The Electron

" The Fundamental Particle of Electromagnetism "

Edited by Paul F. Kisak

Contents

   10.6 Physical origins   86

         10.6.1 Lattice waves   86

         10.6.2 Electronic thermal conductivity   88

   10.7 Equations   89

   10.8 Simple kinetic picture   89

   10.9 See also   90

   10.10 References   90

   10.11 Further reading   91

   10.12 External links   91

11 **Lorentz force**   **92**

   11.1 Equation (SI units)   92

         11.1.1 Charged particle   92

         11.1.2 Continuous charge distribution   93

   11.2 History   93

   11.3 Trajectories of particles due to the Lorentz force   94

   11.4 Significance of the Lorentz force   94

   11.5 Lorentz force law as the definition of E and B   95

   11.6 Force on a current-carrying wire   95

   11.7 EMF   96

   11.8 Lorentz force and Faraday's law of induction   96

   11.9 Lorentz force in terms of potentials   97

   11.10 Lorentz force and analytical mechanics   98

   11.11 Equation (cgs units)   98

   11.12 Relativistic form of the Lorentz force   98

         11.12.1 Covariant form of the Lorentz force   98

         11.12.2 STA form of the Lorentz force   99

   11.13 Applications 100

   11.14 See also 100

   11.15 Footnotes 100

   11.16 References 101

   11.17 External links 101

12 **Electric field**   **102**

   12.1 Definition 102

   12.2 Sources of electric field 102

         12.2.1 Causes and description 102

         12.2.2 Continuous vs. discrete charge repartition 102

Chapter 1

Electron

For other uses, see Electron (disambiguation).

The **electron** is a subatomic particle, symbol e− or β−, with a negative elementary electric charge.[7] Electrons belong to the first generation of the lepton particle family,[8] and are generally thought to be elementary particles because they have no known components or substructure.[1] The electron has a mass that is approximately 1/1836 that of the proton.[9] Quantum mechanical properties of the electron include an intrinsic angular momentum (spin) of a half-integer value in units of h, which means that it is a fermion. Being fermions, no two electrons can occupy the same quantum state, in accordance with the Pauli exclusion principle.[8] Like all matter, electrons have properties of both particles and waves, and so can collide with other particles and can be diffracted like light. The wave properties of electrons are easier to observe with experiments than those of other particles like neutrons and protons because electrons have a lower mass and hence a higher De Broglie wavelength for typical energies.

Many physical phenomena involve electrons in an essential role, such as electricity, magnetism, and thermal conductivity, and they also participate in gravitational, electromagnetic and weak interactions.[10] An electron generates an electric field surrounding it. An electron moving relative to an observer generates a magnetic field. External magnetic fields deflect an electron. Electrons radiate or absorb energy in the form of photons when accelerated. Laboratory instruments are capable of containing and observing individual electrons as well as electron plasma using electromagnetic fields, whereas dedicated telescopes can detect electron plasma in outer space. Electrons have many applications, including electronics, welding, cathode ray tubes, electron microscopes, radiation therapy, lasers, gaseous ionization detectors and particle accelerators.

Interactions involving electrons and other subatomic particles are of interest in fields such as chemistry and nuclear physics. The Coulomb force interaction between positive protons inside atomic nuclei and negative electrons composes atoms. Ionization or changes in the proportions of particles changes the binding energy of the system. The exchange or sharing of the electrons between two or more atoms is the main cause of chemical bonding.[11] British natural philosopher Richard Laming first hypothesized the concept of an indivisible quantity of electric charge to explain the chemical properties of atoms in 1838;[3] Irish physicist George Johnstone Stoney named this charge 'electron' in 1891, and J. J. Thomson and his team of British physicists identified it as a particle in 1897.[5][12][13] Electrons can also participate in nuclear reactions, such as nucleosynthesis in stars, where they are known as beta particles. Electrons may be created through beta decay of radioactive isotopes and in high-energy collisions, for instance when cosmic rays enter the atmosphere. The antiparticle of the electron is called the positron; it is identical to the electron except that it carries electrical and other charges of the opposite sign. When an electron collides with a positron, both particles may be totally annihilated, producing gamma ray photons.

1.1 History

See also: History of electromagnetism

The ancient Greeks noticed that amber attracted small objects when rubbed with fur. Along with lightning, this phenomenon is one of humanity's earliest recorded experiences with electricity.[14] In his 1600 treatise *De Magnete*, the English scientist William Gilbert coined the New Latin term *electricus*, to refer to this property of attracting small objects after being rubbed.[15] Both *electric* and *electricity* are derived from the Latin *ēlectrum* (also the root of the alloy of the same name), which came from the Greek word for amber, ἤλεκτρον (*ēlektron*).

In the early 1700s, Francis Hauksbee and French chemist Charles François de Fay independently discovered what they believed were two kinds of frictional electricity— one generated from rubbing glass, the other from rubbing resin. From this, Du Fay theorized that electricity consists of two electrical fluids, *vitreous* and *resinous*, that are separated by friction, and that neutralize each other when combined.[16] A decade later Benjamin Franklin proposed that electricity was not from different types of electrical fluid, but the same electrical fluid under different pressures. He gave them the modern charge nomenclature of positive and negative respectively.[17] Franklin thought of the charge carrier as being positive, but he did not correctly identify which situation was a surplus of the charge carrier, and which situation was a deficit.[18]

Between 1838 and 1851, British natural philosopher Richard Laming developed the idea that an atom is composed of a core of matter surrounded by subatomic particles that had unit electric charges.[2] Beginning in 1846, German physicist William Weber theorized that electricity was composed of positively and negatively charged fluids, and their interaction was governed by the inverse square law. After studying the phenomenon of electrolysis in 1874, Irish physicist George Johnstone Stoney suggested that there existed a "single definite quantity of electricity", the charge of a monovalent ion. He was able to estimate the value of this elementary charge *e* by means of Faraday's laws of electrolysis.[19] However, Stoney believed these charges were permanently attached to atoms and could not be removed. In 1881, German physicist Hermann von Helmholtz argued that both positive and negative charges were divided into elementary parts, each of which "behaves like atoms of electricity".[3]

Stoney initially coined the term *electrolion* in 1881. Ten years later, he switched to *electron* to describe these elementary charges, writing in 1894: "... an estimate was made of the actual amount of this most remarkable fundamental unit of electricity, for which I have since ventured to suggest the name *electron*". A 1906 proposal to change to *electrion* failed because Hendrik Lorentz preferred to keep *electron*.[20][21] The word *electron* is a combination of the words *electric* and *ion*.[22] The suffix *-on* which is now used to designate other subatomic particles, such as a proton or neutron, is in turn derived from electron.[23][24]

1.1.1 Discovery

The German physicist Johann Wilhelm Hittorf studied electrical conductivity in rarefied gases: in 1869, he discovered a glow emitted from the cathode that increased in size with decrease in gas pressure. In 1876, the German physicist Eugen Goldstein showed that the rays from this glow cast a shadow, and he dubbed the rays cathode rays.[26] During the 1870s, the English chemist and physicist Sir William Crookes developed the first cathode ray tube to have a high vacuum inside.[27] He then showed that the luminescence rays appearing within the tube carried energy and moved

A beam of electrons deflected in a circle by a magnetic field[25]

from the cathode to the anode. Furthermore, by applying a magnetic field, he was able to deflect the rays, thereby demonstrating that the beam behaved as though it were negatively charged.[28][29] In 1879, he proposed that these properties could be explained by what he termed 'radiant matter'. He suggested that this was a fourth state of matter, consisting of negatively charged molecules that were being projected with high velocity from the cathode.[30]

The German-born British physicist Arthur Schuster expanded upon Crookes' experiments by placing metal plates parallel to the cathode rays and applying an electric potential between the plates. The field deflected the rays toward the positively charged plate, providing further evidence that the rays carried negative charge. By measuring the amount of deflection for a given level of current, in 1890 Schuster was able to estimate the charge-to-mass ratio of the ray components. However, this produced a value that was more than a thousand times greater than what was expected, so little credence was given to his calculations at the time.[28][31]

In 1892 Hendrik Lorentz suggested that the mass of these particles (electrons) could be a consequence of their electric charge.[32]

In 1896, the British physicist J. J. Thomson, with his colleagues John S. Townsend and H. A. Wilson,[12] performed experiments indicating that cathode rays really were unique particles, rather than waves, atoms or molecules as was believed earlier.[5] Thomson made good estimates of both the charge *e* and the mass *m*, finding that cathode ray particles, which he called "corpuscles," had perhaps one thousandth of the mass of the least massive ion known: hydrogen.[5][13] He showed that their charge to mass ratio, *e/m*, was independent of cathode material. He further showed that the negatively charged particles produced by radioactive materials, by heated materials and by illuminated materials were universal.[5][33] The name electron was again proposed for these particles by the Irish physicist

George F. Fitzgerald, and the name has since gained universal acceptance.[28]

Robert Millikan

While studying naturally fluorescing minerals in 1896, the French physicist Henri Becquerel discovered that they emitted radiation without any exposure to an external energy source. These radioactive materials became the subject of much interest by scientists, including the New Zealand physicist Ernest Rutherford who discovered they emitted particles. He designated these particles alpha and beta, on the basis of their ability to penetrate matter.[34] In 1900, Becquerel showed that the beta rays emitted by radium could be deflected by an electric field, and that their mass-to-charge ratio was the same as for cathode rays.[35] This evidence strengthened the view that electrons existed as components of atoms.[36][37]

The electron's charge was more carefully measured by the American physicists Robert Millikan and Harvey Fletcher in their oil-drop experiment of 1909, the results of which were published in 1911. This experiment used an electric field to prevent a charged droplet of oil from falling as a result of gravity. This device could measure the electric charge from as few as 1–150 ions with an error margin of less than 0.3%. Comparable experiments had been done earlier by Thomson's team,[5] using clouds of charged wa-

ter droplets generated by electrolysis,[12] and in 1911 by Abram Ioffe, who independently obtained the same result as Millikan using charged microparticles of metals, then published his results in 1913.[38] However, oil drops were more stable than water drops because of their slower evaporation rate, and thus more suited to precise experimentation over longer periods of time.[39]

Around the beginning of the twentieth century, it was found that under certain conditions a fast-moving charged particle caused a condensation of supersaturated water vapor along its path. In 1911, Charles Wilson used this principle to devise his cloud chamber so he could photograph the tracks of charged particles, such as fast-moving electrons.[40]

1.1.2 Atomic theory

See also: The proton–electron model of the nucleus
  By 1914, experiments by physicists Ernest Rutherford,

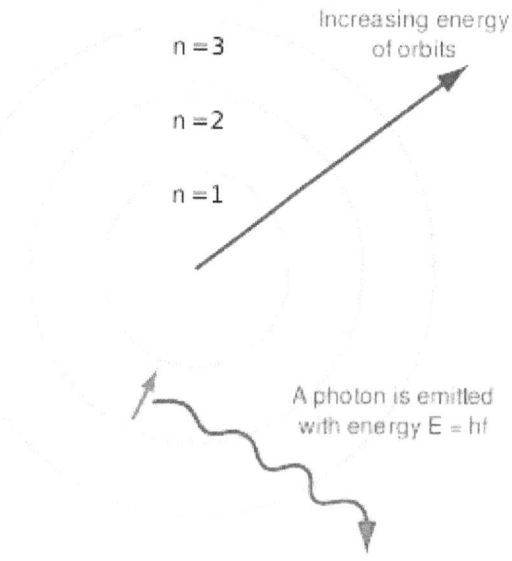

The Bohr model of the atom, showing states of electron with energy quantized by the number n. An electron dropping to a lower orbit emits a photon equal to the energy difference between the orbits.

Henry Moseley, James Franck and Gustav Hertz had largely established the structure of an atom as a dense nucleus of positive charge surrounded by lower-mass electrons.[41] In 1913, Danish physicist Niels Bohr postulated that electrons resided in quantized energy states, with the energy determined by the angular momentum of the electron's orbits about the nucleus. The electrons could move between these states, or orbits, by the emission or absorption of photons at specific frequencies. By means of these quantized orbits, he accurately explained the spectral lines of the hydrogen atom.[42] However, Bohr's model failed to account for the relative intensities of the spectral lines and it was unsuccessful in explaining the spectra of more complex atoms.[41]

Chemical bonds between atoms were explained by Gilbert Newton Lewis, who in 1916 proposed that a covalent bond between two atoms is maintained by a pair of electrons shared between them.[43] Later, in 1927, Walter Heitler and Fritz London gave the full explanation of the electron-pair formation and chemical bonding in terms of quantum mechanics.[44] In 1919, the American chemist Irving Langmuir elaborated on the Lewis' static model of the atom and suggested that all electrons were distributed in successive "concentric (nearly) spherical shells, all of equal thickness".[45] The shells were, in turn, divided by him in a number of cells each containing one pair of electrons. With this model Langmuir was able to qualitatively explain the chemical properties of all elements in the periodic table,[44] which were known to largely repeat themselves according to the periodic law.[46]

In 1924, Austrian physicist Wolfgang Pauli observed that the shell-like structure of the atom could be explained by a set of four parameters that defined every quantum energy state, as long as each state was inhabited by no more than a single electron. (This prohibition against more than one electron occupying the same quantum energy state became known as the Pauli exclusion principle.)[47] The physical mechanism to explain the fourth parameter, which had two distinct possible values, was provided by the Dutch physicists Samuel Goudsmit and George Uhlenbeck. In 1925, Goudsmit and Uhlenbeck suggested that an electron, in addition to the angular momentum of its orbit, possesses an intrinsic angular momentum and magnetic dipole moment.[41][48] The intrinsic angular momentum became known as spin, and explained the previously mysterious splitting of spectral lines observed with a high-resolution spectrograph; this phenomenon is known as fine structure splitting.[49]

1.1.3   Quantum mechanics

See also: History of quantum mechanics

In his 1924 dissertation *Recherches sur la théorie des quanta* (Research on Quantum Theory), French physicist Louis de Broglie hypothesized that all matter possesses a de Broglie wave similar to light.[50] That is, under the appropriate conditions, electrons and other matter would show properties of either particles or waves. The corpuscular properties of a particle are demonstrated when it is shown to have a localized position in space along its trajectory at any given moment.[51] Wave-like nature is observed, for example, when a beam of light is passed through parallel slits and creates interference patterns. In 1927, the interference effect was found in a beam of electrons by English physicist

George Paget Thomson with a thin metal film and by American physicists Clinton Davisson and Lester Germer using a crystal of nickel.[52]

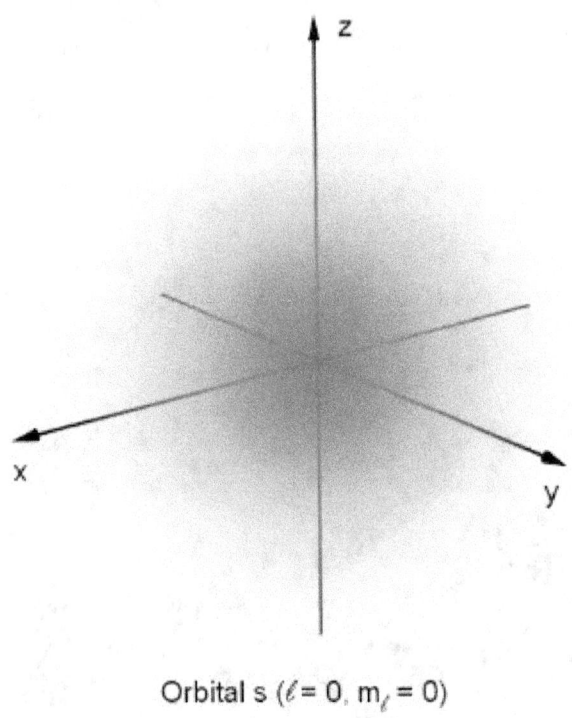

Orbital s ($\ell = 0$, $m_\ell = 0$)

In quantum mechanics, the behavior of an electron in an atom is described by an orbital, which is a probability distribution rather than an orbit. In the figure, the shading indicates the relative probability to "find" the electron, having the energy corresponding to the given quantum numbers, at that point.

De Broglie's prediction of a wave nature for electrons led Erwin Schrödinger to postulate a wave equation for electrons moving under the influence of the nucleus in the atom. In 1926, this equation, the Schrödinger equation, successfully described how electron waves propagated.[53] Rather than yielding a solution that determined the location of an electron over time, this wave equation also could be used to predict the probability of finding an electron near a position, especially a position near where the electron was bound in space, for which the electron wave equations did not change in time. This approach led to a second formulation of quantum mechanics (the first being by Heisenberg in 1925), and solutions of Schrödinger's equation, like Heisenberg's, provided derivations of the energy states of an electron in a hydrogen atom that were equivalent to those that had been derived first by Bohr in 1913, and that were known to reproduce the hydrogen spectrum.[54] Once spin and the interaction between multiple electrons were considered, quantum mechanics later made it possible to predict the configuration of electrons in atoms with higher atomic numbers than hydrogen.[55]

In 1928, building on Wolfgang Pauli's work, Paul Dirac produced a model of the electron – the Dirac equation, consistent with relativity theory, by applying relativistic and symmetry considerations to the hamiltonian formulation of the quantum mechanics of the electro-magnetic field.[56] To resolve some problems within his relativistic equation, in 1930 Dirac developed a model of the vacuum as an infinite sea of particles having negative energy, which was dubbed the Dirac sea. This led him to predict the existence of a positron, the antimatter counterpart of the electron.[57] This particle was discovered in 1932 by Carl Anderson, who proposed calling standard electrons *negatrons*, and using *electron* as a generic term to describe both the positively and negatively charged variants.

In 1947 Willis Lamb, working in collaboration with graduate student Robert Retherford, found that certain quantum states of hydrogen atom, which should have the same energy, were shifted in relation to each other, the difference being the Lamb shift. About the same time, Polykarp Kusch, working with Henry M. Foley, discovered the magnetic moment of the electron is slightly larger than predicted by Dirac's theory. This small difference was later called anomalous magnetic dipole moment of the electron. This difference was later explained by the theory of quantum electrodynamics, developed by Sin-Itiro Tomonaga, Julian Schwinger and Richard Feynman in the late 1940s.[58]

1.1.4   Particle accelerators

With the development of the particle accelerator during the first half of the twentieth century, physicists began to delve deeper into the properties of subatomic particles.[59] The first successful attempt to accelerate electrons using electromagnetic induction was made in 1942 by Donald Kerst. His initial betatron reached energies of 2.3 MeV, while subsequent betatrons achieved 300 MeV. In 1947, synchrotron radiation was discovered with a 70 MeV electron synchrotron at General Electric. This radiation was caused by the acceleration of electrons, moving near the speed of light, through a magnetic field.[60]

With a beam energy of 1.5 GeV, the first high-energy particle collider was ADONE, which began operations in 1968.[61] This device accelerated electrons and positrons in opposite directions, effectively doubling the energy of their collision when compared to striking a static target with an electron.[62] The Large Electron–Positron Collider (LEP) at CERN, which was operational from 1989 to 2000, achieved collision energies of 209 GeV and made important measurements for the Standard Model of particle physics.[63][64]

1.1.5   Confinement of individual electrons

Individual electrons can now be easily confined in ultra small (L = 20 nm, W = 20 nm) CMOS transistors operated at cryogenic temperature over a range of −269 °C (4 K) to about −258 °C (15 K).[65] The electron wavefunction spreads in a semiconductor lattice and negligibly interacts with the valence band electrons, so it can be treated in the single particle formalism, by replacing its mass with the effective mass tensor.

1.2   Characteristics

1.2.1   Classification

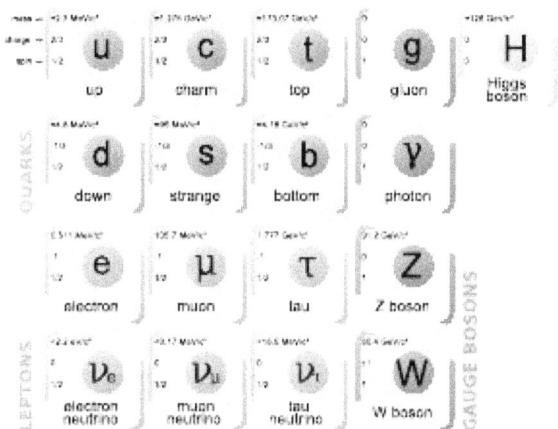

Standard Model of elementary particles. The electron (symbol e) is on the left.

In the Standard Model of particle physics, electrons belong to the group of subatomic particles called leptons, which are believed to be fundamental or elementary particles. Electrons have the lowest mass of any charged lepton (or electrically charged particle of any type) and belong to the first-generation of fundamental particles.[66] The second and third generation contain charged leptons, the muon and the tau, which are identical to the electron in charge, spin and interactions, but are more massive. Leptons differ from the other basic constituent of matter, the quarks, by their lack of strong interaction. All members of the lepton group are fermions, because they all have half-odd integer spin; the electron has spin $\frac{1}{2}$.[67]

1.2.2   Fundamental properties

The invariant mass of an electron is approximately 9.109×10^{-31} kilograms,[68] or 5.489×10^{-4} atomic mass units. On the basis of Einstein's principle of mass–energy equivalence, this mass corresponds to a rest energy of 0.511 MeV. The ratio between the mass of a proton and that of an electron is about 1836.[9][69] Astronomical measurements show that the proton-to-electron mass ratio has held the same value for at least half the age of the universe, as is predicted by the Standard Model.[70]

Electrons have an electric charge of -1.602×10^{-19} coulomb,[68] which is used as a standard unit of charge for subatomic particles, and is also called the elementary charge. This elementary charge has a relative standard uncertainty of 2.2×10^{-8}.[68] Within the limits of experimental accuracy, the electron charge is identical to the charge of a proton, but with the opposite sign.[71] As the symbol e is used for the elementary charge, the electron is commonly symbolized by e−, where the minus sign indicates the negative charge. The positron is symbolized by e+ because it has the same properties as the electron but with a positive rather than negative charge.[67][68]

The electron has an intrinsic angular momentum or spin of $\frac{1}{2}$.[68] This property is usually stated by referring to the electron as a spin-$\frac{1}{2}$ particle.[67] For such particles the spin magnitude is $\sqrt{3}/2\ \hbar$.[note 3] while the result of the measurement of a projection of the spin on any axis can only be $\pm\hbar/2$. In addition to spin, the electron has an intrinsic magnetic moment along its spin axis.[68] It is approximately equal to one Bohr magneton,[72][note 4] which is a physical constant equal to $9.27400915(23)\times10^{-24}$ joules per tesla.[68] The orientation of the spin with respect to the momentum of the electron defines the property of elementary particles known as helicity.[73]

The electron has no known substructure.[1][74] and it is assumed to be a point particle with a point charge and no spatial extent.[8] In classical physics, the angular momentum and magnetic moment of an object depend upon its physical dimensions. Hence, the concept of a dimensionless electron possessing these properties might seem paradoxical and inconsistent to experimental observations in Penning traps which point to finite non-zero radius of the electron. A possible explanation of this paradoxical situation is given below in the "Virtual particles" subsection by taking into consideration the Foldy-Wouthuysen transformation. The issue of the radius of the electron is a challenging problem of the modern theoretical physics. The admission of the hypothesis of a finite radius of the electron is incompatible to the premises of the theory of relativity. On the other hand, a point-like electron (zero radius) generates serious mathematical difficulties due to the self-energy of the electron tending to infinity.[75] These aspects have been analyzed in detail by Dmitri Ivanenko and Arseny Sokolov.

Observation of a single electron in a Penning trap shows the upper limit of the particle's radius is 10^{-22} meters.[76]

There *is* a physical constant called the "classical electron radius", with the much larger value of 2.8179×10^{-15} m, greater than the radius of the proton. However, the terminology comes from a simplistic calculation that ignores the effects of quantum mechanics; in reality, the so-called classical electron radius has little to do with the true fundamental structure of the electron.[77][note 5]

There are elementary particles that spontaneously decay into less massive particles. An example is the muon, which decays into an electron, a neutrino and an antineutrino, with a mean lifetime of 2.2×10^{-6} seconds. However, the electron is thought to be stable on theoretical grounds: the electron is the least massive particle with non-zero electric charge, so its decay would violate charge conservation.[78] The experimental lower bound for the electron's mean lifetime is 4.6×10^{26} years, at a 90% confidence level.[79][80]

1.2.3 Quantum properties

As with all particles, electrons can act as waves. This is called the wave–particle duality and can be demonstrated using the double-slit experiment.

The wave-like nature of the electron allows it to pass through two parallel slits simultaneously, rather than just one slit as would be the case for a classical particle. In quantum mechanics, the wave-like property of one particle can be described mathematically as a complex-valued function, the wave function, commonly denoted by the Greek letter psi (ψ). When the absolute value of this function is squared, it gives the probability that a particle will be observed near a location—a probability density.[81]:162–218

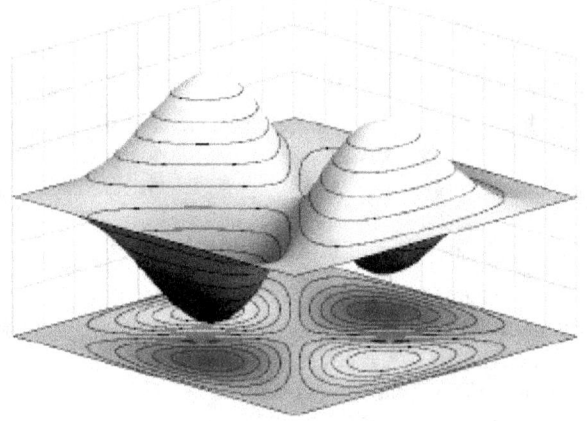

Example of an antisymmetric wave function for a quantum state of two identical fermions in a 1-dimensional box. If the particles swap position, the wave function inverts its sign.

Electrons are identical particles because they cannot be distinguished from each other by their intrinsic physical properties. In quantum mechanics, this means that a pair of interacting electrons must be able to swap positions without an observable change to the state of the system. The wave function of fermions, including electrons, is antisymmetric, meaning that it changes sign when two electrons are swapped; that is, $\psi(r_1, r_2) = -\psi(r_2, r_1)$, where the variables r_1 and r_2 correspond to the first and second electrons, respectively. Since the absolute value is not changed by a sign swap, this corresponds to equal probabilities. Bosons, such as the photon, have symmetric wave functions instead.[81]:162–218

In the case of antisymmetry, solutions of the wave equation for interacting electrons result in a zero probability that each pair will occupy the same location or state. This is responsible for the Pauli exclusion principle, which precludes any two electrons from occupying the same quantum state. This principle explains many of the properties of electrons. For example, it causes groups of bound electrons to occupy different orbitals in an atom, rather than all overlapping each other in the same orbit.[81]:162–218

1.2.4 Virtual particles

Main article: Virtual particle

In a simplified picture, every photon spends some time as a combination of a virtual electron plus its antiparticle, the virtual positron, which rapidly annihilate each other shortly thereafter.[82] The combination of the energy variation needed to create these particles, and the time during which they exist, fall under the threshold of detectability expressed by the Heisenberg uncertainty relation, $\Delta E \cdot \Delta t \geq \hbar$. In effect, the energy needed to create these virtual particles, ΔE, can be "borrowed" from the vacuum for a period of time, Δt, so that their product is no more than the reduced Planck constant, $\hbar \approx 6.6 \times 10^{-16}$ eV·s. Thus, for a virtual electron, Δt is at most 1.3×10^{-21} s.[83]

While an electron–positron virtual pair is in existence, the coulomb force from the ambient electric field surrounding an electron causes a created positron to be attracted to the original electron, while a created electron experiences a repulsion. This causes what is called vacuum polarization. In effect, the vacuum behaves like a medium having a dielectric permittivity more than unity. Thus the effective charge of an electron is actually smaller than its true value, and the charge decreases with increasing distance from the electron.[84][85] This polarization was confirmed experimentally in 1997 using the Japanese TRISTAN particle accelerator.[86] Virtual particles cause a comparable shielding effect for the mass of the electron.[87]

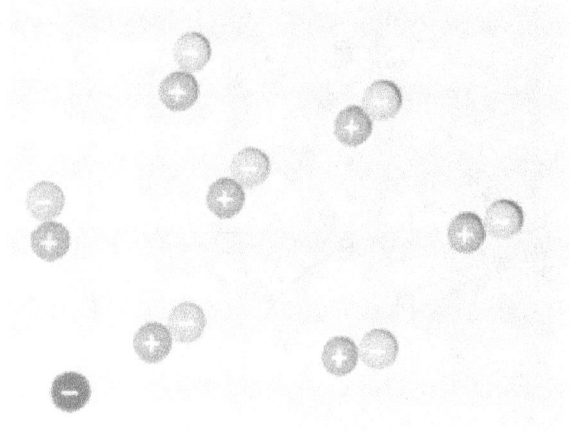

A schematic depiction of virtual electron–positron pairs appearing at random near an electron (at lower left)

The interaction with virtual particles also explains the small (about 0.1%) deviation of the intrinsic magnetic moment of the electron from the Bohr magneton (the anomalous magnetic moment).[72][88] The extraordinarily precise agreement of this predicted difference with the experimentally determined value is viewed as one of the great achievements of quantum electrodynamics.[89]

The apparent paradox (mentioned above in the properties subsection) of a point particle electron having intrinsic angular momentum and magnetic moment can be explained by the formation of virtual photons in the electric field generated by the electron. These photons cause the electron to shift about in a jittery fashion (known as zitterbewegung),[90] which results in a net circular motion with precession. This motion produces both the spin and the magnetic moment of the electron.[8][91] In atoms, this creation of virtual photons explains the Lamb shift observed in spectral lines.[84]

1.2.5 Interaction

An electron generates an electric field that exerts an attractive force on a particle with a positive charge, such as the proton, and a repulsive force on a particle with a negative charge. The strength of this force is determined by Coulomb's inverse square law.[92] When an electron is in motion, it generates a magnetic field.[81]:140 The Ampère-Maxwell law relates the magnetic field to the mass motion of electrons (the current) with respect to an observer. This property of induction supplies the magnetic field that drives an electric motor.[93] The electromagnetic field of an arbitrary moving charged particle is expressed by the Liénard–Wiechert potentials, which are valid even when the particle's speed is close to that of light (relativistic).

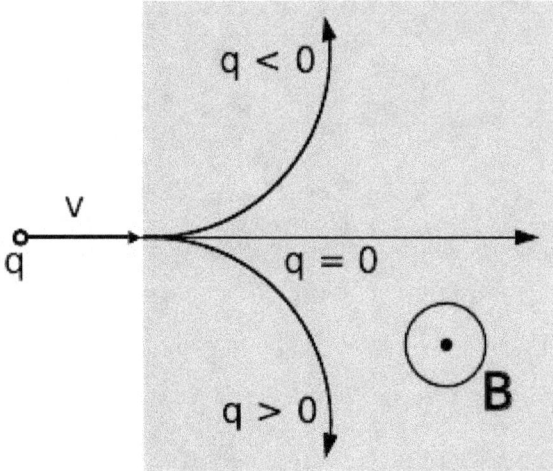

A particle with charge q (at left) is moving with velocity v through a magnetic field B that is oriented toward the viewer. For an electron, q is negative so it follows a curved trajectory toward the top.

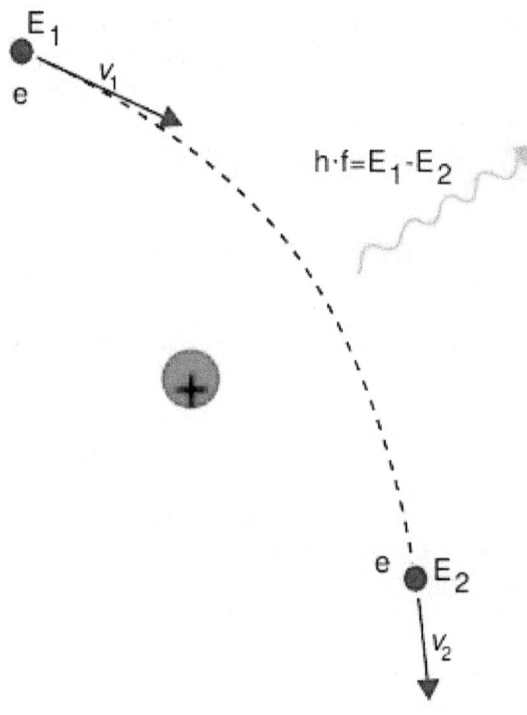

Here, Bremsstrahlung is produced by an electron e deflected by the electric field of an atomic nucleus. The energy change $E_2 - E_1$ determines the frequency f of the emitted photon.

When an electron is moving through a magnetic field, it is subject to the Lorentz force that acts perpendicularly to the plane defined by the magnetic field and the electron velocity. This centripetal force causes the electron to follow a helical trajectory through the field at a radius called the gyroradius. The acceleration from this curving motion induces the electron to radiate energy in the form of synchrotron radiation.[81][:160][94][note 6] The energy emission in turn causes a recoil of the electron, known as the Abraham–Lorentz–Dirac Force, which creates a friction that slows the electron. This force is caused by a back-reaction of the electron's own field upon itself.[95]

Photons mediate electromagnetic interactions between particles in quantum electrodynamics. An isolated electron at a constant velocity cannot emit or absorb a real photon; doing so would violate conservation of energy and momentum. Instead, virtual photons can transfer momentum between two charged particles. This exchange of virtual photons, for example, generates the Coulomb force.[96] Energy emission can occur when a moving electron is deflected by a charged particle, such as a proton. The acceleration of the electron results in the emission of Bremsstrahlung radiation.[97]

An inelastic collision between a photon (light) and a solitary (free) electron is called Compton scattering. This collision results in a transfer of momentum and energy between the particles, which modifies the wavelength of the photon by an amount called the Compton shift.[note 7] The maximum magnitude of this wavelength shift is $h/m_e c$, which is known as the Compton wavelength.[98] For an electron, it has a value of 2.43×10^{-12} m.[68] When the wavelength of the light is long (for instance, the wavelength of the visible light is 0.4–0.7 μm) the wavelength shift becomes negligible. Such interaction between the light and free electrons is called Thomson scattering or Linear Thomson scattering.[99]

The relative strength of the electromagnetic interaction between two charged particles, such as an electron and a proton, is given by the fine-structure constant. This value is a dimensionless quantity formed by the ratio of two energies: the electrostatic energy of attraction (or repulsion) at a separation of one Compton wavelength, and the rest energy of the charge. It is given by $\alpha \approx 7.297353 \times 10^{-3}$, which is approximately equal to $1/137$.[68]

When electrons and positrons collide, they annihilate each other, giving rise to two or more gamma ray photons. If the electron and positron have negligible momentum, a positronium atom can form before annihilation results in two or three gamma ray photons totalling 1.022 MeV.[100][101] On the other hand, high-energy photons may transform into an electron and a positron by a process called pair production, but only in the presence of a nearby charged particle, such as a nucleus.[102][103]

In the theory of electroweak interaction, the left-handed component of electron's wavefunction forms a weak isospin doublet with the electron neutrino. This means that during weak interactions, electron neutrinos behave like electrons. Either member of this doublet can undergo a charged current interaction by emitting or absorbing a W and be converted into the other member. Charge is conserved during this reaction because the W boson also carries a charge, canceling out any net change during the transmutation. Charged current interactions are responsible for the phenomenon of beta decay in a radioactive atom. Both the electron and electron neutrino can undergo a neutral current interaction via a Z0 exchange, and this is responsible for neutrino-electron elastic scattering.[104]

1.2.6 Atoms and molecules

Main article: Atom

An electron can be *bound* to the nucleus of an atom by the

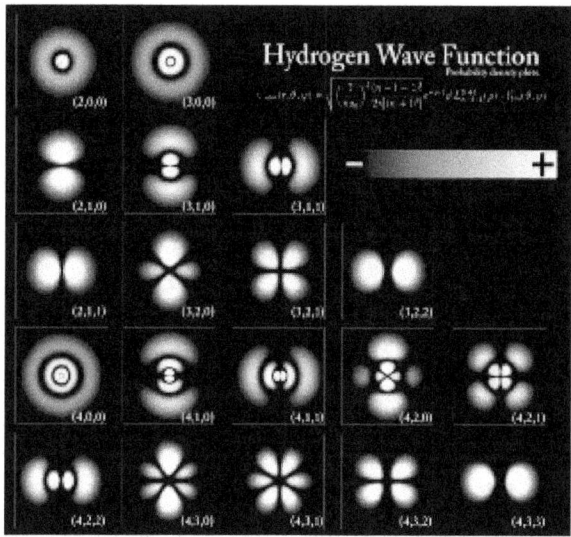

Probability densities for the first few hydrogen atom orbitals, seen in cross-section. The energy level of a bound electron determines the orbital it occupies, and the color reflects the probability of finding the electron at a given position.

attractive Coulomb force. A system of one or more electrons bound to a nucleus is called an atom. If the number of electrons is different from the nucleus' electrical charge, such an atom is called an ion. The wave-like behavior of a bound electron is described by a function called an atomic orbital. Each orbital has its own set of quantum numbers such as energy, angular momentum and projection of angular momentum, and only a discrete set of these orbitals exist around the nucleus. According to the Pauli exclusion principle each orbital can be occupied by up to two electrons, which must differ in their spin quantum number.

Electrons can transfer between different orbitals by the emission or absorption of photons with an energy that matches the difference in potential.[105] Other methods of orbital transfer include collisions with particles, such as electrons, and the Auger effect.[106] To escape the

atom, the energy of the electron must be increased above its binding energy to the atom. This occurs, for example, with the photoelectric effect, where an incident photon exceeding the atom's ionization energy is absorbed by the electron.[107]

The orbital angular momentum of electrons is quantized. Because the electron is charged, it produces an orbital magnetic moment that is proportional to the angular momentum. The net magnetic moment of an atom is equal to the vector sum of orbital and spin magnetic moments of all electrons and the nucleus. The magnetic moment of the nucleus is negligible compared with that of the electrons. The magnetic moments of the electrons that occupy the same orbital (so called, paired electrons) cancel each other out.[108]

The chemical bond between atoms occurs as a result of electromagnetic interactions, as described by the laws of quantum mechanics.[109] The strongest bonds are formed by the sharing or transfer of electrons between atoms, allowing the formation of molecules.[11] Within a molecule, electrons move under the influence of several nuclei, and occupy molecular orbitals; much as they can occupy atomic orbitals in isolated atoms.[110] A fundamental factor in these molecular structures is the existence of electron pairs. These are electrons with opposed spins, allowing them to occupy the same molecular orbital without violating the Pauli exclusion principle (much like in atoms). Different molecular orbitals have different spatial distribution of the electron density. For instance, in bonded pairs (i.e. in the pairs that actually bind atoms together) electrons can be found with the maximal probability in a relatively small volume between the nuclei. On the contrary, in non-bonded pairs electrons are distributed in a large volume around nuclei.[111]

1.2.7 Conductivity

If a body has more or fewer electrons than are required to balance the positive charge of the nuclei, then that object has a net electric charge. When there is an excess of electrons, the object is said to be negatively charged. When there are fewer electrons than the number of protons in nuclei, the object is said to be positively charged. When the number of electrons and the number of protons are equal, their charges cancel each other and the object is said to be electrically neutral. A macroscopic body can develop an electric charge through rubbing, by the triboelectric effect.[115]

Independent electrons moving in vacuum are termed *free* electrons. Electrons in metals also behave as if they were free. In reality the particles that are commonly termed electrons in metals and other solids are quasi-electrons— quasiparticles, which have the same electrical charge, spin

A lightning discharge consists primarily of a flow of electrons.[112] *The electric potential needed for lightning may be generated by a triboelectric effect.*[113]*[114]*

and magnetic moment as real electrons but may have a different mass.[116] When free electrons — both in vacuum and metals — move, they produce a net flow of charge called an electric current, which generates a magnetic field. Likewise a current can be created by a changing magnetic field. These interactions are described mathematically by Maxwell's equations.[117]

At a given temperature, each material has an electrical conductivity that determines the value of electric current when an electric potential is applied. Examples of good conductors include metals such as copper and gold, whereas glass and Teflon are poor conductors. In any dielectric material, the electrons remain bound to their respective atoms and the material behaves as an insulator. Most semiconductors have a variable level of conductivity that lies between the extremes of conduction and insulation.[118] On the other hand, metals have an electronic band structure containing partially filled electronic bands. The presence of such bands allows electrons in metals to behave as if they were free or delocalized electrons. These electrons are not associated with specific atoms, so when an electric field is applied, they are free to move like a gas (called Fermi gas)[119] through the material much like free electrons.

Because of collisions between electrons and atoms, the drift velocity of electrons in a conductor is on the order of millimeters per second. However, the speed at which a change of current at one point in the material causes changes in currents in other parts of the material, the velocity of prop-

agation, is typically about 75% of light speed.[120] This occurs because electrical signals propagate as a wave, with the velocity dependent on the dielectric constant of the material.[121]

Metals make relatively good conductors of heat, primarily because the delocalized electrons are free to transport thermal energy between atoms. However, unlike electrical conductivity, the thermal conductivity of a metal is nearly independent of temperature. This is expressed mathematically by the Wiedemann–Franz law,[119] which states that the ratio of thermal conductivity to the electrical conductivity is proportional to the temperature. The thermal disorder in the metallic lattice increases the electrical resistivity of the material, producing a temperature dependence for electric current.[122]

When cooled below a point called the critical temperature, materials can undergo a phase transition in which they lose all resistivity to electric current, in a process known as superconductivity. In BCS theory, this behavior is modeled by pairs of electrons entering a quantum state known as a Bose–Einstein condensate. These Cooper pairs have their motion coupled to nearby matter via lattice vibrations called phonons, thereby avoiding the collisions with atoms that normally create electrical resistance.[123] (Cooper pairs have a radius of roughly 100 nm, so they can overlap each other.)[124] However, the mechanism by which higher temperature superconductors operate remains uncertain.

Electrons inside conducting solids, which are quasiparticles themselves, when tightly confined at temperatures close to absolute zero, behave as though they had split into three other quasiparticles: spinons, orbitons and holons.[125][126] The former carries spin and magnetic moment, the next carries its orbital location while the latter electrical charge.

1.2.8   Motion and energy

According to Einstein's theory of special relativity, as an electron's speed approaches the speed of light, from an observer's point of view its relativistic mass increases, thereby making it more and more difficult to accelerate it from within the observer's frame of reference. The speed of an electron can approach, but never reach, the speed of light in a vacuum, c. However, when relativistic electrons — that is, electrons moving at a speed close to c — are injected into a dielectric medium such as water, where the local speed of light is significantly less than c, the electrons temporarily travel faster than light in the medium. As they interact with the medium, they generate a faint light called Cherenkov radiation.[127]

The effects of special relativity are based on a quantity

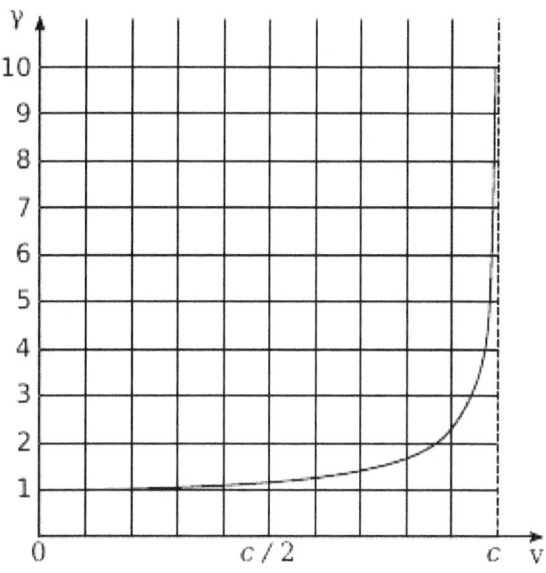

Lorentz factor as a function of velocity. It starts at value 1 and goes to infinity as v approaches c.

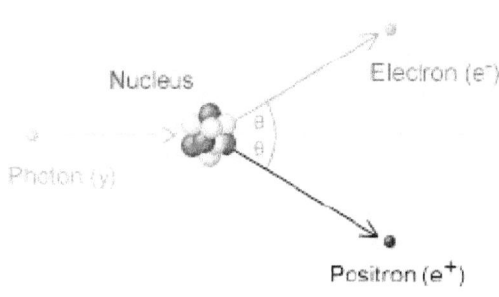

Pair production caused by the collision of a photon with an atomic nucleus

known as the Lorentz factor, defined as $\gamma = 1/\sqrt{1 - v^2/c^2}$ where v is the speed of the particle. The kinetic energy K_e of an electron moving with velocity v is:

$$K_e = (\gamma - 1)m_e c^2.$$

where m_e is the mass of electron. For example, the Stanford linear accelerator can accelerate an electron to roughly 51 GeV.[128] Since an electron behaves as a wave, at a given velocity it has a characteristic de Broglie wavelength. This is given by $\lambda_e = h/p$ where h is the Planck constant and p is the momentum.[50] For the 51 GeV electron above, the wavelength is about 2.4×10^{-17} m, small enough to explore structures well below the size of an atomic nucleus.[129]

1.3 Formation

The Big Bang theory is the most widely accepted scientific theory to explain the early stages in the evolution of the Universe.[130] For the first millisecond of the Big Bang, the temperatures were over 10 billion Kelvin and photons had mean energies over a million electronvolts. These photons were sufficiently energetic that they could react with each other to form pairs of electrons and positrons. Likewise, positron-electron pairs annihilated each other and emitted energetic photons:

$$\gamma + \gamma \leftrightarrow e+ + e-$$

An equilibrium between electrons, positrons and photons was maintained during this phase of the evolution of the Universe. After 15 seconds had passed, however, the temperature of the universe dropped below the threshold where electron-positron formation could occur. Most of the surviving electrons and positrons annihilated each other, releasing gamma radiation that briefly reheated the universe.[131]

For reasons that remain uncertain, during the process of leptogenesis there was an excess in the number of electrons over positrons.[132] Hence, about one electron in every billion survived the annihilation process. This excess matched the excess of protons over antiprotons, in a condition known as baryon asymmetry, resulting in a net charge of zero for the universe.[133][134] The surviving protons and neutrons began to participate in reactions with each other — in the process known as nucleosynthesis, forming isotopes of hydrogen and helium, with trace amounts of lithium. This process peaked after about five minutes.[135] Any leftover neutrons underwent negative beta decay with a half-life of about a thousand seconds, releasing a proton and electron in the process,

$$n \rightarrow p + e- + \nu_e$$

For about the next 300000–400000 years, the excess electrons remained too energetic to bind with atomic nuclei.[136] What followed is a period known as recombination, when neutral atoms were formed and the expanding universe became transparent to radiation.[137]

Roughly one million years after the big bang, the first generation of stars began to form.[137] Within a star, stellar nucleosynthesis results in the production of positrons from the fusion of atomic nuclei. These antimatter particles immediately annihilate with electrons, releasing gamma rays. The net result is a steady reduction in the number of electrons, and a matching increase in the number of neutrons. However, the process of stellar evolution can result in the synthesis of radioactive isotopes. Selected isotopes can subsequently undergo negative beta decay, emitting an electron and antineutrino from the nucleus.*[138] An example is the cobalt-60 (^{60}Co) isotope, which decays to form nickel-60 (60Ni).*[139]

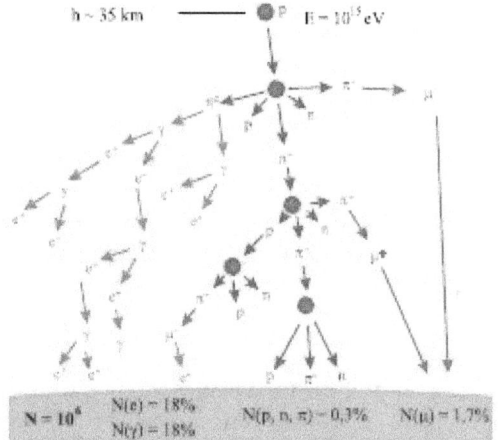

An extended air shower generated by an energetic cosmic ray striking the Earth's atmosphere

At the end of its lifetime, a star with more than about 20 solar masses can undergo gravitational collapse to form a black hole.*[140] According to classical physics, these massive stellar objects exert a gravitational attraction that is strong enough to prevent anything, even electromagnetic radiation, from escaping past the Schwarzschild radius. However, quantum mechanical effects are believed to potentially allow the emission of Hawking radiation at this distance. Electrons (and positrons) are thought to be created at the event horizon of these stellar remnants.

When pairs of virtual particles (such as an electron and positron) are created in the vicinity of the event horizon, the random spatial distribution of these particles may permit one of them to appear on the exterior; this process is called quantum tunnelling. The gravitational potential of the black hole can then supply the energy that transforms this virtual particle into a real particle, allowing it to radiate away into space.*[141] In exchange, the other member of the pair is given negative energy, which results in a net loss of mass-energy by the black hole. The rate of Hawking radiation increases with decreasing mass, eventually causing the black hole to evaporate away until, finally, it explodes.*[142]

Cosmic rays are particles traveling through space with high energies. Energy events as high as 3.0×10^{20} eV have been

recorded.*[143] When these particles collide with nucleons in the Earth's atmosphere, a shower of particles is generated, including pions.*[144] More than half of the cosmic radiation observed from the Earth's surface consists of muons. The particle called a muon is a lepton produced in the upper atmosphere by the decay of a pion.

$$\pi- \rightarrow \mu- + \nu$$
$$\mu$$

A muon, in turn, can decay to form an electron or positron.*[145]

$$\mu- \rightarrow e- + \nu$$
$$e + \nu$$
$$\mu$$

1.4   Observation

Aurorae are mostly caused by energetic electrons precipitating into the atmosphere.[146]

Remote observation of electrons requires detection of their radiated energy. For example, in high-energy environments such as the corona of a star, free electrons form a plasma that radiates energy due to Bremsstrahlung radiation. Electron gas can undergo plasma oscillation, which is waves caused by synchronized variations in electron density, and these produce energy emissions that can be detected by using radio telescopes.*[147]

The frequency of a photon is proportional to its energy. As a bound electron transitions between different energy levels of an atom, it absorbs or emits photons at characteristic frequencies. For instance, when atoms are irradiated by a source with a broad spectrum, distinct absorption lines appear in the spectrum of transmitted radiation. Each element or molecule displays a characteristic set of spectral lines, such as the hydrogen spectral series. Spectroscopic measurements of the strength and width of these lines allow the composition and physical properties of a substance to be determined.[148][149]

In laboratory conditions, the interactions of individual electrons can be observed by means of particle detectors, which allow measurement of specific properties such as energy, spin and charge.[107] The development of the Paul trap and Penning trap allows charged particles to be contained within a small region for long durations. This enables precise measurements of the particle properties. For example, in one instance a Penning trap was used to contain a single electron for a period of 10 months.[150] The magnetic moment of the electron was measured to a precision of eleven digits, which, in 1980, was a greater accuracy than for any other physical constant.[151]

The first video images of an electron's energy distribution were captured by a team at Lund University in Sweden, February 2008. The scientists used extremely short flashes of light, called attosecond pulses, which allowed an electron's motion to be observed for the first time.[152][153]

The distribution of the electrons in solid materials can be visualized by angle-resolved photoemission spectroscopy (ARPES). This technique employs the photoelectric effect to measure the reciprocal space — a mathematical representation of periodic structures that is used to infer the original structure. ARPES can be used to determine the direction, speed and scattering of electrons within the material.[154]

1.5   Plasma applications

1.5.1   Particle beams

During a NASA wind tunnel test, a model of the Space Shuttle is targeted by a beam of electrons, simulating the effect of ionizing gases during re-entry.[155]

Electron beams are used in welding.[156] They allow energy densities up to 10^7 W·cm^{-2} across a narrow focus diameter of 0.1–1.3 mm and usually require no filler material. This welding technique must be performed in a vacuum to prevent the electrons from interacting with the gas before reaching their target, and it can be used to join conductive materials that would otherwise be considered unsuitable for welding.[157][158]

Electron-beam lithography (EBL) is a method of etching semiconductors at resolutions smaller than a micrometer.[159] This technique is limited by high costs, slow performance, the need to operate the beam in the vacuum and the tendency of the electrons to scatter in solids. The last problem limits the resolution to about 10 nm. For this reason, EBL is primarily used for the production of small numbers of specialized integrated circuits.[160]

Electron beam processing is used to irradiate materials in order to change their physical properties or sterilize medical and food products.[161] Electron beams fluidise or quasi-melt glasses without significant increase of temperature on intensive irradiation: e.g. intensive electron radiation causes a many orders of magnitude decrease of viscosity and stepwise decrease of its activation energy.[162]

Linear particle accelerators generate electron beams for treatment of superficial tumors in radiation therapy. Electron therapy can treat such skin lesions as basal-cell carcinomas because an electron beam only penetrates to a limited depth before being absorbed, typically up to 5 cm for electron energies in the range 5–20 MeV. An electron beam can be used to supplement the treatment of areas that have been irradiated by X-rays.[163][164]

Particle accelerators use electric fields to propel electrons and their antiparticles to high energies. These particles emit synchrotron radiation as they pass through magnetic fields. The dependency of the intensity of this radiation upon spin polarizes the electron beam — a process known as the Sokolov–Ternov effect.[note 8] Polarized electron beams can be useful for various experiments. Synchrotron radiation can also cool the electron beams to reduce the momentum spread of the particles. Electron and positron beams are collided upon the particles' accelerating to the required energies; particle detectors observe the resulting energy emissions, which particle physics studies .[165]

1.5.2   Imaging

Low-energy electron diffraction (LEED) is a method of bombarding a crystalline material with a collimated beam of electrons and then observing the resulting diffraction patterns to determine the structure of the material. The required energy of the electrons is typically in the range 20–200 eV.*[166] The reflection high-energy electron diffraction (RHEED) technique uses the reflection of a beam of electrons fired at various low angles to characterize the surface of crystalline materials. The beam energy is typically in the range 8–20 keV and the angle of incidence is 1–4°.*[167]*[168]

The electron microscope directs a focused beam of electrons at a specimen. Some electrons change their properties, such as movement direction, angle, and relative phase and energy as the beam interacts with the material. Microscopists can record these changes in the electron beam to produce atomically resolved images of the material.*[169] In blue light, conventional optical microscopes have a diffraction-limited resolution of about 200 nm.*[170] By comparison, electron microscopes are limited by the de Broglie wavelength of the electron. This wavelength, for example, is equal to 0.0037 nm for electrons accelerated across a 100,000-volt potential.*[171] The Transmission Electron Aberration-Corrected Microscope is capable of sub-0.05 nm resolution, which is more than enough to resolve individual atoms.*[172] This capability makes the electron microscope a useful laboratory instrument for high resolution imaging. However, electron microscopes are expensive instruments that are costly to maintain.

Two main types of electron microscopes exist: transmission and scanning. Transmission electron microscopes function like overhead projectors, with a beam of electrons passing through a slice of material then being projected by lenses on a photographic slide or a charge-coupled device. Scanning electron microscopes rasteri a finely focused electron beam, as in a TV set, across the studied sample to produce the image. Magnifications range from 100× to 1,000,000× or higher for both microscope types. The scanning tunneling microscope uses quantum tunneling of electrons from a sharp metal tip into the studied material and can produce atomically resolved images of its surface.*[173]*[174]*[175]

1.5.3   Other applications

In the free-electron laser (FEL), a relativistic electron beam passes through a pair of undulators that contain arrays of dipole magnets whose fields point in alternating directions. The electrons emit synchrotron radiation that coherently interacts with the same electrons to strongly amplify the radiation field at the resonance frequency. FEL can emit a coherent high-brilliance electromagnetic radiation with a wide range of frequencies, from microwaves to soft X-rays. These devices may find manufacturing, communication and various medical applications, such as soft tissue

surgery.*[176]

Electrons are important in cathode ray tubes, which have been extensively used as display devices in laboratory instruments, computer monitors and television sets.*[177] In a photomultiplier tube, every photon striking the photocathode initiates an avalanche of electrons that produces a detectable current pulse.*[178] Vacuum tubes use the flow of electrons to manipulate electrical signals, and they played a critical role in the development of electronics technology. However, they have been largely supplanted by solid-state devices such as the transistor.*[179]

1.6   See also

- Anyon
- Electride
- Electron bubble
- Exoelectron emission
- g-factor
- Periodic systems of small molecules
- Spintronics
- Stern–Gerlach experiment
- Townsend discharge
- Zeeman effect
- List of particles
- Lepton

1.7   Notes

[1] The fractional version's denominator is the inverse of the decimal value (along with its relative standard uncertainty of $4.2 \times 10^{*}-13$ u).

[2] The electron's charge is the negative of elementary charge, which has a positive value for the proton.

[3] This magnitude is obtained from the spin quantum number as

$$S = \sqrt{s(s+1)} \cdot \frac{h}{2\pi}$$

$$= \frac{\sqrt{3}}{2} \hbar$$

for quantum number $s = \frac{1}{2}$.

See: Gupta, M.C. (2001). *Atomic and Molecular Spectroscopy*. New Age Publishers. p. 81. ISBN 81-224-1300-5.

[4] Bohr magneton:

$$\mu_B = \frac{e\hbar}{2m_e}.$$

[5] The classical electron radius is derived as follows. Assume that the electron's charge is spread uniformly throughout a spherical volume. Since one part of the sphere would repel the other parts, the sphere contains electrostatic potential energy. This energy is assumed to equal the electron's rest energy, defined by special relativity ($E = mc^2$).

From electrostatics theory, the potential energy of a sphere with radius r and charge e is given by:

$$E_p = \frac{e^2}{8\pi\varepsilon_0 r},$$

where ε_0 is the vacuum permittivity. For an electron with rest mass m_0, the rest energy is equal to:

$$E_p = m_0 c^2,$$

where c is the speed of light in a vacuum. Setting them equal and solving for r gives the classical electron radius.

See: Haken, H.; Wolf, H.C.; Brewer, W.D. (2005). *The Physics of Atoms and Quanta: Introduction to Experiments and Theory*. Springer. p. 70. ISBN 3-540-67274-5.

[6] Radiation from non-relativistic electrons is sometimes termed cyclotron radiation.

[7] The change in wavelength, $\Delta\lambda$, depends on the angle of the recoil, θ, as follows,

$$\Delta\lambda = \frac{h}{m_e c}(1 - \cos\theta),$$

where c is the speed of light in a vacuum and m_e is the electron mass. See Zombeck (2007: 393, 396).

[8] The polarization of an electron beam means that the spins of all electrons point into one direction. In other words, the projections of the spins of all electrons onto their momentum vector have the same sign.

1.8 References

[1] Eichten, E.J.; Peskin, M.E.; Peskin, M. (1983). "New Tests for Quark and Lepton Substructure". *Physical Review Letters* **50** (11): 811–814. Bibcode:1983PhRvL..50..811E. doi:10.1103/PhysRevLett.50.811.

[2] Farrar, W.V. (1969). "Richard Laming and the Coal-Gas Industry, with His Views on the Structure of Matter". *Annals of Science* **25** (3): 243–254. doi:10.1080/00033796900200141.

[3] Arabatzis, T. (2006). *Representing Electrons: A Biographical Approach to Theoretical Entities*. University of Chicago Press. pp. 70–74. ISBN 0-226-02421-0.

[4] Buchwald, J.Z.; Warwick, A. (2001). *Histories of the Electron: The Birth of Microphysics*. MIT Press. pp. 195–203. ISBN 0-262-52424-4.

[5] Thomson, J.J. (1897). "Cathode Rays". *Philosophical Magazine* **44** (269): 293. doi:10.1080/14786449708621070.

[6] P.J. Mohr, B.N. Taylor, and D.B. Newell (2011). "The 2010 CODATA Recommended Values of the Fundamental Physical Constants" (Web Version 6.0). This database was developed by J. Baker, M. Douma, and S. Kotochigova. Available: http://physics.nist.gov/constants [Thursday, 02-Jun-2011 21:00:12 EDT]. National Institute of Standards and Technology, Gaithersburg, MD 20899.

[7] "JERRY COFF". Retrieved 10 September 2010.

[8] Curtis, L.J. (2003). *Atomic Structure and Lifetimes: A Conceptual Approach*. Cambridge University Press. p. 74. ISBN 0-521-53635-9.

[9] "CODATA value: proton-electron mass ratio". *2006 CODATA recommended values*. National Institute of Standards and Technology. Retrieved 2009-07-18.

[10] Anastopoulos, C. (2008). *Particle Or Wave: The Evolution of the Concept of Matter in Modern Physics*. Princeton University Press. pp. 236–237. ISBN 0-691-13512-6.

[11] Pauling, L.C. (1960). *The Nature of the Chemical Bond and the Structure of Molecules and Crystals: an introduction to modern structural chemistry* (3rd ed.). Cornell University Press. pp. 4–10. ISBN 0-8014-0333-2.

[12] Dahl (1997:122–185).

[13] Wilson, R. (1997). *Astronomy Through the Ages: The Story of the Human Attempt to Understand the Universe*. CRC Press. p. 138. ISBN 0-7484-0748-0.

[14] Shipley, J.T. (1945). *Dictionary of Word Origins*. The Philosophical Library. p. 133. ISBN 0-88029-751-4.

[15] Baigrie, B. (2006). *Electricity and Magnetism: A Historical Perspective*. Greenwood Press. pp. 7–8. ISBN 0-313-33358-0.

[16] Keithley, J.F. (1999). *The Story of Electrical and Magnetic Measurements: From 500 B.C. to the 1940s*. IEEE Press. pp. 15, 20. ISBN 0-7803-1193-0.

[17] "Benjamin Franklin (1706–1790)". *Eric Weisstein's World of Biography*. Wolfram Research. Retrieved 2010-12-16.

[18] Myers, R.L. (2006). *The Basics of Physics*. Greenwood Publishing Group. p. 242. ISBN 0-313-32857-9.

[19] Barrow, J.D. (1983). "Natural Units Before Planck". *Quarterly Journal of the Royal Astronomical Society* **24**: 24–26. Bibcode:1983QJRAS..24...24B.

[20] Sōgo Okamura (1994). *History of Electron Tubes*. IOS Press. p. 11. ISBN 978-90-5199-145-1. Retrieved 29 May 2015. In 1881, Stoney named this electromagnetic 'electrolion'. It came to be called 'electron' from 1891. [...] In 1906, the suggestion to call cathode ray particles 'electrions' was brought up but through the opinion of Lorentz of Holland 'electrons' came to be widely used.

[21] Stoney, G.J. (1894). "Of the "Electron," or Atom of Electricity". *Philosophical Magazine* **38** (5): 418–420. doi:10.1080/14786449408620653.

[22] "electron, n.2". OED Online. March 2013. Oxford University Press. Accessed 12 April 2013

[23] Soukhanov, A.H. ed. (1986). *Word Mysteries & Histories*. Houghton Mifflin Company. p. 73. ISBN 0-395-40265-4.

[24] Guralnik, D.B. ed. (1970). *Webster's New World Dictionary*. Prentice Hall. p. 450.

[25] Born, M.; Blin-Stoyle. R.J.; Radcliffe, J.M. (1989). *Atomic Physics*. Courier Dover. p. 26. ISBN 0-486-65984-4.

[26] Dahl (1997:55–58).

[27] DeKosky, R.K. (1983). "William Crookes and the quest for absolute vacuum in the 1870s". *Annals of Science* **40** (1): 1–18. doi:10.1080/00033798300200101.

[28] Leicester. H.M. (1971). *The Historical Background of Chemistry*. Courier Dover. pp. 221–222. ISBN 0-486-61053-5.

[29] Dahl (1997:64–78).

[30] Zeeman. P.; Zeeman, P. (1907). "Sir William Crookes, F.R.S". *Nature* **77** (1984): 1–3. Bibcode:1907Natur..77....1C. doi:10.1038/077001a0.

[31] Dahl (1997:99).

[32] Frank Wilczek: "Happy Birthday, Electron" *Scientific American*. June 2012.

[33] Thomson, J.J. (1906). "Nobel Lecture: Carriers of Negative Electricity" (PDF). The Nobel Foundation. Retrieved 2008-08-25.

[34] Trenn, T.J. (1976). "Rutherford on the Alpha-Beta-Gamma Classification of Radioactive Rays". *Isis* **67** (1): 61–75. doi:10.1086/351545. JSTOR 231134.

[35] Becquerel, H. (1900). "Déviation du Rayonnement du Radium dans un Champ Électrique". *Comptes rendus de l'Académie des sciences* (in French) **130**: 809–815.

[36] Buchwald and Warwick (2001:90–91).

[37] Myers, W.G. (1976). "Becquerel's Discovery of Radioactivity in 1896". *Journal of Nuclear Medicine* **17** (7): 579–582. PMID 775027.

[38] Kikoin, I.K.; Sominskiĭ, I.S. (1961). "Abram Fedorovich Ioffe (on his eightieth birthday)". *Soviet Physics Uspekhi* **3** (5): 798–809. Bibcode:1961SvPhU...3..798K. doi:10.1070/PU1961v003n05ABEH005812. Original publication in Russian: Кикоин, И.К.; Соминский, М.С. (1960). "Академик А.Ф. Иоффе" (PDF). *Успехи Физических Наук* **72** (10): 303–321.

[39] Millikan, R.A. (1911). "The Isolation of an Ion, a Precision Measurement of its Charge, and the Correction of Stokes' Law". *Physical Review* **32** (2): 349–397. Bibcode:1911PhRvI..32..349M. doi:10.1103/PhysRevSeriesI.32.349.

[40] Das Gupta, N.N.; Ghosh, S.K. (1999). "A Report on the Wilson Cloud Chamber and Its Applications in Physics". *Reviews of Modern Physics* **18** (2): 225–290. Bibcode:1946RvMP...18..225G. doi:10.1103/RevModPhys.18.225.

[41] Smirnov, B.M. (2003). *Physics of Atoms and Ions*. Springer. pp. 14–21. ISBN 0-387-95550-X.

[42] Bohr, N. (1922). "Nobel Lecture: The Structure of the Atom" (PDF). The Nobel Foundation. Retrieved 2008-12-03.

[43] Lewis, G.N. (1916). "The Atom and the Molecule". *Journal of the American Chemical Society* **38** (4): 762–786. doi:10.1021/ja02261a002.

[44] Arabatzis, T.; Gavroglu, K. (1997). "The chemists' electron". *European Journal of Physics* **18** (3): 150–163. Bibcode:1997EJPh...18..150A. doi:10.1088/0143-0807/18/3/005.

[45] Langmuir, I. (1919). "The Arrangement of Electrons in Atoms and Molecules". *Journal of the American Chemical Society* **41** (6): 868–934. doi:10.1021/ja02227a002.

[46] Scerri, E.R. (2007). *The Periodic Table*. Oxford University Press. pp. 205–226. ISBN 0-19-530573-6.

[47] Massimi, M. (2005). *Pauli's Exclusion Principle, The Origin and Validation of a Scientific Principle*. Cambridge University Press. pp. 7–8. ISBN 0-521-83911-4.

[48] Uhlenbeck. G.E.; Goudsmith, S. (1925). "Ersetzung der Hypothese vom unmechanischen Zwang durch eine Forderung bezüglich des inneren Verhaltens jedes einzelnen Elektrons". *Die Naturwissenschaften* (in German) **13** (47): 953. Bibcode:1925NW.....13..953E. doi:10.1007/BF01558878.

[49] Pauli, W. (1923). "Über die Gesetzmäßigkeiten des anomalen Zeemaneffektes". *Zeitschrift für Physik* (in German) **16** (1): 155–164. Bibcode:1923ZPhy...16..155P. doi:10.1007/BF01327386.

[50] de Broglie, L. (1929). "Nobel Lecture: The Wave Nature of the Electron" (PDF). The Nobel Foundation. Retrieved 2008-08-30.

[51] Falkenburg, B. (2007). *Particle Metaphysics: A Critical Account of Subatomic Reality*. Springer. p. 85. ISBN 3-540-33731-8.

[52] Davisson, C. (1937). "Nobel Lecture: The Discovery of Electron Waves" (PDF). The Nobel Foundation. Retrieved 2008-08-30.

[53] Schrödinger. E. (1926). "Quantisierung als Eigenwertproblem". *Annalen der Physik* (in German) **385** (13): 437–490. Bibcode:1926AnP...385..437S. doi:10.1002/andp.19263851302.

[54] Rigden, J.S. (2003). *Hydrogen*. Harvard University Press. pp. 59–86. ISBN 0-674-01252-6.

[55] Reed, B.C. (2007). *Quantum Mechanics*. Jones & Bartlett Publishers. pp. 275–350. ISBN 0-7637-4451-4.

[56] Dirac, P.A.M. (1928). "The Quantum Theory of the Electron". *Proceedings of the Royal Society A* **117** (778): 610–624. Bibcode:1928RSPSA.117..610D. doi:10.1098/rspa.1928.0023.

[57] Dirac, P.A.M. (1933). "Nobel Lecture: Theory of Electrons and Positrons" (PDF). The Nobel Foundation. Retrieved 2008-11-01.

[58] "The Nobel Prize in Physics 1965". The Nobel Foundation. Retrieved 2008-11-04.

[59] Panofsky, W.K.H. (1997). "The Evolution of Particle Accelerators & Colliders" (PDF). *Beam Line* (Stanford University) **27** (1): 36–44. Retrieved 2008-09-15.

[60] Elder, F.R.; et al. (1947). "Radiation from Electrons in a Synchrotron". *Physical Review* **71** (11): 829–830. Bibcode:1947PhRv...71..829E. doi:10.1103/PhysRev.71.829.5.

[61] Hoddeson, L.; et al. (1997). *The Rise of the Standard Model: Particle Physics in the 1960s and 1970s*. Cambridge University Press. pp. 25–26. ISBN 0-521-57816-7.

[62] Bernardini, C. (2004). "AdA: The First Electron–Positron Collider". *Physics in Perspective* **6** (2): 156–183. Bibcode:2004PhP.....6..156B. doi:10.1007/s00016-003-0202-y.

[63] "Testing the Standard Model: The LEP experiments". CERN. 2008. Retrieved 2008-09-15.

[64] "LEP reaps a final harvest". *CERN Courier* **40** (10). 2000.

[65] Prati, E.; De Michielis, M.; Belli, M.; Cocco, S.; Fanciulli, M.; Kotekar-Patil, D.; Ruoff, M.; Kern, D. P.; Wharam, D. A.; Verduijn, J.; Tettamanzi, G. C.; Rogge, S.; Roche, B.; Wacquez, R.; Jehl, X.; Vinet, M.; Sanquer, M. (2012). "Few electron limit of n-type metal oxide semiconductor single electron transistors". *Nanotechnology* **23** (21): 215204. arXiv:1203.4811. Bibcode:2012Nanot..23u5204P. doi:10.1088/0957-4484/23/21/215204. PMID 22552118.

[66] Frampton, P.H.; Hung, P.Q.; Sher, Marc (2000). "Quarks and Leptons Beyond the Third Generation". *Physics Reports* **330** (5–6): 263–348. arXiv:hep-ph/9903387. Bibcode:2000PhR...330..263F. doi:10.1016/S0370-1573(99)00095-2.

[67] Raith, W.; Mulvey, T. (2001). *Constituents of Matter: Atoms, Molecules, Nuclei and Particles*. CRC Press. pp. 777–781. ISBN 0-8493-1202-7.

[68] The original source for CODATA is Mohr, P.J.; Taylor, B.N.; Newell, D.B. (2006). "CODATA recommended values of the fundamental physical constants". *Reviews of Modern Physics* **80** (2): 633–730. arXiv:0801.0028. Bibcode:2008RvMP...80..633M. doi:10.1103/RevModPhys.80.633.

Individual physical constants from the CODATA are available at: "The NIST Reference on Constants, Units and Uncertainty". National Institute of Standards and Technology. Retrieved 2009-01-15.

[69] Zombeck, M.V. (2007). *Handbook of Space Astronomy and Astrophysics* (3rd ed.). Cambridge University Press. p. 14. ISBN 0-521-78242-2.

[70] Murphy, M.T.; et al. (2008). "Strong Limit on a Variable Proton-to-Electron Mass Ratio from Molecules in the Distant Universe". *Science* **320** (5883): 1611–1613. arXiv:0806.3081. Bibcode:2008Sci...320.1611M. doi:10.1126/science.1156352. PMID 18566280.

[71] Zorn, J.C.; Chamberlain, G.E.; Hughes, V.W. (1963). "Experimental Limits for the Electron-Proton Charge Difference and for the Charge of the Neutron". *Physical Review* **129** (6): 2566–2576. Bibcode:1963PhRv..129.2566Z. doi:10.1103/PhysRev.129.2566.

[72] Odom, B.; et al. (2006). "New Measurement of the Electron Magnetic Moment Using a One-Electron Quantum Cyclotron". *Physical Review Letters* **97** (3): 030801. Bibcode:2006PhRvL..97c0801O. doi:10.1103/PhysRevLett.97.030801. PMID 16907490.

[73] Anastopoulos, C. (2008). *Particle Or Wave: The Evolution of the Concept of Matter in Modern Physics*. Princeton University Press. pp. 261–262. ISBN 0-691-13512-6.

[74] Gabrielse, G.; et al. (2006). "New Determination of the Fine Structure Constant from the Electron g Value and QED". *Physical Review Letters* **97** (3): 030802(1–4). Bibcode:2006PhRvL..97c0802G. doi:10.1103/PhysRevLett.97.030802.

[75] Eduard Shpolsky. Atomic physics (Atomnaia fizika).second edition. 1951

[76] Dehmelt, H. (1988). "A Single Atomic Particle Forever Floating at Rest in Free Space: New Value for Electron Radius". *Physica Scripta* **T22**: 102–10. Bibcode:1988PhST...22..102D. doi:10.1088/0031-8949/1988/T22/016.

[77] Meschede, D. (2004). *Optics, light and lasers: The Practical Approach to Modern Aspects of Photonics and Laser Physics*. Wiley-VCH. p. 168. ISBN 3-527-40364-7.

[78] Steinberg, R.I.; et al. (1999). "Experimental test of charge conservation and the stability of the electron". *Physical Review D* **61** (2): 2582–2586. Bibcode:1975PhRvD..12.2582S. doi:10.1103/PhysRevD.12.2582.

[79] J. Beringer (Particle Data Group); et al. (2012). "Review of Particle Physics: [electron properties]" (PDF). *Physical Review D* **86** (1): 010001. Bibcode:2012PhRvD..86a0001B. doi:10.1103/PhysRevD.86.010001.

[80] Back, H. O.; et al. (2002). "Search for electron decay mode e → γ + ν with prototype of Borexino detector". *Physics Letters B* **525**: 29–40. Bibcode:2002PhLB..525...29B. doi:10.1016/S0370-2693(01)01440-X.

[81] Munowitz, M. (2005). *Knowing, The Nature of Physical Law*. Oxford University Press. ISBN 0-19-516737-6.

[82] Kane, G. (October 9, 2006). "Are virtual particles really constantly popping in and out of existence? Or are they merely a mathematical bookkeeping device for quantum mechanics?". Scientific American. Retrieved 2008-09-19.

[83] Taylor, J. (1989). "Gauge Theories in Particle Physics". In Davies, Paul. *The New Physics*. Cambridge University Press. p. 464. ISBN 0-521-43831-4.

[84] Genz, H. (2001). *Nothingness: The Science of Empty Space*. Da Capo Press. pp. 241–243, 245–247. ISBN 0-7382-0610-5.

[85] Gribbin, J. (January 25, 1997). "More to electrons than meets the eye". *New Scientist*. Retrieved 2008-09-17.

[86] Levine, I.; et al. (1997). "Measurement of the Electromagnetic Coupling at Large Momentum Transfer". *Physical Review Letters* **78** (3): 424–427. Bibcode:1997PhRvL..78..424L. doi:10.1103/PhysRevLett.78.424.

[87] Murayama, H. (March 10–17, 2006). *Supersymmetry Breaking Made Easy, Viable and Generic*. Proceedings of the XLIInd Rencontres de Moriond on Electroweak Interactions and Unified Theories (La Thuile, Italy). arXiv:0709.3041. —lists a 9% mass difference for an electron that is the size of the Planck distance.

[88] Schwinger, J. (1948). "On Quantum-Electrodynamics and the Magnetic Moment of the Electron". *Physical Review* **73** (4): 416–417. Bibcode:1948PhRv...73..416S. doi:10.1103/PhysRev.73.416.

[89] Huang, K. (2007). *Fundamental Forces of Nature: The Story of Gauge Fields*. World Scientific. pp. 123–125. ISBN 981-270-645-3.

[90] Foldy, L.L.; Wouthuysen, S. (1950). "On the Dirac Theory of Spin 1/2 Particles and Its Non-Relativistic Limit". *Physical Review* **78**: 29–36. Bibcode:1950PhRv...78...29F. doi:10.1103/PhysRev.78.29.

[91] Sidharth, B.G. (2008). "Revisiting Zitterbewegung". *International Journal of Theoretical Physics* **48** (2): 497–506. arXiv:0806.0985. Bibcode:2009IJTP...48..497S. doi:10.1007/s10773-008-9825-8.

[92] Elliott, R.S. (1978). "The History of Electromagnetics as Hertz Would Have Known It". *IEEE Transactions on Microwave Theory and Techniques* **36** (5): 806–823. Bibcode:1988ITMTT..36..806E. doi:10.1109/22.3600.

[93] Crowell, B. (2000). *Electricity and Magnetism*. Light and Matter. pp. 129–152. ISBN 0-9704670-4-4.

[94] Mahadevan, R.; Narayan, R.; Yi, I. (1996). "Harmony in Electrons: Cyclotron and Synchrotron Emission by Thermal Electrons in a Magnetic Field". *The Astrophysical Journal* **465**: 327–337. arXiv:astro-ph/9601073. Bibcode:1996ApJ...465..327M. doi:10.1086/177422.

[95] Rohrlich, F. (1999). "The Self-Force and Radiation Reaction". *American Journal of Physics* **68** (12): 1109–1112. Bibcode:2000AmJPh..68.1109R. doi:10.1119/1.1286430.

[96] Georgi, H. (1989). "Grand Unified Theories". In Davies, Paul. *The New Physics*. Cambridge University Press. p. 427. ISBN 0-521-43831-4.

[97] Blumenthal, G.J.; Gould, R. (1970). "Bremsstrahlung, Synchrotron Radiation, and Compton Scattering of High-Energy Electrons Traversing Dilute Gases". *Reviews of Modern Physics* **42** (2): 237–270. Bibcode:1970RvMP...42..237B. doi:10.1103/RevModPhys.42.237.

[98] Staff (2008). "The Nobel Prize in Physics 1927". The Nobel Foundation. Retrieved 2008-09-28.

[99] Chen, S.-Y.; Maksimchuk, A.; Umstadter, D. (1998). "Experimental observation of relativistic nonlinear Thomson scattering". *Nature* **396** (6712): 653–655. arXiv:physics/9810036. Bibcode:1998Natur.396..653C. doi:10.1038/25303.

[100] Beringer, R.; Montgomery, C.G. (1942). "The Angular Distribution of Positron Annihilation Radiation". *Physical Review* **61** (5–6): 222–224. Bibcode:1942PhRv...61..222B. doi:10.1103/PhysRev.61.222.

[101] Buffa, A. (2000). *College Physics* (4th ed.). Prentice Hall. p. 888. ISBN 0-13-082444-5.

[102] Eichler, J. (2005). "Electron–positron pair production in relativistic ion–atom collisions". *Physics Letters A* **347** (1–3): 67–72. Bibcode:2005PhLA..347...67E. doi:10.1016/j.physleta.2005.06.105.

[103] Hubbell, J.H. (2006). "Electron positron pair production by photons: A historical overview". *Radiation Physics and Chemistry* **75** (6): 614–623. Bibcode:2006RaPC...75..614H. doi:10.1016/j.radphyschem.2005.10.008.

[104] Quigg, C. (June 4–30, 2000). *The Electroweak Theory*. TASI 2000: Flavor Physics for the Millennium (Boulder, Colorado): 80. arXiv:hep-ph/0204104.

[105] Mulliken, R.S. (1967). "Spectroscopy, Molecular Orbitals, and Chemical Bonding". *Science* **157** (3784): 13–24. Bibcode:1967Sci...157...13M. doi:10.1126/science.157.3784.13. PMID 5338306.

[106] Burhop, E.H.S. (1952). *The Auger Effect and Other Radiationless Transitions*. Cambridge University Press. pp. 2–3. ISBN 0-88275-966-3.

[107] Grupen, C. (2000). "Physics of Particle Detection". *AIP Conference Proceedings* **536**: 3–34. arXiv:physics/9906063. doi:10.1063/1.1361756.

[108] Jiles, D. (1998). *Introduction to Magnetism and Magnetic Materials*. CRC Press. pp. 280–287. ISBN 0-412-79860-3.

[109] Löwdin, P.O.; Erkki Brändas, E.; Kryachko, E.S. (2003). *Fundamental World of Quantum Chemistry: A Tribute to the Memory of Per-Olov Löwdin*. Springer. pp. 393–394. ISBN 1-4020-1290-X.

[110] McQuarrie, D.A.; Simon, J.D. (1997). *Physical Chemistry: A Molecular Approach*. University Science Books. pp. 325–361. ISBN 0-935702-99-7.

[111] Daudel, R.; et al. (1973). "The Electron Pair in Chemistry". *Canadian Journal of Chemistry* **52** (8): 1310–1320. doi:10.1139/v74-201.

[112] Rakov, V.A.; Uman, M.A. (2007). *Lightning: Physics and Effects*. Cambridge University Press. p. 4. ISBN 0-521-03541-4.

[113] Freeman, G.R.; March, N.H. (1999). "Triboelectricity and some associated phenomena". *Materials Science and Technology* **15** (12): 1454–1458. doi:10.1179/026708399101505464.

[114] Forward, K.M.; Lacks, D.J.; Sankaran, R.M. (2009). "Methodology for studying particle–particle triboelectrification in granular materials". *Journal of Electrostatics* **67** (2–3): 178–183. doi:10.1016/j.elstat.2008.12.002.

[115] Weinberg, S. (2003). *The Discovery of Subatomic Particles*. Cambridge University Press. pp. 15–16. ISBN 0-521-82351-X.

[116] Lou, L.-F. (2003). *Introduction to phonons and electrons*. World Scientific. pp. 162, 164. ISBN 978-981-238-461-4.

[117] Guru, B.S.; Hızıroğlu, H.R. (2004). *Electromagnetic Field Theory*. Cambridge University Press. pp. 138, 276. ISBN 0-521-83016-8.

[118] Achuthan, M.K.; Bhat, K.N. (2007). *Fundamentals of Semiconductor Devices*. Tata McGraw-Hill. pp. 49–67. ISBN 0-07-061220-X.

[119] Ziman, J.M. (2001). *Electrons and Phonons: The Theory of Transport Phenomena in Solids*. Oxford University Press. p. 260. ISBN 0-19-850779-8.

[120] Main, P. (June 12, 1993). "When electrons go with the flow: Remove the obstacles that create electrical resistance, and you get ballistic electrons and a quantum surprise". *New Scientist* **1887**: 30. Retrieved 2008-10-09.

[121] Blackwell, G.R. (2000). *The Electronic Packaging Handbook*. CRC Press. pp. 6.39–6.40. ISBN 0-8493-8591-1.

[122] Durrant, A. (2000). *Quantum Physics of Matter: The Physical World*. CRC Press. pp. 43, 71–78. ISBN 0-7503-0721-8.

[123] Staff (2008). "The Nobel Prize in Physics 1972". The Nobel Foundation. Retrieved 2008-10-13.

[124] Kadin, A.M. (2007). "Spatial Structure of the Cooper Pair". *Journal of Superconductivity and Novel Magnetism* **20** (4): 285–292. arXiv:cond-mat/0510279. doi:10.1007/s10948-006-0198-z.

[125] "Discovery About Behavior Of Building Block Of Nature Could Lead To Computer Revolution". *ScienceDaily*. July 31, 2009. Retrieved 2009-08-01.

[126] Jompol, Y.; et al. (2009). "Probing Spin-Charge Separation in a Tomonaga-Luttinger Liquid". *Science* **325** (5940): 597–601. arXiv:1002.2782. Bibcode:2009Sci...325..597J. doi:10.1126/science.1171769. PMID 19644117.

[127] Staff (2008). "The Nobel Prize in Physics 1958, for the discovery and the interpretation of the Cherenkov effect". The Nobel Foundation. Retrieved 2008-09-25.

[128] Staff (August 26, 2008). "Special Relativity". Stanford Linear Accelerator Center. Retrieved 2008-09-25.

[129] Adams, S. (2000). *Frontiers: Twentieth Century Physics*. CRC Press. p. 215. ISBN 0-7484-0840-1.

[130] Lurquin, P.F. (2003). *The Origins of Life and the Universe*. Columbia University Press. p. 2. ISBN 0-231-12655-7.

[131] Silk, J. (2000). *The Big Bang: The Creation and Evolution of the Universe* (3rd ed.). Macmillan. pp. 110–112, 134–137. ISBN 0-8050-7256-X.

[132] Christianto, V. (2007). "Thirty Unsolved Problems in the Physics of Elementary Particles" (PDF). *Progress in Physics* **4**: 112–114.

[133] Kolb, E.W.; Wolfram, Stephen (1980). "The Development of Baryon Asymmetry in the Early Universe". *Physics Letters B* **91** (2): 217–221. Bibcode:1980PhLB...91..217K. doi:10.1016/0370-2693(80)90435-9.

[134] Sather, E. (Spring–Summer 1996). "The Mystery of Matter Asymmetry" (PDF). *Beam Line*. University of Stanford. Retrieved 2008-11-01.

[135] Burles, S.; Nollett, K.M.; Turner, M.S. (1999). "Big-Bang Nucleosynthesis: Linking Inner Space and Outer Space". arXiv:astro-ph/9903300 [astro-ph].

[136] Boesgaard, A.M.; Steigman, G. (1985). "Big bang nucleosynthesis – Theories and observations". *Annual Review of Astronomy and Astrophysics* **23** (2): 319–378. Bibcode:1985ARA&A..23..319B. doi:10.1146/annurev.aa.23.090185.001535.

[137] Barkana, R. (2006). "The First Stars in the Universe and Cosmic Reionization". *Science* **313** (5789): 931–934. arXiv:astro-ph/0608450. Bibcode:2006Sci...313..931B. doi:10.1126/science.1125644. PMID 16917052.

[138] Burbidge, E.M.; et al. (1957). "Synthesis of Elements in Stars". *Reviews of Modern Physics* **29** (4): 548–647. Bibcode:1957RvMP...29..547B. doi:10.1103/RevModPhys.29.547.

[139] Rodberg, L.S.; Weisskopf, V. (1957). "Fall of Parity: Recent Discoveries Related to Symmetry of Laws of Nature". *Science* **125** (3249): 627–633. Bibcode:1957Sci...125..627R. doi:10.1126/science.125.3249.627. PMID 17810563.

[140] Fryer, C.L. (1999). "Mass Limits For Black Hole Formation". *The Astrophysical Journal* **522** (1): 413–418. arXiv:astro-ph/9902315. Bibcode:1999ApJ...522..413F. doi:10.1086/307647.

[141] Parikh, M.K.; Wilczek, F. (2000). "Hawking Radiation As Tunneling". *Physical Review Letters* **85** (24): 5042–5045. arXiv:hep-th/9907001. Bibcode:2000PhRvL..85.5042P. doi:10.1103/PhysRevLett.85.5042. PMID 11102182.

[142] Hawking, S.W. (1974). "Black hole explosions?". *Nature* **248** (5443): 30–31. Bibcode:1974Natur.248...30H. doi:10.1038/248030a0.

[143] Halzen, F.; Hooper, D. (2002). "High-energy neutrino astronomy: the cosmic ray connection". *Reports on Progress in Physics* **66** (7): 1025–1078. arXiv:astro-ph/0204527. Bibcode:2002astro.ph..4527H. doi:10.1088/0034-4885/65/7/201.

[144] Ziegler, J.F. (1998). "Terrestrial cosmic ray intensities". *IBM Journal of Research and Development* **42** (1): 117–139. doi:10.1147/rd.421.0117.

[145] Sutton, C. (August 4, 1990). "Muons, pions and other strange particles". *New Scientist*. Retrieved 2008-08-28.

[146] Wolpert, S. (July 24, 2008). "Scientists solve 30-year-old aurora borealis mystery". University of California. Retrieved 2008-10-11.

[147] Gurnett, D.A.; Anderson, R. (1976). "Electron Plasma Oscillations Associated with Type III Radio Bursts". *Science* **194** (4270): 1159–1162. Bibcode:1976Sci...194.1159G. doi:10.1126/science.194.4270.1159. PMID 17790910.

[148] Martin, W.C.; Wiese, W.L. (2007). "Atomic Spectroscopy: A Compendium of Basic Ideas, Notation, Data, and Formulas". National Institute of Standards and Technology. Retrieved 2007-01-08.

[149] Fowles, G.R. (1989). *Introduction to Modern Optics*. Courier Dover. pp. 227–233. ISBN 0-486-65957-7.

[150] Staff (2008). "The Nobel Prize in Physics 1989". The Nobel Foundation. Retrieved 2008-09-24.

[151] Ekstrom, P.; Wineland, David (1980). "The isolated Electron" (PDF). *Scientific American* **243** (2): 91–101. doi:10.1038/scientificamerican0880-104. Retrieved 2008-09-24.

[152] Mauritsson, J. "Electron filmed for the first time ever" (PDF). Lund University. Archived from the original (PDF) on March 25, 2009. Retrieved 2008-09-17.

[153] Mauritsson, J.; et al. (2008). "Coherent Electron Scattering Captured by an Attosecond Quantum Stroboscope". *Physical Review Letters* **100** (7): 073003. arXiv:0708.1060. Bibcode:2008PhRvL.100g3003M. doi:10.1103/PhysRevLett.100.073003. PMID 18352546.

[154] Damascelli, A. (2004). "Probing the Electronic Structure of Complex Systems by ARPES". *Physica Scripta* **T109**: 61–74. arXiv:cond-mat/0307085. Bibcode:2004PhST..109...61D. doi:10.1238/Physica.Topical.109a00061.

[155] Staff (April 4, 1975). "Image # L-1975-02972". Langley Research Center, NASA. Retrieved 2008-09-20.

[156] Elmer, J. (March 3, 2008). "Standardizing the Art of Electron-Beam Welding". Lawrence Livermore National Laboratory. Retrieved 2008-10-16.

[157] Schultz, H. (1993). *Electron Beam Welding*. Woodhead Publishing. pp. 2–3. ISBN 1-85573-050-2.

[158] Benedict, G.F. (1987). *Nontraditional Manufacturing Processes*. Manufacturing engineering and materials processing **19**. CRC Press. p. 273. ISBN 0-8247-7352-7.

[159] Ozdemir, F.S. (June 25–27, 1979). *Electron beam lithography*. Proceedings of the 16th Conference on Design automation (San Diego, CA, USA: IEEE Press): 383–391. Retrieved 2008-10-16.

[160] Madou, M.J. (2002). *Fundamentals of Microfabrication: the Science of Miniaturization* (2nd ed.). CRC Press. pp. 53–54. ISBN 0-8493-0826-7.

[161] Jongen, Y.; Herer, A. (May 2–5, 1996). *Electron Beam Scanning in Industrial Applications*. APS/AAPT Joint Meeting (American Physical Society). Bibcode:1996APS..MAY.H9902J.

[162] Mobus G. et al. (2010). Journal of Nuclear Materials, v. 396, 264–271, doi:10.1016/j.jnucmat.2009.11.020

[163] Beddar, A.S.; Domanovic, Mary Ann; Kubu, Mary Lou; Ellis, Rod J.; Sibata, Claudio H.; Kinsella, Timothy J. (2001). "Mobile linear accelerators for intraoperative radiation therapy". *AORN Journal* **74** (5): 700. doi:10.1016/S0001-2092(06)61769-9.

[164] Gazda, M.J.; Coia, L.R. (June 1, 2007). "Principles of Radiation Therapy" (PDF). Retrieved 2013-10-31.

[165] Chao, A.W.; Tigner, M. (1999). *Handbook of Accelerator Physics and Engineering*. World Scientific. pp. 155, 188. ISBN 981-02-3500-3.

[166] Oura, K.; et al. (2003). *Surface Science: An Introduction*. Springer. pp. 1–45. ISBN 3-540-00545-5.

[167] Ichimiya, A.; Cohen, P.I. (2004). *Reflection High-energy Electron Diffraction*. Cambridge University Press. p. 1. ISBN 0-521-45373-9.

[168] Heppell, T.A. (1967). "A combined low energy and reflection high energy electron diffraction apparatus". *Journal of Scientific Instruments* **44** (9): 686–688. Bibcode:1967JScI...44..686H. doi:10.1088/0950-7671/44/9/311.

[169] McMullan, D. (1993). "Scanning Electron Microscopy: 1928–1965". University of Cambridge. Retrieved 2009-03-23.

[170] Slayter, H.S. (1992). *Light and electron microscopy*. Cambridge University Press. p. 1. ISBN 0-521-33948-0.

[171] Cember, H. (1996). *Introduction to Health Physics*. McGraw-Hill Professional. pp. 42–43. ISBN 0-07-105461-8.

[172] Erni, R.; et al. (2009). "Atomic-Resolution Imaging with a Sub-50-pm Electron Probe". *Physical Review Letters* **102** (9): 096101. Bibcode:2009PhRvL.102i6101E. doi:10.1103/PhysRevLett.102.096101. PMID 19392535.

[173] Bozzola, J.J.; Russell, L.D. (1999). *Electron Microscopy: Principles and Techniques for Biologists*. Jones & Bartlett Publishers. pp. 12, 197–199. ISBN 0-7637-0192-0.

[174] Flegler, S.L.; Heckman Jr., J.W.; Klomparens, K.L. (1995). *Scanning and Transmission Electron Microscopy: An Introduction* (Reprint ed.). Oxford University Press. pp. 43–45. ISBN 0-19-510751-9.

[175] Bozzola, J.J.; Russell, L.D. (1999). *Electron Microscopy: Principles and Techniques for Biologists* (2nd ed.). Jones & Bartlett Publishers. p. 9. ISBN 0-7637-0192-0.

[176] Freund, H.P.; Antonsen, T. (1996). *Principles of Free-Electron Lasers*. Springer. pp. 1–30. ISBN 0-412-72540-1.

[177] Kitzmiller, J.W. (1995). *Television Picture Tubes and Other Cathode-Ray Tubes: Industry and Trade Summary*. DIANE Publishing. pp. 3–5. ISBN 0-7881-2100-6.

[178] Sclater, N. (1999). *Electronic Technology Handbook*. McGraw-Hill Professional. pp. 227–228. ISBN 0-07-058048-0.

[179] Staff (2008). "The History of the Integrated Circuit". The Nobel Foundation. Retrieved 2008-10-18.

1.9   External links

- "The Discovery of the Electron". American Institute of Physics, Center for History of Physics.

- "Particle Data Group". University of California.

- Bock, R.K.; Vasilescu, A. (1998). *The Particle Detector BriefBook* (14th ed.). Springer. ISBN 3-540-64120-3.

- Copeland, Ed. "Spherical Electron". *Sixty Symbols*. Brady Haran for the University of Nottingham.

Chapter 2

Subatomic particle

In the physical sciences, **subatomic particles** are particles much smaller than atoms.[1] There are two types of subatomic particles: elementary particles, which according to current theories are not made of other particles; and *composite* particles.[2] Particle physics and nuclear physics study these particles and how they interact.[3]

In particle physics, the concept of a particle is one of several concepts inherited from classical physics. But it also reflects the modern understanding that at the quantum scale matter and energy behave very differently from what much of everyday experience would lead us to expect.

The idea of a particle underwent serious rethinking when experiments showed that light could behave like a stream of particles (called photons) as well as exhibit wave-like properties. This led to the new concept of wave–particle duality to reflect that quantum-scale "particles" behave like both particles and waves (also known as wavicles). Another new concept, the uncertainty principle, states that some of their properties taken together, such as their simultaneous position and momentum, cannot be measured exactly.[4] In more recent times, wave–particle duality has been shown to apply not only to photons but to increasingly massive particles as well.[5]

Interactions of particles in the framework of quantum field theory are understood as creation and annihilation of *quanta* of corresponding fundamental interactions. This blends particle physics with field theory.

2.1 Classification

2.1.1 By statistics

Main article: Spin–statistics theorem

Any subatomic particle, like any particle in the 3-dimensional space that obeys laws of quantum mechanics, can be either a boson (an integer spin) or a fermion (a half-integer spin).

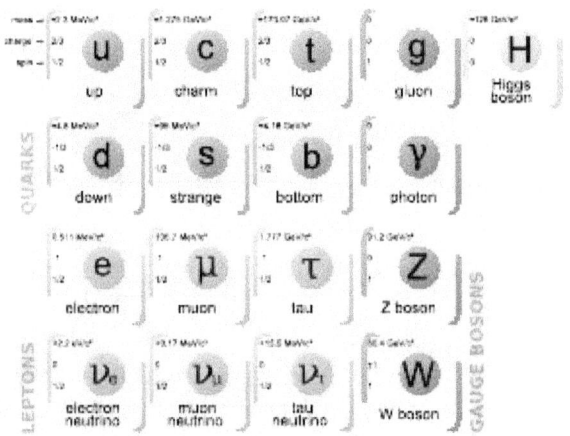

The Standard Model classification of particles

2.1.2 By composition

The elementary particles of the Standard Model include:[6]

- Six "flavors" of quarks: up, down, bottom, top, strange, and charm;

- Six types of leptons: electron, electron neutrino, muon, muon neutrino, tau, tau neutrino;

- Twelve gauge bosons (force carriers): the photon of electromagnetism, the three W and Z bosons of the weak force, and the eight gluons of the strong force;

- The Higgs boson.

Various extensions of the Standard Model predict the existence of an elementary graviton particle and many other elementary particles.

Composite subatomic particles (such as protons or atomic nuclei) are bound states of two or more elementary particles. For example, a proton is made of two up quarks and one down quark, while the atomic nucleus of helium-4 is composed of two protons and two neutrons. The neutron is made of two down quarks and one up quark. Composite particles include all hadrons: these include baryons (such as protons and neutrons) and mesons (such as pions and kaons).

2.1.3 By mass

In special relativity, the energy of a particle at rest equals its mass times the speed of light squared ($E = mc^2$). That is, mass can be expressed in terms of energy and vice versa. If a particle has a frame of reference where it lies at rest, then it has a positive rest mass and is referred to as *massive*.

All composite particles are massive. Baryons (meaning "heavy") tend to have greater mass than mesons (meaning "intermediate"), which in turn tend to be heavier than leptons (meaning "lightweight"), but the heaviest lepton (the tau particle) is heavier than the two lightest flavours of baryons (nucleons). It is also certain that any particle with an electric charge is massive.

All massless particles (particles whose invariant mass is zero) are elementary. These include the photon and gluon, although the latter cannot be isolated.

2.2 Other properties

Through the work of Albert Einstein, Louis de Broglie, and many others, current scientific theory holds that *all* particles also have a wave nature.[7] This has been verified not only for elementary particles but also for compound particles like atoms and even molecules. In fact, according to traditional formulations of non-relativistic quantum mechanics, wave–particle duality applies to all objects, even macroscopic ones; although the wave properties of macroscopic objects cannot be detected due to their small wavelengths.[8]

Interactions between particles have been scrutinized for many centuries, and a few simple laws underpin how particles behave in collisions and interactions. The most fundamental of these are the laws of conservation of energy and conservation of momentum, which let us make calculations of particle interactions on scales of magnitude that range from stars to quarks.[9] These are the prerequisite basics of Newtonian mechanics, a series of statements and equations in *Philosophiae Naturalis Principia Mathematica*, originally published in 1687.

2.3 Dividing an atom

The negatively charged electron has a mass equal to $\frac{1}{1836}$ of that of a hydrogen atom. The remainder of the hydrogen atom's mass comes from the positively charged proton. The atomic number of an element is the number of protons in its nucleus. Neutrons are neutral particles having a mass slightly greater than that of the proton. Different isotopes of the same element contain the same number of protons but differing numbers of neutrons. The mass number of an isotope is the total number of nucleons (neutrons and protons collectively).

Chemistry concerns itself with how electron sharing binds atoms into structures such as crystals and molecules. Nuclear physics deals with how protons and neutrons arrange themselves in nuclei. The study of subatomic particles, atoms and molecules, and their structure and interactions, requires quantum mechanics. Analyzing processes that change the numbers and types of particles requires quantum field theory. The study of subatomic particles *per se* is called particle physics. The term *high-energy physics* is nearly synonymous to "particle physics" since creation of particles requires high energies: it occurs only as a result of cosmic rays, or in particle accelerators. Particle phenomenology systematizes the knowledge about subatomic particles obtained from these experiments.

2.4 History

Main articles: History of subatomic physics and Timeline of particle discoveries

The term "*subatomic* particle" is largely a retronym of 1960s made to distinguish a big number of baryons and mesons (that comprise hadrons) from particles that are now thought to be truly elementary. Before that hadrons were usually classified as "elementary" because their composition was unknown.

A list of important discoveries follows:

2.5 See also

- *Atom: Journey Across the Subatomic Cosmos* (book)

- *Atom: An Odyssey from the Big Bang to Life on Earth...and Beyond* (book)

- CPT invariance

- Dark Matter

- Hot spot effect in subatomic physics

- List of fictional elements, materials, isotopes and atomic particles

- List of particles

- Poincaré symmetry

- Ylem

2.6   References

[1] "Subatomic particles". NTD. Retrieved 5 June 2012.

[2] Bolonkin, Alexander (2011). *Universe, Human Immortality and Future Human Evaluation*. Elsevier. p. 25. ISBN 9780124158016.

[3] Fritzsch, Harald (2005). *Elementary Particles*. World Scientific. pp. 11–20. ISBN 978-981-256-141-1.

[4] Heisenberg, W. (1927). "Über den anschaulichen Inhalt der quantentheoretischen Kinematik und Mechanik", *Zeitschrift für Physik* (in German) **43** (3–4): 172–198. Bibcode:1927ZPhy...43..172H. doi:10.1007/BF01397280.

[5] Arndt, Markus; Nairz, Olaf; Vos-Andreae, Julian; Keller, Claudia; Van Der Zouw, Gerbrand; Zeilinger, Anton (2000). "Wave-particle duality of C60 molecules". *Nature* **401** (6754): 680–682. Bibcode:1999Natur.401..680A. doi:10.1038/44348. PMID 18494170.

[6] Cottingham, W. N.; Greenwood, D. A. (2007). *An introduction to the standard model of particle physics*. Cambridge University Press. p. 1. ISBN 978-0-521-85249-4.

[7] Walter Greiner (2001). *Quantum Mechanics: An Introduction*. Springer. p. 29. ISBN 3-540-67458-6.

[8] R. Eisberg & R. Resnick (1985). *Quantum Physics of Atoms, Molecules, Solids, Nuclei, and Particles* (2nd ed.). John Wiley & Sons. pp. 59–60. ISBN 0-471-87373-X. For both large and small wavelengths, both matter and radiation have both particle and wave aspects. [...] But the wave aspects of their motion become more difficult to observe as their wavelengths become shorter. [...] For ordinary macroscopic particles the mass is so large that the momentum is always sufficiently large to make the de Broglie wavelength small enough to be beyond the range of experimental detection, and classical mechanics reigns supreme.

[9] Isaac Newton (1687).   Newton's Laws of Motion (*Philosophiae Naturalis Principia Mathematica*)

[10] Klemperer, Otto (1959). *Electron Physics: The Physics of the Free Electron*. Academic Press.

[11] Some sources such as The Strange Quark indicate 1947.

[12] http://press.web.cern.ch/press-releases/2014/06/cern-experiments-report-new-higgs-boson-measurements

2.7   Further reading

General readers

- Feynman, R.P. & Weinberg, S. (1987). *Elementary Particles and the Laws of Physics: The 1986 Dirac Memorial Lectures*. Cambridge Univ. Press.

- Brian Greene (1999). *The Elegant Universe*. W.W. Norton & Company. ISBN 0-393-05858-1.

- Oerter, Robert (2006). *The Theory of Almost Everything: The Standard Model, the Unsung Triumph of Modern Physics*. Plume.

- Schumm, Bruce A. (2004). *Deep Down Things: The Breathtaking Beauty of Particle Physics*. Johns Hopkins University Press. ISBN 0-8018-7971-X.

- Martinus Veltman (2003). *Facts and Mysteries in Elementary Particle Physics*. World Scientific. ISBN 981-238-149-X.

Textbooks

- Coughlan, G. D., J. E. Dodd, and B. M. Gripaios (2006). *The Ideas of Particle Physics: An Introduction for Scientists*, 3rd ed. Cambridge Univ. Press. An undergraduate text for those not majoring in physics.

- Griffiths, David J. (1987). *Introduction to Elementary Particles*. Wiley, John & Sons, Inc. ISBN 0-471-60386-4.

- Kane, Gordon L. (1987). *Modern Elementary Particle Physics*. Perseus Books. ISBN 0-201-11749-5.

2.8   External links

- particleadventure.org: The Standard Model.

- cpepweb.org: Particle chart.

- University of California: Particle Data Group.

- Annotated Physics Encyclopædia: Quantum Field Theory.

- Jose Galvez: Chapter 1 Electrodynamics (pdf).

Chapter 3

Electric charge

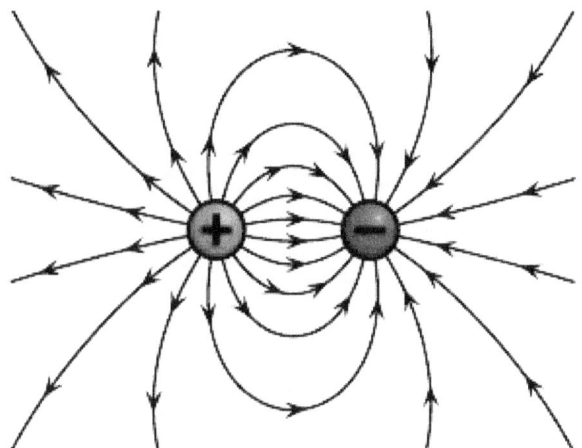

Electric field of a positive and a negative point charge.

Electric charge is the physical property of matter that causes it to experience a force when placed in an electromagnetic field. There are two types of electric charges: positive and negative. Positively charged substances are repelled from other positively charged substances, but attracted to negatively charged substances; negatively charged substances are repelled from negative and attracted to positive. An object is negatively charged if it has an excess of electrons, and is otherwise positively charged or uncharged. The SI derived unit of electric charge is the coulomb (C), although in electrical engineering it is also common to use the ampere-hour (Ah), and in chemistry it is common to use the elementary charge (e) as a unit. The symbol Q is often used to denote charge. The early knowledge of how charged substances interact is now called classical electrodynamics, and is still very accurate if quantum effects do not need to be considered.

The *electric charge* is a fundamental conserved property of some subatomic particles, which determines their electromagnetic interaction. Electrically charged matter is influenced by, and produces, electromagnetic fields. The interaction between a moving charge and an electromagnetic field is the source of the electromagnetic force, which is one of the four fundamental forces (See also: magnetic field).

Twentieth-century experiments demonstrated that electric charge is *quantized*; that is, it comes in integer multiples of individual small units called the elementary charge, e, approximately equal to $1.602 \times 10^{6} - 19$ coulombs (except for particles called quarks, which have charges that are integer multiples of $e/3$). The proton has a charge of $+e$, and the electron has a charge of $-e$. The study of charged particles, and how their interactions are mediated by photons, is called quantum electrodynamics.

3.1   Overview

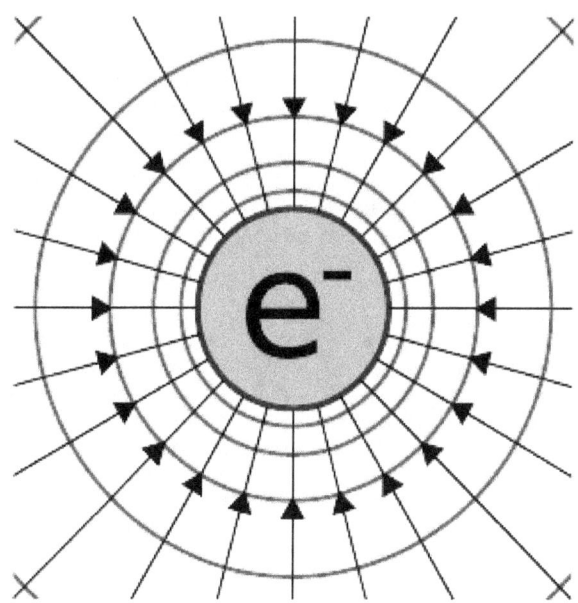

Diagram showing field lines and equipotentials around an electron, a negatively charged particle. In an electrically neutral atom, the number of electrons is equal to the number of protons (which are positively charged), resulting in a net zero overall charge

Charge is the fundamental property of forms of matter that exhibit electrostatic attraction or repulsion in the presence of other matter. Electric charge is a characteristic property of many subatomic particles. The charges of free-standing particles are integer multiples of the elementary charge e; we say that electric charge is *quantized*. Michael Faraday, in his electrolysis experiments, was the first to note the discrete nature of electric charge. Robert Millikan's oil-drop experiment demonstrated this fact directly, and measured the elementary charge.

By convention, the charge of an electron is -1, while that of a proton is $+1$. Charged particles whose charges have the same sign repel one another, and particles whose charges have different signs attract. Coulomb's law quantifies the electrostatic force between two particles by asserting that the force is proportional to the product of their charges, and inversely proportional to the square of the distance between them.

The charge of an antiparticle equals that of the corresponding particle, but with opposite sign. Quarks have fractional charges of either $-\frac{1}{3}$ or $+\frac{2}{3}$, but free-standing quarks have never been observed (the theoretical reason for this fact is asymptotic freedom).

The electric charge of a macroscopic object is the sum of the electric charges of the particles that make it up. This charge is often small, because matter is made of atoms, and atoms typically have equal numbers of protons and electrons, in which case their charges cancel out, yielding a net charge of zero, thus making the atom neutral.

An *ion* is an atom (or group of atoms) that has lost one or more electrons, giving it a net positive charge (cation), or that has gained one or more electrons, giving it a net negative charge (anion). *Monatomic ions* are formed from single atoms, while *polyatomic ions* are formed from two or more atoms that have been bonded together, in each case yielding an ion with a positive or negative net charge.

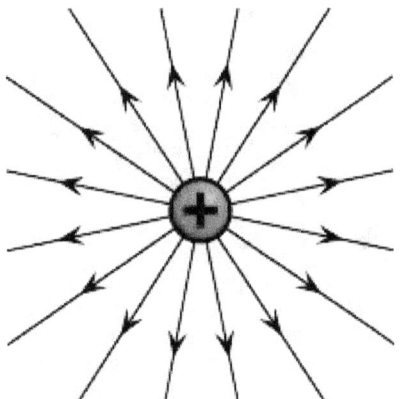

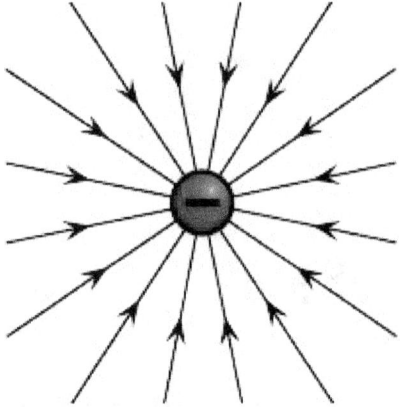

Electric field induced by a positive electric charge (left) and a field induced by a negative electric charge (right).

During formation of macroscopic objects, constituent atoms and ions usually combine to form structures composed of neutral *ionic compounds* electrically bound to neutral atoms. Thus macroscopic objects tend toward being neutral overall, but macroscopic objects are rarely perfectly net neutral.

Sometimes macroscopic objects contain ions distributed throughout the material, rigidly bound in place, giving an overall net positive or negative charge to the object. Also, macroscopic objects made of conductive elements, can more or less easily (depending on the element) take on or give off electrons, and then maintain a net negative or positive charge indefinitely. When the net electric charge of an object is non-zero and motionless, the phenomenon is known as static electricity. This can easily be produced by rubbing two dissimilar materials together, such as rubbing amber with fur or glass with silk. In this way nonconductive materials can be charged to a significant degree, either positively or negatively. Charge taken from one material is moved to the other material, leaving an opposite charge of the same magnitude behind. The law of *conservation of charge* always applies, giving the object from which a negative charge has been taken a positive charge of the same magnitude, and vice versa.

Even when an object's net charge is zero, charge can be distributed non-uniformly in the object (e.g., due to an external electromagnetic field, or bound polar molecules). In such cases the object is said to be polarized. The charge due to polarization is known as bound charge, while charge on an object produced by electrons gained or lost from outside the object is called *free charge*. The motion of electrons in conductive metals in a specific direction is known as electric current.

3.2 Units

The SI unit of quantity of electric charge is the coulomb, which is equivalent to about 6.242×10^{18} e (e is the charge of a proton). Hence, the charge of an electron is approximately -1.602×10^{-19} C. The coulomb is defined as the quantity of charge that has passed through the cross section of an electrical conductor carrying one ampere within one second. The symbol Q is often used to denote a quantity of electricity or charge. The quantity of electric charge can be directly measured with an electrometer, or indirectly measured with a ballistic galvanometer.

After finding the quantized character of charge, in 1891 George Stoney proposed the unit 'electron' for this fundamental unit of electrical charge. This was before the discovery of the particle by J.J. Thomson in 1897. The unit is today treated as nameless, referred to as "elementary charge", "fundamental unit of charge", or simply as "e". A measure of charge should be a multiple of the elementary charge e, even if at large scales charge seems to behave as a real quantity. In some contexts it is meaningful to speak of fractions of a charge; for example in the charging of a capacitor, or in the fractional quantum Hall effect.

In systems of units other than SI such as cgs, electric charge is expressed as combination of only three fundamental quantities such as length, mass and time and not four as in SI where electric charge is a combination of length, mass, time and electric current.

3.3 History

As reported by the ancient Greek mathematician Thales of Miletus around 600 BC, charge (or *electricity*) could be accumulated by rubbing fur on various substances, such as amber. The Greeks noted that the charged amber buttons could attract light objects such as hair. They also noted that if they rubbed the amber for long enough, they could even get an electric spark to jump. This property derives from the triboelectric effect.

In 1600, the English scientist William Gilbert returned to the subject in *De Magnete*, and coined the New Latin word *electricus* from ἤλεκτρον (ēlektron), the Greek word for *amber*, which soon gave rise to the English words "electric" and "electricity." He was followed in 1660 by Otto von Guericke, who invented what was probably the first electrostatic generator. Other European pioneers were Robert Boyle, who in 1675 stated that electric attraction and repulsion can act across a vacuum; Stephen Gray, who in 1729 classified materials as conductors and insulators; and C. F. du Fay, who proposed in 1733 [1] that electricity comes in two varieties that cancel each other, and ex-

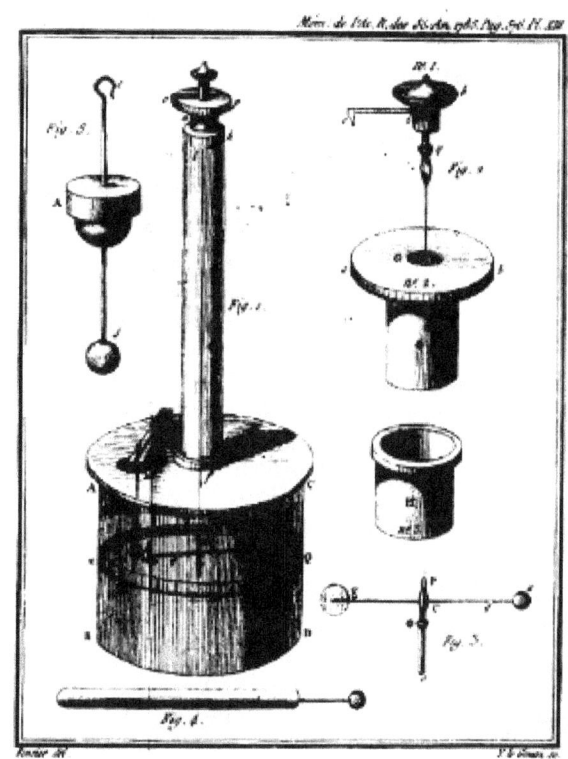

Coulomb's torsion balance

pressed this in terms of a two-fluid theory. When glass was rubbed with silk, du Fay said that the glass was charged with *vitreous electricity*, and, when amber was rubbed with fur, the amber was said to be charged with *resinous electricity*. In 1839, Michael Faraday showed that the apparent division between static electricity, current electricity, and bioelectricity was incorrect, and all were a consequence of the behavior of a single kind of electricity appearing in opposite polarities. It is arbitrary which polarity is called positive and which is called negative. Positive charge can be defined as the charge left on a glass rod after being rubbed with silk. [2]

One of the foremost experts on electricity in the 18th century was Benjamin Franklin, who argued in favour of a one-fluid theory of electricity. Franklin imagined electricity as being a type of invisible fluid present in all matter; for example, he believed that it was the glass in a Leyden jar that held the accumulated charge. He posited that rubbing insulating surfaces together caused this fluid to change location, and that a flow of this fluid constitutes an electric current. He also posited that when matter contained too little of the fluid it was "negatively" charged, and when it had an excess it was "positively" charged. For a reason that was not recorded, he identified the term "positive" with vitreous electricity and "negative" with resinous electricity. William Watson arrived at the same explanation at about the same time.

3.4 Static electricity and electric current

Static electricity and electric current are two separate phenomena. They both involve electric charge, and may occur simultaneously in the same object. Static electricity refers to the electric charge of an object and the related electrostatic discharge when two objects are brought together that are not at equilibrium. An electrostatic discharge creates a change in the charge of each of the two objects. In contrast, electric current is the flow of electric charge through an object, which produces no net loss or gain of electric charge.

3.4.1 Electrification by friction

Further information: triboelectric effect

When a piece of glass and a piece of resin—neither of which exhibit any electrical properties—are rubbed together and left with the rubbed surfaces in contact, they still exhibit no electrical properties. When separated, they attract each other.

A second piece of glass rubbed with a second piece of resin, then separated and suspended near the former pieces of glass and resin causes these phenomena:

- The two pieces of glass repel each other.

- Each piece of glass attracts each piece of resin.

- The two pieces of resin repel each other.

This attraction and repulsion is an *electrical phenomena*, and the bodies that exhibit them are said to be *electrified*, or *electrically charged*. Bodies may be electrified in many other ways, as well as by friction. The electrical properties of the two pieces of glass are similar to each other but opposite to those of the two pieces of resin: The glass attracts what the resin repels and repels what the resin attracts.

If a body electrified in any manner whatsoever behaves as the glass does, that is, if it repels the glass and attracts the resin, the body is said to be 'vitreously' electrified, and if it attracts the glass and repels the resin it is said to be 'resinously' electrified. All electrified bodies are found to be either vitreously or resinously electrified.

It is the established convention of the scientific community to define the vitreous electrification as positive, and the resinous electrification as negative. The exactly opposite properties of the two kinds of electrification justify our indicating them by opposite signs, but the application of the positive sign to one rather than to the other kind must be considered as a matter of arbitrary convention, just as it is a matter of convention in mathematical diagram to reckon positive distances towards the right hand.

No force, either of attraction or of repulsion, can be observed between an electrified body and a body not electrified.[3]

Actually, all bodies are electrified, but may appear not to be so by the relative similar charge of neighboring objects in the environment. An object further electrified + or – creates an equivalent or opposite charge by default in neighboring objects, until those charges can equalize. The effects of attraction can be observed in high-voltage experiments, while lower voltage effects are merely weaker and therefore less obvious. The attraction and repulsion forces are codified by Coulomb's Law (attraction falls off at the square of the distance, which has a corollary for acceleration in a gravitational field, suggesting that gravitation may be merely electrostatic phenomenon between relatively weak charges in terms of scale). See also the Casimir effect.

It is now known that the Franklin/Watson model was fundamentally correct. There is only one kind of electrical charge, and only one variable is required to keep track of the amount of charge.[4] On the other hand, just knowing the charge is not a complete description of the situation. Matter is composed of several kinds of electrically charged particles, and these particles have many properties, not just charge.

The most common charge carriers are the positively charged proton and the negatively charged electron. The movement of any of these charged particles constitutes an electric current. In many situations, it suffices to speak of the *conventional current* without regard to whether it is carried by positive charges moving in the direction of the conventional current or by negative charges moving in the opposite direction. This macroscopic viewpoint is an approximation that simplifies electromagnetic concepts and calculations.

At the opposite extreme, if one looks at the microscopic situation, one sees there are many ways of carrying an electric current, including: a flow of electrons; a flow of electron "holes" that act like positive particles; and both negative and positive particles (ions or other charged particles) flowing in opposite directions in an electrolytic solution or a plasma.

Beware that, in the common and important case of metallic wires, the direction of the conventional current is opposite to the drift velocity of the actual charge carriers, i.e., the electrons. This is a source of confusion for beginners.

3.5 Properties

Aside from the properties described in articles about electromagnetism, charge is a relativistic invariant. This means that any particle that has charge Q, no matter how fast it goes, always has charge Q. This property has been experimentally verified by showing that the charge of *one* helium nucleus (two protons and two neutrons bound together in a nucleus and moving around at high speeds) is the same as *two* deuterium nuclei (one proton and one neutron bound together, but moving much more slowly than they would if they were in a helium nucleus).

3.6 Conservation of electric charge

Main article: Charge conservation

The total electric charge of an isolated system remains constant regardless of changes within the system itself. This law is inherent to all processes known to physics and can be derived in a local form from gauge invariance of the wave function. The conservation of charge results in the charge-current continuity equation. More generally, the net change in charge density ρ within a volume of integration V is equal to the area integral over the current density $\mathbf{J}$ through the closed surface $S = \partial V$, which is in turn equal to the net current I:

$$-\frac{d}{dt} \int_V \rho \, dV = \oiint_{\partial V} \mathbf{J} \cdot d\mathbf{S} = \int J dS \cos\theta = I.$$

Thus, the conservation of electric charge, as expressed by the continuity equation, gives the result:

$$I = \frac{dQ}{dt}.$$

The charge transferred between times t_i and t_f is obtained by integrating both sides:

$$Q = \int_{t_i}^{t_f} I \, dt$$

where I is the net outward current through a closed surface and Q is the electric charge contained within the volume defined by the surface.

3.7 See also

- Quantity of electricity
- SI electromagnetism units

3.8 References

[1] Two Kinds of Electrical Fluid: Vitreous and Resinous – 1733

[2] Electromagnetic Fields (2nd Edition), Roald K. Wangsness, Wiley, 1986. ISBN 0-471-81186-6 (intermediate level textbook)

[3] James Clerk Maxwell A *Treatise on Electricity and Magnetism*, pp. 32-33, Dover Publications Inc., 1954 ASIN: B000HFDK0K, 3rd ed. of 1891

[4] One Kind of Charge

3.9 External links

- How fast does a charge decay?
- Science Aid: Electrostatic charge Easy-to-understand page on electrostatic charge.
- History of the electrical units.

Chapter 4

Fermion

Enrico Fermi

Antisymmetric wavefunction for a (fermionic) 2-particle state in an infinite square well potential.

In particle physics, a **fermion** (a name coined by Paul Dirac°[1] from the surname of Enrico Fermi) is any particle characterized by Fermi–Dirac statistics. These particles obey the Pauli exclusion principle. Fermions include all quarks and leptons, as well as any composite particle made of an odd number of these, such as all baryons and many atoms and nuclei. Fermions differ from bosons, which obey Bose–Einstein statistics.

A fermion can be an elementary particle, such as the electron, or it can be a composite particle, such as the proton. According to the spin-statistics theorem in any reasonable relativistic quantum field theory, particles with integer spin are bosons, while particles with half-integer spin are fermions.

Besides this spin characteristic, fermions have another spe-

cific property: they possess conserved baryon or lepton quantum numbers. Therefore what is usually referred as the spin statistics relation is in fact a spin statistics-quantum number relation.°[2]

As a consequence of the Pauli exclusion principle, only one fermion can occupy a particular quantum state at any given time. If multiple fermions have the same spatial probability distribution, then at least one property of each fermion, such as its spin, must be different. Fermions are usually associated with matter, whereas bosons are generally force carrier particles, although in the current state of particle physics the distinction between the two concepts is unclear. At low temperature fermions show superfluidity for uncharged particles and superconductivity for charged particles. Composite fermions, such as protons and neutrons, are the key building blocks of everyday matter. Weakly interacting fermions can also display bosonic behavior under extreme conditions, such as superconductivity.

4.1 Elementary fermions

The Standard Model recognizes two types of elementary fermions, quarks and leptons. In all, the model distinguishes 24 different fermions. There are six quarks (up, down, strange, charm, bottom and top quarks), and six leptons (electron, electron neutrino, muon, muon neutrino, tau particle and tau neutrino), along with the corresponding antiparticle of each of these.

Mathematically, fermions come in three types - Weyl fermions (massless), Dirac fermions (massive), and Majorana fermions (each its own antiparticle). Most Standard Model fermions are believed to be Dirac fermions, although it is unknown at this time whether the neutrinos are Dirac or Majorana fermions. Dirac fermions can be treated as a combination of two Weyl fermions.[3]:106 In July 2015, Weyl fermions have been experimentally realized in Weyl semimetals.

4.2 Composite fermions

See also: List of particles § Composite particles

Composite particles (such as hadrons, nuclei, and atoms) can be bosons or fermions depending on their constituents. More precisely, because of the relation between spin and statistics, a particle containing an odd number of fermions is itself a fermion. It will have half-integer spin.

Examples include the following:

- A baryon, such as the proton or neutron, contains three fermionic quarks and thus it is a fermion.

- The nucleus of a carbon-13 atom contains six protons and seven neutrons and is therefore a fermion.

- The atom helium-3 (^{3}He) is made of two protons, one neutron, and two electrons, and therefore it is a fermion.

The number of bosons within a composite particle made up of simple particles bound with a potential has no effect on whether it is a boson or a fermion.

Fermionic or bosonic behavior of a composite particle (or system) is only seen at large (compared to size of the system) distances. At proximity, where spatial structure begins to be important, a composite particle (or system) behaves according to its constituent makeup.

Fermions can exhibit bosonic behavior when they become loosely bound in pairs. This is the origin of superconductivity and the superfluidity of helium-3: in superconduct-

ing materials, electrons interact through the exchange of phonons, forming Cooper pairs, while in helium-3, Cooper pairs are formed via spin fluctuations.

The quasiparticles of the fractional quantum Hall effect are also known as composite fermions, which are electrons with an even number of quantized vortices attached to them.

4.2.1 Skyrmions

Main article: Skyrmion

In a quantum field theory, there can be field configurations of bosons which are topologically twisted. These are coherent states (or solitons) which behave like a particle, and they can be fermionic even if all the constituent particles are bosons. This was discovered by Tony Skyrme in the early 1960s, so fermions made of bosons are named skyrmions after him.

Skyrme's original example involved fields which take values on a three-dimensional sphere, the original nonlinear sigma model which describes the large distance behavior of pions. In Skyrme's model, reproduced in the large N or string approximation to quantum chromodynamics (QCD), the proton and neutron are fermionic topological solitons of the pion field.

Whereas Skyrme's example involved pion physics, there is a much more familiar example in quantum electrodynamics with a magnetic monopole. A bosonic monopole with the smallest possible magnetic charge and a bosonic version of the electron will form a fermionic dyon.

The analogy between the Skyrme field and the Higgs field of the electroweak sector has been used[4] to postulate that all fermions are skyrmions. This could explain why all known fermions have baryon or lepton quantum numbers and provide a physical mechanism for the Pauli exclusion principle.

4.3 See also

4.4 Notes

[1] Notes on Dirac's lecture *Developments in Atomic Theory* at Le Palais de la Découverte, 6 December 1945, UK-NATARCHI Dirac Papers BW83/2/257889. See note 64 on page 331 in "The Strangest Man: The Hidden Life of Paul Dirac, Mystic of the Atom" by Graham Farmelo

[2] Physical Review D volume 87, page 0550003, year 2013, author Weiner, Richard M., title "Spin-statistics-quantum number connection and supersymmetry" arxiv:1302.0969

[3] T. Morii; C. S. Lim; S. N. Mukherjee (1 January 2004). *The Physics of the Standard Model and Beyond*. World Scientific. ISBN 978-981-279-560-1.

[4] Weiner, Richard M. (2010). "The Mysteries of Fermions" . *International Journal of Theoretical Physics* **49** (5): 1174–1180. arXiv:0901.3816. Bibcode:2010IJTP...49.1174W. doi:10.1007/s10773-010-0292-7.

Chapter 5

Wave–particle duality

Wave–particle duality is the fact that every elementary particle or quantic entity exhibits the properties of not only particles, but also waves. It addresses the inability of the classical concepts "particle" or "wave" to fully describe the behavior of quantum-scale objects. As Einstein wrote: *"It seems as though we must use sometimes the one theory and sometimes the other, while at times we may use either. We are faced with a new kind of difficulty. We have two contradictory pictures of reality; separately neither of them fully explains the phenomena of light, but together they do".*[1]

Various opinions have arisen about this.

Initiated by Louis de Broglie, before the discovery of quantum mechanics, and developed later as the de Broglie-Bohm theory, the pilot wave interpretation does not regard the duality as paradoxical, seeing both particle and wave aspects as always coexisting. According to Schrödinger the domain of the de Broglie waves is ordinary physical space-time.[2] This formal feature exhibits the pilot wave theory as non-local, which is considered by many physicists to be a grave defect in a theory.[3]

Still in the days of the old quantum theory, another pre-quantum-mechanical version of wave–particle duality was pioneered by William Duane,[4] and developed by others including Alfred Landé.[5] Duane explained diffraction of x-rays by a crystal in terms solely of their particle aspect. The deflection of the trajectory of each diffracted photon was due to quantal translative momentum transfer from the spatially regular structure of the diffracting crystal.[6] Fourier analysis reveals the wave–particle duality as a simply mathematical equivalence, always present, and universal for all quanta. The same reasoning applies for example to diffraction of electrons by a crystal.

In the light of de Broglie's ideas, Erwin Schrödinger developed his wave mechanics by referring the universal wave aspect not to ordinary physical space-time, but rather to a profoundly different and more abstract 'space'. The domain of Schrödinger's wave function is configuration space.[2] Ordinary physical space-time allows more or less direct visualization of cause and effect relations. In contrast, config-

uration space does not directly display cause and effect linkages. Sometimes, nevertheless, it seemed as if Schrödinger visualized his own waves as referring to ordinary space-time, and there was much debate about this.[7]

Niels Bohr regarded the "duality paradox" as a fundamental or metaphysical fact of nature. A given kind of quantum object, will exhibit sometimes wave, sometimes particle, character, in respectively different physical settings. He saw such duality as one aspect of the concept of complementarity.[8]:242, 375–376 Bohr regarded renunciation of the cause-effect relation, or complementarily, of the space-time picture, as essential to the quantum mechanical account.[9]

Werner Heisenberg considered the question further. He saw the duality as present for all quantic entities, but not quite in the usual quantum mechanical account considered by Bohr. He saw it in what is called second quantization, which generates an entirely new concept of fields which exist in ordinary space-time, causality still being visualizable. Classical field values (e.g. the electric and magnetic field strengths of Maxwell) are replaced by an entirely new kind of field value, as considered in quantum field theory. Turning the reasoning around, ordinary quantum mechanics can be deduced as a specialized consequence of quantum field theory.[10][11]

Because of the difference of views of Bohr and Heisenberg, the main sources of the so-called Copenhagen interpretation, the position of that interpretation on wave–particle duality is ill-defined.

In a modern perspective, wave functions arise naturally in relativistic quantum field theory in the formulation of free quantum fields. They are necessary for the Lorentz invariance of the theory. Their form and the equations of motion they obey are dictated by under which representation of the Lorentz group they transform.[12]

5.1   Origin of theory

The idea of duality originated in a debate over the nature of light and matter that dates back to the 17th century, when Christiaan Huygens and Isaac Newton proposed competing theories of light: light was thought either to consist of waves (Huygens) or of particles (Newton). Through the work of Max Planck, Albert Einstein, Louis de Broglie, Arthur Compton, Niels Bohr, and many others, current scientific theory holds that *all* particles *also* have a wave nature (and vice versa).[13] This phenomenon has been verified not only for elementary particles, but also for compound particles like atoms and even molecules. For macroscopic particles, because of their extremely short wavelengths, wave properties usually cannot be detected.[14]

5.2   Brief history of wave and particle viewpoints

Aristotle was one of the first to publicly hypothesize about the nature of light, proposing that light is a disturbance in the element aether (that is, it is a wave-like phenomenon). On the other hand, Democritus—the original *atomist*—argued that all things in the universe, including light, are composed of indivisible sub-components (light being some form of solar atom).[15] At the beginning of the 11th Century, the Arabic scientist Alhazen wrote the first comprehensive treatise on optics; describing refraction, reflection, and the operation of a pinhole lens via rays of light traveling from the point of emission to the eye. He asserted that these rays were composed of particles of light. In 1630, René Descartes popularized and accredited the opposing wave description in his treatise on light, showing that the behavior of light could be re-created by modeling wave-like disturbances in a universal medium ("plenum"). Beginning in 1670 and progressing over three decades, Isaac Newton developed and championed his corpuscular hypothesis, arguing that the perfectly straight lines of reflection demonstrated light's particle nature; only particles could travel in such straight lines. He explained refraction by positing that particles of light accelerated laterally upon entering a denser medium. Around the same time, Newton's contemporaries Robert Hooke and Christiaan Huygens—and later Augustin-Jean Fresnel—mathematically refined the wave viewpoint, showing that if light traveled at different speeds in different media (such as water and air), refraction could be easily explained as the medium-dependent propagation of light waves. The resulting Huygens–Fresnel principle was extremely successful at reproducing light's behavior and, subsequently supported by Thomas Young's 1803 discovery of double-slit interference, was the beginning of the end for the particle light camp.[16][17]

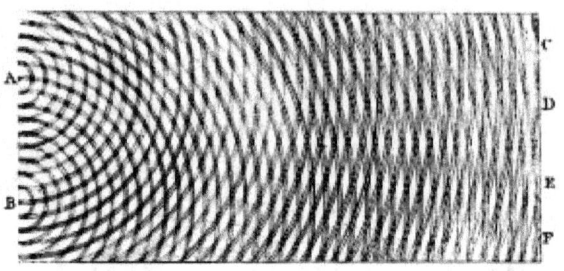

Thomas Young's sketch of two-slit diffraction of waves, 1803

The final blow against corpuscular theory came when James Clerk Maxwell discovered that he could combine four simple equations, which had been previously discovered, along with a slight modification to describe self-propagating waves of oscillating electric and magnetic fields. When the propagation speed of these electromagnetic waves was calculated, the speed of light fell out. It quickly became apparent that visible light, ultraviolet light, and infrared light (phenomena thought previously to be unrelated) were all electromagnetic waves of differing frequency. The wave theory had prevailed—or at least it seemed to.

While the 19th century had seen the success of the wave theory at describing light, it had also witnessed the rise of the atomic theory at describing matter. In 1789, Antoine Lavoisier securely differentiated chemistry from alchemy by introducing rigor and precision into his laboratory techniques; allowing him to deduce the conservation of mass and categorize many new chemical elements and compounds. However, the nature of these essential chemical elements remained unknown. In 1799, Joseph Louis Proust advanced chemistry towards the atom by showing that elements combined in definite proportions. This led John Dalton to resurrect Democritus' atom in 1803, when he proposed that elements were invisible sub components; which explained why the varying oxides of metals (e.g. stannous oxide and cassiterite, SnO and SnO_2 respectively) possess a 1:2 ratio of oxygen to one another. But Dalton and other chemists of the time had not considered that some elements occur in monatomic form (like Helium) and others in diatomic form (like Hydrogen), or that water was H_2O, not the simpler and more intuitive HO—thus the atomic weights presented at the time were varied and often incorrect. Additionally, the formation of H_2O by two parts of hydrogen gas and one part of oxygen gas would require an atom of oxygen to split in half (or two half-atoms of hydrogen to come together). This problem was solved by Amedeo Avogadro, who studied the reacting volumes of gases as they formed liquids and solids. By postulating that equal volumes of elemental gas contain an equal number of atoms, he was able to show that H_2O was formed from two parts H_2 and one part O_2. By discovering diatomic gases, Avogadro completed the basic atomic theory, allowing the correct molecular formulae of most known compounds — as well as the correct weights of atoms — to be deduced and categorized in a consistent manner. The final stroke in classical atomic theory came when Dimitri Mendeleev saw an order in recurring chemical properties, and created a table presenting the elements in unprecedented order and symmetry. But there were holes in Mendeleev's table, with no element to fill them in. His critics initially cited this as a fatal flaw, but were silenced when new elements were discovered that perfectly fit into these holes. The success of the periodic table effectively converted any remaining opposition to atomic theory; even though no single atom had ever been observed in the laboratory, chemistry was now an atomic science.

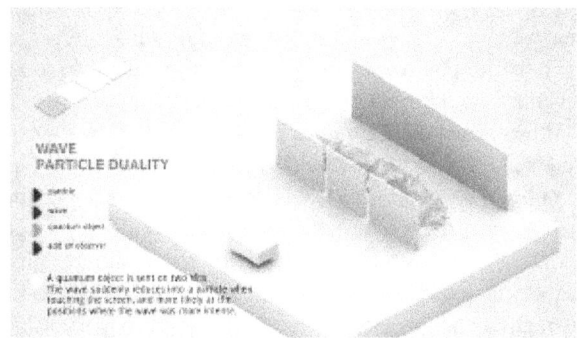

*Animation showing the wave-particle duality with a double slit experiment and effect of an observer. Increase size to see explanations in the video itself. See also **quiz based on this animation**.*

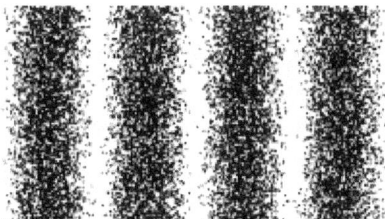

Particle impacts make visible the interference pattern of waves.

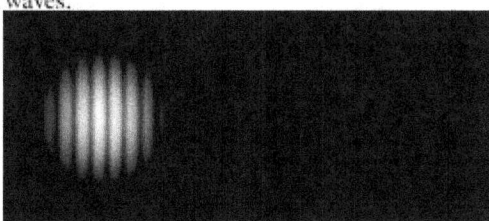

A quantum particle is represented by a wave packet.

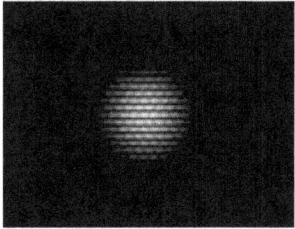

Interference of a quantum particle with itself.
Click images for animations.

5.3 Turn of the 20th century and the paradigm shift

5.3.1 Particles of electricity

At the close of the 19th century, the reductionism of atomic theory began to advance into the atom itself; determining, through physics, the nature of the atom and the operation of chemical reactions. Electricity, first thought to be a fluid, was now understood to consist of particles called electrons. This was first demonstrated by J. J. Thomson in 1897 when, using a cathode ray tube, he found that an electrical charge would travel across a vacuum (which would possess infinite resistance in classical theory). Since the vacuum offered no medium for an electric fluid to travel, this discovery could only be explained via a particle carrying a negative charge and moving through the vacuum. This *electron* flew in the face of classical electrodynamics, which had successfully treated electricity as a fluid for many years (leading to the invention of batteries, electric motors, dynamos, and arc lamps). More importantly, the intimate relation between electric charge and electromagnetism had been well documented following the discoveries of Michael Faraday and James Clerk Maxwell. Since electromagnetism was *known* to be a wave generated by a changing electric or magnetic *field* (a continuous, wave-like entity itself) an atomic/particle description of electricity and charge was a non sequitur. Furthermore, classical electrodynamics was not the only classical theory rendered incomplete.

5.3.2 Radiation quantization

Black-body radiation, the emission of electromagnetic energy due to an object's heat, could not be explained from classical arguments alone. The equipartition theorem of classical mechanics, the basis of all classical thermodynamic theories, stated that an object's energy is partitioned equally among the object's vibrational modes. This worked well when describing thermal objects, whose vibrational modes were defined as the speeds of their constituent atoms, and the speed distribution derived from egalitarian partitioning of these vibrational modes closely matched experimental results. Speeds much higher than the average speed were suppressed by the fact that kinetic energy is quadratic — doubling the speed requires four times the energy — thus the number of atoms occupying high energy modes (high

speeds) quickly drops off because the constant, equal partition can excite successively fewer atoms. Low speed modes
would *ostensibly* dominate the distribution, since low speed
modes would require ever less energy, and *prima facie* a
zero-speed mode would require zero energy and its energy
partition would contain an infinite number of atoms. *But*
this would only occur in the absence of atomic interaction;
when collisions are allowed, the low speed modes are immediately suppressed by jostling from the higher energy atoms,
exciting them to higher energy modes. An equilibrium is
swiftly reached where most atoms occupy a speed proportional to the temperature of the object (thus defining temperature as the average kinetic energy of the object).

But applying the same reasoning to the electromagnetic
emission of such a thermal object was not so successful.
It had been long known that thermal objects emit light. Hot
metal glows red, and upon further heating, white (this is the
underlying principle of the incandescent bulb). Since light
was known to be waves of electromagnetism, physicists
hoped to describe this emission via classical laws. This became known as the black body problem. Since the equipartition theorem worked so well in describing the vibrational
modes of the thermal object itself, it was trivial to assume
that it would perform equally well in describing the radiative
emission of such objects. But a problem quickly arose when
determining the vibrational modes of light. To simplify the
problem (by limiting the vibrational modes) a longest allowable wavelength was defined by placing the thermal object
in a cavity. Any electromagnetic mode at equilibrium (i.e.
any standing wave) could only exist if it used the walls of
the cavities as nodes. Thus there were no waves/modes with
a wavelength larger than twice the length (L) of the cavity.

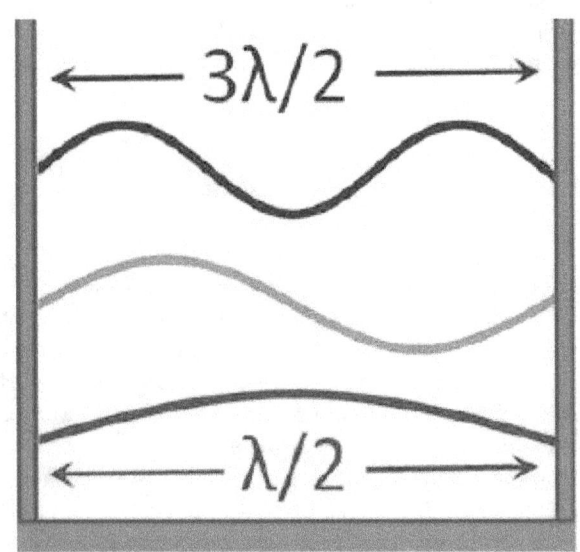

Standing waves in a cavity

The first few allowable modes would therefore have wave-

lengths of : $2L$, L, $2L/3$, $L/2$, etc. (each successive wavelength adding one node to the wave). However, while the
wavelength could never exceed $2L$, there was no such limit
on decreasing the wavelength, and adding nodes to reduce
the wavelength could proceed *ad infinitum*. Suddenly it became apparent that the short wavelength modes completely
dominated the distribution, since ever shorter wavelength
modes could be crammed into the cavity. If each mode received an equal partition of energy, the short wavelength
modes would consume all the energy. This became clear
when plotting the Rayleigh–Jeans law which, while correctly predicting the intensity of long wavelength emissions,
predicted infinite total energy as the intensity diverges to
infinity for short wavelengths. This became known as the
ultraviolet catastrophe.

The solution arrived in 1900 when Max Planck hypothesized that the frequency of light emitted by the black body
depended on the frequency of the *oscillator* that emitted it,
and the energy of these oscillators increased linearly with
frequency (according to his constant h, where $E = hv$). This
was not an unsound proposal considering that macroscopic
oscillators operate similarly: when studying five simple
harmonic oscillators of equal amplitude but different frequency, the oscillator with the highest frequency possesses
the highest energy (though this relationship is not linear like
Planck's). By demanding that high-frequency light must be
emitted by an oscillator of equal frequency, and further requiring that this oscillator occupy higher energy than one of
a lesser frequency, Planck avoided any catastrophe; giving
an equal partition to high-frequency oscillators produced
successively fewer oscillators and less emitted light. And as
in the Maxwell–Boltzmann distribution, the low-frequency,
low-energy oscillators were suppressed by the onslaught of
thermal jiggling from higher energy oscillators, which necessarily increased their energy and frequency.

The most revolutionary aspect of Planck's treatment of the
black body is that it inherently relies on an integer number
of oscillators in thermal equilibrium with the electromagnetic field. These oscillators *give* their entire energy to the
electromagnetic field, creating a quantum of light, as often
as they are *excited* by the electromagnetic field, absorbing
a quantum of light and beginning to oscillate at the corresponding frequency. Planck had intentionally created an
atomic theory of the black body, but had unintentionally
generated an atomic theory of light, where the black body
never generates quanta of light at a given frequency with an
energy less than **hv**. However, once realizing that he had
quantized the electromagnetic field, he denounced particles
of light as a limitation of his approximation, not a property
of reality.

5.3.3 Photoelectric effect illuminated

While Planck had solved the ultraviolet catastrophe by using atoms and a quantized electromagnetic field, most contemporary physicists agreed that Planck's "light quanta" represented only flaws in his model. A more-complete derivation of black body radiation would yield a fully continuous and 'wave-like' electromagnetic field with no quantization. However, in 1905 Albert Einstein took Planck's black body model to produce his solution to another outstanding problem of the day: the photoelectric effect, wherein electrons are emitted from atoms when they absorb energy from light. Since their discovery eight years previously, electrons had been *the* thing to study in physics laboratories worldwide.

In 1902 Philipp Lenard discovered that the energy of these ejected electrons did *not* depend on the intensity of the incoming light, but instead on its *frequency*. So if one shines a little low-frequency light upon a metal, a few low energy electrons are ejected. If one now shines a very intense beam of low-frequency light upon the same metal, a whole slew of electrons are ejected; however they possess the same low energy, there are merely *more of them*. The more light there is, the more electrons are ejected. Whereas in order to get high energy electrons, one must illuminate the metal with high-frequency light. Like blackbody radiation, this was at odds with a theory invoking continuous transfer of energy between radiation and matter. However, it can still be explained using a fully classical description of light, as long as matter is quantum mechanical in nature.[18]

If one used Planck's energy quanta, and demanded that electromagnetic radiation at a given frequency could only transfer energy to matter in integer multiples of an energy quantum **hv**, then the photoelectric effect could be explained very simply. Low-frequency light only ejects low-energy electrons because each electron is excited by the absorption of a single photon. Increasing the intensity of the low-frequency light (increasing the number of photons) only increases the number of excited electrons, not their energy, because the energy of each photon remains low. Only by increasing the frequency of the light, and thus increasing the energy of the photons, can one eject electrons with higher energy. Thus, using Planck's constant h to determine the energy of the photons based upon their frequency, the energy of ejected electrons should also increase linearly with frequency; the gradient of the line being Planck's constant. These results were not confirmed until 1915, when Robert Andrews Millikan, who had previously determined the charge of the electron, produced experimental results in perfect accord with Einstein's predictions. While the energy of ejected electrons reflected Planck's constant, the existence of photons was not explicitly proven until the discovery of the photon antibunching effect, of which a modern experiment can be performed in undergraduate-level labs.[19] This phenomenon could only be explained via photons, and not through any semi-classical theory (which could alternatively explain the photoelectric effect). When Einstein received his Nobel Prize in 1921, it was not for his more difficult and mathematically laborious special and general relativity, but for the simple, yet totally revolutionary, suggestion of quantized light. Einstein's "light quanta" would not be called photons until 1925, but even in 1905 they represented the quintessential example of wave-particle duality. Electromagnetic radiation propagates following linear wave equations, but can only be emitted or absorbed as discrete elements, thus acting as a wave and a particle simultaneously.

5.4 Developmental milestones

5.4.1 Huygens and Newton

The earliest comprehensive theory of light was advanced by Christiaan Huygens, who proposed a wave theory of light, and in particular demonstrated how waves might interfere to form a wavefront, propagating in a straight line. However, the theory had difficulties in other matters, and was soon overshadowed by Isaac Newton's corpuscular theory of light. That is, Newton proposed that light consisted of small particles, with which he could easily explain the phenomenon of reflection. With considerably more difficulty, he could also explain refraction through a lens, and the splitting of sunlight into a rainbow by a prism. Newton's particle viewpoint went essentially unchallenged for over a century.[20]

5.4.2 Young, Fresnel, and Maxwell

In the early 19th century, the double-slit experiments by Young and Fresnel provided evidence for Huygens' wave theories. The double-slit experiments showed that when light is sent through a grid, a characteristic interference pattern is observed, very similar to the pattern resulting from the interference of water waves; the wavelength of light can be computed from such patterns. The wave view did not immediately displace the ray and particle view, but began to dominate scientific thinking about light in the mid 19th century, since it could explain polarization phenomena that the alternatives could not.[21]

In the late 19th century, James Clerk Maxwell explained light as the propagation of electromagnetic waves according to the Maxwell equations. These equations were verified by experiment by Heinrich Hertz in 1887, and the wave theory became widely accepted.

5.4.3 Planck's formula for black-body radiation

Main article: Planck's law

In 1901, Max Planck published an analysis that succeeded in reproducing the observed spectrum of light emitted by a glowing object. To accomplish this, Planck had to make an ad hoc mathematical assumption of quantized energy of the oscillators (atoms of the black body) that emit radiation. It was Einstein who later proposed that it is the electromagnetic radiation itself that is quantized, and not the energy of radiating atoms.

5.4.4 Einstein's explanation of the photoelectric effect

Main article: Photoelectric effect

In 1905, Albert Einstein provided an explanation of the

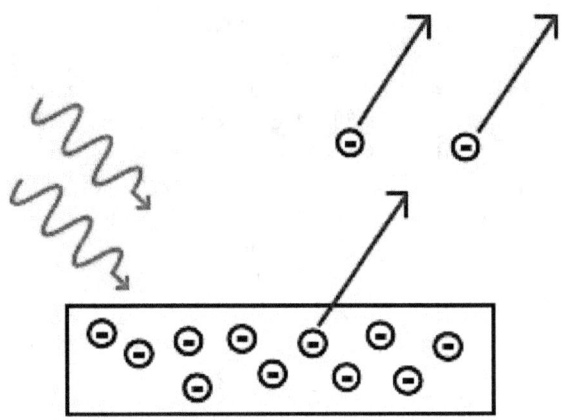

The photoelectric effect. Incoming photons on the left strike a metal plate (bottom), and eject electrons, depicted as flying off to the right.

photoelectric effect, a hitherto troubling experiment that the wave theory of light seemed incapable of explaining. He did so by postulating the existence of photons, quanta of light energy with particulate qualities.

In the photoelectric effect, it was observed that shining a light on certain metals would lead to an electric current in a circuit. Presumably, the light was knocking electrons out of the metal, causing current to flow. However, using the case of potassium as an example, it was also observed that while a dim blue light was enough to cause a current, even the strongest, brightest red light available with the technology of the time caused no current at all. According to the classical theory of light and matter, the strength or amplitude of a light wave was in proportion to its brightness: a bright light should have been easily strong enough to create a large

current. Yet, oddly, this was not so.

Einstein explained this conundrum by postulating that the electrons can receive energy from electromagnetic field only in discrete portions (quanta that were called photons): an amount of energy E that was related to the frequency f of the light by

$$E = hf$$

where h is Planck's constant (6.626×10^{-34} J seconds). Only photons of a high enough frequency (above a certain *threshold* value) could knock an electron free. For example, photons of blue light had sufficient energy to free an electron from the metal, but photons of red light did not. More intense light above the threshold frequency could release more electrons, but no amount of light (using technology available at the time) below the threshold frequency could release an electron. To "violate" this law would require extremely high intensity lasers which had not yet been invented. Intensity-dependent phenomena have now been studied in detail with such lasers.[22]

Einstein was awarded the Nobel Prize in Physics in 1921 for his discovery of the law of the photoelectric effect.

5.4.5 De Broglie's wavelength

Main article: Matter wave

In 1924, Louis-Victor de Broglie formulated the de Broglie hypothesis, claiming that *all* matter,[23][24] not just light, has a wave-like nature; he related wavelength (denoted as λ), and momentum (denoted as p):

$$\lambda = \frac{h}{p}$$

This is a generalization of Einstein's equation above, since the momentum of a photon is given by $p = \frac{E}{c}$ and the wavelength (in a vacuum) by $\lambda = \frac{c}{f}$, where c is the speed of light in vacuum.

De Broglie's formula was confirmed three years later for electrons (which differ from photons in having a rest mass) with the observation of electron diffraction in two independent experiments. At the University of Aberdeen, George Paget Thomson passed a beam of electrons through a thin metal film and observed the predicted interference patterns. At Bell Labs Clinton Joseph Davisson and Lester Halbert Germer guided their beam through a crystalline grid.

De Broglie was awarded the Nobel Prize for Physics in 1929 for his hypothesis. Thomson and Davisson shared the Nobel Prize for Physics in 1937 for their experimental work.

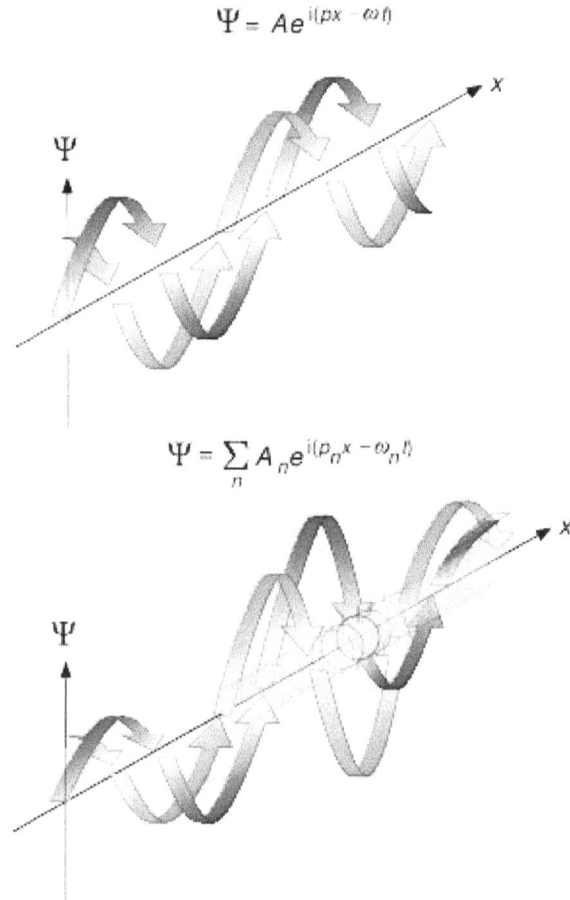

$$\Psi = Ae^{i(px - \omega t)}$$

$$\Psi = \sum_n A_n e^{i(p_n x - \omega_n t)}$$

Propagation of de Broglie waves in 1d — real part of the complex amplitude is blue, imaginary part is green. The probability (shown as the colour opacity) of finding the particle at a given point x is spread out like a waveform: there is no definite position of the particle. As the amplitude increases above zero the curvature decreases, so the amplitude decreases again, and vice versa — the result is an alternating amplitude: a wave. Top: Plane wave. Bottom: Wave packet.

5.4.6 Heisenberg's uncertainty principle

Main article: Heisenberg uncertainty principle

In his work on formulating quantum mechanics, Werner Heisenberg postulated his uncertainty principle, which states:

$$\Delta x \Delta p \geq \frac{\hbar}{2}$$

where

> Δ here indicates standard deviation, a measure of spread or uncertainty;

x and **p** are a particle's position and linear momentum respectively.

$\hbar$ is the reduced Planck's constant (Planck's constant divided by 2π).

Heisenberg originally explained this as a consequence of the process of measuring: Measuring position accurately would disturb momentum and vice versa, offering an example (the "gamma-ray microscope") that depended crucially on the de Broglie hypothesis. It is now thought, however, that this only partly explains the phenomenon, but that the uncertainty also exists in the particle itself, even before the measurement is made.

In fact, the modern explanation of the uncertainty principle, extending the Copenhagen interpretation first put forward by Bohr and Heisenberg, depends even more centrally on the wave nature of a particle: Just as it is nonsensical to discuss the precise location of a wave on a string, particles do not have perfectly precise positions; likewise, just as it is nonsensical to discuss the wavelength of a "pulse" wave traveling down a string, particles do not have perfectly precise momenta (which corresponds to the inverse of wavelength). Moreover, when position is relatively well defined, the wave is pulse-like and has a very ill-defined wavelength (and thus momentum). And conversely, when momentum (and thus wavelength) is relatively well defined, the wave looks long and sinusoidal, and therefore it has a very ill-defined position.

5.4.7 de Broglie–Bohm theory

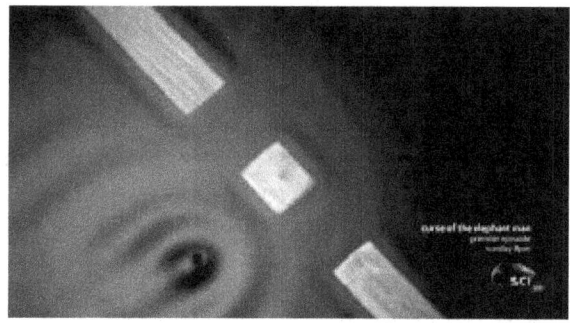

Couder experiments,[25] "materializing" the pilot wave model.

De Broglie himself had proposed a pilot wave construct to explain the observed wave-particle duality. In this view, each particle has a well-defined position and momentum, but is guided by a wave function derived from Schrödinger's equation. The pilot wave theory was initially rejected because it generated non-local effects when applied to systems involving more than one particle. Non-locality, however, soon became established as an integral feature of quantum theory (see EPR paradox), and David Bohm extended de Broglie's model to explicitly include it.

In the resulting representation, also called the de Broglie–Bohm theory or Bohmian mechanics,[26] the wave-particle duality vanishes, and explains the wave behaviour as a scattering with wave appearance, because the particle's motion is subject to a guiding equation or quantum potential. *"This idea seems to me so natural and simple, to resolve the wave-particle dilemma in such a clear and ordinary way, that it is a great mystery to me that it was so generally ignored"* ,[27] J.S.Bell.

The best illustration of the *pilot-wave model* was given by Couder's 2010 "walking droplets" experiments,[28] demonstrating the pilot-wave behaviour in a macroscopic mechanical analog.[25]

5.5   Wave behavior of large objects

Since the demonstrations of wave-like properties in photons and electrons, similar experiments have been conducted with neutrons and protons. Among the most famous experiments are those of Estermann and Otto Stern in 1929.[29] Authors of similar recent experiments with atoms and molecules, described below, claim that these larger particles also act like waves. A wave is basically a group of particles which moves in a particular form of motion i.e. to and fro, if we break that flow by an object it will convert into radiants.

A dramatic series of experiments emphasizing the action of gravity in relation to wave–particle duality was conducted in the 1970s using the neutron interferometer.[30] Neutrons, one of the components of the atomic nucleus, provide much of the mass of a nucleus and thus of ordinary matter. In the neutron interferometer, they act as quantum-mechanical waves directly subject to the force of gravity. While the results were not surprising since gravity was known to act on everything, including light (see tests of general relativity and the Pound–Rebka falling photon experiment), the self-interference of the quantum mechanical wave of a massive fermion in a gravitational field had never been experimentally confirmed before.

In 1999, the diffraction of C_{60} fullerenes by researchers from the University of Vienna was reported.[31] Fullerenes are comparatively large and massive objects, having an atomic mass of about 720 u. The de Broglie wavelength is 2.5 pm, whereas the diameter of the molecule is about 1 nm, about 400 times larger. In 2012, these far-field diffraction experiments could be extended to phthalocyanine molecules and their heavier derivatives, which are composed of 58 and 114 atoms respectively. In these experiments the build-up of such interference patterns could be recorded in real time and with single molecule sensitivity.[32][33]

In 2003, the Vienna group also demonstrated the wave nature of tetraphenylporphyrin[34]—a flat biodye with an extension of about 2 nm and a mass of 614 u. For this demonstration they employed a near-field Talbot Lau interferometer.[35][36] In the same interferometer they also found interference fringes for $C_{60}F_{48}$, a fluorinated buckyball with a mass of about 1600 u, composed of 108 atoms.[34] Large molecules are already so complex that they give experimental access to some aspects of the quantum-classical interface, i.e., to certain decoherence mechanisms.[37][38] In 2011, the interference of molecules as heavy as 6910 u could be demonstrated in a Kapitza–Dirac–Talbot–Lau interferometer.[39] In 2013, the interference of molecules beyond 10,000 u has been demonstrated.[40]

Whether objects heavier than the Planck mass (about the weight of a large bacterium) have a de Broglie wavelength is theoretically unclear and experimentally unreachable; above the Planck mass a particle's Compton wavelength would be smaller than the Planck length and its own Schwarzschild radius, a scale at which current theories of physics may break down or need to be replaced by more general ones.[41]

Recently Couder, Fort, *et al.* showed[42] that we can use macroscopic oil droplets on a vibrating surface as a model of wave–particle duality—localized droplet creates periodical waves around and interaction with them leads to quantum-like phenomena: interference in double-slit experiment,[43] unpredictable tunneling[44] (depending in complicated way on practically hidden state of field), orbit quantization[45] (that particle has to 'find a resonance' with field perturbations it creates—after one orbit, its internal phase has to return to the initial state) and Zeeman effect.[46]

5.6   Treatment in modern quantum mechanics

Wave–particle duality is deeply embedded into the foundations of quantum mechanics. In the formalism of the theory, all the information about a particle is encoded in its *wave function*, a complex-valued function roughly analogous to the amplitude of a wave at each point in space. This function evolves according to a differential equation (generically called the Schrödinger equation). For particles with mass this equation has solutions that follow the form of the wave equation. Propagation of such waves leads to wave-like phenomena such as interference and diffraction. Particles without mass, like photons, have no solutions of the Schrödinger equation so have another wave.

The particle-like behavior is most evident due to phenomena associated with measurement in quantum mechanics. Upon measuring the location of the particle, the particle will be forced into a more localized state as given by the uncertainty principle. When viewed through this formalism, the measurement of the wave function will randomly "collapse", or rather "decohere", to a sharply peaked function at some location. For particles with mass the likelihood of detecting the particle at any particular location is equal to the squared amplitude of the wave function there. The measurement will return a well-defined position, (subject to uncertainty), a property traditionally associated with particles. It is important to note that a measurement is only a particular type of interaction where some data is recorded and the measured quantity is forced into a particular eigenstate. The act of measurement is therefore not fundamentally different from any other interaction.

Following the development of quantum field theory the ambiguity disappeared. The field permits solutions that follow the wave equation, which are referred to as the wave functions. The term particle is used to label the irreducible representations of the Lorentz group that are permitted by the field. An interaction as in a Feynman diagram is accepted as a calculationally convenient approximation where the outgoing legs are known to be simplifications of the propagation and the internal lines are for some order in an expansion of the field interaction. Since the field is non-local and quantized, the phenomena which previously were thought of as paradoxes are explained. Within the limits of the wave-particle duality the quantum field theory gives the same results.

5.6.1 Visualization

There are two ways to visualize the wave-particle behaviour: by the "standard model", described below; and by the Broglie–Bohm model, where no duality is perceived.

Below is an illustration of wave–particle duality as it relates to De Broglie's hypothesis and Heisenberg's uncertainty principle (above), in terms of the position and momentum space wavefunctions for one spinless particle with mass in one dimension. These wavefunctions are Fourier transforms of each other.

The more localized the position-space wavefunction, the more likely the particle is to be found with the position coordinates in that region, and correspondingly the momentum-space wavefunction is less localized so the possible momentum components the particle could have are more widespread.

Conversely the more localized the momentum-space wavefunction, the more likely the particle is to be found with

those values of momentum components in that region, and correspondingly the less localized the position-space wavefunction, so the position coordinates the particle could occupy are more widespread.

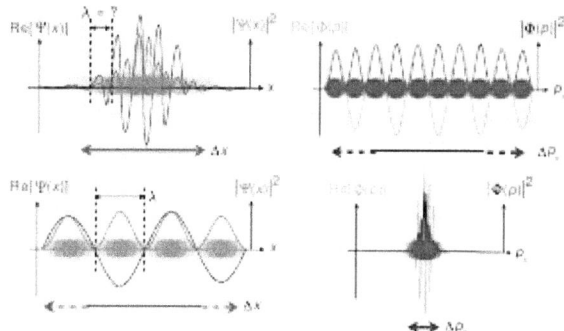

Position x and momentum p wavefunctions corresponding to quantum particles. The colour opacity (%) of the particles corresponds to the probability density of finding the particle with position x or momentum component p.

Top: If wavelength λ is unknown, so are momentum p, wave-vector k and energy E (de Broglie relations). As the particle is more localized in position space, Δx is smaller than for $Δp_x$.

Bottom: If λ is known, so are p, k, and E. As the particle is more localized in momentum space, Δp is smaller than for Δx.

5.7 Alternative views

Wave–particle duality is an ongoing conundrum in modern physics. Most physicists accept wave-particle duality as the best explanation for a broad range of observed phenomena; however, it is not without controversy. Alternative views are also presented here. These views are not generally accepted by mainstream physics, but serve as a basis for valuable discussion within the community.

5.7.1 Both-particle-and-wave view

The pilot wave model, originally developed by Louis de Broglie and further developed by David Bohm into the hidden variable theory proposes that there is no duality, but rather a system exhibits both particle properties and wave properties simultaneously, and particles are guided, in a deterministic fashion, by the pilot wave (or its "quantum potential") which will direct them to areas of constructive interference in preference to areas of destructive interference. This idea is held by a significant minority within the physics community. [47]

At least one physicist considers the "wave-duality" as not being an incomprehensible mystery. L.E. Ballentine, *Quantum Mechanics, A Modern Development*, p. 4, explains:

When first discovered, particle diffraction was a source of great puzzlement. Are "particles" really "waves?" In the early experiments, the diffraction patterns were detected holistically by means of a photographic plate, which could not detect individual particles. As a result, the notion grew that particle and wave properties were mutually incompatible, or complementary, in the sense that different measurement apparatuses would be required to observe them. That idea, however, was only an unfortunate generalization from a technological limitation. Today it is possible to detect the arrival of individual electrons, and to see the diffraction pattern emerge as a statistical pattern made up of many small spots (Tonomura et al., 1989). Evidently, quantum particles are indeed particles, but whose behaviour is very different from classical physics would have us to expect.

It has been claimed that the Afshar experiment [48] (2007) shows that it is possible to simultaneously observe both wave and particle properties of photons. This claim is, however, rejected by other scientists.

5.7.2   Wave-only view

At least one scientist proposes that the duality can be replaced by a "wave-only" view. In his book *Collective Electrodynamics: Quantum Foundations of Electromagnetism* (2000), Carver Mead purports to analyze the behavior of electrons and photons purely in terms of electron wave functions, and attributes the apparent particle-like behavior to quantization effects and eigenstates. According to reviewer David Haddon: [49]

Mead has cut the Gordian knot of quantum complementarity. He claims that atoms, with their neutrons, protons, and electrons, are not particles at all but pure waves of matter. Mead cites as the gross evidence of the exclusively wave nature of both light and matter the discovery between 1933 and 1996 of ten examples of pure wave phenomena, including the ubiquitous laser of CD players, the self-propagating electrical currents of superconductors, and the Bose–Einstein condensate of atoms.

Albert Einstein, who, in his search for a Unified Field Theory, did not accept wave-particle duality, wrote: [50]

This double nature of radiation (and of material corpuscles)...has been interpreted by quantum-mechanics in an ingenious and amazingly successful fashion.   This interpretation...appears to me as only a temporary way out...

The many-worlds interpretation (MWI) is sometimes presented as a waves-only theory, including by its originator, Hugh Everett who referred to MWI as "the wave interpretation". [51]

The *Three Wave Hypothesis* of R. Horodecki relates the particle to wave. [52] [53] The hypothesis implies that a massive particle is an intrinsically spatially as well as temporally extended wave phenomenon by a nonlinear law.

5.7.3   Neither-wave-nor-particle view

It has been argued that there are never exact particles or waves, but only some compromise or intermediate between them.   For this reason, in 1928 Arthur Eddington [54] coined the name "*wavicle*" to describe the objects although it is not regularly used today.   One consideration is that zero-dimensional mathematical points cannot be observed. Another is that the formal representation of such points, the Dirac delta function is unphysical, because it cannot be normalized.   Parallel arguments apply to pure wave states. Roger Penrose states: [55]

"Such 'position states' are idealized wavefunctions in the opposite sense from the momentum states. Whereas the momentum states are infinitely spread out, the position states are infinitely concentrated.   Neither is normalizable [...]."

5.7.4   Relational approach to wave–particle duality

Relational quantum mechanics is developed which regards the detection event as establishing a relationship between the quantized field and the detector.   The inherent ambiguity associated with applying Heisenberg's uncertainty principle and thus wave–particle duality is subsequently avoided. [56]

5.8   Image of Wave-Particle nature of light has been captured

Researchers at the Ecole polytechnique federale de Lausanne have experimentally been able to image the wave-particle nature of light waves. [57] [58] The full length technical paper published on 2 March 2015. is available for access.*[59]

5.9 Applications

Although it is difficult to draw a line separating wave–particle duality from the rest of quantum mechanics. it is nevertheless possible to list some applications of this basic idea.

- Wave–particle duality is exploited in electron microscopy. where the small wavelengths associated with the electron can be used to view objects much smaller than what is visible using visible light.

- Similarly. neutron diffraction uses neutrons with a wavelength of about 0.1 nm. the typical spacing of atoms in a solid. to determine the structure of solids.

5.10 See also

- Arago spot

- Afshar experiment

- Basic concepts of quantum mechanics

- Complementarity (physics)

- Englert–Greenberger–Yasin duality relation

- Kapitsa–Dirac effect

- Electron wave-packet interference

- Faraday wave

- Hanbury Brown and Twiss effect

- Photon polarization

- Scattering theory

- Wavelet

- Wheeler's delayed choice experiment

5.11 Notes and references

[1] Harrison, David (2002). "Complementarity and the Copenhagen Interpretation of Quantum Mechanics". *UPSCALE*. Dept. of Physics, U. of Toronto. Retrieved 2008-06-21.

[2] Schrödinger, E. (1928). Wave mechanics, pp. 185–206 of *Électrons et Photons: Rapports et Discussions du Cinquième Conseil de Physique, tenu à Bruxelles du 24 au 29 Octobre 1927, sous les Auspices de l'Institut International de Physique Solvay*. Gauthier-Villars, Paris, pp. 185–186; translation at p. 447 of Bacciagaluppi, G., Valentini, A. (2009). *Quantum Theory at the Crossroads: Reconsidering the 1927 Solvay Conference*, Cambridge University Press, Cambridge UK, ISBN 978-0-521-81421-8.

[3] Bransden, B.H., Joachain, C.J. (1989/2000). *Quantum Mechanics*, second edition, Pearson, Prentice Hall, Harlow UK, ISBN 978-0-582-35691-7, p. 760.

[4] Duane, W. (1923). The transfer in quanta of radiation momentum to matter. *Proc. Natl. Acad. Sci.* 9(5): 158–164.

[5] Landé, A. (1951). *Quantum Mechanics*, Sir Isaac Pitman and Sons, London, pp. 19–22.

[6] Heisenberg, W. (1930). *The Physical Principles of the Quantum Theory*, translated by C. Eckart and F.C. Hoyt, University of Chicago Press, Chicago, pp. 77–78.

[7] Heisenberg, W., (1967). Quantum theory and its interpretation, quoted on p. 56 by eds. J.A. Wheeler, W.H. Zurek, (1983), *Quantum Theory and Measurement*, Princeton University Press, Princeton NJ, from ed. S. Rozental, *Niels Bohr: his Life and Work as seen by his Friends and Colleagues*, North Holland, Amsterdam.

[8] Kumar, Manjit (2011). *Quantum: Einstein, Bohr, and the Great Debate about the Nature of Reality* (Reprint ed.). W. W. Norton & Company. ISBN 978-0393339888.

[9] Bohr, N. (1927/1928). The quantum postulate and the recent development of atomic theory. *Nature* Supplement April 14 1928, **121**: 580–590.

[10] Camilleri, K. (2009). *Heisenberg and the Interpretation of Quantum Mechanics: the Physicist as Philosopher*, Cambridge University Press, Cambridge UK, ISBN 978-0-521-88484-6.

[11] Preparata, G. (2002). *An Introduction to a Realistic Quantum Physics*, World Scientific, River Edge NJ, ISBN 978-981-238-176-7.

[12] Weinberg, S. (2002). *The Quantum Theory of Fields* **1**, Cambridge University Press, ISBN 0-521-55001-7 Chapter 5.

[13] Walter Greiner (2001). *Quantum Mechanics: An Introduction*. Springer. ISBN 3-540-67458-6.

[14] R. Eisberg and R. Resnick (1985). *Quantum Physics of Atoms, Molecules, Solids, Nuclei, and Particles* (2nd ed.). John Wiley & Sons. pp. 59–60. ISBN 047187373X. For both large and small wavelengths, both matter and radiation have both particle and wave aspects.... But the wave aspects of their motion become more difficult to observe as their wavelengths become shorter.... For ordinary macroscopic

particles the mass is so large that the momentum is always sufficiently large to make the de Broglie wavelength small enough to be beyond the range of experimental detection, and classical mechanics reigns supreme.

[15] Nathaniel Page Stites, M.A./M.S. "Light I: Particle or Wave?," Visionlearning Vol. PHY-1 (3), 2005. http://www.visionlearning.com/library/module_viewer.php?mid=132

[16] Young, Thomas (1804). "Bakerian Lecture: Experiments and calculations relative to physical optics". *Philosophical Transactions of the Royal Society* **94**: 1–16. Bibcode:1804RSPT...94....1Y. doi:10.1098/rstl.1804.0001.

[17] Thomas Young: The Double Slit Experiment

[18] Lamb, Willis E.; Scully, Marlan O. (1968). "The photoelectric effect without photons" (PDF).

[19] "Observing the quantum behavior of light in an undergraduate laboratory". *American Journal of Physics* **72**: 1210. Bibcode:2004AmJPh..72.1210T. doi:10.1119/1.1737397.

[20] "light", The Columbia Encyclopedia, Sixth Edition. 2001–05.

[21] Buchwald, Jed (1989). *The Rise of the Wave Theory of Light: Optical Theory and Experiment in the Early Nineteenth Century*. Chicago: University of Chicago Press. ISBN 0-226-07886-8. OCLC 18069573 59210058.

[22] Zhang, Q (1996). "Intensity dependence of the photoelectric effect induced by a circularly polarized laser beam". *Physics Letters A* **216** (1-5): 125–128. Bibcode:1996PhLA..216..125Z. doi:10.1016/0375-9601(96)00259-9.

[23] Donald H Menzel, *"Fundamental formulas of Physics"*, volume 1, page 153; Gives the de Broglie wavelengths for composite particles such as protons and neutrons.

[24] Brian Greene, The Elegant Universe, page 104 "all matter has a wave-like character"

[25] See this Science Channel production (Season II, Episode VI "How Does The Universe Work?"), presented by Morgan Freeman, https://www.youtube.com/watch?v=W9yWv5dqSKk

[26] Bohmian Mechanics, *Stanford Encyclopedia of Philosophy*.

[27] Bell, J. S., "Speakable and Unspeakable in Quantum Mechanics". Cambridge: Cambridge University Press, 1987.

[28] Y. Couder, A. Boudaoud, S. Protière, Julien Moukhtar, E. Fort: *Walking droplets: a form of wave-particle duality at macroscopic level?*, doi:10.1051/epn/2010101, (PDF)

[29] Estermann, I.; Stern O. (1930). "Beugung von Molekularstrahlen". *Zeitschrift für Physik* **61** (1-2): 95–125. Bibcode:1930ZPhy...61...95E. doi:10.1007/BF01340293.

[30] R. Colella, A. W. Overhauser and S. A. Werner, Observation of Gravitationally Induced Quantum Interference, *Phys. Rev. Lett.* **34**, 1472–1474 (1975).

[31] Arndt, Markus; O. Nairz; J. Voss-Andreae, C. Keller, G. van der Zouw, A. Zeilinger (14 October 1999). "Wave-particle duality of C_{60}". *Nature* **401** (6754): 680–682. Bibcode:1999Natur.401..680A. doi:10.1038/44348. PMID 18494170.

[32] Juffmann, Thomas; et al. (25 March 2012). "Real-time single-molecule imaging of quantum interference". Nature Nanotechnology. Retrieved 27 March 2012.

[33] Quantumnanovienna. "Single molecules in a quantum interference movie". Retrieved 2012-04-21.

[34] Hackermüller, Lucia; Stefan Uttenthaler; Klaus Hornberger; Elisabeth Reiger; Björn Brezger; Anton Zeilinger; Markus Arndt (2003). "The wave nature of biomolecules and fluorofullerenes". *Phys. Rev. Lett.* **91** (9): 090408. arXiv:quant-ph/0309016. Bibcode:2003PhRvL..91i0408H. doi:10.1103/PhysRevLett.91.090408. PMID 14525169.

[35] Clauser, John F.; S. Li (1994). "Talbot von Lau interefometry with cold slow potassium atoms.". *Phys. Rev. A* **49** (4): R2213–17. Bibcode:1994PhRvA..49.2213C. doi:10.1103/PhysRevA.49.R2213. PMID 9910609.

[36] Brezger, Björn; Lucia Hackermüller; Stefan Uttenthaler; Julia Petschinka; Markus Arndt; Anton Zeilinger (2002). "Matter-wave interferometer for large molecules". *Phys. Rev. Lett.* **88** (10): 100404. arXiv:quant-ph/0202158. Bibcode:2002PhRvL..88j0404B. doi:10.1103/PhysRevLett.88.100404. PMID 11909334.

[37] Hornberger, Klaus; Stefan Uttenthaler; Björn Brezger; Lucia Hackermüller; Markus Arndt; Anton Zeilinger (2003). "Observation of Collisional Decoherence in Interferometry". *Phys. Rev. Lett.* **90** (16): 160401. arXiv:quant-ph/0303093. Bibcode:2003PhRvL..90p0401H. doi:10.1103/PhysRevLett.90.160401. PMID 12731960.

[38] Hackermüller, Lucia; Klaus Hornberger; Björn Brezger; Anton Zeilinger; Markus Arndt (2004). "Decoherence of matter waves by thermal emission of radiation". *Nature* **427** (6976): 711–714. arXiv:quant-ph/0402146. Bibcode:2004Natur.427..711H. doi:10.1038/nature02276. PMID 14973478.

[39] Gerlich, Stefan; et al. (2011). "Quantum interference of large organic molecules". *Nature Communications* **2** (263). Bibcode:2011NatCo...2E.263G. doi:10.1038/ncomms1263. PMC 3104521. PMID 21468015.

[40] Eibenberger, S.; Gerlich, S.; Arndt, M.; Mayor, M.; Tüxen, J. (2013). "Matter–wave interference of particles selected from a molecular library with masses exceeding 10 000 amu". *Physical Chemistry Chemical Physics* **15** (35): 14696–14700. doi:10.1039/c3cp51500a. PMID 23900710.

[41] Peter Gabriel Bergmann, *The Riddle of Gravitation*, Courier Dover Publications, 1993 ISBN 0-486-27378-4 online

[42] http://www.youtube.com/watch?v=W9yWv5dqSKk - You Tube video - Yves Couder Explains Wave/Particle Duality via Silicon Droplets

[43] Y. Couder, E. Fort, *Single-Particle Diffraction and Interference at a Macroscopic Scale*, PRL 97, 154101 (2006) online

[44] A. Eddi, E. Fort, F. Moisy, Y. Couder, *Unpredictable Tunneling of a Classical Wave–Particle Association*, PRL 102, 240401 (2009)

[45] Fort, E.; Eddi, A.; Boudaoud, A.; Moukhtar, J.; Couder, Y. (2010). "Path-memory induced quantization of classical orbits". *PNAS* 107 (41): 17515–17520. doi:10.1073/pnas.1007386107.

[46] http://prl.aps.org/abstract/PRL/v108/i26/e264503 - Level Splitting at Macroscopic Scale

[47] (Buchanan pp. 29–31)

[48] Afshar S.S. et al: Paradox in Wave Particle Duality. Found. Phys. 37, 295 (2007) http://arxiv.org/abs/quant-ph/0702188 arXiv:quant-ph/0702188

[49] David Haddon. "Recovering Rational Science". *Touchstone*. Retrieved 2007-09-12.

[50] Paul Arthur Schilpp, ed. *Albert Einstein: Philosopher-Scientist*, Open Court (1949), ISBN 0-87548-133-7, p 51.

[51] See section VI(e) of Everett's thesis: *The Theory of the Universal Wave Function*, in Bryce Seligman DeWitt, R. Neill Graham, eds, *The Many-Worlds Interpretation of Quantum Mechanics*, Princeton Series in Physics, Princeton University Press (1973), ISBN 0-691-08131-X, pp 3–140.

[52] Horodecki, R. (1981). "De broglie wave and its dual wave". *Phys. Lett. A* 87 (3): 95–97. Bibcode:1981PhLA...87...95H. doi:10.1016/0375-9601(81)90571-5.

[53] Horodecki, R. (1983). "Superluminal singular dual wave". *Lett. Novo Cimento* 38: 509–511.

[54] Eddington, Arthur Stanley (1928). *The Nature of the Physical World*. Cambridge, UK.: MacMillan. p. 201.

[55] Penrose, Roger (2007). *The Road to Reality: A Complete Guide to the Laws of the Universe*. Vintage. p. 521, §21.10. ISBN 978-0-679-77631-4.

[56] http://www.quantum-relativity.org/Quantum-Relativity.pdf. See Q. Zheng and T. Kobayashi, *Quantum Optics as a Relativistic Theory of Light*; Physics Essays 9 (1996) 447. Annual Report, Department of Physics, School of Science, University of Tokyo (1992) 240.

[57]

[58] EPFL News 2015-02-03 The first ever photograph of light as both a particle and wave

[59]

5.12 External links

- Animation, applications and research linked to the wave-particle duality and other basic quantum phenomena (Université Paris Sud)

- H. Nikolic. "Quantum mechanics: Myths and facts". arXiv:quant-ph/0609163.

- Young & Geller. "College Physics".

- B. Crowell. "Light as a Particle" (Web page). Retrieved December 10, 2006.

- E.H. Carlson, *Wave–Particle Duality: Light* on Project PHYSNET

- R. Nave. "Wave–Particle Duality" (Web page). *HyperPhysics*. Georgia State University, Department of Physics and Astronomy. Retrieved December 12, 2005.

- Juffmann, Thomas; et al. (25 March 2012). "Real-time single-molecule imaging of quantum interference". Nature Nanotechnology. Retrieved 21 January 2014.

Chapter 6

Pauli exclusion principle

Wolfgang Pauli

The **Pauli exclusion principle** is the quantum mechanical principle that states that two identical fermions (particles with half-integer spin) cannot occupy the same quantum state simultaneously. In the case of electrons, it can be stated as follows: it is impossible for two electrons of a poly-electron atom to have the same values of the four quantum numbers (n, ℓ, m_ℓ and m_s). For two electrons residing in the same orbital, n, ℓ, and m_ℓ are the same, so m_s must be different and the electrons have opposite spins. This principle was formulated by Austrian physicist Wolfgang Pauli in 1925.

A more rigorous statement is that the total wave function for two identical fermions is antisymmetric with respect to exchange of the particles. This means that the wave function changes its sign if the space *and* spin co-ordinates of any two particles are interchanged.

Integer spin particles, bosons, are not subject to the Pauli exclusion principle: any number of identical bosons can occupy the same quantum state, as with, for instance, photons produced by a laser and Bose–Einstein condensate.

6.1 Overview

The Pauli exclusion principle governs the behavior of all fermions (particles with "half-integer spin"), while bosons (particles with "integer spin") are not subject to it. Fermions include elementary particles such as quarks (the constituent particles of protons and neutrons), electrons and neutrinos. In addition, protons and neutrons (subatomic particles composed from three quarks) and some atoms are fermions, and are therefore subject to the Pauli exclusion principle as well. Atoms can have different overall "spin", which determines whether they are fermions or bosons —for example helium-3 has spin 1/2 and is therefore a fermion, in contrast to helium-4 which has spin 0 and is a boson.[1]:123–125 As such, the Pauli exclusion principle underpins many properties of everyday matter, from its large-scale stability, to the chemical behavior of atoms.

"Half-integer spin" means that the intrinsic angular momentum value of fermions is $\hbar = h/2\pi$ (reduced Planck's constant) times a half-integer (1/2, 3/2, 5/2, etc.). In the theory of quantum mechanics fermions are described by antisymmetric states. In contrast, particles with integer spin (called bosons) have symmetric wave functions; unlike fermions they may share the same quantum states. Bosons include the photon, the Cooper pairs which are responsible for superconductivity, and the W and Z bosons. (Fermions take their name from the Fermi–Dirac statistical distribution that they obey, and bosons from their Bose–Einstein distribution).

6.2 History

In the early 20th century it became evident that atoms and molecules with even numbers of electrons are more chemically stable than those with odd numbers of electrons. In the 1916 article "The Atom and the Molecule" by Gilbert N. Lewis, for example, the third of his six postulates of chemical behavior states that the atom tends to hold an even number of electrons in the shell and especially to hold eight electrons which are normally arranged symmetrically at the eight corners of a cube (see: cubical atom).[2] In 1919 chemist Irving Langmuir suggested that the periodic table could be explained if the electrons in an atom were connected or clustered in some manner. Groups of electrons were thought to occupy a set of electron shells around the nucleus.[3] In 1922, Niels Bohr updated his model of the atom by assuming that certain numbers of electrons (for example 2, 8 and 18) corresponded to stable "closed shells".[4]:203

Pauli looked for an explanation for these numbers, which were at first only empirical. At the same time he was trying to explain experimental results of the Zeeman effect in atomic spectroscopy and in ferromagnetism. He found an essential clue in a 1924 paper by Edmund C. Stoner, which pointed out that for a given value of the principal quantum number (n), the number of energy levels of a single electron in the alkali metal spectra in an external magnetic field, where all degenerate energy levels are separated, is equal to the number of electrons in the closed shell of the noble gases for the same value of n. This led Pauli to realize that the complicated numbers of electrons in closed shells can be reduced to the simple rule of *one* electron per state, if the electron states are defined using four quantum numbers. For this purpose he introduced a new two-valued quantum number, identified by Samuel Goudsmit and George Uhlenbeck as electron spin.[5]

6.3 Connection to quantum state symmetry

The Pauli exclusion principle with a single-valued many-particle wavefunction is equivalent to requiring the wavefunction to be antisymmetric. An antisymmetric two-particle state is represented as a sum of states in which one particle is in state $|x\rangle$ and the other in state $|y\rangle$:

$$|\psi\rangle = \sum_{x,y} A(x,y)|x,y\rangle.$$

and antisymmetry under exchange means that $A(x,y) = -A(y,x)$. This implies $A(x,y) = 0$ when $x=y$, which is Pauli

exclusion. It is true in any basis, since unitary changes of basis keep antisymmetric matrices antisymmetric, although strictly speaking, the quantity $A(x,y)$ is not a matrix but an antisymmetric rank-two tensor.

Conversely, if the diagonal quantities $A(x,x)$ are zero *in every basis*, then the wavefunction component

$$A(x,y) = \langle\psi|x,y\rangle = \langle\psi|(|x\rangle \otimes |y\rangle)$$

is necessarily antisymmetric. To prove it, consider the matrix element

$$\langle\psi|\Big((|x\rangle + |y\rangle) \otimes (|x\rangle + |y\rangle)\Big).$$

This is zero, because the two particles have zero probability to both be in the superposition state $|x\rangle + |y\rangle$. But this is equal to

$$\langle\psi|x,x\rangle + \langle\psi|x,y\rangle + \langle\psi|y,x\rangle + \langle\psi|y,y\rangle.$$

The first and last terms on the right side are diagonal elements and are zero, and the whole sum is equal to zero. So the wavefunction matrix elements obey:

$$\langle\psi|x,y\rangle + \langle\psi|y,x\rangle = 0.$$

or

$$A(x,y) = -A(y,x).$$

6.3.1 Pauli principle in advanced quantum theory

According to the spin-statistics theorem, particles with integer spin occupy symmetric quantum states, and particles with half-integer spin occupy antisymmetric states; furthermore, only integer or half-integer values of spin are allowed by the principles of quantum mechanics. In relativistic quantum field theory, the Pauli principle follows from applying a rotation operator in imaginary time to particles of half-integer spin.

In one dimension, bosons, as well as fermions, can obey the exclusion principle. A one-dimensional Bose gas with delta-function repulsive interactions of infinite strength is equivalent to a gas of free fermions. The reason for this is that, in one dimension, exchange of particles requires that they pass through each other; for infinitely strong repulsion this cannot happen. This model is described by a quantum nonlinear Schrödinger equation. In momentum space the exclusion principle is valid also for finite repulsion in a Bose gas with delta-function interactions,[6] as well as for interacting spins and Hubbard model in one dimension, and for other models solvable by Bethe ansatz. The ground state in models solvable by Bethe ansatz is a Fermi sphere.

6.4 Consequences

6.4.1 Atoms and the Pauli principle

The Pauli exclusion principle helps explain a wide variety of physical phenomena. One particularly important consequence of the principle is the elaborate electron shell structure of atoms and the way atoms share electrons, explaining the variety of chemical elements and their chemical combinations. An electrically neutral atom contains bound electrons equal in number to the protons in the nucleus. Electrons, being fermions, cannot occupy the same quantum state as other electrons, so electrons have to "stack" within an atom, i.e. have different spins while at the same electron orbital as described below.

An example is the neutral [helium] atom, which has two bound electrons, both of which can occupy the lowest-energy ($1s$) states by acquiring opposite spin; as spin is part of the quantum state of the electron, the two electrons are in different quantum states and do not violate the Pauli principle. However, the spin can take only two different values (eigenvalues). In a lithium atom, with three bound electrons, the third electron cannot reside in a $1s$ state, and must occupy one of the higher-energy $2s$ states instead. Similarly, successively larger elements must have shells of successively higher energy. The chemical properties of an element largely depend on the number of electrons in the outermost shell; atoms with different numbers of occupied electron shells but the same number of electrons in the outermost shell have similar properties, which gives rise to the periodic table of the elements.[7]:214–218

6.4.2 Solid state properties and the Pauli principle

In conductors and semiconductors, there are very large numbers of molecular orbitals which effectively form a continuous band structure of energy levels. In strong conductors (metals) electrons are so degenerate that they cannot even contribute much to the thermal capacity of a metal.[8]:133–147 Many mechanical, electrical, magnetic, optical and chemical properties of solids are the direct consequence of Pauli exclusion.

6.4.3 Stability of matter

The stability of the electrons in an atom itself is unrelated to the exclusion principle, but is described by the quantum theory of the atom. The underlying idea is that close approach of an electron to the nucleus of the atom necessarily increases its kinetic energy, an application of the uncertainty principle of Heisenberg.[9] However, stability of large systems with many electrons and many nucleons is a different matter, and requires the Pauli exclusion principle.[10]

It has been shown that the Pauli exclusion principle is responsible for the fact that ordinary bulk matter is stable and occupies volume. This suggestion was first made in 1931 by Paul Ehrenfest, who pointed out that the electrons of each atom cannot all fall into the lowest-energy orbital and must occupy successively larger shells. Atoms therefore occupy a volume and cannot be squeezed too closely together.[11]

A more rigorous proof was provided in 1967 by Freeman Dyson and Andrew Lenard, who considered the balance of attractive (electron–nuclear) and repulsive (electron–electron and nuclear–nuclear) forces and showed that ordinary matter would collapse and occupy a much smaller volume without the Pauli principle.[12][13]

The consequence of the Pauli principle here is that electrons of the same spin are kept apart by a repulsive exchange interaction, which is a short-range effect, acting simultaneously with the long-range electrostatic or Coulombic force. This effect is partly responsible for the everyday observation in the macroscopic world that two solid objects cannot be in the same place at the same time.

6.4.4 Astrophysics and the Pauli principle

Dyson and Lenard did not consider the extreme magnetic or gravitational forces which occur in some astronomical objects. In 1995 Elliott Lieb and coworkers showed that the Pauli principle still leads to stability in intense magnetic fields such as in neutron stars, although at a much higher density than in ordinary matter.[14] It is a consequence of general relativity that, in sufficiently intense gravitational fields, matter collapses to form a black hole.

Astronomy provides a spectacular demonstration of the effect of the Pauli principle, in the form of white dwarf and neutron stars. In both types of body, atomic structure is disrupted by large gravitational forces, leaving the constituents supported by "degeneracy pressure" alone. This exotic form of matter is known as degenerate matter. In white dwarfs atoms are held apart by electron degeneracy pressure. In neutron stars, subject to even stronger gravitational forces, electrons have merged with protons to form neutrons. Neutrons are capable of producing an even higher degeneracy pressure, albeit over a shorter range. This can stabilize neutron stars from further collapse, but at a smaller size and higher density than a white dwarf. Neutrons are the most "rigid" objects known; their Young modulus (or more accurately, bulk modulus) is 20 orders of magnitude larger than that of diamond. However, even this enormous rigidity can be overcome by the gravitational field of a massive star or by the pressure of a supernova, leading to the formation of a black hole."[15]":286–287

6.5 See also

- Exchange force

- Exchange interaction

- Exchange symmetry

- Hund's rule

- Fermi hole

- Pauli effect

6.6 References

[1] Kenneth S. Krane (5 November 1987). *Introductory Nuclear Physics*. Wiley. ISBN 978-0-471-80553-3.

[2]

[3] Langmuir, Irving (1919). "The Arrangement of Electrons in Atoms and Molecules" (PDF). *Journal of the American Chemical Society* **41** (6): 868–934. doi:10.1021/ja02227a002. Retrieved 2008-09-01.

[4] Shaviv, Glora. *The Life of Stars: The Controversial Inception and Emergence of the Theory of Stellar Structure* (2010 ed.). Springer. ISBN 978-3642020872.

[5] Straumann, Norbert (2004). "The Role of the Exclusion Principle for Atoms to Stars: A Historical Account". *Invited talk at the 12th Workshop on Nuclear Astrophysics*.

[6] A. Izergin and V. Korepin. Letter in Mathematical Physics vol 6, page 283, 1982

[7] Griffiths, David J. (2004). *Introduction to Quantum Mechanics (2nd ed.)*. Prentice Hall. ISBN 0-13-111892-7

[8] Kittel, Charles (2005). *Introduction to Solid State Physics* (8th ed.). USA: John Wiley & Sons, Inc.. ISBN 978-0-471-41526-8

[9] Elliot J. Lieb *The Stability of Matter and Quantum Electrodynamics*

[10] This realization is attributed by Lieb and by GL Sewell (2002). *Quantum Mechanics and Its Emergent Macrophysics*. Princeton University Press. ISBN 0-691-05832-6. to FJ Dyson and A Lenard: *Stability of Matter, Parts I and II (J. Math. Phys.*, **8**, 423–434 (1967); *J. Math. Phys.*, **9**, 698–711 (1968)).

[11] As described by FJ Dyson (J.Math.Phys. **8**, 1538–1545 (1967)), Ehrenfest made this suggestion in his address on the occasion of the award of the Lorentz Medal to Pauli.

[12] FJ Dyson and A Lenard: *Stability of Matter, Parts I and II (J. Math. Phys.*, **8**, 423–434 (1967); *J. Math. Phys.*, **9**, 698–711 (1968))

[13] Dyson, Freeman (1967). "Ground-State Energy of a Finite System of Charged Particles". *J. Math. Phys.* **8** (8): 1538–1545. Bibcode:1967JMP.....8.1538D. doi:10.1063/1.1705389.

[14] Lieb, E. H.; Loss, M.; Solovej, J. P. (1995). "Stability of Matter in Magnetic Fields". *Phys. Rev. Letters* **75** (6): 985–9. arXiv:cond-mat/9506047. Bibcode:1995PhRvL..75..985L. doi:10.1103/PhysRevLett.75.985.

[15] Martin Bojowald (5 November 2012). *The Universe: A View from Classical and Quantum Gravity*. John Wiley & Sons. ISBN 978-3-527-66769-7.

- Dill, Dan (2006). "Chapter 3.5, Many-electron atoms: Fermi holes and Fermi heaps". *Notes on General Chemistry (2nd ed.)*. W. H. Freeman. ISBN 1-4292-0068-5.

- Liboff, Richard L. (2002). *Introductory Quantum Mechanics*. Addison-Wesley. ISBN 0-8053-8714-5.

- Massimi, Michela (2005). *Pauli's Exclusion Principle*. Cambridge University Press. ISBN 0-521-83911-4.

- Tipler, Paul; Llewellyn, Ralph (2002). *Modern Physics (4th ed.)*. W. H. Freeman. ISBN 0-7167-4345-0.

6.7 External links

- Nobel Lecture: Exclusion Principle and Quantum Mechanics Pauli's own account of the development of the Exclusion Principle.

Chapter 7

Matter wave

The de Broglie relations redirect here.
This article is about wave-like phenomena exhibited by particles of matter. For the ordinary type of wave propagating through material media, see Mechanical wave.

All matter can exhibit wave-like behaviour. For example a beam of electrons can be diffracted just like a beam of light or a water wave. **Matter waves** are a central part of the theory of quantum mechanics, being an example of wave–particle duality. The concept that matter behaves like a wave is also referred to as the **de Broglie hypothesis** (/dəˈbrɔɪ/) due to having been proposed by Louis de Broglie in 1924.*[1] Matter waves are often referred to as **de Broglie waves**.

The **de Broglie wavelength** is the wavelength, λ, associated with a massive particle and is related to its momentum, p, through the Planck constant, h:

$$\lambda = \frac{h}{p}.$$

Wave-like behaviour of matter was first experimentally demonstrated in the Davisson–Germer experiment using electrons, and it has also been confirmed for other elementary particles, neutral atoms and even molecules. The wave-like behaviour of matter is crucial to the modern theory of atomic structure and particle physics.

7.1 Historical context

At the end of the 19th century, light was thought to consist of waves of electromagnetic fields which propagated according to Maxwell's equations, while matter was thought to consist of localized particles (See history of wave and particle viewpoints). In 1900, this division was exposed to doubt, when, investigating the theory of black body thermal radiation, Max Planck proposed that light is emitted in discrete quanta of energy. It was thoroughly challenged in

1905. Extending Planck's investigation in several ways, including its connection with the photoelectric effect, Albert Einstein proposed that light is also propagated and absorbed in quanta. Light quanta are now called photons. These quanta would have an energy given by the Planck–Einstein relation:

$$E = h\nu$$

and a momentum

$$p = \frac{E}{c} = \frac{h}{\lambda}$$

where ν (lowercase Greek letter nu) and λ (lowercase Greek letter lambda) denote the frequency and wavelength of the light, c the speed of light, and h Planck's constant.*[2] In the modern convention, frequency is symbolized by f as is done in the rest of this article. Einstein's postulate was confirmed experimentally by Robert Millikan and Arthur Compton over the next two decades.

7.2 The de Broglie hypothesis

De Broglie, in his 1924 PhD thesis, proposed that just as light has both wave-like and particle-like properties, electrons also have wave-like properties. By rearranging the momentum equation stated in the above section, we find a relationship between the wavelength, λ associated with an electron and its momentum, p, through the Planck constant, h:*[3]

$$\lambda = \frac{h}{p}.$$

The relationship is now known to hold for all types of matter: all matter exhibits properties of both particles and waves.

$$\Psi = Ae^{i(px - \omega t)}$$

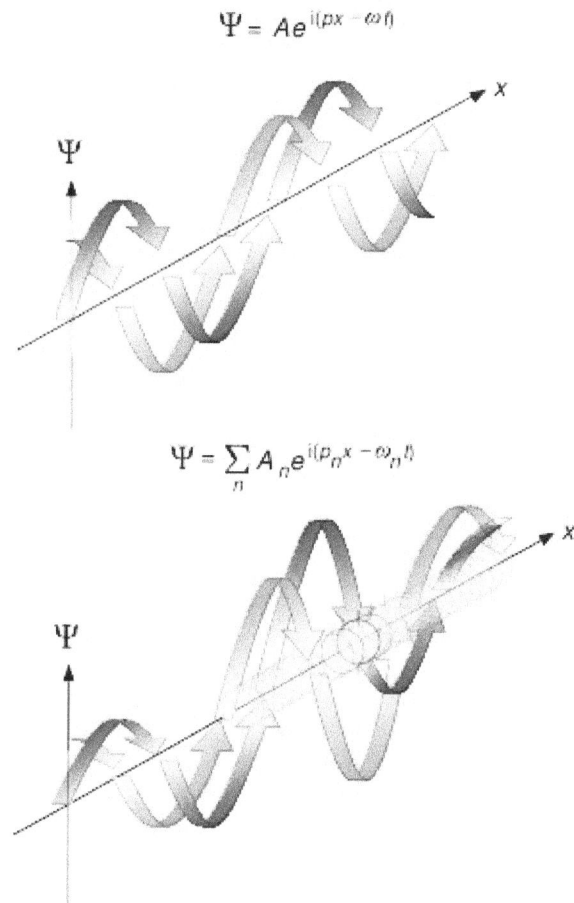

$$\Psi = \sum_n A_n e^{i(p_n x - \omega_n t)}$$

*Propagation of **de Broglie waves** in 1d – real part of the complex amplitude is blue, imaginary part is green. The probability (shown as the colour opacity) of finding the particle at a given point x is spread out like a waveform, there is no definite position of the particle. As the amplitude increases above zero the curvature decreases, so the amplitude decreases again, and vice versa – the result is an alternating amplitude: a wave. Top: plane wave. Bottom: wave packet.*

In 1926, Erwin Schrödinger published an equation describing how a matter wave should evolve—the matter wave analogue of Maxwell's equations—and used it to derive the energy spectrum of hydrogen.

7.3   Experimental confirmation

Matter waves were first experimentally confirmed to occur in the Davisson–Germer experiment for electrons, and the de Broglie hypothesis has been confirmed for other elementary particles. Furthermore, neutral atoms and even molecules have been shown to be wave-like.

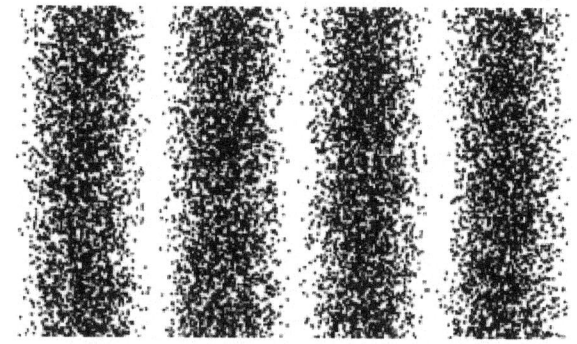

Demonstration of a matter wave in diffraction of electrons

7.3.1   Electrons

Further information: Davisson–Germer experiment and Electron diffraction

In 1927 at Bell Labs, Clinton Davisson and Lester Germer fired slow-moving electrons at a crystalline nickel target. The angular dependence of the diffracted electron intensity was measured, and was determined to have the same diffraction pattern as those predicted by Bragg for x-rays. Before the acceptance of the de Broglie hypothesis, diffraction was a property that was thought to be only exhibited by waves. Therefore, the presence of any diffraction effects by matter demonstrated the wave-like nature of matter. When the de Broglie wavelength was inserted into the Bragg condition, the observed diffraction pattern was predicted, thereby experimentally confirming the de Broglie hypothesis for electrons.[5]

This was a pivotal result in the development of quantum mechanics. Just as the photoelectric effect demonstrated the particle nature of light, the Davisson–Germer experiment showed the wave-nature of matter, and completed the theory of wave-particle duality. For physicists this idea was important because it meant that not only could any particle exhibit wave characteristics, but that one could use wave equations to describe phenomena in matter if one used the de Broglie wavelength.

7.3.2   Neutral atoms

Further information: Atom optics

Experiments with Fresnel diffraction[6] and an atomic mirror for specular reflection[7][8] of neutral atoms confirm the application of the de Broglie hypothesis to atoms, i.e. the existence of atomic waves which undergo diffraction, interference and allow quantum reflection by the tails of the attractive potential.[9] Advances in laser cooling have allowed cooling of neutral atoms down to nanokelvin temperatures. At these temperatures, the thermal de Broglie wavelengths come into the micrometre range. Using Bragg diffraction of atoms and a Ramsey interferometry technique, the de Broglie wavelength of cold sodium atoms was explicitly measured and found to be consistent with the temperature measured by a different method.[10]

This effect has been used to demonstrate atomic holography, and it may allow the construction of an atom probe imaging system with nanometer resolution.[11][12] The description of these phenomena is based on the wave properties of neutral atoms, confirming the de Broglie hypothesis.

The effect has also been used to explain the spatial version of the quantum Zeno effect, in which an otherwise unstable object may be stabilised by rapidly-repeated observations.[8]

7.3.3   Molecules

Recent experiments even confirm the relations for molecules and even macromolecules that otherwise might be supposed too large to undergo quantum mechanical effects. In 1999, a research team in Vienna demonstrated diffraction for molecules as large as fullerenes.[13] The researchers calculated a De Broglie wavelength of the most probable C_{60} velocity as 2.5 pm. More recent experiments prove the quantum nature of molecules with a mass up to 6910 amu.[14]

7.4   de Broglie relations

The de Broglie equations relate the wavelength λ to the momentum p, and frequency f to the total energy E of a particle:[15]

where h is Planck's constant. The equations can also be written as

where $\hbar$ is the reduced Planck's constant, $\mathbf{k}$ is the wave vector, and ω is the angular frequency.

In each pair, the second equation is also referred to as the Planck-Einstein relation, since it was also proposed by Planck and Einstein.

7.4.1   Special relativity

Using two formulas from special relativity, one for the relativistic momentum and one for the energy

$$E = mc^2 = \gamma m_0 c^2$$

$$\vec{p} = m\vec{v} = \gamma m_0 \vec{v}$$

allows the equations to be written as

$$\lambda = \frac{h}{\gamma m_0 v} = \frac{h}{m_0 v}\sqrt{1 - \frac{v^2}{c^2}}$$
$$f = \frac{\gamma m_0 c^2}{h} = \frac{m_0 c^2}{h} \Big/ \sqrt{1 - \frac{v^2}{c^2}}$$

where m_0 denotes the particle's rest mass, v its velocity, γ the Lorentz factor, and c the speed of light in a vacuum.[16][17][18] See below for details of the derivation of the de Broglie relations. Group velocity (equal to the particle's speed) should not be confused with phase velocity (equal to the product of the particle's frequency and its wavelength). In the case of a non-dispersive medium, they happen to be equal, but otherwise they are not.

Group velocity

Albert Einstein first explained the wave–particle duality of light in 1905. Louis de Broglie hypothesized that any particle should also exhibit such a duality. The velocity of a particle, he concluded, should always equal the group velocity of the corresponding wave. The magnitude of the group velocity is equal to the particle's speed.

Both in relativistic and non-relativistic quantum physics, we can identify the group velocity of a particle's wave function with the particle velocity. Quantum mechanics has very accurately demonstrated this hypothesis, and the relation has been shown explicitly for particles as large as molecules.

De Broglie deduced that if the duality equations already known for light were the same for any particle, then his hypothesis would hold. This means that

$$v_g = \frac{\partial \omega}{\partial k} = \frac{\partial (E/\hbar)}{\partial (p/\hbar)} = \frac{\partial E}{\partial p}$$

where E is the total energy of the particle, p is its momentum, $\hbar$ is the reduced Planck constant. For a free non-relativistic particle it follows that

$$v_g = \frac{\partial E}{\partial p} = \frac{\partial}{\partial p}\left(\frac{1}{2}\frac{p^2}{m}\right)$$
$$= \frac{p}{m}$$
$$= v$$

where m is the mass of the particle and v its velocity.

Also in special relativity we find that

$$v_g = \frac{\partial E}{\partial p} = \frac{\partial}{\partial p}\left(\sqrt{p^2 c^2 + m_0^2 c^4}\right)$$
$$= \frac{pc^2}{\sqrt{p^2 c^2 + m_0^2 c^4}}$$
$$= \frac{pc^2}{E}$$

where m_0 is the rest mass of the particle and c is the speed of light in a vacuum. But (see below), using that the phase velocity is $v_p = E/p = c^2/v$, therefore

$$v_g = \frac{pc^2}{E}$$
$$= \frac{c^2}{v_p}$$
$$= v$$

where v is the velocity of the particle regardless of wave behavior.

Phase velocity

In quantum mechanics, particles also behave as waves with complex phases. The phase velocity is equal to the product of the frequency multiplied by the wavelength.

By the de Broglie hypothesis, we see that

$$v_p = \frac{\omega}{k} = \frac{E/\hbar}{p/\hbar} = \frac{E}{p}.$$

Using relativistic relations for energy and momentum, we have

$$v_p = \frac{E}{p} = \frac{\gamma m_0 c^2}{\gamma m_0 v} = \frac{c^2}{v} = \frac{c}{\beta}$$

where E is the total energy of the particle (i.e. rest energy plus kinetic energy in kinematic sense), p the momentum,

γ the Lorentz factor, c the speed of light, and β the speed as a fraction of c. The variable v can either be taken to be the speed of the particle or the group velocity of the corresponding matter wave. Since the particle speed $v < c$ for any particle that has mass (according to special relativity), the phase velocity of matter waves always exceeds c, i.e.

$$v_p > c.$$

and as we can see, it approaches c when the particle speed is in the relativistic range. The superluminal phase velocity does not violate special relativity, because phase propagation carries no energy. See the article on *Dispersion (optics)* for details.

7.4.2 Four-vectors

Main article: Four-vector

Using 4-Vectors, the De Broglie relations form a single equation:

which is frame-independent.

Likewise, the relation between group/particle velocity and phase velocity is given in frame-independent form by:

where

4-Momentum $\mathbf{P} = \left(\frac{E}{c}, \mathbf{p}\right)$

4-WaveVector $\mathbf{K} = \left(\frac{\omega}{c}, \mathbf{k}\right) = \left(\frac{\omega}{c}, \frac{\omega}{v_p}\hat{\mathbf{n}}\right)$

4-Velocity $\mathbf{U} = \gamma(c, \mathbf{u}) = \gamma(c, v_g\hat{\mathbf{n}})$

7.5 Interpretations

The physical reality underlying de Broglie waves is a subject of ongoing debate. Some theories treat either the particle or the wave aspect as its fundamental nature, seeking to explain the other as an emergent property. Some, such as the hidden variable theory, treat the wave and the particle as distinct entities. Yet others propose some intermediate entity that is neither quite wave nor quite particle but only appears as such when we measure one or the other property.

The Copenhagen interpretation states that the nature of the underlying reality is unknowable and beyond the bounds of scientific enquiry.

Schrödinger's quantum mechanical waves are conceptually different from ordinary physical waves such as water or sound. Ordinary physical waves are characterized by undulating real-number 'displacements' of dimensioned physical variables at each point of ordinary physical space at each instant of time. Schrödinger's "waves" are characterized by the undulating value of a dimensionless complex number at each point of an abstract multi-dimensional space, for example of configuration space.

At the Fifth Solvay Conference in 1927, Max Born and Werner Heisenberg reported as follows

At the same conference, Erwin Schrödinger reported likewise.

In 1955, Heisenberg reiterated this.

It is mentioned above that the "displaced quantity" of the Schrödinger wave has values that are dimensionless complex numbers. One may ask what is the physical meaning of those numbers. According to Heisenberg, rather than being of some ordinary physical quantity such as for example Maxwell's electric field intensity, or for example mass density, the Schrödinger-wave packet's "displaced quantity" is probability amplitude. He wrote that instead of using the term 'wave packet', it is preferable to speak of a probability packet. [22] The probability amplitude supports calculation of probability of location or momentum of discrete particles. Heisenberg recites Duane's account of particle diffraction by probabilistic quantal translation momentum transfer, which allows, for example in Young's two-slit experiment, each diffracted particle probabilistically to pass discretely through a particular slit. [23] Thus one does not need necessarily think of the matter wave, as it were, as 'composed of smeared matter'.

These ideas may be expressed in ordinary language as follows. In the account of ordinary physical waves, a 'point' refers to a position in ordinary physical space at an instant of time, at which there is specified a 'displacement' of some physical quantity. But in the account of quantum mechanics, a 'point' refers to a configuration of the system at an instant of time, every particle of the system being in a sense present in every 'point' of configuration space, each particle at such a 'point' being located possibly at a different position in ordinary physical space. There is no explicit definite indication that, at an instant, this particle is 'here' and that particle is 'there' in some separate 'location' in configuration space. This conceptual difference entails that, in contrast to de Broglie's pre-quantum mechanical wave description, the quantum mechanical probability packet description does not directly and explicitly express the Aristotelian

idea, referred to by Newton, that causal efficacy propagates through ordinary space by contact, nor the Einsteinian idea that such propagation is no faster than light. In contrast, these ideas are so expressed in the classical wave account, through the Green's function, though it is inadequate for the observed quantal phenomena. The physical reasoning for this was first recognized by Einstein.[24][25]

7.6 De Broglie's phase wave and periodic phenomenon

De Broglie's thesis started from the hypothesis, "that to each portion of energy with a proper mass m_0 one may associate a periodic phenomenon of the frequency ν_0, such that one finds: $h\nu_0 = m_0 c^2$. The frequency ν_0 is to be measured, of course, in the rest frame of the energy packet. This hypothesis is the basis of our theory." [26][27][28][29][30][31]

De Broglie followed his initial hypothesis of a periodic phenomenon, with frequency ν_0, associated with the energy packet. He used the special theory of relativity to find, in the frame of the observer of the electron energy packet that is moving with velocity v, that its frequency was apparently reduced to

$$f = \nu_0 \sqrt{1 - \frac{v^2}{c^2}}.$$

Then

$$\lambda f = E/p = v_{\mathrm{p}}.$$

using the same notation as above. The quantity v_{p} is the velocity of what de Broglie called the "phase wave". Its wavelength is λ and frequency f. De Broglie reasoned that his hypothetical intrinsic particle periodic phenomenon is in phase with that phase wave. This was his basic matter wave conception. He noted, as above, that $v_{\mathrm{p}} > c$, and the phase wave does not transfer energy.[28][32]

While the concept of waves being associated with matter is correct, de Broglie did not leap directly to the final understanding of quantum mechanics with no missteps. There are conceptual problems with the approach that de Broglie took in his thesis that he was not able to resolve, despite trying a number of different fundamental hypotheses in different papers published while working on, and shortly after publishing, his thesis.[29][33] These difficulties were resolved by Erwin Schrödinger, who developed the wave mechanics approach, starting from a somewhat different basic hypothesis.

7.7   See also

- Bohr model
- Faraday wave
- Kapitsa–Dirac effect
- Matter wave clock
- Schrödinger equation
- Theoretical and experimental justification for the Schrödinger equation
- Thermal de Broglie wavelength
- De Broglie–Bohm theory

7.8   References

[1] Feynman, R.: *QED the Strange Theory of Light and matter.* Penguin 1990 Edition, page 84.

[2] Einstein, A. (1917). Zur Quantentheorie der Strahlung. *Physicalische Zeitschrift* **18**: 121–128. Translated in ter Haar, D. (1967). *The Old Quantum Theory.* Pergamon Press. pp. 167–183. LCCN 66029628.

[3] J. P. McEvoy & Oscar Zarate (2004). *Introducing Quantum Theory.* Totem Books. pp. 110–114. ISBN 1-84046-577-8.

[4] Louis de Broglie "The Reinterpretation of Wave Mechanics" Foundations of Physics, Vol. 1 No. 1 (1970)

[5] Mauro Dardo, *Nobel Laureates and Twentieth-Century Physics,* Cambridge University Press 2004, pp. 156–157

[6] R.B.Doak; R.E.Grisenti; S.Rehbein; G.Schmahl; J.P.Toennies; Ch. Wöll (1999). "Towards Realization of an Atomic de Broglie Microscope: Helium Atom Focusing Using Fresnel Zone Plates". *Physical Review Letters* **83** (21): 4229–4232. Bibcode:1999PhRvL..83.4229D. doi:10.1103/PhysRevLett.83.4229.

[7] F. Shimizu (2000). "Specular Reflection of Very Slow Metastable Neon Atoms from a Solid Surface". *Physical Review Letters* **86** (6): 987–990. Bibcode:2001PhRvL..86..987S. doi:10.1103/PhysRevLett.86.987. PMID 11177991.

[8] D. Kouznetsov; H. Oberst (2005). "Reflection of Waves from a Ridged Surface and the Zeno Effect". *Optical Review* **12** (5): 1605–1623. Bibcode:2005OptRv..12..363K. doi:10.1007/s10043-005-0363-9.

[9] H.Friedrich; G.Jacoby; C.G.Meister (2002). "quantum reflection by Casimir–van der Waals potential tails". *Physical Review A* **65** (3): 032902. Bibcode:2002PhRvA..65c2902F. doi:10.1103/PhysRevA.65.032902.

[10] Pierre Cladé; Changhyun Ryu; Anand Ramanathan; Kristian Helmerson; William D. Phillips (2008). "Observation of a 2D Bose Gas: From thermal to quasi-condensate to superfluid". arXiv:0805.3519.

[11] Shimizu; J.Fujita (2002). "Reflection-Type Hologram for Atoms". *Physical Review Letters* **88** (12): 123201. Bibcode:2002PhRvL..88l3201S. doi:10.1103/PhysRevLett.88.123201. PMID 11909457.

[12] D. Kouznetsov; H. Oberst; K. Shimizu; A. Neumann; Y. Kuznetsova; J.-F. Bisson; K. Ueda; S. R. J. Brueck (2006). "Ridged atomic mirrors and atomic nanoscope". *Journal of Physics B* **39** (7): 1605–1623. Bibcode:2006JPhB...39.1605K. doi:10.1088/0953-4075/39/7/005.

[13] Arndt, M.; O. Nairz; J. Voss-Andreae; C. Keller; G. van der Zouw; A. Zeilinger (14 October 1999). "Wave-particle duality of C60". *Nature* **401** (6754): 680–682. Bibcode:1999Natur.401..680A. doi:10.1038/44348. PMID 18494170.

[14] Gerlich, S.; S. Eibenberger; M. Tomandl; S. Nimmrichter; K. Hornberger; P. J. Fagan; J. Tüxen; M. Mayor & M. Arndt (5 April 2011). "Quantum interference of large organic molecules". *Nature Communications* **2** (263): 263–. Bibcode:2011NatCo...2E.263G. doi:10.1038/ncomms1263. PMC 3104521. PMID 21468015.

[15] Resnick, R.; Eisberg, R. (1985). *Quantum Physics of Atoms, Molecules, Solids, Nuclei and Particles* (2nd ed.). New York: John Wiley & Sons. ISBN 0-471-87373-X.

[16] Holden, Alan (1971). *Stationary states.* New York: Oxford University Press. ISBN 0-19-501497-9.

[17] Williams, W.S.C. (2002). *Introducing Special Relativity,* Taylor & Francis, London. ISBN 0-415-27761-2, p. 192.

[18] de Broglie, L. (1970). The reinterpretation of wave mechanics, *Foundations of Physics* **1**(1): 5–15, p. 9.

[19] Born, M., Heisenberg, W. (1928). Quantum mechanics, pp. 143–181 of *Électrons et Photons: Rapports et Discussions du Cinquième Conseil de Physique, tenu à Bruxelles du 24 au 29 Octobre 1927, sous les Auspices de l'Institut International de Physique Solvay,* Gauthier-Villars, Paris. p. 166; this translation at p. 425 of Bacciagaluppi, G., Valentini, A. (2009), *Quantum Theory at the Crossroads: Reconsidering the 1927 Solvay Conference,* Cambridge University Press, Cambridge UK. ISBN 978-0-521-81421-8.

[20] Schrödinger, E. (1928). Wave mechanics, pp. 185–206 of *Électrons et Photons: Rapports et Discussions du Cinquième Conseil de Physique, tenu à Bruxelles du 24 au 29 Octobre 1927, sous les Auspices de l'Institut International de Physique Solvay,* Gauthier-Villars, Paris. pp. 185–186; this translation at p. 447 of Bacciagaluppi, G., Valentini, A. (2009), *Quantum Theory at the Crossroads: Reconsidering the 1927 Solvay Conference,* Cambridge University Press, Cambridge UK. ISBN 978-0-521-81421-8.

[21] Heisenberg, W. (1955). The development of the interpretation of the quantum theory, pp. 12–29, in *Niels Bohr and the Development of Physics: Essays dedicated to Niels Bohr on the occasion of his seventieth birthday*, edited by W. Pauli, with the assistance of L. Rosenfeld and V. Weisskopf. Pergamon Press, London, p. 13.

[22] Heisenberg, W. (1927). Über den anschaulichen Inhalt der quantentheoretischen Kinematik und Mechanik, *Z. Phys.* **43**: 172–198, translated by eds. Wheeler, J.A., Zurek, W.H. (1983), at pp. 62–84 of *Quantum Theory and Measurement*. Princeton University Press, Princeton NJ, p. 73. Also translated as 'The actual content of quantum theoretical kinematics and mechanics' here

[23] Heisenberg, W. (1930). *The Physical Principles of the Quantum Theory*, translated by C. Eckart, F. C. Hoyt, University of Chicago Press, Chicago IL, pp. 77–78.

[24] Fine, A. (1986). *The Shaky Game: Einstein Realism and the Quantum Theory*, University of Chicago, Chicago, ISBN 0-226-24946-8

[25] Howard, D. (1990). "Nicht sein kann was nicht sein darf", or the prehistory of the EPR, 1909–1935: Einstein's early worries about the quantum mechanics of composite systems, pp. 61–112 in *Sixty-two Years of Uncertainty: Historical Philosophical and Physical Inquiries into the Foundations of Quantum Mechanics*, edited by A.I. Miller, Plenum Press, New York, ISBN 978-1-4684-8773-2.

[26] de Broglie, L. (1923). Waves and quanta, *Nature* **112**: 540.

[27] de Broglie, L. (1924). Thesis, p. 8 of Kracklauer's translation.

[28] Medicus, H.A. (1974). Fifty years of matter waves, *Physics Today* **27**(2): 38–45.

[29] MacKinnon, E. (1976). De Broglie's thesis: a critical retrospective, *Am. J. Phys.* **44**: 1047–1055.

[30] Espinosa, J.M. (1982). Physical properties of de Broglie's phase waves, *Am. J. Phys.* **50**: 357–362.

[31] Brown, H.R., Martins, R.deA. (1984). De Broglie's relativistic phase waves and wave groups. *Am. J. Phys.* **52**: 1130–1140.

[32] Bacciagaluppi, G., Valentini, A. (2009). *Quantum Theory at the Crossroads: Reconsidering the 1927 Solvay Conference*. Cambridge University Press, Cambridge UK, ISBN 978-0-521-81421-8, pp. 30–88.

[33] Martins, Roberto de Andrade (2010). "Louis de Broglie's Struggle with the Wave-Particle Dualism, 1923-1925". Quantum History Project, Fritz Haber Institute of the Max Planck Society and the Max Planck Institute for the History of Science. Retrieved 2015-01-03.

7.9 Further reading

- L. de Broglie, *Recherches sur la théorie des quanta* (Researches on the quantum theory), Thesis (Paris), 1924; L. de Broglie, *Ann. Phys.* (Paris) **3**, 22 (1925). English translation by A.F. Kracklauer. And here.

- Broglie, Louis de. *The wave nature of the electron* Nobel Lecture, 12, 1929

- Tipler, Paul A. and Ralph A. Llewellyn (2003). *Modern Physics*. 4th ed. New York; W. H. Freeman and Co. ISBN 0-7167-4345-0. pp. 203–4, 222–3, 236.

- Zumdahl, Steven S. (2005). *Chemical Principles* (5th ed.). Boston: Houghton Mifflin. ISBN 0-618-37206-7.

- An extensive review article "Optics and interferometry with atoms and molecules" appeared in July 2009: http://www.atomwave.org/rmparticle/RMPLAO.pdf.

- "Scientific Papers Presented to Max Born on his retirement from the Tait Chair of Natural Philosophy in the University of Edinburgh", 1953 (Oliver and Boyd)

7.10 External links

- Bowley, Roger. "de Broglie Waves". *Sixty Symbols*. Brady Haran for the University of Nottingham.

Chapter 8

Electricity

"Electric" redirects here. For other uses, see Electric (disambiguation) and Electricity (disambiguation).

Lightning is one of the most dramatic effects of electricity.

Electricity is the set of physical phenomena associated with the presence and flow of electric charge. Electricity gives a wide variety of well-known effects, such as lightning, static electricity, electromagnetic induction and electric current. In addition, electricity permits the creation and reception of electromagnetic radiation such as radio waves.

In electricity, charges produce electromagnetic fields which act on other charges. Electricity occurs due to several types of physics:

- **electric charge**: a property of some subatomic particles, which determines their electromagnetic interactions. Electrically charged matter is influenced by, and produces, electromagnetic fields.

- **electric field** (see electrostatics): an especially simple type of electromagnetic field produced by an electric charge even when it is not moving (i.e., there is no

electric current). The electric field produces a force on other charges in its vicinity.

- **electric potential**: the capacity of an electric field to do work on an electric charge, typically measured in volts.

- **electric current**: a movement or flow of electrically charged particles, typically measured in amperes.

- **electromagnets**: Moving charges produce a magnetic field. Electric currents generate magnetic fields, and changing magnetic fields generate electric currents.

In electrical engineering, electricity is used for:

- **electric power** where electric current is used to energise equipment;

- **electronics** which deals with electrical circuits that involve active electrical components such as vacuum tubes, transistors, diodes and integrated circuits, and associated passive interconnection technologies.

Electrical phenomena have been studied since antiquity, though progress in theoretical understanding remained slow until the seventeenth and eighteenth centuries. Even then, practical applications for electricity were few, and it would not be until the late nineteenth century that engineers were able to put it to industrial and residential use. The rapid expansion in electrical technology at this time transformed industry and society. Electricity's extraordinary versatility means it can be put to an almost limitless set of applications which include transport, heating, lighting, communications, and computation. Electrical power is now the backbone of modern industrial society."[1]

8.1 History

Main articles: History of electromagnetic theory and History of electrical engineering

Thales, the earliest known researcher into electricity

See also: Etymology of electricity

Long before any knowledge of electricity existed people were aware of shocks from electric fish. Ancient Egyptian texts dating from 2750 BC referred to these fish as the "Thunderer of the Nile", and described them as the "protectors" of all other fish. Electric fish were again reported millennia later by ancient Greek, Roman and Arabic naturalists and physicians.[2] Several ancient writers, such as Pliny the Elder and Scribonius Largus, attested to the numbing effect of electric shocks delivered by catfish and torpedo rays, and knew that such shocks could travel along conducting objects.[3] Patients suffering from ailments such as gout or headache were directed to touch electric fish in the hope that the powerful jolt might cure them.[4] Possibly the earliest and nearest approach to the discovery of the identity of lightning, and electricity from any other source, is to be attributed to the Arabs, who before the 15th century had the Arabic word for lightning (*raad*) applied to the electric ray.[5]

Ancient cultures around the Mediterranean knew that certain objects, such as rods of amber, could be rubbed with cat's fur to attract light objects like feathers. Thales of Miletus made a series of observations on static electricity around 600 BC, from which he believed that friction rendered amber magnetic, in contrast to minerals such as

magnetite, which needed no rubbing.[6][7] Thales was incorrect in believing the attraction was due to a magnetic effect, but later science would prove a link between magnetism and electricity. According to a controversial theory, the Parthians may have had knowledge of electroplating, based on the 1936 discovery of the Baghdad Battery, which resembles a galvanic cell, though it is uncertain whether the artifact was electrical in nature.[8]

Benjamin Franklin conducted extensive research on electricity in the 18th century, as documented by Joseph Priestley (1767) History and Present Status of Electricity, *with whom Franklin carried on extended correspondence.*

Electricity would remain little more than an intellectual curiosity for millennia until 1600, when the English scientist William Gilbert made a careful study of electricity and magnetism, distinguishing the lodestone effect from static electricity produced by rubbing amber.[6] He coined the New Latin word *electricus* ("of amber" or "like amber" , from ἤλεκτρον, *elektron*, the Greek word for "amber") to refer to the property of attracting small objects after being rubbed.[9] This association gave rise to the English words "electric" and "electricity" , which made their first appearance in print in Thomas Browne's *Pseudodoxia Epidemica* of 1646.[10]

Further work was conducted by Otto von Guericke, Robert Boyle, Stephen Gray and C. F. du Fay. In the 18th century, Benjamin Franklin conducted extensive research in electricity, selling his possessions to fund his work. In June 1752 he is reputed to have attached a metal key to the bottom of a dampened kite string and flown the kite in a storm-threatened sky.[11] A succession of sparks jumping from the key to the back of his hand showed that lightning was indeed electrical in nature.[12] He also explained the apparently paradoxical behavior[13] of the Leyden jar as a device for storing large amounts of electrical charge in terms of electricity consisting of both positive and negative charges.

Michael Faraday's discoveries formed the foundation of electric motor technology

In 1791, Luigi Galvani published his discovery of bioelectricity, demonstrating that electricity was the medium by which nerve cells passed signals to the muscles.[14] Alessandro Volta's battery, or voltaic pile, of 1800, made from alternating layers of zinc and copper, provided scientists with a more reliable source of electrical energy than the electrostatic machines previously used.[14] The recognition of electromagnetism, the unity of electric and magnetic phenomena, is due to Hans Christian Ørsted and André-Marie Ampère in 1819-1820; Michael Faraday invented the electric motor in 1821, and Georg Ohm mathematically analysed the electrical circuit in 1827.[14] Electricity and magnetism (and light) were definitively linked by James Clerk Maxwell, in particular in his "On Physical Lines of Force" in 1861 and 1862.[15]

While the early 19th century had seen rapid progress in electrical science, the late 19th century would see the greatest progress in electrical engineering. Through such people as Alexander Graham Bell, Ottó Bláthy, Thomas Edison, Galileo Ferraris, Oliver Heaviside, Ányos Jedlik, Lord Kelvin, Sir Charles Parsons, Ernst Werner von Siemens, Joseph Swan, Nikola Tesla and George Westinghouse, electricity turned from a scientific curiosity into an essential tool for modern life, becoming a driving force of the Second Industrial Revolution.[16]

In 1887, Heinrich Hertz[17]:843–844[18] discovered that electrodes illuminated with ultraviolet light create electric sparks more easily. In 1905 Albert Einstein published a paper that explained experimental data from the photoelectric effect as being the result of light energy being carried in discrete quantized packets, energising electrons. This discovery led to the quantum revolution. Einstein was awarded the Nobel Prize in 1921 for "his discovery of the law of the photoelectric effect".[19] The photoelectric effect is also employed in photocells such as can be found in solar panels and this is frequently used to make electricity commercially.

The first solid-state device was the "cat's whisker" detector, first used in the 1900s in radio receivers. A whisker-like wire is placed lightly in contact with a solid crystal (such as a germanium crystal) in order to detect a radio signal by the contact junction effect.[20] In a solid-state component, the current is confined to solid elements and compounds engineered specifically to switch and amplify it. Current flow can be understood in two forms: as negatively charged electrons, and as positively charged electron deficiencies called holes. These charges and holes are understood in terms of quantum physics. The building material is most often a crystalline semiconductor.[21][22]

The solid-state device came into its own with the invention of the transistor in 1947. Common solid-state devices include transistors, microprocessor chips, and RAM. A specialized type of RAM called flash RAM is used in flash drives and more recently, solid state drives to replace mechanically rotating magnetic disc hard drives. Solid state devices became prevalent in the 1950s and the 1960s, during the transition from vacuum tubes to semiconductor diodes, transistors, integrated circuit (IC) and the light-emitting diode (LED).

8.2 Concepts

8.2.1 Electric charge

Main article: Electric charge
See also: electron, proton and ion

The presence of charge gives rise to an electrostatic force:

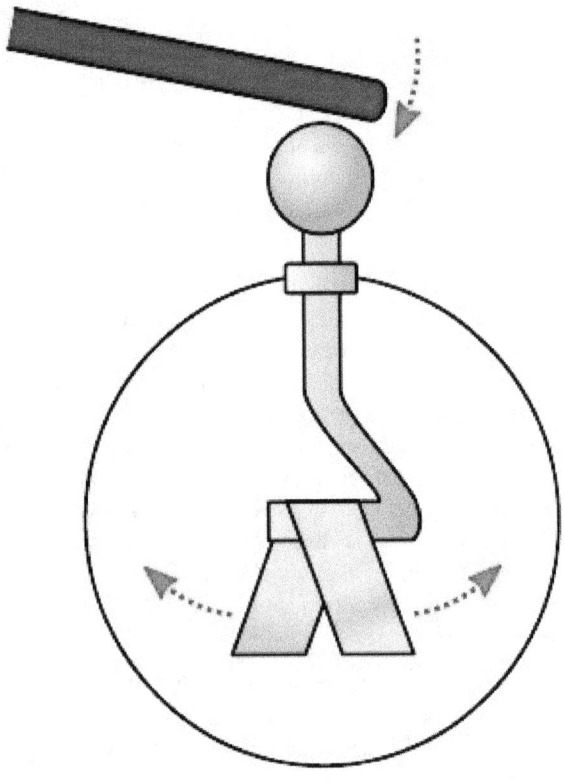

Charge on a gold-leaf electroscope causes the leaves to visibly repel each other

charges exert a force on each other, an effect that was known, though not understood, in antiquity.[17]:457 A lightweight ball suspended from a string can be charged by touching it with a glass rod that has itself been charged by rubbing with a cloth. If a similar ball is charged by the same glass rod, it is found to repel the first: the charge acts to force the two balls apart. Two balls that are charged with a rubbed amber rod also repel each other. However, if one ball is charged by the glass rod, and the other by an amber rod, the two balls are found to attract each other. These phenomena were investigated in the late eighteenth century by Charles-Augustin de Coulomb, who deduced that charge manifests itself in two opposing forms. This discovery led to the well-known axiom: *like-charged objects repel and opposite-charged objects attract.*[17]

The force acts on the charged particles themselves, hence charge has a tendency to spread itself as evenly as possible over a conducting surface. The magnitude of the electromagnetic force, whether attractive or repulsive, is given by Coulomb's law, which relates the force to the product of the charges and has an inverse-square relation to the distance between them.[23][24]:35 The electromagnetic force is very strong, second only in strength to the strong interaction,[25] but unlike that force it operates over

all distances.[26] In comparison with the much weaker gravitational force, the electromagnetic force pushing two electrons apart is 10^{42} times that of the gravitational attraction pulling them together.[27]

Study has shown that the origin of charge is from certain types of subatomic particles which have the property of electric charge. Electric charge gives rise to and interacts with the electromagnetic force, one of the four fundamental forces of nature. The most familiar carriers of electrical charge are the electron and proton. Experiment has shown charge to be a conserved quantity, that is, the net charge within an isolated system will always remain constant regardless of any changes taking place within that system.[28] Within the system, charge may be transferred between bodies, either by direct contact, or by passing along a conducting material, such as a wire.[24]:2–5 The informal term static electricity refers to the net presence (or 'imbalance') of charge on a body, usually caused when dissimilar materials are rubbed together, transferring charge from one to the other.

The charge on electrons and protons is opposite in sign, hence an amount of charge may be expressed as being either negative or positive. By convention, the charge carried by electrons is deemed negative, and that by protons positive, a custom that originated with the work of Benjamin Franklin.[29] The amount of charge is usually given the symbol Q and expressed in coulombs;[30] each electron carries the same charge of approximately -1.6022×10^{-19} coulomb. The proton has a charge that is equal and opposite, and thus $+1.6022\times10^{-19}$ coulomb. Charge is possessed not just by matter, but also by antimatter, each antiparticle bearing an equal and opposite charge to its corresponding particle.[31]

Charge can be measured by a number of means, an early instrument being the gold-leaf electroscope, which although still in use for classroom demonstrations, has been superseded by the electronic electrometer.[24]:2–5

8.2.2   Electric current

Main article: Electric current

The movement of electric charge is known as an electric current, the intensity of which is usually measured in amperes. Current can consist of any moving charged particles; most commonly these are electrons, but any charge in motion constitutes a current.

By historical convention, a positive current is defined as having the same direction of flow as any positive charge it contains, or to flow from the most positive part of a circuit to the most negative part. Current defined in this manner is called conventional current. The motion of negatively charged electrons around an electric circuit, one of the most familiar forms of current, is thus deemed positive in the *opposite* direction to that of the electrons.[32] However, depending on the conditions, an electric current can consist of a flow of charged particles in either direction, or even in both directions at once. The positive-to-negative convention is widely used to simplify this situation.

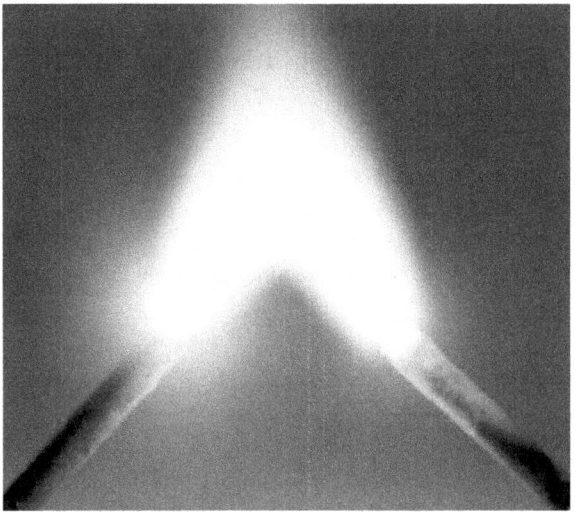

An electric arc provides an energetic demonstration of electric current

The process by which electric current passes through a material is termed electrical conduction, and its nature varies with that of the charged particles and the material through which they are travelling. Examples of electric currents include metallic conduction, where electrons flow through a conductor such as metal, and electrolysis, where ions (charged atoms) flow through liquids, or through plasmas such as electrical sparks. While the particles themselves can move quite slowly, sometimes with an average drift velocity only fractions of a millimetre per second,[24]:17 the electric field that drives them itself propagates at close to the speed of light, enabling electrical signals to pass rapidly along wires.[33]

Current causes several observable effects, which historically were the means of recognising its presence. That water could be decomposed by the current from a voltaic pile was discovered by Nicholson and Carlisle in 1800, a process now known as electrolysis. Their work was greatly expanded upon by Michael Faraday in 1833. Current through a resistance causes localised heating, an effect James Prescott Joule studied mathematically in 1840.[24]:23–24 One of the most important discoveries relating to current was made accidentally by Hans Christian Ørsted in 1820, when, while preparing a lecture, he witnessed the current in a wire disturbing the needle of a magnetic compass.[34]

He had discovered electromagnetism, a fundamental interaction between electricity and magnetics. The level of electromagnetic emissions generated by electric arcing is high enough to produce electromagnetic interference, which can be detrimental to the workings of adjacent equipment.[35]

In engineering or household applications, current is often described as being either direct current (DC) or alternating current (AC). These terms refer to how the current varies in time. Direct current, as produced by example from a battery and required by most electronic devices, is a unidirectional flow from the positive part of a circuit to the negative.[36]:11 If, as is most common, this flow is carried by electrons, they will be travelling in the opposite direction. Alternating current is any current that reverses direction repeatedly; almost always this takes the form of a sine wave.[36]:206–207 Alternating current thus pulses back and forth within a conductor without the charge moving any net distance over time. The time-averaged value of an alternating current is zero, but it delivers energy in first one direction, and then the reverse. Alternating current is affected by electrical properties that are not observed under steady state direct current, such as inductance and capacitance.[36]:223–225 These properties however can become important when circuitry is subjected to transients, such as when first energised.

8.2.3 Electric field

Main article: Electric field
See also: Electrostatics

The concept of the electric field was introduced by Michael Faraday. An electric field is created by a charged body in the space that surrounds it, and results in a force exerted on any other charges placed within the field. The electric field acts between two charges in a similar manner to the way that the gravitational field acts between two masses, and like it, extends towards infinity and shows an inverse square relationship with distance.[26] However, there is an important difference. Gravity always acts in attraction, drawing two masses together, while the electric field can result in either attraction or repulsion. Since large bodies such as planets generally carry no net charge, the electric field at a distance is usually zero. Thus gravity is the dominant force at distance in the universe, despite being much weaker.[27]

An electric field generally varies in space,[37] and its strength at any one point is defined as the force (per unit charge) that would be felt by a stationary, negligible charge if placed at that point.[17]:469–470 The conceptual charge, termed a 'test charge', must be vanishingly small to prevent its own electric field disturbing the main field and must also be stationary to prevent the effect of magnetic

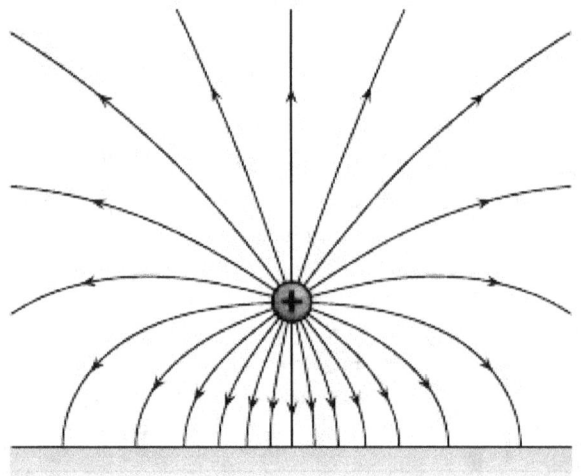

Field lines emanating from a positive charge above a plane conductor

fields. As the electric field is defined in terms of force, and force is a vector, so it follows that an electric field is also a vector, having both magnitude and direction. Specifically, it is a vector field.[17]:469–470

The study of electric fields created by stationary charges is called electrostatics. The field may be visualised by a set of imaginary lines whose direction at any point is the same as that of the field. This concept was introduced by Faraday,[38] whose term 'lines of force' still sometimes sees use. The field lines are the paths that a point positive charge would seek to make as it was forced to move within the field; they are however an imaginary concept with no physical existence, and the field permeates all the intervening space between the lines.[38] Field lines emanating from stationary charges have several key properties: first, that they originate at positive charges and terminate at negative charges; second, that they must enter any good conductor at right angles, and third, that they may never cross nor close in on themselves.[17]:479

A hollow conducting body carries all its charge on its outer surface. The field is therefore zero at all places inside the body.[24]:88 This is the operating principal of the Faraday cage, a conducting metal shell which isolates its interior from outside electrical effects.

The principles of electrostatics are important when designing items of high-voltage equipment. There is a finite limit to the electric field strength that may be withstood by any medium. Beyond this point, electrical breakdown occurs and an electric arc causes flashover between the charged parts. Air, for example, tends to arc across small gaps at electric field strengths which exceed 30 kV per centimetre. Over larger gaps, its breakdown strength is weaker, perhaps 1 kV per centimetre.[39] The most visible natural occur-

rence of this is lightning, caused when charge becomes separated in the clouds by rising columns of air, and raises the electric field in the air to greater than it can withstand. The voltage of a large lightning cloud may be as high as 100 MV and have discharge energies as great as 250 kWh.[40]

The field strength is greatly affected by nearby conducting objects, and it is particularly intense when it is forced to curve around sharply pointed objects. This principle is exploited in the lightning conductor, the sharp spike of which acts to encourage the lightning stroke to develop there, rather than to the building it serves to protect[41]:155

8.2.4   Electric potential

Main article: Electric potential
See also: Voltage and Battery (electricity)
  The concept of electric potential is closely linked to that

A pair of AA cells. The + sign indicates the polarity of the potential difference between the battery terminals.

of the electric field. A small charge placed within an electric field experiences a force, and to have brought that charge to that point against the force requires work. The electric potential at any point is defined as the energy required to bring a unit test charge from an infinite distance slowly to that point. It is usually measured in volts, and one volt is the potential for which one joule of work must be expended to bring a charge of one coulomb from infinity.[17]:494–498 This definition of potential, while formal, has little practical application, and a more useful concept is that of electric potential difference, and is the energy required to move a unit charge between two specified points. An electric field has the special property that it is *conservative*, which means that the path taken by the test charge is irrelevant: all paths between two specified points expend the same energy, and thus a unique value for potential difference may be stated.[17]:494–498 The volt is so strongly identified as the unit of choice for measurement and description of electric potential difference that the term voltage sees greater everyday usage.

For practical purposes, it is useful to define a common reference point to which potentials may be expressed and compared. While this could be at infinity, a much more useful reference is the Earth itself, which is assumed to be at the same potential everywhere. This reference point naturally takes the name earth or ground. Earth is assumed to be an infinite source of equal amounts of positive and negative charge, and is therefore electrically uncharged — and unchargeable.[42]

Electric potential is a scalar quantity, that is, it has only magnitude and not direction. It may be viewed as analogous to height: just as a released object will fall through a difference in heights caused by a gravitational field, so a charge will 'fall' across the voltage caused by an electric field.[43] As relief maps show contour lines marking points of equal height, a set of lines marking points of equal potential (known as equipotentials) may be drawn around an electrostatically charged object. The equipotentials cross all lines of force at right angles. They must also lie parallel to a conductor's surface, otherwise this would produce a force that will move the charge carriers to even the potential of the surface.

The electric field was formally defined as the force exerted per unit charge, but the concept of potential allows for a more useful and equivalent definition: the electric field is the local gradient of the electric potential. Usually expressed in volts per metre, the vector direction of the field is the line of greatest slope of potential, and where the equipotentials lie closest together.[24]:60

8.2.5 Electromagnets

Main article: Electromagnets

Ørsted's discovery in 1821 that a magnetic field existed around all sides of a wire carrying an electric current indicated that there was a direct relationship between electricity and magnetism. Moreover, the interaction seemed different from gravitational and electrostatic forces, the two forces of nature then known. The force on the compass needle did not direct it to or away from the current-carrying wire, but acted at right angles to it.[34] Ørsted's slightly

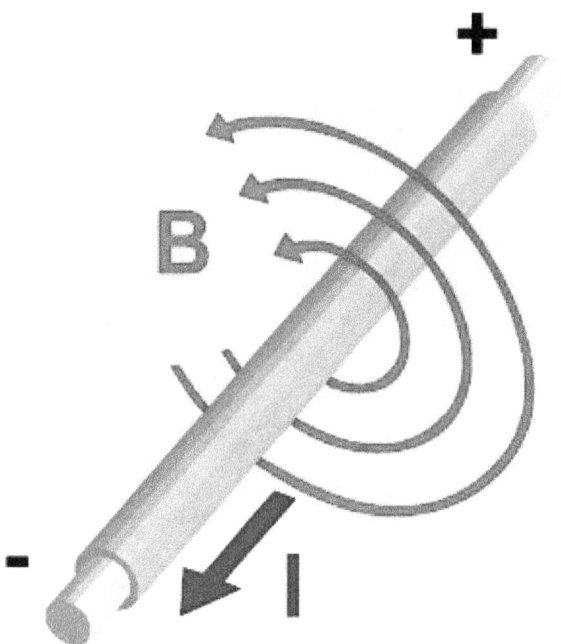

Magnetic field circles around a current

obscure words were that "the electric conflict acts in a revolving manner." The force also depended on the direction of the current, for if the flow was reversed, then the force did too.[44]

Ørsted did not fully understand his discovery, but he observed the effect was reciprocal: a current exerts a force on a magnet, and a magnetic field exerts a force on a current. The phenomenon was further investigated by Ampère, who discovered that two parallel current-carrying wires exerted a force upon each other: two wires conducting currents in the same direction are attracted to each other, while wires containing currents in opposite directions are forced apart.[45] The interaction is mediated by the magnetic field each current produces and forms the basis for the international definition of the ampere.[45]

This relationship between magnetic fields and currents is extremely important, for it led to Michael Faraday's invention of the electric motor in 1821. Faraday's homopolar motor consisted of a permanent magnet sitting in a pool of mercury. A current was allowed through a wire suspended from a pivot above the magnet and dipped into the mercury. The magnet exerted a tangential force on the wire, making it circle around the magnet for as long as the current was maintained.[46]

Experimentation by Faraday in 1831 revealed that a wire moving perpendicular to a magnetic field developed a potential difference between its ends. Further analysis of this process, known as electromagnetic induction, enabled him to state the principle, now known as Faraday's law of induc-

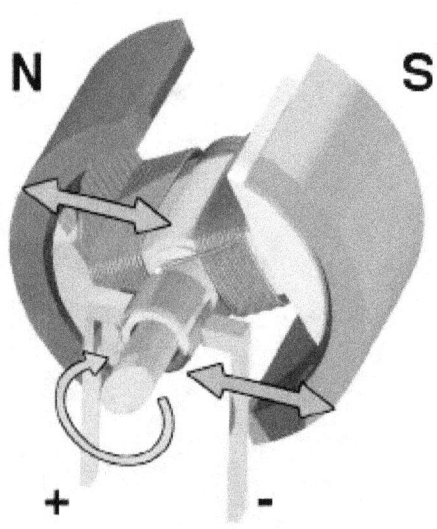

The electric motor exploits an important effect of electromagnetism: a current through a magnetic field experiences a force at right angles to both the field and current

Italian physicist Alessandro Volta showing his "battery" to French emperor Napoleon Bonaparte in the early 19th century.

8.2.7 Electric circuits

Main article: Electric circuit
 An electric circuit is an interconnection of electric compo-

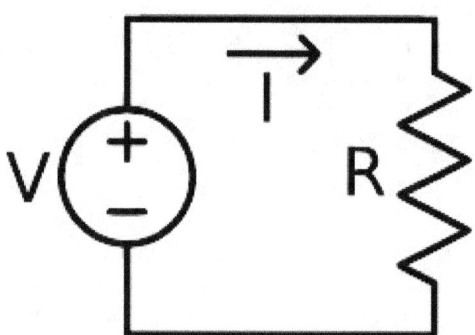

A basic electric circuit. The voltage source V on the left drives a current I around the circuit, delivering electrical energy into the resistor R. From the resistor, the current returns to the source, completing the circuit.

tion, that the potential difference induced in a closed circuit is proportional to the rate of change of magnetic flux through the loop. Exploitation of this discovery enabled him to invent the first electrical generator in 1831, in which he converted the mechanical energy of a rotating copper disc to electrical energy.*[46] Faraday's disc was inefficient and of no use as a practical generator, but it showed the possibility of generating electric power using magnetism, a possibility that would be taken up by those that followed on from his work.

8.2.6 Electrochemistry

Main article: Electrochemistry

The ability of chemical reactions to produce electricity, and conversely the ability of electricity to drive chemical reactions has a wide array of uses.

Electrochemistry has always been an important part of electricity.   From the initial invention of the Voltaic pile, electrochemical cells have evolved into the many different types of batteries, electroplating and electrolysis cells. Aluminium is produced in vast quantities this way, and many portable devices are electrically powered using rechargeable cells.

nents such that electric charge is made to flow along a closed path (a circuit), usually to perform some useful task.

The components in an electric circuit can take many forms, which can include elements such as resistors, capacitors, switches, transformers and electronics. Electronic circuits contain active components, usually semiconductors, and typically exhibit non-linear behaviour, requiring complex analysis. The simplest electric components are those that are termed passive and linear: while they may temporarily store energy, they contain no sources of it, and exhibit linear responses to stimuli.[47]:15–16

The resistor is perhaps the simplest of passive circuit elements: as its name suggests, it resists the current through it, dissipating its energy as heat. The resistance is a consequence of the motion of charge through a conductor: in metals, for example, resistance is primarily due to collisions between electrons and ions. Ohm's law is a basic law of circuit theory, stating that the current passing through a resistance is directly proportional to the potential difference across it. The resistance of most materials is relatively constant over a range of temperatures and currents; materials under these conditions are known as 'ohmic'. The ohm, the unit of resistance, was named in honour of Georg Ohm, and is symbolised by the Greek letter Ω. 1 Ω is the resistance that will produce a potential difference of one volt in response to a current of one amp.[47]:30–35

The capacitor is a development of the Leyden jar and is a device that can store charge, and thereby storing electrical energy in the resulting field. It consists of two conducting plates separated by a thin insulating dielectric layer; in practice, thin metal foils are coiled together, increasing the surface area per unit volume and therefore the capacitance. The unit of capacitance is the farad, named after Michael Faraday, and given the symbol F: one farad is the capacitance that develops a potential difference of one volt when it stores a charge of one coulomb. A capacitor connected to a voltage supply initially causes a current as it accumulates charge; this current will however decay in time as the capacitor fills, eventually falling to zero. A capacitor will therefore not permit a steady state current, but instead blocks it.[47]:216–220

The inductor is a conductor, usually a coil of wire, that stores energy in a magnetic field in response to the current through it. When the current changes, the magnetic field does too, inducing a voltage between the ends of the conductor. The induced voltage is proportional to the time rate of change of the current. The constant of proportionality is termed the inductance. The unit of inductance is the henry, named after Joseph Henry, a contemporary of Faraday. One henry is the inductance that will induce a potential difference of one volt if the current through it changes at a rate of one ampere per second. The inductor's behaviour is in some regards converse to that of the capacitor: it will freely allow an unchanging current, but opposes a rapidly changing one.[47]:226–229

8.2.8 Electric power

Main article: electric power

Electric power is the rate at which electric energy is transferred by an electric circuit. The SI unit of power is the watt, one joule per second.

Electric power, like mechanical power, is the rate of doing work, measured in watts, and represented by the letter P. The term *wattage* is used colloquially to mean "electric power in watts." The electric power in watts produced by an electric current I consisting of a charge of Q coulombs every t seconds passing through an electric potential (voltage) difference of V is

$$P = \text{time unit per done work} = \frac{QV}{t} = IV$$

where

Q is electric charge in coulombs

t is time in seconds

I is electric current in amperes

V is electric potential or voltage in volts

Electricity generation is often done with electric generators, but can also be supplied by chemical sources such as electric batteries or by other means from a wide variety of sources of energy. Electric power is generally supplied to businesses and homes by the electric power industry. Electricity is usually sold by the kilowatt hour (3.6 MJ) which is the product of power in kilowatts multiplied by running time in hours. Electric utilities measure power using electricity meters, which keep a running total of the electric energy delivered to a customer.

8.2.9 Electronics

Main article: electronics

Electronics deals with electrical circuits that involve active electrical components such as vacuum tubes, transistors, diodes and integrated circuits, and associated passive interconnection technologies. The nonlinear behaviour of active components and their ability to control electron flows makes amplification of weak signals possible and electronics is widely used in information processing, telecommunications, and signal processing. The ability of electronic devices to act as switches makes digital information processing possible. Interconnection technologies such as circuit boards, electronics packaging technology, and

Surface mount electronic components

other varied forms of communication infrastructure complete circuit functionality and transform the mixed components into a regular working system.

Today, most electronic devices use semiconductor components to perform electron control. The study of semiconductor devices and related technology is considered a branch of solid state physics, whereas the design and construction of electronic circuits to solve practical problems come under electronics engineering.

8.2.10   Electromagnetic wave

Main article: Electromagnetic wave

Faraday's and Ampère's work showed that a time-varying magnetic field acted as a source of an electric field, and a time-varying electric field was a source of a magnetic field. Thus, when either field is changing in time, then a field of the other is necessarily induced.[17]:696–700 Such a phenomenon has the properties of a wave, and is naturally referred to as an electromagnetic wave. Electromagnetic waves were analysed theoretically by James Clerk Maxwell in 1864. Maxwell developed a set of equations that could unambiguously describe the interrelationship between electric field, magnetic field, electric charge, and electric current. He could moreover prove that such a wave would necessarily travel at the speed of light, and thus light itself was a form of electromagnetic radiation. Maxwell's Laws, which unify light, fields, and charge are one of the great milestones of theoretical physics.[17]:696–700

Thus, the work of many researchers enabled the use of electronics to convert signals into high frequency oscillating currents, and via suitably shaped conductors, electricity permits the transmission and reception of these signals via radio waves over very long distances.

8.3   Production and uses

8.3.1   Generation and transmission

Main article: Electricity generation
See also: Electric power transmission and Mains electricity
In the 6th century BC, the Greek philosopher Thales of

Early 20th-century alternator made in Budapest, Hungary, in the power generating hall of a hydroelectric station (photograph by Prokudin-Gorsky, 1905–1915).

Miletus experimented with amber rods and these experiments were the first studies into the production of electrical energy. While this method, now known as the triboelectric effect, can lift light objects and generate sparks, it is extremely inefficient.[48] It was not until the invention of the voltaic pile in the eighteenth century that a viable source of electricity became available. The voltaic pile, and its modern descendant, the electrical battery, store energy chemically and make it available on demand in the form of electrical energy.[48] The battery is a versatile and very common power source which is ideally suited to many applications, but its energy storage is finite, and once discharged it must be disposed of or recharged. For large electrical demands electrical energy must be generated and transmitted continuously over conductive transmission lines.

Electrical power is usually generated by electro-mechanical generators driven by steam produced from fossil fuel combustion, or the heat released from nuclear reactions; or from other sources such as kinetic energy extracted from wind or flowing water. The modern steam turbine invented by Sir Charles Parsons in 1884 today generates about 80 percent of the electric power in the world using a variety of heat sources. Such generators bear no resemblance to Faraday's homopolar disc generator of 1831, but they still rely on his electromagnetic principle that a conductor linking a changing magnetic field induces a potential difference across its ends.[49] The invention in the late nineteenth century of the transformer meant that electrical power could be transmitted more efficiently at a higher voltage but lower current. Efficient electrical transmission meant in turn that electricity could be generated at centralised power stations, where it benefited from economies of scale, and then be despatched relatively long distances to where it was needed.[50][51]

Wind power is of increasing importance in many countries

The light bulb, an early application of electricity, operates by Joule heating: the passage of current through resistance generating heat

Since electrical energy cannot easily be stored in quantities large enough to meet demands on a national scale, at all times exactly as much must be produced as is required.[50] This requires electricity utilities to make careful predictions of their electrical loads, and maintain constant coordination with their power stations. A certain amount of generation must always be held in reserve to cushion an electrical grid against inevitable disturbances and losses.

Demand for electricity grows with great rapidity as a nation modernises and its economy develops. The United States showed a 12% increase in demand during each year of the first three decades of the twentieth century,[52] a rate of growth that is now being experienced by emerging economies such as those of India or China.[53][54] Historically, the growth rate for electricity demand has outstripped that for other forms of energy.[55]:16

Environmental concerns with electricity generation have led to an increased focus on generation from renewable sources, in particular from wind and hydropower. While debate can be expected to continue over the environmental impact of different means of electricity production, its final form is relatively clean[55]:89

8.3.2 Applications

Electricity is a very convenient way to transfer energy, and it has been adapted to a huge, and growing, number of uses.[56] The invention of a practical incandescent light bulb in the 1870s led to lighting becoming one of the first publicly available applications of electrical power. Although electrification brought with it its own dangers, replacing the naked flames of gas lighting greatly reduced fire hazards within homes and factories.[57] Public utilities were set up in many cities targeting the burgeoning market for electrical lighting.

The Joule heating effect employed in the light bulb also sees more direct use in electric heating. While this is versatile and controllable, it can be seen as wasteful, since most electrical generation has already required the production of heat at a power station.[58] A number of countries, such as Denmark, have issued legislation restricting or banning the use of electric heating in new buildings.[59] Electricity is however a highly practical energy source for refrigeration.[60] with air conditioning representing a growing sector for electricity demand, the effects of which electricity utilities are increasingly obliged to accommodate.[61]

Electricity is used within telecommunications, and indeed the electrical telegraph, demonstrated commercially in 1837 by Cooke and Wheatstone, was one of its earliest applications. With the construction of first intercontinental, and then transatlantic, telegraph systems in the 1860s, electricity had enabled communications in minutes across the globe. Optical fibre and satellite communication have taken a share of the market for communications systems, but electricity can be expected to remain an essential part of the process.

The effects of electromagnetism are most visibly employed in the electric motor, which provides a clean and efficient means of motive power. A stationary motor such as a winch is easily provided with a supply of power, but a motor that moves with its application, such as an electric vehicle, is obliged to either carry along a power source such as a battery, or to collect current from a sliding contact such as a pantograph.

Electronic devices make use of the transistor, perhaps one of the most important inventions of the twentieth century,[62] and a fundamental building block of all modern circuitry. A modern integrated circuit may contain several billion miniaturised transistors in a region only a few centimetres square.[63]

Electricity is also used to fuel public transportation, including electric buses and trains.[64]

8.4   Electricity and the natural world

8.4.1   Physiological effects

Main article: Electric shock

A voltage applied to a human body causes an electric current through the tissues, and although the relationship is non-linear, the greater the voltage, the greater the current.[65] The threshold for perception varies with the supply frequency and with the path of the current, but is about 0.1 mA to 1 mA for mains-frequency electricity, though a current as low as a microamp can be detected as an electrovibration effect under certain conditions.[66] If the current is sufficiently high, it will cause muscle contraction, fibrillation of the heart, and tissue burns.[65] The lack of any visible sign that a conductor is electrified makes electricity a particular hazard. The pain caused by an electric

shock can be intense, leading electricity at times to be employed as a method of torture. Death caused by an electric shock is referred to as electrocution. Electrocution is still the means of judicial execution in some jurisdictions, though its use has become rarer in recent times.[67]

8.4.2   Electrical phenomena in nature

The electric eel, Electrophorus electricus

Main article: Electrical phenomena

Electricity is not a human invention, and may be observed in several forms in nature, a prominent manifestation of which is lightning. Many interactions familiar at the macroscopic level, such as touch, friction or chemical bonding, are due to interactions between electric fields on the atomic scale. The Earth's magnetic field is thought to arise from a natural dynamo of circulating currents in the planet's core.[68] Certain crystals, such as quartz, or even sugar, generate a potential difference across their faces when subjected to external pressure.[69] This phenomenon is known as piezoelectricity, from the Greek *piezein* (πιέζειν), meaning to press, and was discovered in 1880 by Pierre and Jacques Curie. The effect is reciprocal, and when a piezoelectric material is subjected to an electric field, a small change in physical dimensions takes place.[69]

Some organisms, such as sharks, are able to detect and respond to changes in electric fields, an ability known as electroreception,[70] while others, termed electrogenic, are able to generate voltages themselves to serve as a predatory or defensive weapon.[3] The order Gymnotiformes, of which the best known example is the electric eel, detect or stun their prey via high voltages generated from modified muscle cells called electrocytes.[3][4] All animals transmit information along their cell membranes with voltage pulses called action potentials, whose functions include communication by the nervous system between neurons and muscles.[71] An electric shock stimulates this system, and causes muscles to contract.[72] Action potentials are also responsible for coordinating activities in certain plants.[71]

8.5 Cultural perception

In 1850, William Gladstone asked the scientist Michael Faraday why electricity was valuable. Faraday answered, "One day sir, you may tax it." [73]

In the 19th and early 20th century, electricity was not part of the everyday life of many people, even in the industrialised Western world. The popular culture of the time accordingly often depicts it as a mysterious, quasi-magical force that can slay the living, revive the dead or otherwise bend the laws of nature.[74] This attitude began with the 1771 experiments of Luigi Galvani in which the legs of dead frogs were shown to twitch on application of animal electricity. "Revitalization" or resuscitation of apparently dead or drowned persons was reported in the medical literature shortly after Galvani's work. These results were known to Mary Shelley when she authored *Frankenstein* (1819), although she does not name the method of revitalization of the monster. The revitalization of monsters with electricity later became a stock theme in horror films.

As the public familiarity with electricity as the lifeblood of the Second Industrial Revolution grew, its wielders were more often cast in a positive light,[75] such as the workers who "finger death at their gloves' end as they piece and repiece the living wires" in Rudyard Kipling's 1907 poem *Sons of Martha*.[75] Electrically powered vehicles of every sort featured large in adventure stories such as those of Jules Verne and the *Tom Swift* books.[75] The masters of electricity, whether fictional or real—including scientists such as Thomas Edison, Charles Steinmetz or Nikola Tesla—were popularly conceived of as having wizard-like powers.[75]

With electricity ceasing to be a novelty and becoming a necessity of everyday life in the later half of the 20th century, it required particular attention by popular culture only when it *stops* flowing,[75] an event that usually signals disaster.[75] The people who *keep* it flowing, such as the nameless hero of Jimmy Webb's song "Wichita Lineman" (1968),[75] are still often cast as heroic, wizard-like figures.[75]

8.6 See also

- Ampère's circuital law, connects the direction of an

electric current and its associated magnetic currents.

- Electric potential energy, the potential energy of a system of charges

- Electricity market, the sale of electrical energy

- Hydraulic analogy, an analogy between the flow of water and electric current

8.7 Notes

[1] Jones, D.A. (1991). "Electrical engineering: the backbone of society". *Proceedings of the IEE: Science, Measurement and Technology* **138** (1): 1–10. doi:10.1049/ip-a-3.1991.0001

[2] Moller, Peter; Kramer, Bernd (December 1991). "Review: Electric Fish". *BioScience* (American Institute of Biological Sciences) **41** (11): 794–6 [794]. doi:10.2307/1311732. JSTOR 1311732

[3] Bullock, Theodore H. (2005). *Electroreception*. Springer. pp. 5–7. ISBN 0-387-23192-7

[4] Morris, Simon C. (2003). *Life's Solution: Inevitable Humans in a Lonely Universe*. Cambridge University Press. pp. 182–185. ISBN 0-521-82704-3

[5] *The Encyclopedia Americana; a library of universal knowledge* (1918). New York: Encyclopedia Americana Corp

[6] Stewart, Joseph (2001). *Intermediate Electromagnetic Theory*. World Scientific. p. 50. ISBN 981-02-4471-1

[7] Simpson, Brian (2003). *Electrical Stimulation and the Relief of Pain*. Elsevier Health Sciences. pp. 6–7. ISBN 0-444-51258-6

[8] Frood, Arran (27 February 2003). *Riddle of 'Baghdad's batteries'*. BBC. retrieved 2008-02-16

[9] Baigrie, Brian (2006). *Electricity and Magnetism: A Historical Perspective*. Greenwood Press. pp. 7–8. ISBN 0-313-33358-0

[10] Chalmers, Gordon (1937). "The Lodestone and the Understanding of Matter in Seventeenth Century England". *Philosophy of Science* **4** (1): 75–95. doi:10.1086/286445

[11] Srodes, James (2002). *Franklin: The Essential Founding Father*. Regnery Publishing. pp. 92–94. ISBN 0-89526-163-4 It is uncertain if Franklin personally carried out this experiment, but it is popularly attributed to him.

[12] Uman, Martin (1987). *All About Lightning* (PDF). Dover Publications. ISBN 0-486-25237-X

[13] Riskin, Jessica (1998). *Poor Richard's Leyden Jar: Electricity and economy in Franklinist France* (PDF). p. 327

[14] Kirby, Richard S. (1990), *Engineering in History*, Courier Dover Publications, pp. 331–333, ISBN 0-486-26412-2

[15] Berkson, William (1974) Fields of force: the development of a world view from Faraday to Einstein p.148. Routledge, 1974

[16] Marković, Dragana, *The Second Industrial Revolution*, retrieved 2007-12-09

[17] Sears, Francis; et al. (1982), *University Physics, Sixth Edition*, Addison Wesley, ISBN 0-201-07199-1

[18] Hertz, Heinrich (1887). "Ueber den Einfluss des ultravioletten Lichtes auf die electrische Entladung". *Annalen der Physik* **267** (8): S. 983–1000. Bibcode:1887AnP...267..983H. doi:10.1002/andp.18872670827.

[19] "The Nobel Prize in Physics 1921". Nobel Foundation. Retrieved 2013-03-16.

[20] "Solid state", *The Free Dictionary*

[21] John Sydney Blakemore, *Solid state physics*, pp.1-3, Cambridge University Press, 1985 ISBN 0-521-31391-0.

[22] Richard C. Jaeger, Travis N. Blalock, *Microelectronic circuit design*, pp.46-47, McGraw-Hill Professional, 2003 ISBN 0-07-250503-6.

[23] "The repulsive force between two small spheres charged with the same type of electricity is inversely proportional to the square of the distance between the centres of the two spheres." Charles-Augustin de Coulomb, *Histoire de l'Academie Royal des Sciences*, Paris 1785.

[24] Duffin, W.J. (1980), *Electricity and Magnetism, 3rd edition*, McGraw-Hill, ISBN 0-07-084111-X

[25] National Research Council (1998), *Physics Through the 1990s*, National Academies Press, pp. 215–216, ISBN 0-309-03576-7

[26] Umashankar, Korada (1989), *Introduction to Engineering Electromagnetic Fields*, World Scientific, pp. 77–79, ISBN 9971-5-0921-0

[27] Hawking, Stephen (1988), *A Brief History of Time*, Bantam Press. p. 77, ISBN 0-553-17521-1

[28] Trefil, James (2003), *The Nature of Science: An A–Z Guide to the Laws and Principles Governing Our Universe*, Houghton Mifflin Books. p. 74, ISBN 0-618-31938-7

[29] Shectman, Jonathan (2003), *Groundbreaking Scientific Experiments, Inventions, and Discoveries of the 18th Century*, Greenwood Press, pp. 87–91, ISBN 0-313-32015-2

[30] Sewell, Tyson (1902), *The Elements of Electrical Engineering*, Lockwood, p. 18. The Q originally stood for 'quantity of electricity', the term 'electricity' now more commonly expressed as 'charge'.

[31] Close, Frank (2007), *The New Cosmic Onion: Quarks and the Nature of the Universe*, CRC Press, p. 51, ISBN 1-58488-798-2

[32] Ward, Robert (1960), *Introduction to Electrical Engineering*, Prentice-Hall, p. 18

[33] Solymar, L. (1984), *Lectures on electromagnetic theory*, Oxford University Press, p. 140, ISBN 0-19-856169-5

[34] Berkson, William (1974), *Fields of Force: The Development of a World View from Faraday to Einstein*, Routledge, p. 370, ISBN 0-7100-7626-6 Accounts differ as to whether this was before, during, or after a lecture.

[35] "Lab Note #105 *EMI Reduction - Unsuppressed vs. Suppressed*". Arc Suppression Technologies. April 2011. Retrieved March 7, 2012.

[36] Bird, John (2007), *Electrical and Electronic Principles and Technology, 3rd edition*, Newnes, ISBN 9781417505432

[37] Almost all electric fields vary in space. An exception is the electric field surrounding a planar conductor of infinite extent, the field of which is uniform.

[38] Morely & Hughes, *Principles of Electricity, Fifth edition*, p. 73, ISBN 0-582-42629-4

[39] Naidu, M.S.; Kamataru, V. (1982), *High Voltage Engineering*, Tata McGraw-Hill, p. 2, ISBN 0-07-451786-4

[40] Naidu, M.S.; Kamataru, V. (1982), *High Voltage Engineering*, Tata McGraw-Hill, pp. 201–202, ISBN 0-07-451786-4

[41] Paul J. Nahin (9 October 2002). *Oliver Heaviside: The Life, Work, and Times of an Electrical Genius of the Victorian Age*. JHU Press. ISBN 978-0-8018-6909-9.

[42] Serway, Raymond A. (2006), *Serway's College Physics*, Thomson Brooks, p. 500, ISBN 0-534-99724-4

[43] Saeli, Sue; MacIsaac, Dan (2007), "Using Gravitational Analogies To Introduce Elementary Electrical Field Theory Concepts", *The Physics Teacher* **45** (2): 104. Bibcode:2007PhTea..45..104S, doi:10.1119/1.2432088, retrieved 2007-12-09

[44] Thompson, Silvanus P. (2004), *Michael Faraday: His Life and Work*, Elibron Classics. p. 79, ISBN 1-4212-7387-X

[45] Morely & Hughes, *Principles of Electricity, Fifth edition*, pp. 92–93

[46] Institution of Engineering and Technology, *Michael Faraday: Biography*, retrieved 2007-12-09

[47] Alexander, Charles; Sadiku, Matthew (2006), *Fundamentals of Electric Circuits* (3, revised ed.), McGraw-Hill, ISBN 9780073301150

[48] Dell, Ronald; Rand, David (2001), "Understanding Batteries", *Unknown* (Royal Society of Chemistry) **86**: 2–4, Bibcode:1985STIN...8619754M, ISBN 0-85404-605-4

[49] McLaren, Peter G. (1984), *Elementary Electric Power and Machines*, Ellis Horwood, pp. 182–183, ISBN 0-85312-269-5

[50] Patterson, Walter C. (1999), *Transforming Electricity: The Coming Generation of Change*, Earthscan, pp. 44–48, ISBN 1-85383-341-X

[51] Edison Electric Institute, *History of the Electric Power Industry*, archived from the original on November 13, 2007, retrieved 2007-12-08

[52] Edison Electric Institute, *History of the U.S. Electric Power Industry, 1882-1991*, retrieved 2007-12-08

[53] Carbon Sequestration Leadership Forum, *An Energy Summary of India*, archived from the original on 2007-12-05, retrieved 2007-12-08

[54] IndexMundi, *China Electricity - consumption*, retrieved 2007-12-08

[55] National Research Council (1986), *Electricity in Economic Growth*, National Academies Press, ISBN 0-309-03677-1

[56] Wald, Matthew (21 March 1990), "Growing Use of Electricity Raises Questions on Supply", *New York Times*, retrieved 2007-12-09

[57] d'Alroy Jones, Peter, *The Consumer Society: A History of American Capitalism*, Penguin Books, p. 211

[58] ReVelle, Charles and Penelope (1992), *The Global Environment: Securing a Sustainable Future*, Jones & Bartlett, p. 298, ISBN 0-86720-321-8

[59] Danish Ministry of Environment and Energy, "F.2 The Heat Supply Act", *Denmark's Second National Communication on Climate Change*, retrieved 2007-12-09

[60] Brown, Charles E. (2002), *Power resources*, Springer, ISBN 3-540-42634-5

[61] Hojjati, B.; Battles, S., *The Growth in Electricity Demand in U.S. Households, 1981-2001: Implications for Carbon Emissions* (PDF), retrieved 2007-12-09

[62] Herrick, Dennis F. (2003), *Media Management in the Age of Giants: Business Dynamics of Journalism*, Blackwell Publishing, ISBN 0-8138-1699-8

[63] Das, Saswato R. (2007-12-15), "The tiny, mighty transistor", *Los Angeles Times*

[64] "Public Transportation", *Alternative Energy News*, 2010-03-10

[65] Tleis, Nasser (2008), *Power System Modelling and Fault Analysis*, Elsevier, pp. 552–554, ISBN 978-0-7506-8074-5

[66] Grimnes, Sverre (2000), *Bioimpedance and Bioelectricity Basic*, Academic Press, pp. 301–309, ISBN 0-12-303260-1

[67] Lipschultz, J.H.; Hilt, M.L.J.H. (2002), *Crime and Local Television News*, Lawrence Erlbaum Associates, p. 95, ISBN 0-8058-3620-9

[68] Encrenaz, Thérèse (2004), *The Solar System*, Springer, p. 217, ISBN 3-540-00241-3

[69] Lima-de-Faria, José; Buerger, Martin J. (1990), *Historical Atlas of Crystallography*, Springer, p. 67, ISBN 0-7923-0649-X

[70] Ivancevic, Vladimir & Tijana (2005), *Natural Biodynamics*, World Scientific, p. 602, ISBN 981-256-534-5

[71] Kandel, E.; Schwartz, J.; Jessell, T. (2000), *Principles of Neural Science*, McGraw-Hill Professional, pp. 27–28, ISBN 0-8385-7701-6

[72] Davidovits, Paul (2007), *Physics in Biology and Medicine*, Academic Press, pp. 204–205, ISBN 978-0-12-369411-9

[73] Jackson, Mark (4 November 2013), *Theoretical physics – like sex, but with no need to experiment*, The Conversation

[74] Van Riper, A. Bowdoin (2002), *Science in popular culture: a reference guide*, Westport: Greenwood Press, p. 69, ISBN 0-313-31822-0

[75] Van Riper, op.cit., p. 71.

8.8 References

- Nahvi, Mahmood; Joseph, Edminister (1965), *Electric Circuits*, McGraw-Hill, ISBN 9780071422413

- Hammond, Percy (1981), "Electromagnetism for Engineers", *Nature* (Pergamon) **168** (4262): 4, Bibcode:1951Natur.168....4G, doi:10.1038/168004b0, ISBN 0-08-022104-1

- Morely, A.; Hughes, E. (1994), *Principles of Electricity* (5th ed.), Longman, ISBN 0-582-22874-3

- Naidu, M.S.; Kamataru, V. (1982), *High Voltage Engineering*, Tata McGraw-Hill, ISBN 0-07-451786-4

- Nilsson, James; Riedel, Susan (2007), *Electric Circuits*, Prentice Hall, ISBN 978-0-13-198925-2

- Patterson, Walter C. (1999), *Transforming Electricity: The Coming Generation of Change*, Earthscan, ISBN 1-85383-341-X

- Benjamin, P. (1898). A history of electricity (The intellectual rise in electricity) from antiquity to the days of Benjamin Franklin. New York: J. Wiley & Sons.

8.9   External links

- "One-Hundred Years of Electricity", May 1931, Popular Mechanics

- Illustrated view of how an American home's electrical system works

- Electricity around the world

- Electricity Misconceptions

- Electricity and Magnetism

- Understanding Electricity and Electronics in about 10 Minutes

- World Bank report on Water, Electricity and Utility subsidies

Chapter 9

Magnetism

"Magnetic" redirects here. For other uses, see Magnetic (disambiguation) and Magnetism (disambiguation).

Magnetism is a class of physical phenomena that are

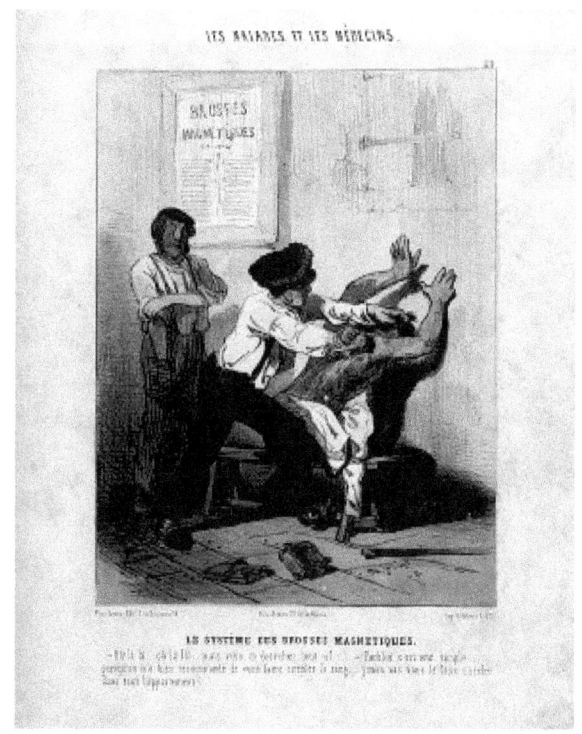

A magnetic quadrupole

mediated by magnetic fields. Electric currents and the magnetic moments of elementary particles give rise to a magnetic field, which acts on other currents and magnetic moments. Every material is influenced to some extent by a magnetic field. The most familiar effect is on permanent magnets, which have persistent magnetic moments caused by ferromagnetism. Most materials do not have permanent moments. Some are attracted to a magnetic field (paramagnetism); others are repulsed by a magnetic field (diamagnetism); others have a more complex relationship with an applied magnetic field (spin glass behavior and antiferromagnetism). Substances that are negligibly affected by magnetic fields are known as *non-magnetic* substances. These include copper, aluminium, gases, and plastic. Pure oxygen exhibits magnetic properties when cooled to a liquid state.

The magnetic state (or magnetic phase) of a material depends on temperature and other variables such as pressure and the applied magnetic field. A material may exhibit more than one form of magnetism as these variables change.

9.1 History

Main article: History of electromagnetism

Aristotle attributed the first of what could be called a sci-

Drawing of a medical treatment using magnetic brushes. Charles Jacque 1843, France.

entific discussion on magnetism to Thales of Miletus, who lived from about 625 BC to about 545 BC.[1] Around the same time, in ancient India, the Indian surgeon, Sushruta, was the first to make use of the magnet for surgical purposes.[2]

In ancient China, the earliest literary reference to magnetism lies in a 4th-century BC book named after its author, *The Master of Demon Valley* (鬼谷子): "The lodestone makes iron come or it attracts it." [3] The earliest mention of the attraction of a needle appears in a work composed between AD 20 and 100 (*Louen-heng*): "A lodestone attracts a needle." [4] The Chinese scientist Shen Kuo (1031–1095) was the first person to write of the magnetic needle compass and that it improved the accuracy of navigation by employing the astronomical concept of true north (*Dream Pool Essays*, AD 1088), and by the 12th century the Chinese were known to use the lodestone compass for navigation. They sculpted a directional spoon from lodestone in such a way that the handle of the spoon always pointed south.

Alexander Neckam, by 1187, was the first in Europe to describe the compass and its use for navigation. In 1269, Peter Peregrinus de Maricourt wrote the *Epistola de magnete*, the first extant treatise describing the properties of magnets. In 1282, the properties of magnets and the dry compass were discussed by Al-Ashraf, a Yemeni physicist, astronomer, and geographer.[5]

Michael Faraday, 1842

In 1600, William Gilbert published his *De Magnete, Magneticisque Corporibus, et de Magno Magnete Tellure* (*On the Magnet and Magnetic Bodies, and on the Great Magnet the Earth*). In this work he describes many of his experiments with his model earth called the terrella. From his experiments, he concluded that the Earth was itself magnetic and that this was the reason compasses pointed north (previously, some believed that it was the pole star (Polaris) or a large magnetic island on the north pole that attracted the compass).

An understanding of the relationship between electricity and magnetism began in 1819 with work by Hans Christian Ørsted, a professor at the University of Copenhagen, who discovered more or less by accident that an electric current could influence a compass needle. This landmark experiment is known as Ørsted's Experiment. Several other experiments followed, with André-Marie Ampère, who in 1820 discovered that the magnetic field circulating in a closed-path was related to the current flowing through the perimeter of the path; Carl Friedrich Gauss; Jean-Baptiste Biot and Félix Savart, both of whom in 1820 came up with the Biot–Savart law giving an equation for the magnetic field from a current-carrying wire; Michael Faraday, who in 1831 found that a time-varying magnetic flux through a loop of wire induced a voltage, and others finding further links between magnetism and electricity. James Clerk Maxwell synthesized and expanded these insights into Maxwell's equations, unifying electricity, magnetism, and optics into the field of electromagnetism. In 1905, Einstein used these laws in motivating his theory of special relativity,[6] requiring that the laws held true in all inertial reference frames.

Electromagnetism has continued to develop into the 21st century, being incorporated into the more fundamental theories of gauge theory, quantum electrodynamics, electroweak theory, and finally the standard model.

9.2   Sources of magnetism

See also: Magnetic moment

Magnetism, at its root, arises from two sources:

1. Electric current (see *Electron magnetic moment*).

2. Spin magnetic moments of elementary particles. The magnetic moments of the nuclei of atoms are typically thousands of times smaller than the electrons' magnetic moments, so they are negligible in the context of the magnetization of materials. Nuclear magnetic moments are very important in other contexts, particularly in nuclear magnetic resonance (NMR) and magnetic resonance imaging (MRI).

Ordinarily, the enormous number of electrons in a material are arranged such that their magnetic moments (both orbital and intrinsic) cancel out. This is due, to some extent, to electrons combining into pairs with opposite intrinsic magnetic moments as a result of the Pauli exclusion principle (see *electron configuration*), or combining into filled subshells with zero net orbital motion. In both cases, the electron arrangement is so as to exactly cancel the magnetic moments from each electron. Moreover, even when the electron configuration *is* such that there are unpaired electrons and/or non-filled subshells, it is often the case that the various electrons in the solid will contribute magnetic moments that point in different, random directions, so that the material will not be magnetic.

However, sometimes—either spontaneously, or owing to an applied external magnetic field—each of the electron magnetic moments will be, on average, lined up. Then the material can produce a net total magnetic field, which can potentially be quite strong.

The magnetic behavior of a material depends on its structure, particularly its electron configuration, for the reasons mentioned above, and also on the temperature. At high temperatures, random thermal motion makes it more difficult for the electrons to maintain alignment.

9.3   Materials

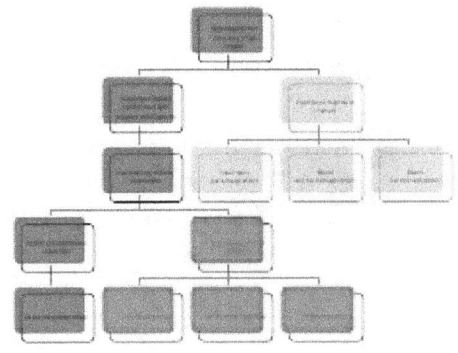

Hierarchy of types of magnetism. [7]

9.3.1   Diamagnetism

Main article: Diamagnetism

Diamagnetism appears in all materials, and is the tendency of a material to oppose an applied magnetic field, and therefore, to be repelled by a magnetic field. However, in a material with paramagnetic properties (that is, with a tendency to enhance an external magnetic field), the paramagnetic behavior dominates. [8] Thus, despite its universal occurrence, diamagnetic behavior is observed only in a purely diamagnetic material. In a diamagnetic material, there are no unpaired electrons, so the intrinsic electron magnetic moments cannot produce any bulk effect. In these cases, the magnetization arises from the electrons' orbital motions, which can be understood classically as follows:

> When a material is put in a magnetic field, the electrons circling the nucleus will experience, in addition to their Coulomb attraction to the nucleus, a Lorentz force from the magnetic field. Depending on which direction the electron is orbiting, this force may increase the centripetal force on the electrons, pulling them in towards the nucleus, or it may decrease the force, pulling them away from the nucleus. This effect systematically increases the orbital magnetic moments that were aligned opposite the field, and decreases the ones aligned parallel to the field (in accordance with Lenz's law). This results in a small bulk magnetic moment, with an opposite direction to the applied field.

Note that this description is meant only as an heuristic; a proper understanding requires a quantum-mechanical description.

Note that all materials undergo this orbital response. However, in paramagnetic and ferromagnetic substances, the diamagnetic effect is overwhelmed by the much stronger effects caused by the unpaired electrons.

9.3.2   Paramagnetism

Main article: Paramagnetism

In a paramagnetic material there are *unpaired electrons*, i.e. atomic or molecular orbitals with exactly one electron in them. While paired electrons are required by the Pauli exclusion principle to have their intrinsic ('spin') magnetic moments pointing in opposite directions, causing their magnetic fields to cancel out, an unpaired electron is free to align its magnetic moment in any direction. When an external magnetic field is applied, these magnetic moments will tend to align themselves in the same direction as the applied field, thus reinforcing it.

9.3.3   Ferromagnetism

Main article: Ferromagnetism

A permanent magnet holding up several coins

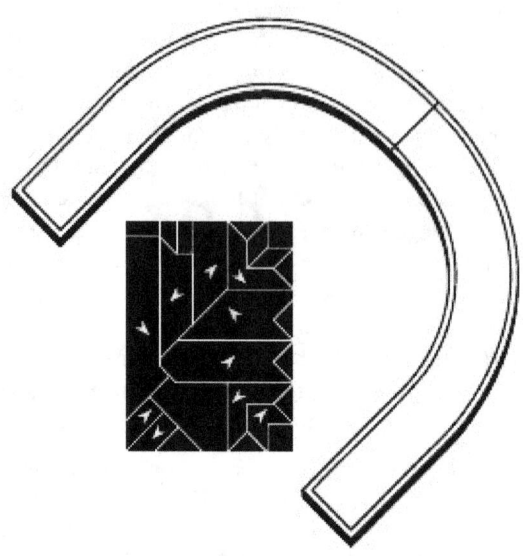

Magnetic domains boundaries (white lines) in ferromagnetic material (black rectangle).

A ferromagnet, like a paramagnetic substance, has unpaired electrons. However, in *addition* to the electrons' intrinsic magnetic moment's tendency to be parallel to an *applied field*, there is also in these materials a tendency for these magnetic moments to orient parallel to *each other* to maintain a lowered-energy state. Thus, even in the absence of an applied field, the magnetic moments of the electrons in the material spontaneously line up parallel to one another.

Every ferromagnetic substance has its own individual temperature, called the Curie temperature, or Curie point, above which it loses its ferromagnetic properties. This is because the thermal tendency to disorder overwhelms the energy-lowering due to ferromagnetic order.

Ferromagnetism only occurs in a few substances; the common ones are iron, nickel, cobalt, their alloys, and some alloys of rare earth metals.

Magnetic domains

Main article: Magnetic domains

The magnetic moments of atoms in a ferromagnetic material cause them to behave something like tiny permanent magnets. They stick together and align themselves into small regions of more or less uniform alignment called magnetic domains or Weiss domains. Magnetic domains can be observed with a magnetic force microscope to reveal magnetic domain boundaries that resemble white lines in the sketch. There are many scientific experiments that can physically show magnetic fields.

When a domain contains too many molecules, it becomes unstable and divides into two domains aligned in opposite directions so that they stick together more stably as shown at the right.

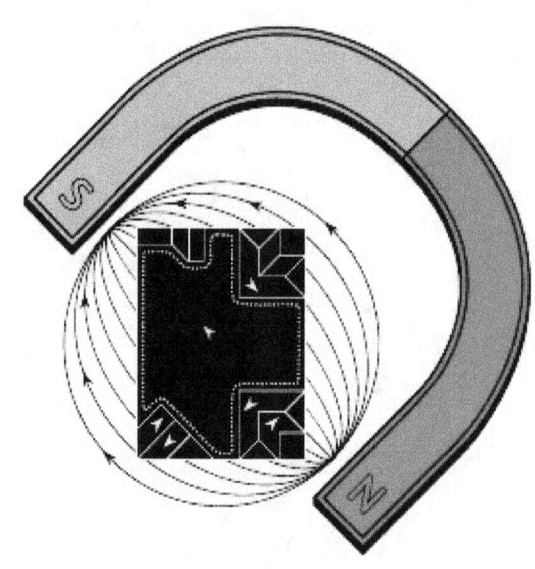

Effect of a magnet on the domains.

When exposed to a magnetic field, the domain boundaries move so that the domains aligned with the magnetic field grow and dominate the structure (dotted yellow area) as shown at the left. When the magnetizing field is removed, the domains may not return to an unmagnetized state. This results in the ferromagnetic material's being magnetized, forming a permanent magnet.

When magnetized strongly enough that the prevailing domain overruns all others to result in only one single domain, the material is magnetically saturated. When a magnetized ferromagnetic material is heated to the Curie point temperature, the molecules are agitated to the point that the magnetic domains lose the organization and the magnetic properties they cause cease. When the material is cooled, this domain alignment structure spontaneously returns, in a manner roughly analogous to how a liquid can freeze into a crystalline solid.

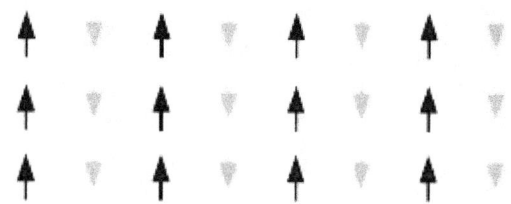

Ferrimagnetic ordering

9.3.4 Antiferromagnetism

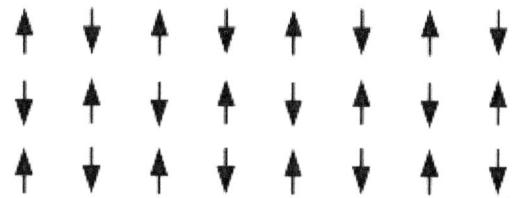

Antiferromagnetic ordering

Main article: Antiferromagnetism

In an antiferromagnet, unlike a ferromagnet, there is a tendency for the intrinsic magnetic moments of neighboring valence electrons to point in *opposite* directions. When all atoms are arranged in a substance so that each neighbor is 'anti-aligned', the substance is **antiferromagnetic**. Antiferromagnets have a zero net magnetic moment, meaning no field is produced by them. Antiferromagnets are less common compared to the other types of behaviors, and are mostly observed at low temperatures. In varying temperatures, antiferromagnets can be seen to exhibit diamagnetic and ferrimagnetic properties.

In some materials, neighboring electrons want to point in opposite directions, but there is no geometrical arrangement in which *each* pair of neighbors is anti-aligned. This is called a spin glass, and is an example of geometrical frustration.

9.3.5 Ferrimagnetism

Main article: Ferrimagnetism

Like ferromagnetism, **ferrimagnets** retain their magnetization in the absence of a field. However, like antiferromagnets, neighboring pairs of electron spins like to point in opposite directions. These two properties are not contradictory, because in the optimal geometrical arrangement,

there is more magnetic moment from the sublattice of electrons that point in one direction, than from the sublattice that points in the opposite direction.

Most ferrites are ferrimagnetic. The first discovered magnetic substance, magnetite, is a ferrite and was originally believed to be a ferromagnet; Louis Néel disproved this, however, after discovering ferrimagnetism.

9.3.6 Superparamagnetism

Main article: Superparamagnetism

When a ferromagnet or ferrimagnet is sufficiently small, it acts like a single magnetic spin that is subject to Brownian motion. Its response to a magnetic field is qualitatively similar to the response of a paramagnet, but much larger.

9.3.7 Other types of magnetism

- Metamagnetism
- Molecule-based magnet
- Spin glass

9.4 Electromagnet

An *electromagnet* is a type of magnet whose magnetism is produced by the flow of electric current. The magnetic field disappears when the current ceases.

9.5 Magnetism, electricity, and special relativity

Main article: Classical electromagnetism and special relativity

As a consequence of Einstein's theory of special relativity,

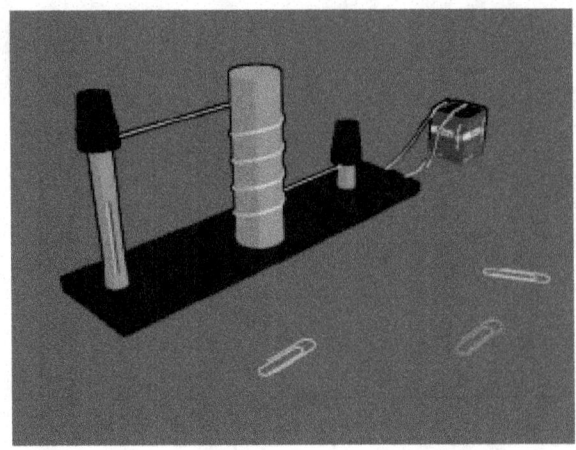

An electromagnet attracts paper clips when current is applied creating a magnetic field. The electromagnet loses them when current and magnetic field are removed.

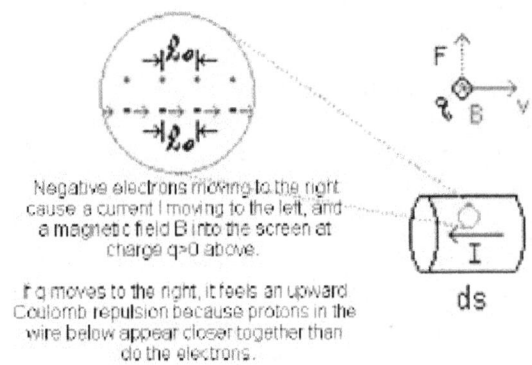

Negative electrons moving to the right cause a current I moving to the left, and a magnetic field B into the screen at charge q>0 above.

If q moves to the right, it feels an upward Coulomb repulsion because protons in the wire below appear closer together than do the electrons.

Magnetism from length-contraction.

electricity and magnetism are fundamentally interlinked. Both magnetism lacking electricity, and electricity without magnetism, are inconsistent with special relativity, due to such effects as length contraction, time dilation, and the fact that the magnetic force is velocity-dependent. However, when both electricity and magnetism are taken into account, the resulting theory (electromagnetism) is fully consistent with special relativity.[6][9] In particular, a phenomenon that appears purely electric or purely magnetic to one observer may be a mix of both to another, or more generally the relative contributions of electricity and magnetism are dependent on the frame of reference. Thus, special relativity "mixes" electricity and magnetism into a single, inseparable phenomenon called electromagnetism, analogous to how relativity "mixes" space and time into spacetime.

All observations on electromagnetism apply to what might be considered to be primarily magnetism, e.g. perturbations in the magnetic field are necessarily accompanied by a nonzero electric field, and propagate at the speed of light.

9.6   Magnetic fields in a material

See also: Magnetic field § H and B inside and outside of magnetic materials

In a vacuum,

$$\mathbf{B} = \mu_0 \mathbf{H}.$$

where μ_0 is the vacuum permeability.

In a material,

$$\mathbf{B} = \mu_0(\mathbf{H} + \mathbf{M}).$$

The quantity $\mu_0 \mathbf{M}$ is called *magnetic polarization*.

If the field $\mathbf{H}$ is small, the response of the magnetization $\mathbf{M}$ in a diamagnet or paramagnet is approximately linear:

$$\mathbf{M} = \chi \mathbf{H}.$$

the constant of proportionality being called the magnetic susceptibility. If so,

$$\mu_0(\mathbf{H} + \mathbf{M}) = \mu_0(1 + \chi)\mathbf{H} = \mu_r \mu_0 \mathbf{H} = \mu \mathbf{H}.$$

In a hard magnet such as a ferromagnet, $\mathbf{M}$ is not proportional to the field and is generally nonzero even when $\mathbf{H}$ is zero (see Remanence).

9.7   Magnetic force

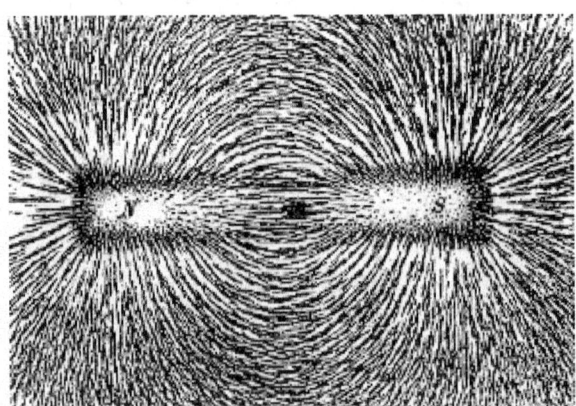

Magnetic lines of force of a bar magnet shown by iron filings on paper

Main article: Magnetic field

The phenomenon of magnetism is "mediated" by the magnetic field. An electric current or magnetic dipole creates a magnetic field, and that field, in turn, imparts magnetic forces on other particles that are in the fields.

Maxwell's equations, which simplify to the Biot–Savart law in the case of steady currents, describe the origin and behavior of the fields that govern these forces. Therefore, magnetism is seen whenever electrically charged particles are in motion — for example, from movement of electrons in an electric current, or in certain cases from the orbital motion of electrons around an atom's nucleus. They also arise from "intrinsic" magnetic dipoles arising from quantum-mechanical spin.

The same situations that create magnetic fields — charge moving in a current or in an atom, and intrinsic magnetic dipoles — are also the situations in which a magnetic field has an effect, creating a force. Following is the formula for moving charge; for the forces on an intrinsic dipole, see magnetic dipole.

When a charged particle moves through a magnetic field **B**, it feels a Lorentz force **F** given by the cross product:[10]

$$\mathbf{F} = q(\mathbf{v} \times \mathbf{B})$$

where

q is the electric charge of the particle, and

$\mathbf{v}$ is the velocity vector of the particle

Because this is a cross product, the force is perpendicular to both the motion of the particle and the magnetic field. It follows that the magnetic force does no work on the particle; it may change the direction of the particle's movement, but it cannot cause it to speed up or slow down. The magnitude of the force is

$$F = qvB \sin\theta$$

where θ is the angle between **v** and **B**.

One tool for determining the direction of the velocity vector of a moving charge, the magnetic field, and the force exerted is labeling the index finger "V", the middle finger "B", and the thumb "F" with your right hand. When making a gun-like configuration, with the middle finger crossing under the index finger, the fingers represent the velocity vector, magnetic field vector, and force vector, respectively. See also right hand rule.

9.8 Magnetic dipoles

Main article: Magnetic dipole

A very common source of magnetic field shown in nature is a dipole, with a "South pole" and a "North pole", terms dating back to the use of magnets as compasses, interacting with the Earth's magnetic field to indicate North and South on the globe. Since opposite ends of magnets are attracted, the north pole of a magnet is attracted to the south pole of another magnet. The Earth's North Magnetic Pole (currently in the Arctic Ocean, north of Canada) is physically a south pole, as it attracts the north pole of a compass. A magnetic field contains energy, and physical systems move toward configurations with lower energy. When diamagnetic material is placed in a magnetic field, a *magnetic dipole* tends to align itself in opposed polarity to that field, thereby lowering the net field strength. When ferromagnetic material is placed within a magnetic field, the magnetic dipoles align to the applied field, thus expanding the domain walls of the magnetic domains.

9.8.1 Magnetic monopoles

Main article: Magnetic monopole

Since a bar magnet gets its ferromagnetism from electrons distributed evenly throughout the bar, when a bar magnet is cut in half, each of the resulting pieces is a smaller bar magnet. Even though a magnet is said to have a north pole and a south pole, these two poles cannot be separated from each other. A monopole — if such a thing exists — would be a new and fundamentally different kind of magnetic object. It would act as an isolated north pole, not attached to a south pole, or vice versa. Monopoles would carry "magnetic charge" analogous to electric charge. Despite systematic searches since 1931, as of 2010, they have never been observed, and could very well not exist.[11]

Nevertheless, some theoretical physics models predict the existence of these magnetic monopoles. Paul Dirac observed in 1931 that, because electricity and magnetism show a certain symmetry, just as quantum theory predicts that individual positive or negative electric charges can be observed without the opposing charge, isolated South or North magnetic poles should be observable. Using quantum theory Dirac showed that if magnetic monopoles exist, then one could explain the quantization of electric charge — that is, why the observed elementary particles carry charges that are multiples of the charge of the electron.

Certain grand unified theories predict the existence of monopoles which, unlike elementary particles, are solitons (localized energy packets). The initial results of using these models to estimate the number of monopoles created in the big bang contradicted cosmological observations — the monopoles would have been so plentiful and massive that they would have long since halted the expansion of the universe. However, the idea of inflation (for which this problem served as a partial motivation) was successful in solving this problem, creating models in which monopoles existed but were rare enough to be consistent with current observations.[12]

9.9   Quantum-mechanical origin of magnetism

In principle all kinds of magnetism originate (similar to superconductivity) from specific quantum-mechanical phenomena (e.g. Mathematical formulation of quantum mechanics, in particular the chapters on spin and on the Pauli principle). A successful model was developed already in 1927, by Walter Heitler and Fritz London, who derived quantum-mechanically, how hydrogen molecules are formed from hydrogen atoms, i.e. from the atomic hydrogen orbitals u_A and u_B centered at the nuclei A and B, see below. That this leads to magnetism is not at all obvious, but will be explained in the following.

According to the Heitler-London theory, so-called two-body molecular σ-orbitals are formed, namely the resulting orbital is:

$$\psi(\mathbf{r}_1, \mathbf{r}_2) = \frac{1}{\sqrt{2}} \ (u_A(\mathbf{r}_1)u_B(\mathbf{r}_2) + u_B(\mathbf{r}_1)u_A(\mathbf{r}_2))$$

Here the last product means that a first electron, $\mathbf{r}_1$, is in an atomic hydrogen-orbital centered at the second nucleus, whereas the second electron runs around the first nucleus. This "exchange" phenomenon is an expression for the quantum-mechanical property that particles with identical properties cannot be distinguished. It is specific not only for the formation of chemical bonds, but as we will see, also for magnetism, i.e. in this connection the term exchange interaction arises, a term which is essential for the origin of magnetism, and which is stronger, roughly by factors 100 and even by 1000, than the energies arising from the electrodynamic dipole-dipole interaction.

As for the *spin function* $\chi(s_1, s_2)$, which is responsible for the magnetism, we have the already mentioned Pauli's principle, namely that a symmetric orbital (i.e. with the + sign as above) must be multiplied with an antisymmetric spin function (i.e. with a − sign), and *vice versa*. Thus:

$$\chi(s_1, s_2) = \frac{1}{\sqrt{2}} \ (\alpha(s_1)\beta(s_2) - \beta(s_1)\alpha(s_2))$$

I.e., not only u_A and u_B must be substituted by α and β, respectively (the first entity means "spin up", the second one "spin down"), but also the sign + by the − sign, and finally $\mathbf{r}_i$ by the discrete values s_i (= ±½); thereby we have $\alpha(+1/2) = \beta(-1/2) = 1$ and $\alpha(-1/2) = \beta(+1/2) = 0$. The "singlet state", i.e. the − sign, means: the spins are *antiparallel*, i.e. for the solid we have antiferromagnetism, and for two-atomic molecules one has diamagnetism. The tendency to form a (homoeopolar) chemical bond (this means: the formation of a *symmetric* molecular orbital, i.e. with the + sign) results through the Pauli principle automatically in an *antisymmetric* spin state (i.e. with the − sign). In contrast, the Coulomb repulsion of the electrons, i.e. the tendency that they try to avoid each other by this repulsion, would lead to an *antisymmetric* orbital function (i.e. with the − sign) of these two particles, and complementary to a *symmetric* spin function (i.e. with the + sign, one of the so-called "triplet functions"). Thus, now the spins would be *parallel* (ferromagnetism in a solid, paramagnetism in two-atomic gases).

The last-mentioned tendency dominates in the metals iron, cobalt and nickel, and in some rare earths, which are *ferromagnetic*. Most of the other metals, where the first-mentioned tendency dominates, are *nonmagnetic* (e.g. sodium, aluminium, and magnesium) or *antiferromagnetic* (e.g. manganese). Diatomic gases are also almost exclusively diamagnetic, and not paramagnetic. However, the oxygen molecule, because of the involvement of π-orbitals, is an exception important for the life-sciences.

The Heitler-London considerations can be generalized to the Heisenberg model of magnetism (Heisenberg 1928).

The explanation of the phenomena is thus essentially based on all subtleties of quantum mechanics, whereas the electrodynamics covers mainly the phenomenology.

9.10   Units

9.10.1   SI

9.10.2   Other

- gauss – the centimeter-gram-second (CGS) unit of magnetic field (denoted **B**).

- oersted – the CGS unit of magnetizing field (denoted **H**).

- maxwell – the CGS unit for magnetic flux.

- gamma – a unit of *magnetic flux density* that was commonly used before the tesla came into use (1.0 gamma = 1.0 nanotesla)

- μ_0 – common symbol for the permeability of free space ($4\pi \times 10^{-7}$ newton/(ampere-turn)^2).

9.11   Living things

Some organisms can detect magnetic fields, a phenomenon known as magnetoception. Magnetobiology studies magnetic fields as a medical treatment; fields naturally produced by an organism are known as biomagnetism.

9.12   See also

- Coercivity

- Magnetic hysteresis

- Magnetar

- Magnetic bearing

- Magnetic circuit

- Magnetic cooling

- Magnetic field viewing film

- Magnetic stirrer

- Magnetic structure

- Magnetism and temperature

- Micromagnetism

- Neodymium magnet

- Plastic magnet

- Rare-earth magnet

- Spin wave

- Spontaneous magnetization

- Vibrating sample magnetometer

- Gravitomagnetism

9.13   References

[1] Fowler, Michael (1997). "Historical Beginnings of Theories of Electricity and Magnetism". Retrieved 2008-04-02.

[2] Vowles, Hugh P. (1932). "Early Evolution of Power Engineering". *Isis* (University of Chicago Press) **17** (2): 412–420 [419–20]. doi:10.1086/346662.

[3] Li Shu-hua. "Origine de la Boussole 11. Aimant et Boussole," *Isis*, Vol. 45, No. 2. (Jul., 1954), p.175

[4] Li Shu-hua. "Origine de la Boussole 11. Aimant et Boussole," *Isis*, Vol. 45, No. 2. (Jul., 1954), p.176

[5] Schmidl, Petra G. (1996–1997). "Two Early Arabic Sources On The Magnetic Compass". *Journal of Arabic and Islamic Studies* **1**: 81–132.

[6] A. Einstein: "On the Electrodynamics of Moving Bodies". June 30, 1905.

[7] HP Meyers (1997). *Introductory solid state physics* (2 ed.). CRC Press. p. 362; Figure 11.1. ISBN 9781420075021.

[8] Catherine Westbrook, Carolyn Kaut, Carolyn Kaut-Roth (1998). *MRI (Magnetic Resonance Imaging) in practice* (2 ed.). Wiley-Blackwell. p. 217. ISBN 0-632-04205-2.

[9] Griffiths 1998, chapter 12

[10] Jackson, John David (1999). *Classical electrodynamics* (3rd ed.). New York: Wiley. ISBN 0-471-30932-X.

[11] Milton mentions some inconclusive events (p.60) and still concludes that "no evidence at all of magnetic monopoles has survived" (p.3). Milton, Kimball A. (June 2006). "Theoretical and experimental status of magnetic monopoles". *Reports on Progress in Physics* **69** (6): 1637–1711. arXiv:hep-ex/0602040. Bibcode:2006RPPh...69.1637M. doi:10.1088/0034-4885/69/6/R02..

[12] Guth, Alan (1997). *The Inflationary Universe: The Quest for a New Theory of Cosmic Origins*. Perseus. ISBN 0-201-32840-2. OCLC 38941224..

[13] International Union of Pure and Applied Chemistry (1993). *Quantities, Units and Symbols in Physical Chemistry*, 2nd edition, Oxford: Blackwell Science. ISBN 0-632-03583-8. pp. 14–15. Electronic version.

9.14   Further reading

- Furlani, Edward P. (2001). *Permanent Magnet and Electromechanical Devices: Materials, Analysis and Applications*. Academic Press. ISBN 0-12-269951-3. OCLC 162129430.

- Griffiths, David J. (1998). *Introduction to Electrodynamics (3rd ed.)*. Prentice Hall. ISBN 0-13-805326-X. OCLC 40251748.

- Kronmüller, Helmut. (2007). *Handbook of Magnetism and Advanced Magnetic Materials, 5 Volume Set*. John Wiley & Sons. ISBN 978-0-470-02217-7. OCLC 124165851.

- Tipler, Paul (2004). *Physics for Scientists and Engineers: Electricity, Magnetism, Light, and Elementary Modern Physics (5th ed.)*. W. H. Freeman. ISBN 0-7167-0810-8. OCLC 51095685.

- David K. Cheng (1992). *Field and Wave Electromagnetics*. Addison-Wesley Publishing Company, Inc. ISBN 0-201-12819-5.

9.15 External links

-

- Magnetism on *In Our Time* at the BBC. (listen now)

- The Exploratorium Science Snacks – Snacks about Magnetism

- Electromagnetism - a chapter from an online textbook

- Video: The physicist Richard Feynman answers the question, Why do bar magnets attract or repel each other? on YouTube

- On the Magnet, 1600 First scientific book on magnetism by the father of electrical engineering. Full English text, full text search.

- Magnetism and magnetization - Astronoo

Chapter 10

Thermal conductivity

For thermal conductivity values, see List of thermal conductivities.

In physics, **thermal conductivity** (often denoted k, λ, or κ) is the property of a material to conduct heat. It is evaluated primarily in terms of Fourier's Law for heat conduction.

Heat transfer occurs at a lower rate across materials of low thermal conductivity than across materials of high thermal conductivity. Correspondingly, materials of high thermal conductivity are widely used in heat sink applications and materials of low thermal conductivity are used as thermal insulation. The thermal conductivity of a material may depend on temperature. The reciprocal of thermal conductivity is called thermal resistivity.

Thermal conductivity is actually a tensor, which means it is possible to have different values in different directions. See #Thermal anisotropy below.

10.1   Units of thermal conductivity

In SI units, thermal conductivity is measured in watts per meter kelvin (W/(m·K)). The dimension of thermal conductivity is $M^1 L^1 T^{-3} \Theta^{-1}$. These variables are (M)mass, (L)length, (T)time, and (Θ)temperature. In Imperial units, thermal conductivity is measured in BTU/(hr·ft·°F).[note 1][1]

Other units which are closely related to the thermal conductivity are in common use in the construction and textile industries. The construction industry makes use of units such as the R-value (resistance) and the U-value (conductivity). Although related to the thermal conductivity of a material used in an insulation product, R and U-values are dependent on the thickness of the product.[note 2]

Likewise the textile industry has several units including the tog and the clo which express thermal resistance of a material in a way analogous to the R-values used in the construction industry.

10.2   Measurement

Main article: Thermal conductivity measurement

There are a number of ways to measure thermal conductivity. Each of these is suitable for a limited range of materials, depending on the thermal properties and the medium temperature. There is a distinction between steady-state and transient techniques.

In general, steady-state techniques are useful when the temperature of the material does not change with time. This makes the signal analysis straightforward (steady state implies constant signals). The disadvantage is that a well-engineered experimental setup is usually needed. The Divided Bar (various types) is the most common device used for consolidated rock solids.

10.3   Experimental values

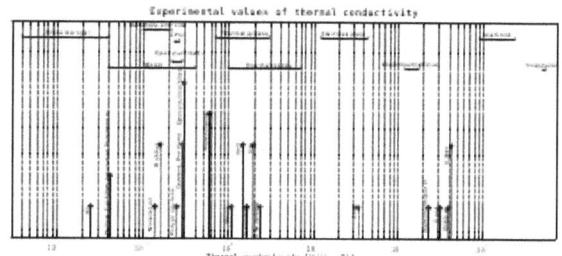

Experimental values of thermal conductivity.

Main article: List of thermal conductivities

Thermal conductivity is important in material science, research, electronics, building insulation and related fields, especially where high operating temperatures are achieved. Several materials are shown in the list of thermal conductivities. These should be considered approximate due to the uncertainties related to material definitions.

High energy generation rates within electronics or turbines require the use of materials with high thermal conductivity such as copper *(see: Copper in heat exchangers)*, aluminium, and silver. On the other hand, materials with low thermal conductance, such as polystyrene and alumina, are used in building construction or in furnaces in an effort to slow the flow of heat, i.e. for insulation purposes.

10.4 Definitions

The reciprocal of thermal conductivity is *thermal resistivity*, usually expressed in kelvin-meters per watt ($K \cdot m \cdot W^{-1}$). For a given thickness of a material, that particular construction's *thermal resistance* and the reciprocal property, *thermal conductance*, can be calculated. Unfortunately, there are differing definitions for these terms.

10.4.1 Conductance

For general scientific use, *thermal conductance* is the quantity of heat that passes in unit time through a plate of *particular area and thickness* when its opposite faces differ in temperature by one kelvin. For a plate of thermal conductivity k, area A and thickness L, the conductance calculated is kA/L, measured in $W \cdot K^{-1}$ (equivalent to: W/°C). The thermal conductance of that particular construction is the inverse of the thermal resistance. Thermal conductivity and conductance are analogous to electrical conductivity ($A \cdot m^{-1} \cdot V^{-1}$) and electrical conductance ($A \cdot V^{-1}$).

There is also a measure known as heat transfer coefficient: the quantity of heat that passes in unit time through a *unit area* of a plate of particular thickness when its opposite faces differ in temperature by one kelvin. The reciprocal is *thermal insulance*. In summary:

- thermal conductance = kA/L, measured in $W \cdot K^{-1}$

  - thermal resistance = $L/(kA)$, measured in $K \cdot W^{-1}$ (equivalent to: °C/W)

- heat transfer coefficient = k/L, measured in $W \cdot K^{-1} \cdot m^{-2}$

  - thermal insulance = L/k, measured in $K \cdot m^{2} \cdot W^{-1}$.

The heat transfer coefficient is also known as *thermal admittance* in the sense that the material may be seen as admitting heat to flow.

10.4.2 Resistance

Main article: Thermal resistance

Thermal resistance is the ability of a material to resist the flow of heat.

Thermal resistance is the reciprocal of thermal conductance, i.e., lowering its value will raise the heat conduction and vice versa.

When thermal resistances occur in series, they are *additive*. Thus, when heat flows consecutively through two components each with a resistance of 3 °C/W, the total resistance is 3+3=6 °C/W.

A common engineering design problem involves the selection of an appropriate sized heat sink for a given heat source. Working in units of thermal resistance greatly simplifies the design calculation. The following formula can be used to estimate the performance:

$$R_{hs} = \frac{\Delta T}{P_{th}} - R_s$$

where:

- R_{hs} is the maximum thermal resistance of the heat sink to ambient, in °C/W (equivalent to K/W)

- ΔT is the required temperature difference (temperature drop), in °C

- P_{th} is the thermal power (heat flow), in watts

- R_s is the thermal resistance of the heat source, in °C/W

For example, if a component produces 100 W of heat, and has a thermal resistance of 0.5 °C/W, what is the maximum thermal resistance of the heat sink? Suppose the maximum temperature is 125 °C, and the ambient temperature is 25 °C; then ΔT is 100 °C. The heat sink's thermal resistance to ambient must then be 0.5 °C/W or less (total resistance component and heat sink is then 1.0 °C/W).

10.4.3 Transmittance

A third term, *thermal transmittance*, quantifies the thermal conductance of a structure along with heat transfer due to convection and radiation. It is measured in the same units as thermal conductance and is sometimes known as the *composite thermal conductance*. The term *U-value* is often used.

10.4.4 Admittance

The thermal admittance of a material, such as a building fabric, is a measure of the ability of a material to transfer heat in the presence of a temperature difference on opposite sides of the material. Thermal admittance is measured in the same units as a heat transfer coefficient, power (watts) per unit area (square meters) per temperature change (kelvin). Thermal admittance of a building fabric affects a building's thermal response to variation in outside temperature.*[2]

10.5 Influencing factors

10.5.1 Temperature

The effect of temperature on thermal conductivity is different for metals and nonmetals. In metals conductivity is primarily due to free electrons. Following the Wiedemann–Franz law, thermal conductivity of metals is approximately proportional to the absolute temperature (in kelvin) times electrical conductivity. In pure metals the electrical conductivity decreases with increasing temperature and thus the product of the two, the thermal conductivity, stays approximately constant. In alloys the change in electrical conductivity is usually smaller and thus thermal conductivity increases with temperature, often proportionally to temperature.

On the other hand, heat conductivity in nonmetals is mainly due to lattice vibrations (phonons). Except for high quality crystals at low temperatures, the phonon mean free path is not reduced significantly at higher temperatures. Thus, the thermal conductivity of nonmetals is approximately constant at high temperatures. At low temperatures well below the Debye temperature, thermal conductivity decreases, as does the heat capacity.

10.5.2 Chemical phase

When a material undergoes a phase change from solid to liquid or from liquid to gas the thermal conductivity may change. An example of this would be the change in thermal conductivity that occurs when ice (thermal conductivity of 2.18 W/(m·K) at 0 °C) melts to form liquid water (thermal conductivity of 0.56 W/(m·K) at 0 °C).*[3]

10.5.3 Thermal anisotropy

Some substances, such as non-cubic crystals, can exhibit different thermal conductivities along different crystal axes, due to differences in phonon coupling along a given crystal axis. Sapphire is a notable example of variable thermal conductivity based on orientation and temperature, with 35 W/(m·K) along the C-axis and 32 W/(m·K) along the A-axis.*[4] Wood generally conducts better along the grain than across it.

When anisotropy is present, the direction of heat flow may not be exactly the same as the direction of the thermal gradient.

10.5.4 Electrical conductivity

In metals, thermal conductivity approximately tracks electrical conductivity according to the Wiedemann–Franz law, as freely moving valence electrons transfer not only electric current but also heat energy. However, the general correlation between electrical and thermal conductance does not hold for other materials, due to the increased importance of phonon carriers for heat in non-metals. Highly electrically conductive silver is less thermally conductive than diamond, which is an electrical insulator, but due to its orderly array of atoms it is conductive of heat via phonons.

10.5.5 Magnetic field

The influence of magnetic fields on thermal conductivity is known as the Righi-Leduc effect.

10.5.6 Convection

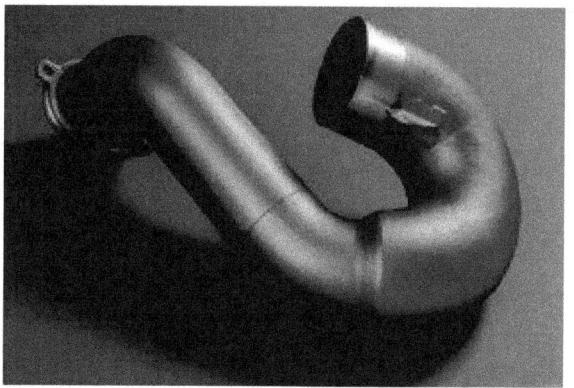

Ceramic coatings with low thermal conductivities are used on exhaust systems to prevent heat from reaching sensitive components

Air and other gases are generally good insulators, in the absence of convection. Therefore, many insulating materials function simply by having a large number of gas-filled pockets which prevent large-scale convection. Examples of these include expanded and extruded polystyrene (popularly referred to as "styrofoam") and silica aerogel, as well as warm clothes. Natural, biological insulators such as fur and feathers achieve similar effects by dramatically inhibiting convection of air or water near an animal's skin.

Light gases, such as hydrogen and helium typically have high thermal conductivity. Dense gases such as xenon and dichlorodifluoromethane have low thermal conductivity. An exception, sulfur hexafluoride, a dense gas, has a relatively high thermal conductivity due to its high heat capacity. Argon, a gas denser than air, is often used in insulated glazing (double paned windows) to improve their insulation characteristics.

10.6 Physical origins

Heat flux is exceedingly difficult to control and isolate in a laboratory setting. At the atomic level, there are no simple, correct expressions for thermal conductivity. Atomically, the thermal conductivity of a system is determined by how atoms composing the system interact. There are two different approaches for calculating the thermal conductivity of a system.

- The first approach employs the Green-Kubo relations. Although this employs analytic expressions, which, in principle, can be solved, calculating the thermal conductivity of a dense fluid or solid using this relation requires the use of molecular dynamics computer simulation.

- The second approach is based on the relaxation time approach. Due to the anharmonicity within the crystal potential, the phonons in the system are known to scatter. There are three main mechanisms for scattering:

  - Boundary scattering, a phonon hitting the boundary of a system;

  - Mass defect scattering, a phonon hitting an impurity within the system and scattering;

  - Phonon-phonon scattering, a phonon breaking into two lower energy phonons or a phonon colliding with another phonon and merging into one higher-energy phonon.

10.6.1 Lattice waves

Heat transport in both amorphous and crystalline dielectric solids is by way of elastic vibrations of the lattice (phonons). This transport mode is limited by the elastic scattering of acoustic phonons at lattice defects. These predictions were confirmed by the experiments of Chang and Jones on commercial glasses and glass ceramics, where the mean free paths were limited by "internal boundary scattering" to length scales of 10^{-2} cm to 10^{-3} cm.[5][6]

The phonon mean free path has been associated directly with the effective relaxation length for processes without directional correlation. If V_g is the group velocity of a phonon wave packet, then the relaxation length l is defined as:

$$l = V_g t$$

where t is the characteristic relaxation time. Since longitudinal waves have a much greater phase velocity than transverse waves, V_{long} is much greater than V_{trans}, and the relaxation length or mean free path of longitudinal phonons will be much greater. Thus, thermal conductivity will be largely determined by the speed of longitudinal phonons.[5][7]

Regarding the dependence of wave velocity on wavelength or frequency (dispersion), low-frequency phonons of long wavelength will be limited in relaxation length by elastic Rayleigh scattering. This type of light scattering from small particles is proportional to the fourth power of the frequency. For higher frequencies, the power of the frequency will decrease until at highest frequencies scattering is almost frequency independent. Similar arguments were subsequently generalized to many glass forming substances using Brillouin scattering.[8][9][10][11]

Phonons in the acoustical branch dominate the phonon heat conduction as they have greater energy dispersion and therefore a greater distribution of phonon velocities. Additional optical modes could also be caused by the presence of internal structure (i.e., charge or mass) at a lattice point; it is implied that the group velocity of these modes is low and therefore their contribution to the lattice thermal conductivity λ_L (κ_L) is small.[12]

Each phonon mode can be split into one longitudinal and two transverse polarization branches. By extrapolating the phenomenology of lattice points to the unit cells it is seen that the total number of degrees of freedom is 3pq when p is the number of primitive cells with q atoms/unit cell. From these only 3p are associated with the acoustic modes, the remaining 3p(q-1) are accommodated through the optical branches. This implies that structures with larger p and q contain a greater number of optical modes and a reduced λ_L.

From these ideas, it can be concluded that increasing crystal complexity, which is described by a complexity factor CF (defined as the number of atoms/primitive unit cell), decreases λ_L. Micheline Roufosse and P.G. Klemens derived the exact proportionality in their article Thermal Conductivity of Complex Dielectric Crystals at Phys. Rev. B 7,

5379–5386 (1973). This was done by assuming that the relaxation time τ decreases with increasing number of atoms in the unit cell and then scaling the parameters of the expression for thermal conductivity in high temperatures accordingly.*[12]

Describing of anharmonic effects is complicated because exact treatment as in the harmonic case is not possible and phonons are no longer exact eigensolutions to the equations of motion. Even if the state of motion of the crystal could be described with a plane wave at a particular time, its accuracy would deteriorate progressively with time. Time development would have to be described by introducing a spectrum of other phonons, which is known as the phonon decay. The two most important anharmonic effects are the thermal expansion and the phonon thermal conductivity.

Only when the phonon number $\langle n \rangle$ deviates from the equilibrium value $\langle n \rangle^0$, can a thermal current arise as stated in the following expression

$$Q_x = \frac{1}{V} \sum_{q,j} \hbar\omega \left(\langle n \rangle - \langle n \rangle^0 \right) v_x.$$

where v is the energy transport velocity of phonons. Only two mechanisms exist that can cause time variation of $\langle n \rangle$ in a particular region. The number of phonons that diffuse into the region from neighboring regions differs from those that diffuse out, or phonons decay inside the same region into other phonons. A special form of the Boltzmann equation

$$\frac{d \langle n \rangle}{dt} = \left(\frac{\partial \langle n \rangle}{\partial t} \right)_{\text{diff.}} + \left(\frac{\partial \langle n \rangle}{\partial t} \right)_{\text{decay}}$$

states this. When steady state conditions are assumed the total time derivate of phonon number is zero, because the temperature is constant in time and therefore the phonon number stays also constant. Time variation due to phonon decay is described with a relaxation time (τ) approximation

$$\left(\frac{\partial \langle n \rangle}{\partial t} \right)_{\text{decay}} = - \frac{\langle n \rangle - \langle n \rangle^0}{\tau}.$$

which states that the more the phonon number deviates from its equilibrium value, the more its time variation increases. At steady state conditions and local thermal equilibrium are assumed we get the following equation

$$\left(\frac{\partial \langle n \rangle}{\partial t} \right)_{\text{diff.}} = -v_x \frac{\partial \langle n \rangle^0}{\partial T} \frac{\partial T}{\partial x}.$$

Using the relaxation time approximation for the Boltzmann equation and assuming steady-state conditions, the phonon

thermal conductivity λ_L can be determined. The temperature dependence for λ_L originates from the variety of processes, whose significance for λ_L depends on the temperature range of interest. Mean free path is one factor that determines the temperature dependence for λ_L, as stated in the following equation

$$\lambda_L = \frac{1}{3V} \sum_{q,j} v(q,j) \Lambda(q,j) \frac{\partial}{\partial T} \epsilon(\omega(q,j),T).$$

where Λ is the mean free path for phonon and $\frac{\partial}{\partial T}\epsilon$ denotes the heat capacity. This equation is a result of combining the four previous equations with each other and knowing that $\langle v_x^2 \rangle = \frac{1}{3}c^2$ for cubic or isotropic systems and $\Lambda = v\tau$.*[13]

At low temperatures (<10 K) the anharmonic interaction does not influence the mean free path and therefore, the thermal resistivity is determined only from processes for which q-conservation does not hold. These processes include the scattering of phonons by crystal defects, or the scattering from the surface of the crystal in case of high quality single crystal. Therefore, thermal conductance depends on the external dimensions of the crystal and the quality of the surface. Thus, temperature dependence of λ_L is determined by the specific heat and is therefore proportional to T^3.*[13]

Phonon quasimomentum is defined as $\hbar q$ and differs from normal momentum because it is only defined within an arbitrary reciprocal lattice vector. At higher temperatures (10 K$<$T$<\Theta$), the conservation of energy $\hbar\omega_1 = \hbar\omega_2 + \hbar\omega_3$ and quasimomentum $q_1 = q_2 + q_3 + G$, where $\mathbf{q}_1$ is wave vector of the incident phonon and $\mathbf{q}_2$, $\mathbf{q}_3$ are wave vectors of the resultant phonons, may also involve a reciprocal lattice vector $\mathbf{G}$ complicating the energy transport process. These processes can also reverse the direction of energy transport.

Therefore, these processes are also known as Umklapp (U) processes and can only occur when phonons with sufficiently large q-vectors are excited, because unless the sum of $\mathbf{q}_2$ and $\mathbf{q}_3$ points outside of the Brillouin zone the momentum is conserved and the process is normal scattering (N-process). The probability of a phonon to have energy E is given by the Boltzmann distribution $P \propto e^{-E/kT}$. To U-process to occur the decaying phonon to have a wave vector $\mathbf{q}_1$ that is roughly half of the diameter of the Brillouin zone, because otherwise quasimomentum would not be conserved.

Therefore, these phonons have to possess energy of $\sim k\Theta/2$, which is a significant fraction of Debye energy that is needed to generate new phonons. The probability for this is proportional to $e^{-\Theta/bT}$, with $b = 2$. Temperature dependence of the mean free path has an exponential form $e^{\Theta/bT}$. The presence of the reciprocal lattice wave vec-

tor implies a net phonon backscattering and a resistance
to phonon and thermal transport resulting finite λ_L.[12]
as it means that momentum is not conserved. Only momentum non-conserving processes can cause thermal resistance.[13]

At high temperatures (T>Θ) the mean free path and therefore λ_L has a temperature dependence T^{-1}, to which one
arrives from formula $e^{\Theta/bT}$ by making the following approximation $e^x \propto x$, $(x) < 1$ and writing $x = \Theta/bT$.
This dependency is known as Eucken's law and originates
from the temperature dependency of the probability for the
U-process to occur.[12][13]

Thermal conductivity is usually described by the Boltzmann
equation with the relaxation time approximation in which
phonon scattering is a limiting factor. Another approach
is to use analytic models or molecular dynamics or Monte
Carlo based methods to describe thermal conductivity in
solids.

Short wavelength phonons are strongly scattered by impurity atoms if an alloyed phase is present, but mid and long
wavelength phonons are less affected. Mid and long wavelength phonons carry significant fraction of heat, so to further reduce lattice thermal conductivity one has to introduce structures to scatter these phonons. This is achieved
by introducing interface scattering mechanism, which requires structures whose characteristic length is longer than
that of impurity atom. Some possible ways to realize these
interfaces are nanocomposites and embedded nanoparti-
cles/structures.[14]

10.6.2   Electronic thermal conductivity

Hot electrons from higher energy states carry more thermal
energy than cold electrons, while electrical conductivity is
rather insensitive to the energy distribution of carriers because the amount of charge that electrons carry, does not
depend on their energy. This is a physical reason for the
greater sensitivity of electronic thermal conductivity to energy dependence of density of states and relaxation time,
respectively.[12]

Mahan and Sofo (*PNAS* 1996 93 (15) 7436-7439) showed
that materials with a certain electron structure have reduced
electron thermal conductivity. Based on their analysis one
can demonstrate that if the electron density of states in the
material is close to the delta-function, the electronic thermal
conductivity drops to zero. By taking the following equa-
tion $\lambda_E = \lambda_0 - T\sigma S^2$, where λ_0 is the electronic thermal conductivity when the electrochemical potential gradi-
ent inside the sample is zero, as a starting point. As next
step the transport coefficients are written as following

$$\sigma = \sigma_0 I_0$$

$$\sigma S = \left(\frac{k}{e}\right) \sigma_0 I_1$$

$$\lambda_0 = \left(\frac{k}{e}\right)^2 \sigma_0 T I_2$$

where $\sigma_0 = e^2/(\hbar a_0)$ and a_0 the Bohr radius. The dimensionless integrals I_n are defined as

$$I_n = \int_{-\infty}^{\infty} \frac{e^x}{(e^x + 1)^2} s(x) x^n dx$$

where $s(x)$ is the dimensionless transport distribution function. The integrals I_n are the moments of the function

$$P(x) = D(x) s(x), \quad D(x) = \frac{e^x}{(e^x + 1)^2}$$

where x is the energy of carriers. By substituting the previous formulas for the transport coefficient to the equation
for λ_E we get the following equation

$$\lambda_E = \left(\frac{k}{e}\right)^2 \sigma_0 T \left(I_2 - \frac{I_1^2}{I_0}\right)$$

From the previous equation we see that λ_E to be zero the
bracketed term containing I_n terms have to be zero. Now if
we assume that

$$s(x) = f(x) \delta(x - b)$$

where δ is the Dirac delta function. I_n terms get the following expressions

$$I_0 = D(b) f(b)$$

$$I_1 = D(b) f(b) b$$

$$I_2 = D(b) f(b) b^2$$

By substituting these expressions to the equation for λ_E, we
see that it goes to zero. Therefore, $P(x)$ has to be delta function.[14]

10.7 Equations

In an isotropic medium the thermal conductivity is the parameter k in the Fourier expression for the heat flux

$$\vec{q} = -k\vec{\nabla}T$$

where $\vec{q}$ is the heat flux (amount of heat flowing per second and per unit area) and $\vec{\nabla}T$ the temperature gradient. The sign in the expression is chosen so that always $k > 0$ as heat always flows from a high temperature to a low temperature. This is a direct consequence of the second law of thermodynamics.

In the one-dimensional case $q = H/A$ with H the amount of heat flowing per second through a surface with area A and the temperature gradient is dT/dx so

$$H = -kA\frac{dT}{dx}.$$

In case of a thermally-insulated bar (except at the ends) in the steady state H is constant. If A is constant as well the expression can be integrated with the result

$$HL = A \int_{T_L}^{T_H} k(T)dT$$

where T_H and T_L are the temperatures at the hot end and the cold end respectively, and L is the length of the bar. It is convenient to introduce the thermal-conductivity integral

$$I_k(T) = \int_0^T k(T')dT'.$$

The heat flow rate is then given by

$$H = \frac{A}{L}[I_k(T_H) - I_k(T_L)].$$

If the temperature difference is small k can be taken as constant. In that case

$$H = kA\frac{T_H - T_L}{L}.$$

10.8 Simple kinetic picture

In this Section we will derive an expression for the thermal conductivity. Consider a gas with hard-core interactions but

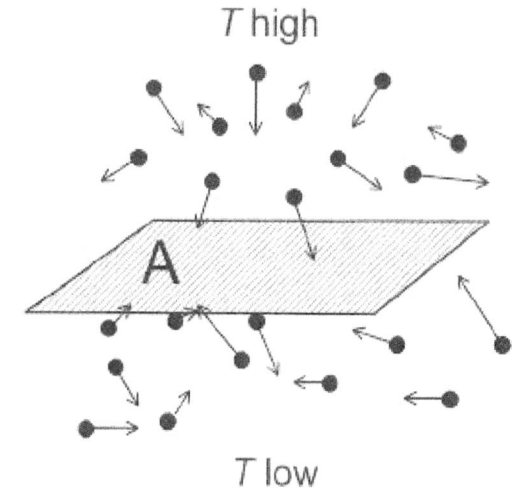

Gas atoms moving randomly through a surface.

negligible volume within a vertical temperature gradient. The upper side is hot and the lower side cold. There is a downward energy flow because the gas atoms, going down, have a higher energy than the atoms going up. The net flow of energy per second is the heat flow H. The heat flow is proportional to the number of particles that cross the area A per second. This number is proportional to the product nvA where n is the particle density and v the mean particle velocity. The magnitude of the heat flow will also be proportional to amount of energy transported per particle so with the heat capacity per particle c and some characteristic temperature difference ΔT. So far we have

$$H \propto nvcA\Delta T.$$

The unit of H is J/s and of the right-hand side it is (particle/m^3)•(m/s)•(J/(K•particle))•(m^2)•(K) = J/s, so this is already of the right dimension. Only a numerical factor is missing. For ΔT we take the temperature difference of the gas between two collisions

$$\Delta T = l\frac{dT}{dz}$$

where l is the mean free path. Detailed kinetic calculations[15] show that the numerical factor is $-1/3$, so, all in all,

$$H = -\frac{1}{3}nvclA\frac{dT}{dz}.$$

Comparison with the one-dimension expression for the heat flow, given above, gives as the final result

$$k = \frac{1}{3} n v c l.$$

The particle density and the heat capacity per particle can be combined as the heat capacity per unit volume

so

$$k = \frac{1}{3} v l \frac{C_V}{V_m}$$

where C_V is the molar heat capacity at constant volume and V_m the molar volume.

For the hard-core gas the mean free path is given by

$$l \propto \frac{1}{n\sigma}.$$

where σ is the collision cross section. So

$$k \propto \frac{c}{\sigma} v.$$

The heat capacity per particle c and the cross section σ both are temperature independent so the temperature dependence of k is determined by the T dependence of v. For a monatomic gas, with atomic mass M, v is given by

$$v = \sqrt{\frac{3RT}{M}}.$$

So

$$k \propto \sqrt{\frac{T}{M}}.$$

This expression also shows why gases with a low mass (hydrogen, helium) have a high thermal conductivity.

For metals at low temperatures the heat is carried mainly by the free electrons. In this case the mean velocity is the Fermi velocity which is temperature independent. The mean free path is determined by the impurities and the crystal imperfections which are temperature independent as well. So the only temperature-dependent quantity is the heat capacity c, which, in this case, is proportional to T. So

$$k = k_0 T \text{ temperature) low at (metal}$$

with k_0 a constant. For pure metals such as copper, silver, etc. l is large, so the thermal conductivity is high. At higher

temperatures the mean free path is limited by the phonons, so the thermal conductivity tends to decrease with temperature. In alloys the density of the impurities is very high, so l and, consequently k, are small. Therefore, alloys, such as stainless steel, can be used for thermal insulation.

10.9   See also

- Copper in heat exchangers
- Heat transfer
- Heat transfer mechanisms
- Insulated pipes
- Interfacial thermal resistance
- Laser flash analysis
- Specific heat
- Thermal bridge
- Thermal conductance quantum
- Thermal contact conductance
- Thermal diffusivity
- Thermal rectifier
- Thermal resistance in electronics
- Thermistor
- Thermocouple

10.10   References

Notes

[1] 1 Btu/(hr·ft·F) = 1.730735 W/(m·K)

[2] R-Values and U-Values quoted in the US (based on the imperial units of measurement) do not correspond with and are not compatible with those used outside the US (based on the SI units of measurement).

References

[1] Perry, R. H.; Green, D. W., eds. (1997). *Perry's Chemical Engineers' Handbook* (7th ed.). McGraw-Hill. Table 1–4. ISBN 978-0-07-049841-9.

[2] "Thermal Mass in Buildings" . Reidsteel. Retrieved 23 January 2013.

[3] NIST: Standard reference data for the thermal conductivity of water

[4] "Sapphire, Al₂O₃". Almaz Optics. Retrieved 2012-08-15.

[5] Klemens, P.G. (1951). "The Thermal Conductivity of Dielectric Solids at Low Temperatures". *Proceedings of the Royal Society of London A* **208** (1092): 108. Bibcode:1951RSPSA.208..108K. doi:10.1098/rspa.1951.0147.

[6] Chan, G. K.; Jones, R. E. (1962). "Low-Temperature Thermal Conductivity of Amorphous Solids". *Physical Review* **126** (6): 2055. Bibcode:1962PhRv..126.2055C. doi:10.1103/PhysRev.126.2055.

[7] Pomeranchuk, I. (1941). "Thermal conductivity of the paramagnetic dielectrics at low temperatures". *Journal of Physics (Moscow)* **4**: 357. ISSN 0368-3400.

[8] Zeller, R. C.; Pohl, R. O. (1971). "Thermal Conductivity and Specific Heat of Non-crystalline Solids". *Physical Review B* **4** (6): 2029. Bibcode:1971PhRvB...4.2029Z. doi:10.1103/PhysRevB.4.2029.

[9] Love, W. F. (1973). "Low-Temperature Thermal Brillouin Scattering in Fused Silica and Borosilicate Glass". *Physical Review Letters* **31** (13): 822. Bibcode:1973PhRvL..31..822L. doi:10.1103/PhysRevLett.31.822.

[10] Zaitlin, M. P.; Anderson, M. C. (1975). "Phonon thermal transport in noncrystalline materials". *Physical Review B* **12** (10): 4475. Bibcode:1975PhRvB..12.4475Z. doi:10.1103/PhysRevB.12.4475.

[11] Zaitlin, M. P.; Scherr, L. M.; Anderson, M. C. (1975). "Boundary scattering of phonons in noncrystalline materials". *Physical Review B* **12** (10): 4487. Bibcode:1975PhRvB..12.4487Z. doi:10.1103/PhysRevB.12.4487.

[12] Pichanusakorn, P.; Bandaru, P. (2010). "Nanostructured thermoelectrics". *Materials Science and Engineering: R: Reports* **67** (2–4): 19–63. doi:10.1016/j.mser.2009.10.001.

[13] Ibach, H.; Luth, H. (2009). *Solid-State Physics: An Introduction to Principles of Materials Science*. Springer. ISBN 978-3-540-93803-3.

[14] Minnich, A. J.; Dresselhaus, M. S.; Ren, Z. F.; Chen, G. (2009). "Bulk nanostructured thermoelectric materials: Current research and future prospects". *Energy & Environmental Science* **2** (5): 466–479. doi:10.1039/b822664b.

[15] Kittel, C.; Kroemer, H. (1980). *Thermal Physics*. W. H. Freeman and Company. Chapter 14. ISBN 978-0716710882.

10.11   Further reading

- Callister, William (2003). "Appendix B". *Materials Science and Engineering - An Introduction*. John Wiley & Sons. p. 757. ISBN 0-471-22471-5.

- Halliday, David; Resnick, Robert; & Walker, Jearl(1997). *Fundamentals of Physics* (5th ed.). John Wiley and Sons, New York ISBN 0-471-10558-9.

- Srivastava G. P (1990), *The Physics of Phonons*. Adam Hilger, IOP Publishing Ltd, Bristol.

- TM 5-852-6 AFR 88-19, Volume 6 (Army Corp of Engineers publication)

- Reid, C. R., Prausnitz, J. M., Poling B. E., *Properties of gases and liquids*, IV edition, Mc Graw-Hill, 1987

- R. Joven, R. Das, A. Ahmed, P. Roozbehjavan, B. Minaie, "Thermal properties of carbon fiber/epoxy composites with different fabric weaves", in: SAMPE International Symposium Proceedings, Charleston, SC: 2012

10.12   External links

- Thermopedia

- J Chem Phys thermal conductivity of electrolyte solutions

- The importance of Soil Thermal Conductivity for power companies

- J Chem Phys gas mixtures

Chapter 11

Lorentz force

In physics, particularly electromagnetism, the **Lorentz force** is the combination of electric and magnetic force on a point charge due to electromagnetic fields. If a particle of charge q moves with velocity $\mathbf{v}$ in the presence of an electric field $\mathbf{E}$ and a magnetic field $\mathbf{B}$, then it will experience a force

$$\mathbf{F} = q\,[\mathbf{E} + (\mathbf{v} \times \mathbf{B})]$$

(in SI units). Variations on this basic formula describe the magnetic force on a current-carrying wire (sometimes called *Laplace force*), the electromotive force in a wire loop moving through a magnetic field (an aspect of Faraday's law of induction), and the force on a charged particle which might be travelling near the speed of light (relativistic form of the Lorentz force).

The first derivation of the Lorentz force is commonly attributed to Oliver Heaviside in 1889,[1] although other historians suggest an earlier origin in an 1865 paper by James Clerk Maxwell.[2] Hendrik Lorentz derived it a few years after Heaviside.

11.1 Equation (SI units)

See also: SI units

11.1.1 Charged particle

The force $\mathbf{F}$ acting on a particle of electric charge q with instantaneous velocity $\mathbf{v}$, due to an external electric field $\mathbf{E}$ and magnetic field $\mathbf{B}$, is given by:[3]

where × is the vector cross product. All boldface quantities are vectors. More explicitly stated:

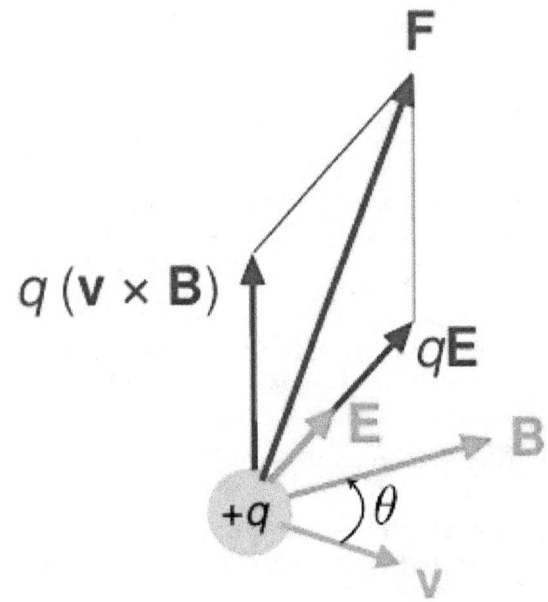

Lorentz force $\mathbf{F}$ on a charged particle (of charge q) in motion (instantaneous velocity $\mathbf{v}$). The $\mathbf{E}$ field and $\mathbf{B}$ field vary in space and time.

$$\mathbf{F}(\mathbf{r}, \dot{\mathbf{r}}, t, q) = q[\mathbf{E}(\mathbf{r}, t) + \dot{\mathbf{r}} \times \mathbf{B}(\mathbf{r}, t)]$$

in which $\mathbf{r}$ is the position vector of the charged particle, t is time, and the overdot is a time derivative.

A positively charged particle will be accelerated in the *same* linear orientation as the $\mathbf{E}$ field, but will curve perpendicularly to both the instantaneous velocity vector $\mathbf{v}$ and the $\mathbf{B}$ field according to the right-hand rule (in detail, if the thumb of the right hand points along $\mathbf{v}$ and the index finger along $\mathbf{B}$, then the middle finger points along $\mathbf{F}$).

The term $q\mathbf{E}$ is called the **electric force**, while the term $q\mathbf{v} \times \mathbf{B}$ is called the **magnetic force**.[4] According to some definitions, the term "Lorentz force" refers specifically to the formula for the magnetic force,[5] with the *total* electromagnetic force (including the electric force) given some other (nonstandard) name. This article will *not* follow this nomenclature: In what follows, the term "Lorentz force" will refer only to the expression for the total force.

The magnetic force component of the Lorentz force manifests itself as the force that acts on a current-carrying wire in a magnetic field. In that context, it is also called the **Laplace force**.

11.1.2 Continuous charge distribution

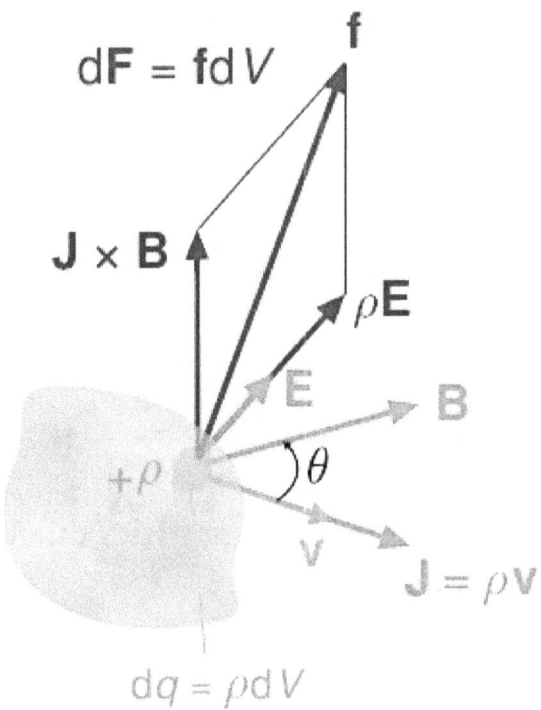

Lorentz force (per unit 3-volume) f on a continuous charge distribution (charge density ρ) in motion. The 3-current density J corresponds to the motion of the charge element dq in volume element dV and varies throughout the continuum.

For a continuous charge distribution in motion, the Lorentz force equation becomes:

$$d\mathbf{F} = dq \, (\mathbf{E} + \mathbf{v} \times \mathbf{B})$$

where $d\mathbf{F}$ is the force on a small piece of the charge distribution with charge dq. If both sides of this equation are divided by the volume of this small piece of the charge distribution dV, the result is:

$$\mathbf{f} = \rho \, (\mathbf{E} + \mathbf{v} \times \mathbf{B})$$

where $\mathbf{f}$ is the *force density* (force per unit volume) and ρ is the charge density (charge per unit volume). Next, the current density corresponding to the motion of the charge continuum is

$$\mathbf{J} = \rho \mathbf{v}$$

so the continuous analogue to the equation is [6]

The total force is the volume integral over the charge distribution:

$$\mathbf{F} = \iiint (\rho \mathbf{E} + \mathbf{J} \times \mathbf{B}) \, dV.$$

By eliminating ρ and $\mathbf{J}$, using Maxwell's equations, and manipulating using the theorems of vector calculus, this form of the equation can be used to derive the Maxwell stress tensor $\boldsymbol{\sigma}$, in turn this can be combined with the Poynting vector $\mathbf{S}$ to obtain the electromagnetic stress–energy tensor $\mathbf{T}$ used in general relativity. [6]

In terms of $\boldsymbol{\sigma}$ and $\mathbf{S}$, another way to write the Lorentz force (per unit 3D volume) is [6]

$$\mathbf{f} = \nabla \cdot \boldsymbol{\sigma} - \frac{1}{c^2} \frac{\partial \mathbf{S}}{\partial t}$$

where c is the speed of light and $\nabla \cdot$ denotes the divergence of a tensor field. Rather than the amount of charge and its velocity in electric and magnetic fields, this equation relates the energy flux (flow of *energy* per unit time per unit distance) in the fields to the force exerted on a charge distribution. See Covariant formulation of classical electromagnetism for more details.

11.2 History

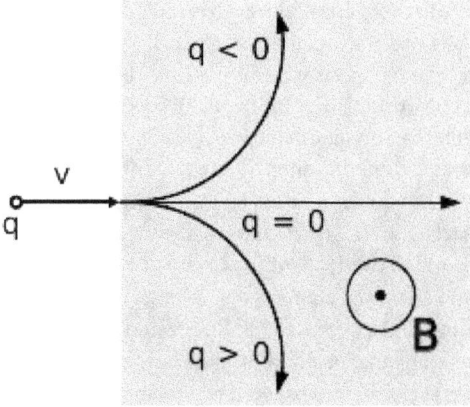

Trajectory of a particle with a positive or negative charge q under the influence of a magnetic field B, which is directed perpendicularly out of the screen.

Beam of electrons moving in a circle, due to the presence of a magnetic field. Purple light is emitted along the electron path, due to the electrons colliding with gas molecules in the bulb. A Teltron tube is used in this example.
Charged particles experiencing the Lorentz force.

Early attempts to quantitatively describe the electromagnetic force were made in the mid-18th century. It was proposed that the force on magnetic poles, by Johann Tobias Mayer and others in 1760, and electrically charged objects, by Henry Cavendish in 1762, obeyed an inverse-square law. However, in both cases the experimental proof was neither complete nor conclusive. It was not until 1784 when Charles-Augustin de Coulomb, using a torsion balance, was able to definitively show through experiment that this was true.[7] Soon after the discovery in 1820 by H. C. Ørsted that a magnetic needle is acted on by a voltaic current, André-Marie Ampère that same year was able to devise through experimentation the formula for the angular dependence of the force between two current elements.[8][9] In all these descriptions, the force was always given in terms of the properties of the objects involved and the distances between them rather than in terms of electric and magnetic fields.[10]

The modern concept of electric and magnetic fields first arose in the theories of Michael Faraday, particularly his idea of lines of force, later to be given full mathematical description by Lord Kelvin and James Clerk Maxwell.[11] From a modern perspective it is possible to identify in Maxwell's 1865 formulation of his field equations a form of the Lorentz force equation in relation to electric currents,[2] however, in the time of Maxwell it was not evident how his equations related to the forces on moving charged objects. J. J. Thomson was the first to attempt to derive from Maxwell's field equations the electromagnetic forces on a moving charged object in terms of the object's properties and external fields. Interested in determining the electromagnetic behavior of the charged particles in cathode rays, Thomson published a paper in 1881 wherein

he gave the force on the particles due to an external magnetic field as[1]

$$\mathbf{F} = \frac{q}{2}\mathbf{v} \times \mathbf{B}.$$

Thomson derived the correct basic form of the formula, but, because of some miscalculations and an incomplete description of the displacement current, included an incorrect scale-factor of a half in front of the formula. It was Oliver Heaviside, who had invented the modern vector notation and applied them to Maxwell's field equations, that in 1885 and 1889 fixed the mistakes of Thomson's derivation and arrived at the correct form of the magnetic force on a moving charged object.[1][12][13] Finally, in 1892, Hendrik Lorentz derived the modern form of the formula for the electromagnetic force which includes the contributions to the total force from both the electric and the magnetic fields. Lorentz began by abandoning the Maxwellian descriptions of the ether and conduction. Instead, Lorentz made a distinction between matter and the luminiferous aether and sought to apply the Maxwell equations at a microscopic scale. Using Heaviside's version of the Maxwell equations for a stationary ether and applying Lagrangian mechanics (see below), Lorentz arrived at the correct and complete form of the force law that now bears his name.[14][15]

11.3   Trajectories of particles due to the Lorentz force

Main article: Guiding center
In many cases of practical interest, the motion in a magnetic field of an electrically charged particle (such as an electron or ion in a plasma) can be treated as the superposition of a relatively fast circular motion around a point called the **guiding center** and a relatively slow **drift** of this point. The drift speeds may differ for various species depending on their charge states, masses, or temperatures, possibly resulting in electric currents or chemical separation.

11.4   Significance of the Lorentz force

While the modern Maxwell's equations describe how electrically charged particles and currents or moving charged particles give rise to electric and magnetic fields, the Lorentz force law completes that picture by describing the force acting on a moving point charge q in the presence of electromagnetic fields.[3][16] The Lorentz force law describes the effect of $\mathbf{E}$ and $\mathbf{B}$ upon a point charge, but such

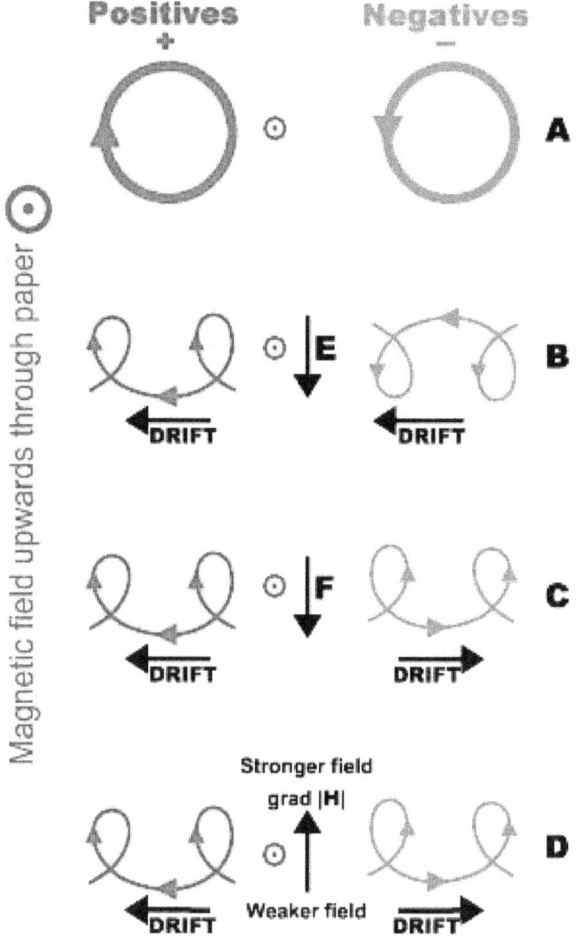

Charged particle drifts in a homogeneous magnetic field. (A) No disturbing force (B) With an electric field, E (C) With an independent force, F (e.g. gravity) (D) In an inhomogeneous magnetic field, grad H

electromagnetic forces are not the entire picture. Charged particles are possibly coupled to other forces, notably gravity and nuclear forces. Thus, Maxwell's equations do not stand separate from other physical laws, but are coupled to them via the charge and current densities. The response of a point charge to the Lorentz law is one aspect; the generation of E and B by currents and charges is another.

In real materials the Lorentz force is inadequate to describe the behavior of charged particles, both in principle and as a matter of computation. The charged particles in a material medium both respond to the E and B fields and generate these fields. Complex transport equations must be solved to determine the time and spatial response of charges, for example, the Boltzmann equation or the Fokker–Planck equation or the Navier–Stokes equations. For example, see magnetohydrodynamics, fluid dynamics, electrohydrodynamics, superconductivity, stellar evolution. An entire physical apparatus for dealing with these matters

has developed. See for example, Green–Kubo relations and Green's function (many-body theory).

11.5   Lorentz force law as the definition of E and B

In many textbook treatments of classical electromagnetism, the Lorentz force Law is used as the *definition* of the electric and magnetic fields E and B.[17][18][19] To be specific, the Lorentz force is understood to be the following empirical statement:

> *The electromagnetic force F on a test charge at a given point and time is a certain function of its charge q and velocity v, which can be parameterized by exactly two vectors E and B, in the functional form:*

$$\mathbf{F} = q(\mathbf{E} + \mathbf{v} \times \mathbf{B})$$

This *is* valid; countless experiments have shown that it is, even for particles approaching the speed of light (that is, magnitude of $\mathbf{v} = |\mathbf{v}| = c$).[20] So the two vector fields E and B are thereby defined throughout space and time, and these are called the "electric field" and "magnetic field". Note that the fields are defined everywhere in space and time with respect to what force a test charge would receive regardless of whether a charge is present to experience the force.

Note also that as a definition of E and B, the Lorentz force is only a definition in principle because a real particle (as opposed to the hypothetical "test charge" of infinitesimally-small mass and charge) would generate its own finite E and B fields, which would alter the electromagnetic force that it experiences. In addition, if the charge experiences acceleration, as if forced into a curved trajectory by some external agency, it emits radiation that causes braking of its motion. See for example Bremsstrahlung and synchrotron light. These effects occur through both a direct effect (called the radiation reaction force) and indirectly (by affecting the motion of nearby charges and currents). Moreover, net force must include gravity, electroweak, and any other forces aside from electromagnetic force.

11.6   Force on a current-carrying wire

When a wire carrying an electric current is placed in a magnetic field, each of the moving charges, which comprise

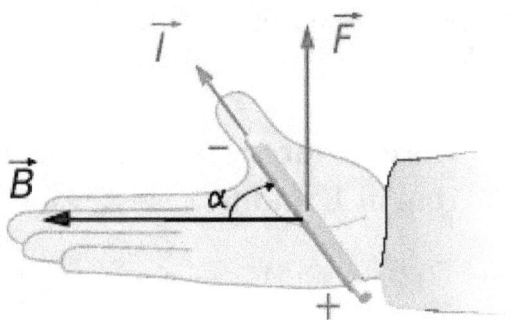

Right-hand rule for a current-carrying wire in a magnetic field B

the current, experiences the Lorentz force, and together they can create a macroscopic force on the wire (sometimes called the **Laplace force**). By combining the Lorentz force law above with the definition of electric current, the following equation results, in the case of a straight, stationary wire:

$$\mathbf{F} = I\boldsymbol{\ell} \times \mathbf{B}$$

where $\boldsymbol{\ell}$ is a vector whose magnitude is the length of wire, and whose direction is along the wire, aligned with the direction of conventional current flow I.

If the wire is not straight but curved, the force on it can be computed by applying this formula to each infinitesimal segment of wire $d\boldsymbol{\ell}$, then adding up all these forces by integration. Formally, the net force on a stationary, rigid wire carrying a steady current I is

$$\mathbf{F} = I \int d\boldsymbol{\ell} \times \mathbf{B}$$

This is the net force. In addition, there will usually be torque, plus other effects if the wire is not perfectly rigid.

One application of this is Ampère's force law, which describes how two current-carrying wires can attract or repel each other, since each experiences a Lorentz force from the other's magnetic field. For more information, see the article: Ampère's force law.

11.7   EMF

The magnetic force ($q\,\mathbf{v} \times \mathbf{B}$) component of the Lorentz force is responsible for *motional* electromotive force (or *motional EMF*), the phenomenon underlying many electrical generators. When a conductor is moved through a magnetic field, the magnetic force tries to push electrons through the wire, and this creates the EMF. The term "motional EMF"

is applied to this phenomenon, since the EMF is due to the *motion* of the wire.

In other electrical generators, the magnets move, while the conductors do not. In this case, the EMF is due to the electric force ($q\mathbf{E}$) term in the Lorentz Force equation. The electric field in question is created by the changing magnetic field, resulting in an *induced* EMF, as described by the Maxwell–Faraday equation (one of the four modern Maxwell's equations). [21]

Both of these EMF's, despite their different origins, can be described by the same equation, namely, the EMF is the rate of change of magnetic flux through the wire. (This is Faraday's law of induction, see below.) Einstein's special theory of relativity was partially motivated by the desire to better understand this link between the two effects. [21] In fact, the electric and magnetic fields are different faces of the same electromagnetic field, and in moving from one inertial frame to another, the solenoidal vector field portion of the E-field can change in whole or in part to a B-field or *vice versa*. [22]

11.8   Lorentz force and Faraday's law of induction

Main article: Faraday's law of induction

Given a loop of wire in a magnetic field, Faraday's law of induction states the induced electromotive force (EMF) in the wire is:

$$\mathcal{E} = -\frac{d\Phi_B}{dt}$$

where

$$\Phi_B = \iint_{\Sigma(t)} d\mathbf{A} \cdot \mathbf{B}(\mathbf{r}, t)$$

is the magnetic flux through the loop, **B** is the magnetic field, $\Sigma(t)$ is a surface bounded by the closed contour $\partial\Sigma(t)$, at all at time t, $d\mathbf{A}$ is an infinitesimal vector area element of $\Sigma(t)$ (magnitude is the area of an infinitesimal patch of surface, direction is orthogonal to that surface patch).

The *sign* of the EMF is determined by Lenz's law. Note that this is valid for not only a *stationary* wire — but also for a *moving* wire.

From Faraday's law of induction (that is valid for a moving wire, for instance in a motor) and the Maxwell Equations, the Lorentz Force can be deduced. The reverse is also true, the Lorentz force and the Maxwell Equations can be used to derive the Faraday Law.

Let $\Sigma(t)$ be the moving wire, moving together without rotation and with constant velocity $\mathbf{v}$ and $\Sigma(t)$ be the internal surface of the wire. The EMF around the closed path $\partial\Sigma(t)$ is given by:[23]

$$\mathcal{E} = \oint_{\partial\Sigma(t)} d\boldsymbol{\ell} \cdot \mathbf{F}/q$$

where

$$\mathbf{E} = \mathbf{F}/q$$

is the electric field and $d\boldsymbol{\ell}$ is an infinitesimal vector element of the contour $\partial\Sigma(t)$.

NB: Both $d\boldsymbol{\ell}$ and $d\mathbf{A}$ have a sign ambiguity; to get the correct sign, the right-hand rule is used, as explained in the article Kelvin-Stokes theorem.

The above result can be compared with the version of Faraday's law of induction that appears in the modern Maxwell's equations, called here the *Maxwell-Faraday equation*:

$$\nabla \times \mathbf{E} = -\frac{\partial\mathbf{B}}{\partial t} .$$

The Maxwell-Faraday equation also can be written in an *integral form* using the Kelvin-Stokes theorem.[24]

So we have, the Maxwell Faraday equation:

$$\oint_{\partial\Sigma(t)} d\boldsymbol{\ell} \cdot \mathbf{E}(\mathbf{r}, t) = -\iint_{\Sigma(t)} d\mathbf{A} \cdot \frac{d\mathbf{B}(\mathbf{r}, t)}{dt}$$

and the Faraday Law,

$$\oint_{\partial\Sigma(t)} d\boldsymbol{\ell} \cdot \mathbf{F}/q(\mathbf{r}, t) = -\frac{d}{dt} \iint_{\Sigma(t)} d\mathbf{A} \cdot \mathbf{B}(\mathbf{r}, t).$$

The two are equivalent if the wire is not moving. Using the Leibniz integral rule and that $div\ \mathbf{B} = 0$, results in,

$$\oint_{\partial\Sigma(t)} d\boldsymbol{\ell}\cdot\mathbf{F}/q(\mathbf{r}, t) = -\iint_{\Sigma(t)} d\mathbf{A}\cdot\frac{\partial}{\partial t}\mathbf{B}(\mathbf{r}, t) + \oint_{\partial\Sigma(t)} \mathbf{v}\times\mathbf{B}\, d\boldsymbol{\ell}$$

and using the Maxwell Faraday equation,

$$\oint_{\partial\Sigma(t)} d\boldsymbol{\ell}\cdot\mathbf{F}/q(\mathbf{r}, t) = \oint_{\partial\Sigma(t)} d\boldsymbol{\ell}\cdot\mathbf{E}(\mathbf{r}, t) + \oint_{\partial\Sigma(t)} \mathbf{v}\times\mathbf{B}(\mathbf{r}, t)\, d\boldsymbol{\ell}$$

since this is valid for any wire position it implies that,

$$\mathbf{F} = q\,\mathbf{E}(\mathbf{r}, t) + q\,\mathbf{v} \times \mathbf{B}(\mathbf{r}, t).$$

Faraday's law of induction holds whether the loop of wire is rigid and stationary, or in motion or in process of deformation, and it holds whether the magnetic field is constant in time or changing. However, there are cases where Faraday's law is either inadequate or difficult to use, and application of the underlying Lorentz force law is necessary. See inapplicability of Faraday's law.

If the magnetic field is fixed in time and the conducting loop moves through the field, the magnetic flux Φ_B linking the loop can change in several ways. For example, if the **B**-field varies with position, and the loop moves to a location with different **B**-field, Φ_B will change. Alternatively, if the loop changes orientation with respect to the **B**-field, the **B** • d**A** differential element will change because of the different angle between **B** and d**A**, also changing Φ_B. As a third example, if a portion of the circuit is swept through a uniform, time-independent **B**-field, and another portion of the circuit is held stationary, the flux linking the entire closed circuit can change due to the shift in relative position of the circuit's component parts with time (surface $\partial\Sigma(t)$ time-dependent). In all three cases, Faraday's law of induction then predicts the EMF generated by the change in Φ_B.

Note that the Maxwell Faraday's equation implies that the Electric Field **E** is non conservative when the Magnetic Field **B** varies in time, and is not expressible as the gradient of a scalar field, and not subject to the gradient theorem since its rotational is not zero.[23][25]

11.9 Lorentz force in terms of potentials

See also: Mathematical descriptions of the electromagnetic field, Maxwell's equations and Helmholtz decomposition

The **E** and **B** fields can be replaced by the magnetic vector potential **A** and (scalar) electrostatic potential φ by

$$\mathbf{E} = -\nabla\varphi - \frac{\partial\mathbf{A}}{\partial t}$$

$$\mathbf{B} = \nabla \times \mathbf{A}$$

where ∇ is the gradient, $\nabla\bullet$ is the divergence, $\nabla \times$ is the curl. The force becomes

$$\mathbf{F} = q \left[-\nabla \phi - \frac{\partial \mathbf{A}}{\partial t} + \mathbf{v} \times (\nabla \times \mathbf{A}) \right]$$

and using an identity for the triple product simplifies to

using the chain rule, the total derivative of $\mathbf{A}$ is:

$$\frac{d\mathbf{A}}{dt} = \frac{\partial \mathbf{A}}{\partial t} + (\mathbf{v} \cdot \nabla)\mathbf{A}$$

so the above expression can be rewritten as:

$$\mathbf{F} = q \left[-\nabla(\phi - \mathbf{v} \cdot \mathbf{A}) - \frac{d\mathbf{A}}{dt} \right]$$

which can take the convenient Euler–Lagrange form

11.10   Lorentz force and analytical mechanics

See also: Momentum

The Lagrangian for a charged particle of mass m and charge q in an electromagnetic field equivalently describes the dynamics of the particle in terms of its *energy*, rather than the force exerted on it. The classical expression is given by:[26]

$$L = \frac{m}{2}\dot{\mathbf{r}} \cdot \dot{\mathbf{r}} + q\mathbf{A} \cdot \dot{\mathbf{r}} - q\phi$$

where $\mathbf{A}$ and ϕ are the potential fields as above. Using Lagrange's equations, the equation for the Lorentz force can be obtained.

The potential energy depends on the velocity of the particle, so the force is velocity dependent, so it is not conservative.

The relativistic Lagrangian is

$$L = -mc^2 \sqrt{1 - \left(\frac{\dot{\mathbf{r}}}{c}\right)^2} + e\mathbf{A}(\mathbf{r}) \cdot \dot{\mathbf{r}} - e\phi(\mathbf{r})$$

The action is the relativistic arclength of the path of the particle in space time, minus the potential energy contribution, plus an extra contribution which quantum mechanically is an extra phase a charged particle gets when it is moving along a vector potential.

11.11   Equation (cgs units)

See also: cgs units

The above-mentioned formulae use SI units which are the most common among experimentalists, technicians, and engineers. In cgs-Gaussian units, which are somewhat more common among theoretical physicists, one has instead

$$\mathbf{F} = q_{cgs} \left(\mathbf{E}_{cgs} + \frac{\mathbf{v}}{c} \times \mathbf{B}_{cgs} \right).$$

where c is the speed of light. Although this equation looks slightly different, it is completely equivalent, since one has the following relations:

$$q_{cgs} = \frac{q_{SI}}{\sqrt{4\pi\epsilon_0}}, \quad \mathbf{E}_{cgs} = \sqrt{4\pi\epsilon_0}\,\mathbf{E}_{SI}, \quad \mathbf{B}_{cgs} = \sqrt{4\pi/\mu_0}\,\mathbf{B}_{SI}$$

where ϵ_0 is the vacuum permittivity and μ_0 the vacuum permeability. In practice, the subscripts "cgs" and "SI" are always omitted, and the unit system has to be assessed from context.

11.12   Relativistic form of the Lorentz force

11.12.1   Covariant form of the Lorentz force

Field tensor

Main articles: Covariant formulation of classical electromagnetism and Mathematical descriptions of the electromagnetic field

Using the metric signature $(-1, 1, 1, 1)$, The Lorentz force for a charge q can be written in covariant form:

where p^α is the four-momentum, defined as

$$p^\alpha = (p_0, p_1, p_2, p_3) = (\gamma mc, p_x, p_y, p_z) .$$

τ the proper time of the particle, $F^{\alpha\beta}$ the contravariant electromagnetic tensor

$$F^{\alpha\beta} = \begin{pmatrix} 0 & E_x/c & E_y/c & E_z/c \\ -E_x/c & 0 & B_z & -B_y \\ -E_y/c & -B_z & 0 & B_x \\ -E_z/c & B_y & -B_x & 0 \end{pmatrix}$$

and U is the covariant 4-velocity of the particle, defined as:

$$U_\beta = (U_0, U_1, U_2, U_3) = \gamma(-c, v_x, v_y, v_z) .$$

in which

$$\gamma(v) = \frac{1}{\sqrt{1 - \frac{v^2}{c^2}}} = \frac{1}{\sqrt{1 - \frac{v_x^2 + v_y^2 + v_z^2}{c^2}}}$$

is the Lorentz factor.

The fields are transformed to a frame moving with constant relative velocity by:

$$F'^{\mu\nu} = \Lambda^\mu{}_\alpha \Lambda^\nu{}_\beta F^{\alpha\beta} .$$

where $\Lambda^\mu{}_\alpha$ is the Lorentz transformation tensor.

Translation to vector notation

The $\alpha = 1$ component (x-component) of the force is

$$\frac{dp^1}{d\tau} = qU_\beta F^{1\beta} = q\left(U_0 F^{10} + U_1 F^{11} + U_2 F^{12} + U_3 F^{13}\right) .$$

Substituting the components of the covariant electromagnetic tensor F yields

$$\frac{dp^1}{d\tau} = q\left[U_0\left(\frac{-E_x}{c}\right) + U_2(B_z) + U_3(-B_y)\right] .$$

Using the components of covariant four-velocity yields

$$\begin{aligned} \frac{dp^1}{d\tau} &= q\gamma\left[-c\left(\frac{-E_x}{c}\right) + v_y B_z + v_z(-B_y)\right] \\ &= q\gamma\left(E_x + v_y B_z - v_z B_y\right) \\ &= q\gamma\left[E_x + (\mathbf{v} \times \mathbf{B})_x\right] . \end{aligned}$$

The calculation for $\alpha = 2, 3$ (force components in the y and z directions) yields similar results, so collecting the 3 equations into one:

$$\frac{d\mathbf{p}}{d\tau} = q\gamma\left(\mathbf{E} + \mathbf{v} \times \mathbf{B}\right) .$$

and since differentials in coordinate time dt and proper time $d\tau$ are related by the Lorentz factor,

$$dt = \gamma(v)d\tau ,$$

so we arrive at

$$\frac{d\mathbf{p}}{dt} = q\left(\mathbf{E} + \mathbf{v} \times \mathbf{B}\right) .$$

This is precisely the Lorentz force law, however, it is important to note that $\mathbf{p}$ is the relativistic expression,

$$\mathbf{p} = \gamma(v)m_0\mathbf{v} .$$

11.12.2 STA form of the Lorentz force

The electric and magnetic fields are dependent on the velocity of an observer, so the relativistic form of the Lorentz force law can best be exhibited starting from a coordinate-independent expression for the electromagnetic and magnetic fields,[27] $\mathcal{F}$, and an arbitrary time-direction, γ_0, where

$$\mathbf{E} = (\mathcal{F} \cdot \gamma_0)\gamma_0$$

and

$$i\mathbf{B} = (\mathcal{F} \wedge \gamma_0)\gamma_0$$

$\mathcal{F}$ is a space-time bivector (an oriented plane segment, just like a vector is an oriented line segment), which has six degrees of freedom corresponding to boosts (rotations in space-time planes) and rotations (rotations in space-space planes). The dot product with the vector γ_0 pulls a vector (in the space algebra) from the translational part, while the wedge-product creates a trivector (in the space algebra) who is dual to a vector which is the usual magnetic field vector. The relativistic velocity is given by the (time-like) changes in a time-position vector $v = \dot{x}$, where

$$v^2 = 1.$$

(which shows our choice for the metric) and the velocity is

$$\mathbf{v} = cv \wedge \gamma_0 / (v \cdot \gamma_0).$$

The proper (invariant is an inadequate term because no transformation has been defined) form of the Lorentz force law is simply

Note that the order is important because between a bivector and a vector the dot product is anti-symmetric. Upon a space time split like one can obtain the velocity, and fields as above yielding the usual expression.

11.13   Applications

The Lorentz force occurs in many devices, including:

- Cyclotrons and other circular path particle accelerators
- Mass spectrometers
- Velocity Filters
- Magnetrons
- Lorentz force velocimetry

In its manifestation as the Laplace force on an electric current in a conductor, this force occurs in many devices including:

11.14   See also

11.15   Footnotes

[1] Oliver Heaviside By Paul J. Nahin, p120

[2] Huray, Paul G. (2009). *Maxwell's Equations*. Wiley-IEEE. p. 22. ISBN 0-470-54276-4.

[3] See Jackson page 2. The book lists the four modern Maxwell's equations, and then states, "Also essential for consideration of charged particle motion is the Lorentz force equation, $\mathbf{F} = q (\mathbf{E} + \mathbf{v} \times \mathbf{B})$, which gives the force acting on a point charge q in the presence of electromagnetic fields."

[4] See Griffiths page 204.

[5] For example, see the website of the "Lorentz Institute": \. or Griffiths.

[6] Griffiths, David J. (1999). *Introduction to electrodynamics*. reprint. with corr. (3rd ed.). Upper Saddle River, New Jersey [u.a.]: Prentice Hall. ISBN 978-0-13-805326-0.

[7] Meyer, Herbert W. (1972). *A History of Electricity and Magnetism*. Norwalk, Connecticut: Burndy Library. pp. 30–31. ISBN 0-262-13070-X.

[8] Verschuur, Gerrit L. (1993). *Hidden Attraction : The History And Mystery Of Magnetism*. New York: Oxford University Press. pp. 78–79. ISBN 0-19-506488-7.

[9] Darrigol, Olivier (2000). *Electrodynamics from Ampère to Einstein*. Oxford, [England]: Oxford University Press. pp. 9, 25. ISBN 0-19-850593-0.

[10] Verschuur, Gerrit L. (1993). *Hidden Attraction : The History And Mystery Of Magnetism*. New York: Oxford University Press. p. 76. ISBN 0-19-506488-7.

[11] Darrigol, Olivier (2000). *Electrodynamics from Ampère to Einstein*. Oxford, [England]: Oxford University Press. pp. 126–131, 139–144. ISBN 0-19-850593-0.

[12] Darrigol, Olivier (2000). *Electrodynamics from Ampère to Einstein*. Oxford, [England]: Oxford University Press. pp. 200, 429–430. ISBN 0-19-850593-0.

[13] Heaviside, Oliver. "On the Electromagnetic Effects due to the Motion of Electrification through a Dielectric". *Philosophical Magazine*, April 1889, p. 324.

[14] Darrigol, Olivier (2000). *Electrodynamics from Ampère to Einstein*. Oxford, [England]: Oxford University Press. p. 327. ISBN 0-19-850593-0.

[15] Whittaker, E. T. (1910). *A History of the Theories of Aether and Electricity: From the Age of Descartes to the Close of the Nineteenth Century*. Longmans, Green and Co. pp. 420–423. ISBN 1-143-01208-9.

[16] See Griffiths page 326, which states that Maxwell's equations, "together with the [Lorentz] force law...summarize the entire theoretical content of classical electrodynamics".

[17] See, for example, Jackson p777-8.

[18] J.A. Wheeler, C. Misner, K.S. Thorne (1973). *Gravitation*. W.H. Freeman & Co. pp. 72–73. ISBN 0-7167-0344-0.. These authors use the Lorentz force in tensor form as definer of the electromagnetic tensor F, in turn the fields $\mathbf{E}$ and $\mathbf{B}$.

[19] I.S. Grant, W.R. Phillips, Manchester Physics (2008). *Electromagnetism* (2nd ed.). John Wiley & Sons. p. 122. ISBN 978-0-471-92712-9.

[20] I.S. Grant, W.R. Phillips, Manchester Physics (2008). *Electromagnetism* (2nd Edition). John Wiley & Sons. p. 123. ISBN 978-0-471-92712-9.

[21] See Griffiths pages 301–3.

[22] Tai L. Chow (2006). *Electromagnetic theory*. Sudbury MA: Jones and Bartlett. p. 395. ISBN 0-7637-3827-1.

[23] Landau, L. D., Lifshifs, E. M., & Pitaevskiĭ, L. P. (1984). *Electrodynamics of continuous media: Volume 8* Course of Theoretical Physics (Second ed.). Oxford: Butterworth-Heinemann. p. §63 (§49 pp. 205–207 in 1960 edition). ISBN 0-7506-2634-8.

[24] Roger F Harrington (2003). *Introduction to electromagnetic engineering*. Mineola, New York: Dover Publications. p. 56. ISBN 0-486-43241-6.

[25] M N O Sadiku (2007). *Elements of elctromagnetics* (Fourth ed.). NY/Oxford: Oxford University Press. p. 391. ISBN 0-19-530048-3.

[26] Classical Mechanics (2nd Edition), T.W.B. Kibble, European Physics Series, Mc Graw Hill (UK), 1973, ISBN 0-07-084018-0.

[27] Hestenes, David. "SpaceTime Calculus".

11.16 References

The numbered references refer in part to the list immediately below.

- Feynman, Richard Phillips; Leighton, Robert B.; Sands, Matthew L. (2006). *The Feynman lectures on physics (3 vol.)*. Pearson / Addison-Wesley. ISBN 0-8053-9047-2.: volume 2.

- Griffiths, David J. (1999). *Introduction to electrodynamics* (3rd ed.). Upper Saddle River, [NJ.]: Prentice-Hall. ISBN 0-13-805326-X.

- Jackson, John David (1999). *Classical electrodynamics* (3rd ed.). New York, [NY.]: Wiley. ISBN 0-471-30932-X.

- Serway, Raymond A.; Jewett, John W., Jr. (2004). *Physics for scientists and engineers, with modern physics*. Belmont, [CA.]: Thomson Brooks/Cole. ISBN 0-534-40846-X.

- Srednicki, Mark A. (2007). *Quantum field theory*. Cambridge, [England] ; New York [NY.]: Cambridge University Press. ISBN 978-0-521-86449-7.

11.17 External links

- Interactive Java tutorial on the Lorentz force National High Magnetic Field Laboratory

- Lorentz force (demonstration)

- Faraday's law: Tankersley and Mosca

- Notes from Physics and Astronomy HyperPhysics at Georgia State University: see also home page

- Interactive Java applet on the magnetic deflection of a particle beam in a homogeneous magnetic field by Wolfgang Bauer

Chapter 12

Electric field

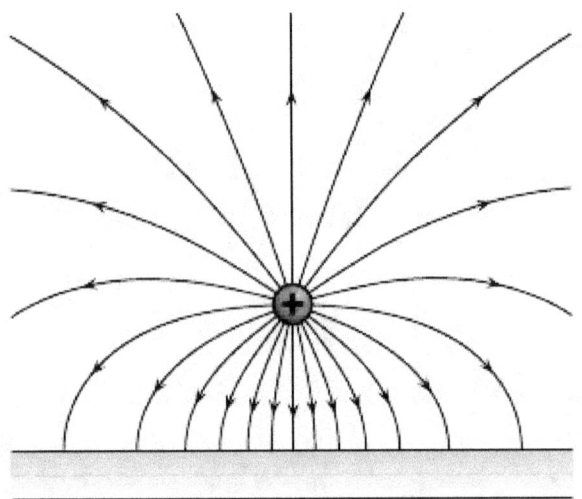

Electric field lines emanating from a point positive electric charge suspended over an infinite sheet of conducting material.

The **electric field** is a component of the electromagnetic field. It is a vector field and is generated by electric charges or time-varying magnetic fields, as described by Maxwell's equations.[1] The concept of an electric field was introduced by Michael Faraday.[2]

12.1 Definition

The electric field **E** at a given point is defined as the (vectorial) force **F** that would be exerted on a stationary test particle of unit charge by electromagnetic forces (i.e. the Lorentz force). A particle of charge q would be subject to a force $\mathbf{F} = q \cdot \mathbf{E}$.

Its SI units are newtons per coulomb ($N \cdot C^{-1}$) or, equivalently, volts per metre ($V \cdot m^{-1}$), which in terms of SI base units are $kg \cdot m \cdot s^{-3} \cdot A^{-1}$.

12.2 Sources of electric field

12.2.1 Causes and description

Electric fields are caused by electric charges or varying magnetic fields. The former effect is described by Gauss's law, the latter by Faraday's law of induction, which together are enough to define the behavior of the electric field as a function of charge repartition and magnetic field. However, since the magnetic field is described as a function of electric field, the equations of both fields are coupled and together form Maxwell's equations that describe both fields as a function of charges and currents.

In the special case of a steady state (stationary charges and currents), the Maxwell-Faraday inductive effect disappears. The resulting two equations (Gauss's law $\nabla \cdot \mathbf{E} = \frac{\rho}{\varepsilon_0}$ and Faraday's law with no induction term $\nabla \times \mathbf{E} = 0$), taken together, are equivalent to Coulomb's law, written as $\boldsymbol{E}(\boldsymbol{r}) = \frac{1}{4\pi\varepsilon_0} \int d r' \rho(\boldsymbol{r}') \frac{\boldsymbol{r}-\boldsymbol{r}'}{|\boldsymbol{r}-\boldsymbol{r}'|^3}$ for a charge density $\rho(\mathbf{r})$ ($\mathbf{r}$ denotes the position in space). Notice that ε_0 , the permittivity of vacuum, must be substituted if charges are considered in non-empty media.

12.2.2 Continuous vs. discrete charge repartition

Main article: Charge density

The equations of electromagnetism are best described in a continuous description. However, charges are sometimes best described as discrete points; for example, some models may describe electrons as punctual sources where charge density is infinite on an infinitesimal section of space.

A charge q located in $\mathbf{r}_0$ can be described mathematically as a charge density $\rho(\mathbf{r}) = q\delta(\mathbf{r} - \mathbf{r}_0)$, where the Dirac delta function (in three dimensions) is used. Conversely, a charge distribution can be approximated by many small punctual charge.

12.3 Superposition principle

Electric fields satisfy the superposition principle, because Maxwell's equations are linear. As a result, if $\mathbf{E}_1$ and $\mathbf{E}_2$ are the electric fields resulting from distribution of charges ρ_1 and ρ_2, a distribution of charges $\rho_1 + \rho_2$ will create an electric field $\mathbf{E}_1 + \mathbf{E}_2$; for instance, Coulomb's law is linear in charge density as well.

This principle is useful to calculate the field created by multiple point charges. If charges $q_1, q_2, ..., q_n$ are stationary in space at $\mathbf{r}_1, \mathbf{r}_2, ...\mathbf{r}_n$, in the absence of currents, the superposition principle proves that the resulting field is the sum of fields generated by each particle as described by Coulomb's law:

$$\mathbf{E}(\mathbf{r}) = \sum_{i=1}^{N} \mathbf{E}_i(\mathbf{r}) = \frac{1}{4\pi\varepsilon_0}\sum_{i=1}^{N} q_i \frac{\mathbf{r} - \mathbf{r}_i}{|\mathbf{r} - \mathbf{r}_i|^3}$$

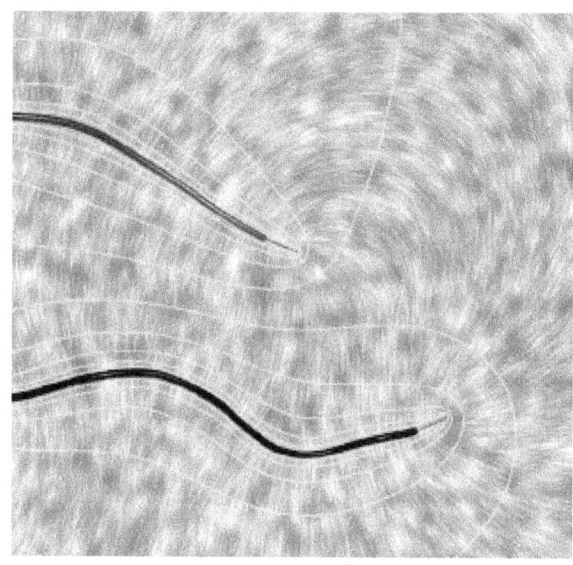

Electric field between two conductors

12.4 Electrostatic fields

Main article: Electrostatics
Electrostatic fields are **E**-fields which do not change with

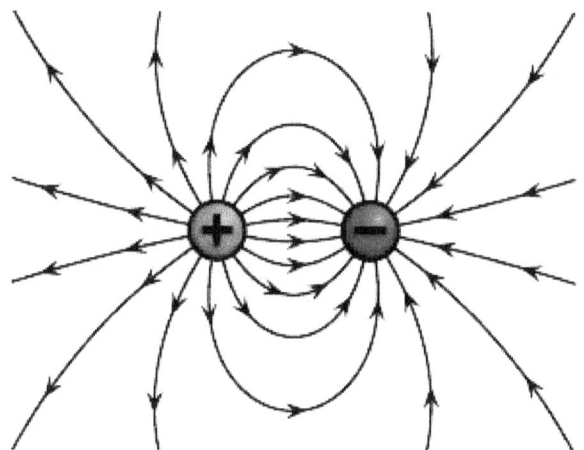

Illustration of the electric field surrounding a positive (red) and a negative (blue) charge.

time, which happens when charges and currents are stationary. In that case, Coulomb's law fully describes the field.

12.4.1 Electric potential

Main article: Conservative vector field § Irrotational vector fields

By Faraday's law, the electric field has zero curl. One can then define an electric potential, that is, a function Φ such

that $\mathbf{E} = -\nabla\Phi$. [3] This is analogous to the gravitational potential.

12.4.2 Parallels between electrostatic and gravitational fields

Coulomb's law, which describes the interaction of electric charges:

$$\mathbf{F} = q\left(\frac{Q}{4\pi\varepsilon_0}\frac{\hat{\mathbf{r}}}{|\mathbf{r}|^2}\right) = q\mathbf{E}$$

is similar to Newton's law of universal gravitation:

$$\mathbf{F} = m\left(-GM\frac{\hat{\mathbf{r}}}{|\mathbf{r}|^2}\right) = m\mathbf{g}$$

(where $\hat{\mathbf{r}} = \frac{\mathbf{r}}{|\mathbf{r}|}$).

This suggests similarities between the electric field **E** and the gravitational field **g**, or their associated potentials. Mass is sometimes called "gravitational charge" because of that similarity.

Electrostatic and gravitational forces both are central, conservative and obey an inverse-square law.

12.4.3 Uniform fields

A uniform field is one in which the electric field is constant at every point. It can be approximated by placing two conducting plates parallel to each other and maintaining a voltage (potential difference) between them; it is only an approximation because of boundary effects (near the edge of the planes, electric field is distorted because the plane does not continue). Assuming infinite planes, the magnitude of the electric field E is:

$$E = -\frac{\Delta\phi}{d}$$

where $\Delta\phi$ is the potential difference between the plates and d is the distance separating the plates. The negative sign arises as positive charges repel, so a positive charge will experience a force away from the positively charged plate, in the opposite direction to that in which the voltage increases. In micro- and nanoapplications, for instance in relation to semiconductors, a typical magnitude of an electric field is in the order of 10^6 V·m*−1, achieved by applying a voltage of the order of 1 volt between conductors spaced 1 μm apart.

12.5   Electrodynamic fields

Main article: Electrodynamics

Electrodynamic fields are E-fields which do change with time, for instance when charges are in motion.

The electric field cannot be described independently of the magnetic field in that case. If $\mathbf{A}$ is the magnetic vector potential, defined so that $\mathbf{B} = \nabla \times \mathbf{A}$, one can still define an electric potential Φ such that:

$$\mathbf{E} = -\nabla\Phi - \frac{\partial\mathbf{A}}{\partial t}$$

One can recover Faraday's law of induction by taking the curl of that equation

*[4]

$$\nabla \times \mathbf{E} = -\frac{\partial(\nabla \times \mathbf{A})}{\partial t} = -\frac{\partial\mathbf{B}}{\partial t}$$

which justifies, a posteriori, the previous form for $\mathbf{E}$.

12.6   Energy in the electric field

If the magnetic field $\mathbf{B}$ is nonzero,

The total energy per unit volume stored by the electromagnetic field is*[5]

$$u_{EM} = \frac{\varepsilon}{2}|\mathbf{E}|^2 + \frac{1}{2\mu}|\mathbf{B}|^2$$

where ε is the permittivity of the medium in which the field exists, μ its magnetic permeability, and $\mathbf{E}$ and $\mathbf{B}$ are the electric and magnetic field vectors.

As $\mathbf{E}$ and $\mathbf{B}$ fields are coupled, it would be misleading to split this expression into "electric" and "magnetic" contributions. However, in the steady-state case, the fields are no longer coupled (see Maxwell's equations). It makes sense in that case to compute the electrostatic energy per unit volume:

$$u_{ES} = \frac{1}{2}\varepsilon|\mathbf{E}|^2 .$$

The total energy U stored in the electric field in a given volume V is therefore

$$U_{ES} = \frac{1}{2}\varepsilon\int_V |\mathbf{E}|^2\, dV .$$

12.7   Further extensions

12.7.1   Definitive equation of vector fields

See also: Defining equation (physics) and List of electromagnetism equations

In the presence of matter, it is helpful in electromagnetism to extend the notion of the electric field into three vector fields, rather than just one:*[6]

$$\mathbf{D} = \varepsilon_0\mathbf{E} + \mathbf{P}$$

where $\mathbf{P}$ is the electric polarization – the volume density of electric dipole moments, and $\mathbf{D}$ is the electric displacement field. Since $\mathbf{E}$ and $\mathbf{P}$ are defined separately, this equation can be used to define $\mathbf{D}$. The physical interpretation of $\mathbf{D}$ is not as clear as $\mathbf{E}$ (effectively the field applied to the material) or $\mathbf{P}$ (induced field due to the dipoles in the material), but still serves as a convenient mathematical simplification, since Maxwell's equations can be simplified in terms of free charges and currents.

12.7.2   Constitutive relation

Main article: Constitutive equation

The **E** and **D** fields are related by the permittivity of the material, ε.[7][8]

For linear, homogeneous, isotropic materials **E** and **D** are proportional and constant throughout the region, there is no position dependence: For inhomogeneous materials, there is a position dependence throughout the material:

$$\mathbf{D}(\mathbf{r}) = \varepsilon \mathbf{E}(\mathbf{r})$$

For anisotropic materials the **E** and **D** fields are not parallel, and so **E** and **D** are related by the permittivity tensor (a 2nd order tensor field), in component form:

$$D_i = \varepsilon_{ij} E_j$$

For non-linear media, **E** and **D** are not proportional. Materials can have varying extents of linearity, homogeneity and isotropy.

12.8   See also

- Classical electromagnetism
- Field strength
- signal strength in telecommunications
- Magnetism
- Teltron Tube
- Teledeltos, a conductive paper that may be used as a simple analog computer for modelling fields.

12.9   References

[1] Richard Feynman (1970). *The Feynman Lectures on Physics Vol II*. Addison Wesley Longman. ISBN 978-0-201-02115-8.

[2] http://public.wsu.edu/~{}jtd/Physics206/michael_faraday.htm

[3] http://physicspages.com/2011/10/08/curl-potential-in-electrostatics/

[4] Huray, Paul G. (2009). *Maxwell's Equations*. Wiley-IEEE. p. 205. ISBN 0-470-54276-4.

[5] Introduction to Electrodynamics (3rd Edition), D.J. Griffiths, Pearson Education, Dorling Kindersley, 2007, ISBN 81-7758-293-3

[6] Electromagnetism (2nd Edition), I.S. Grant, W.R. Phillips, Manchester Physics, John Wiley & Sons, 2008, ISBN 978-0-471-92712-9

[7] Electricity and Modern Physics (2nd Edition), G.A.G. Bennet, Edward Arnold (UK), 1974, ISBN 0-7131-2459-8

[8] Electromagnetism (2nd Edition), I.S. Grant, W.R. Phillips, Manchester Physics, John Wiley & Sons, 2008, ISBN 978-0-471-92712-9

12.10   External links

- Electric field in "Electricity and Magnetism", R Nave – Hyperphysics, Georgia State University
- 'Gauss's Law' – Chapter 24 of Frank Wolfs's lectures at University of Rochester
- 'The Electric Field' – Chapter 23 of Frank Wolfs's lectures at University of Rochester
- – An applet that shows the electric field of a moving point charge.
- Fields – a chapter from an online textbook
- Learning by Simulations Interactive simulation of an electric field of up to four point charges
- Java simulations of electrostatics in 2-D and 3-D
- Interactive Flash simulation picturing the electric field of user-defined or preselected sets of point charges by field vectors, field lines, or equipotential lines. Author: David Chappell

Chapter 13

History of electromagnetic theory

For a chronological guide to this subject, see Timeline of electromagnetic theory.

The **history of electromagnetic theory** begins with ancient measures to deal with atmospheric electricity, in particular lightning.[1] People then had little understanding of electricity, and were unable to scientifically explain the phenomena.[2] In the 19th century there was a unification of the **history of electric theory** with the **history of magnetic theory**. It became clear that electricity should be treated jointly with magnetism, because wherever electricity is in motion, magnetism is also present.[3] Magnetism was not fully explained until the idea of magnetic induction was developed.[4] Electricity was not fully explained until the idea of electric charge was developed.

13.1 Ancient and classical history

The knowledge of static electricity dates back to the earliest civilizations, but for millennia it remained merely an interesting and mystifying phenomenon, without a theory to explain its behavior and often confused with magnetism. The ancients were acquainted with rather curious properties possessed by two minerals, amber (Greek: ἤλεκτρον, *electron*) and magnetic iron ore (Greek: Μάγνης λίθος, *Magnes lithos*, "the Magnesian stone, lodestone"). Amber, when rubbed, attracts light bodies; magnetic iron ore has the power of attracting iron.[5]

Based on his find of an Olmec hematite artifact in Central America, the American astronomer John Carlson has suggested that "the Olmec may have discovered and used the geomagnetic lodestone compass earlier than 1000 BC". If true, this "predates the Chinese discovery of the geomagnetic lodestone compass by more than a millennium".[6][7] Carlson speculates that the Olmecs may have used similar artifacts as a directional device for astrological or geomantic purposes, or to orient their temples, the dwellings of the living or the interments of the dead. The earliest Chinese literature reference to *magnetism* lies

The discovery of the property of magnets.

Magnets were first found in a natural state; certain iron oxides were discovered in various parts of the world, notably in Magnesia in Asia Minor, that had the property of attracting small pieces of iron, which is shown here.

in a 4th-century BC book called *Book of the Devil Valley Master* (鬼谷子): "The lodestone makes iron come or it attracts it."[8]

Long before any knowledge of electromagnetism existed, people were aware of the effects of electricity. Lightning and other manifestations of electricity such as St. Elmo's fire were known in ancient times, but it was not understood that these phenomena had a common origin.[9] Ancient Egyptians were aware of shocks when interacting with electric fish (such as the electric catfish) or other ani-

Electric catfish are found in tropical Africa and the Nile River.

artifacts remains in doubt.[19]

13.2   Middle Ages and the Renaissance

Magnetic attraction was once accounted by Aristotle and Thales for as the working of a soul in the stone.[20]

Shen Kuo wrote Dream Pool Essays *(夢溪筆談); Shen also first described the magnetic needle.*

mals (such as electric eels).[10] The shocks from animals were apparent to observers since pre-history by a variety of peoples that came into contact with them. Texts from 2750 BC by the ancient Egyptians referred to these fish as "thunderer of the Nile" and saw them as the "protectors" of all the other fish.[5] Another possible approach to the discovery of the identity of lightning and electricity from any other source, is to be attributed to the Arabs, who before the 15th century used the same Arabic word for lightning (*barq*) and the electric ray.[9]

Thales of Miletus, writing at around 600 BC, noted that rubbing fur on various substances such as amber would cause them to attract specks of dust and other light objects.[11] Thales wrote on the effect now known as static electricity. The Greeks noted that if they rubbed the amber for long enough they could even get an electric spark to jump.

The electrostatic phenomena was again reported millennia later by Roman and Arabic naturalists and physicians.[12] Several ancient writers, such as Pliny the Elder and Scribonius Largus, attested to the numbing effect of electric shocks delivered by catfish and torpedo rays. Pliny in his books writes: "The ancient Tuscans by their learning hold that there are nine gods that send forth lightning and those of eleven sorts." This was in general the early pagan idea of lightning.[9] The ancients held some concept that shocks could travel along conducting objects.[13] Patients suffering from ailments such as gout or headache were directed to touch electric fish in the hope that the powerful jolt might cure them.[14]

A number of objects found in Iraq in 1938 dated to the early centuries AD (Sassanid Mesopotamia), called the Baghdad Battery, resembles a galvanic cell and is believed by some to have been used for electroplating.[15] The claims are controversial because of supporting evidence and theories for the uses of the artifacts,[16][17] physical evidence on the objects conducive for electrical functions,[18] and if they were electrical in nature. As a result the nature of these objects is based on speculation, and the function of these

In the 11th century, the Chinese scientist Shen Kuo (1031–1095) was the first person to write of the magnetic needle compass and that it improved the accuracy of navigation by employing the astronomical concept of true north (*Dream Pool Essays*, AD 1088), and by the 12th century the Chinese were known to use the lodestone compass for navigation. In 1187, Alexander Neckam was the first in Europe to describe the compass and its use for navigation.

Magnetism was one of the few sciences which progressed in medieval Europe; for in the thirteenth century Peter Peregrinus, a native of Maricourt in Picardy, made a discovery of fundamental importance.[21] The French 13th century scholar conducted experiments on magnetism and wrote the first extant treatise describing the properties of magnets and pivoting compass needles.[5] The dry compass was invented around 1300 by Italian inventor Flavio Gioja.[22]

Archbishop Eustathius of Thessalonica, Greek scholar and writer of the 12th century, records that *Woliver*, king of the Goths, was able to draw sparks from his body. The same writer states that a certain philosopher was able while dressing to draw sparks from his clothes, a result seemingly akin to that obtained by Robert Symmer in his silk stocking experiments, a careful account of which may be found in the

'Philosophical Transactions,' 1759.*[9]

Italian physician Gerolamo Cardano wrote about electricity in *De Subtilitate* (1550) distinguishing, perhaps for the first time, between electrical and magnetic forces.

Toward the late 16th century, a physician of Queen Elizabeth's time, Dr. William Gilbert, in *De Magnete*, expanded on Cardano's work and invented the New Latin word *electricus* from ἤλεκτρον (*elektron*), the Greek word for "amber". Gilbert, a native of Colchester, Fellow of St John's College, Cambridge, and sometime President of the College of Physicians, was one of the earliest and most distinguished English men of science — a man whose work Galileo thought enviably great. He was appointed Court physician, and a pension was settled on him to set him free to continue his researches in Physics and Chemistry.*[23]

Gilbert undertook a number of careful electrical experiments, in the course of which he discovered that many substances other than amber, such as sulphur, wax, glass, etc.,*[24] were capable of manifesting electrical properties. Gilbert also discovered that a heated body lost its electricity and that moisture prevented the electrification of all bodies, due to the now well-known fact that moisture impaired the insulation of such bodies. He also noticed that electrified substances attracted all other substances indiscriminately, whereas a magnet only attracted iron. The many discoveries of this nature earned for Gilbert the title of *founder of the electrical science.*[9] By investigating the forces on a light metallic needle, balanced on a point, he extended the list of electric bodies, and found also that many substances, including metals and natural magnets, showed no attractive forces when rubbed. He noticed that dry weather with north or east wind was the most favourable atmospheric condition for exhibiting electric phenomena — an observation liable to misconception until the difference between conductor and insulator was understood.*[23]

Gilbert's work was followed up by Robert Boyle (1627–1691), the famous natural philosopher who was once described as "father of Chemistry, and uncle of the Earl of Cork." Boyle was one of the founders of the Royal Society when it met privately in Oxford, and became a member of the Council after the Society was incorporated by Charles II. in 1663. He worked frequently at the new science of electricity, and added several substances to Gilbert's list of electrics. He left a detailed account of his researches under the title of *Experiments on the Origin of Electricity.*[23] Boyle, in 1675, stated that electric attraction and repulsion can act across a vacuum. One of his important discoveries was that electrified bodies in a vacuum would attract light substances, thus indicating that the electrical effect did not depend upon the air as a medium. He also added resin to the then known list of electrics.*[9]*[25]*[26]*[27]

This was followed in 1660 by Otto von Guericke, who in-

Robert Boyle.

vented an early electrostatic generator. By the end of the 17th Century, researchers had developed practical means of generating electricity by friction with an electrostatic generator, but the development of electrostatic machines did not begin in earnest until the 18th century, when they became fundamental instruments in the studies about the new science of electricity.

The first usage of the word *electricity* is ascribed to Sir Thomas Browne in his 1646 work, *Pseudodoxia Epidemica*.

The first appearance of the term *electromagnetism* on the other hand comes from an earlier date: 1641. *Magnes.*[28] by the Jesuit luminary Athanasius Kircher, carries on page 640 the provocative chapter-heading: "*Elektromagnetismos* i.e. On the Magnetism of amber, or electrical attractions and their causes" (ηλεκτρο-μαγνητισμος *id est sive De Magnetismo electri, seu electricis attractionibus earumque causis*).

13.3   18th century

13.3.1   Improving the electric machine

Main article: electrostatic machine

The electric machine was subsequently improved by Francis Hauksbee, Litzendorf, and by Prof.   Georg

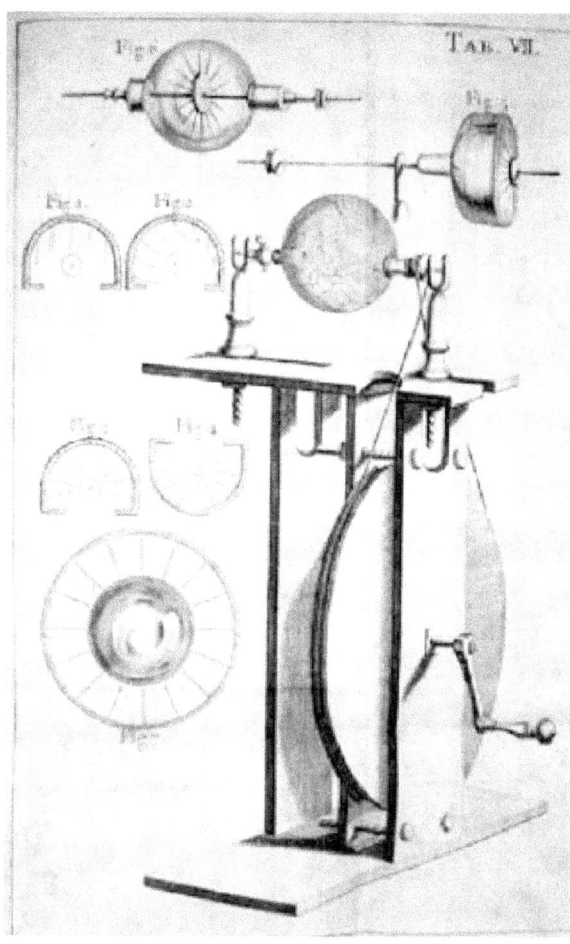

Generator built by Francis Hauksbee.[29]

Matthias Bose, about 1750. Litzendorf, researching for Christian August Hausen, substituted a glass ball for the sulphur ball of Guericke. Bose was the first to employ the "prime conductor" in such machines, this consisting of an iron rod held in the hand of a person whose body was insulated by standing on a block of resin. Ingenhousz, during 1746, invented electric machines made of plate glass.[30] Experiments with the electric machine were largely aided by the discovery of the property of a glass plate, when coated on both sides with tinfoil, of accumulating a charge of electricity when connected with a source of electromotive force. The electric machine was soon further improved by Andrew Gordon, a Scotsman, Professor at Erfurt, who substituted a glass cylinder in place of a glass globe; and by Giessing of Leipzig who added a "rubber" consisting of a cushion of woollen material. The collector, consisting of a series of metal points, was added to the machine by Benjamin Wilson about 1746, and in 1762, John Canton of England (also the inventor of the first pith-ball electroscope) improved the efficiency of electric machines by sprinkling an amalgam of tin over the surface of the rub-

ber.[9]

13.3.2 Electrics and non-electrics

In 1729, Stephen Gray conducted a series of experiments that demonstrated the difference between conductors and non-conductors (insulators), showing amongst other things that a metal wire and even pack thread conducted electricity, whereas silk did not. In one of his experiments he sent an electric current through 800 feet of hempen thread which was suspended at intervals by loops of silk thread. When he tried to conduct the same experiment substituting the silk for finely spun brass wire, he found that the electric current was no longer carried throughout the hemp cord, but instead seemed to vanish into the brass wire. From this experiment he classified substances into two categories: "electrics" like glass, resin and silk and "non-electrics" like metal and water. "Non-electrics" conducted charges while "electrics" held the charge.[9][31]

13.3.3 Vitreous and resinous

Intrigued by Gray's results, in 1732, C. F. du Fay began to conduct several experiments. In his first experiment, Du Fay concluded that all objects except metals, animals, and liquids could be electrified by rubbing and that metals, animals and liquids could be electrified by means of an electric machine, thus discrediting Gray's "electrics" and "non-electrics" classification of substances.

In 1737 Du Fay and Hauksbee independently discovered what they believed to be two kinds of frictional electricity; one generated from rubbing glass, the other from rubbing resin. From this, Du Fay theorized that electricity consists of two electrical fluids, "vitreous" and "resinous", that are separated by friction and that neutralize each other when combined.[32] This two-fluid theory would later give rise to the concept of *positive* and *negative* electrical charges devised by Benjamin Franklin.[9]

13.3.4 Leyden jar

The Leyden jar, a type of capacitor for electrical energy in large quantities, was invented independently by Ewald Georg von Kleist on 11 October 1744 and by Pieter van Musschenbroek in 1745 - 1746 at Leiden University (the latter location giving the device its name).[33] William Watson, when experimenting with the Leyden jar, discovered in 1747 that a discharge of static electricity was equivalent to an electric current. Capacitance was first observed by Von Kleist of Leyden in 1754.[34] Von Kleist happened to hold, near his electric machine, a small bottle, in the neck

Pieter van Musschenbroek.

myrtle trees was quickened by electrification. These myrtles were electrified "during the whole month of October, 1746, and they put forth branches and blossoms sooner than other shrubs of the same kind not electrified." . [35] Abbé Ménon in France tried the effects of a continued application of electricity upon men and birds and found that the subjects experimented on lost weight, thus apparently showing that electricity quickened the excretions. [36] [37] The efficacy of electric shocks in cases of paralysis was tested in the county hospital at Shrewsbury, England, with rather poor success. [38]

13.3.5   Late 18th century

Benjamin Franklin.

of which there was an iron nail. Touching the iron nail accidentally with his other hand he received a severe electric shock. In much the same way Musschenbroeck assisted by Cunaens received a more severe shock from a somewhat similar glass bottle. Sir William Watson of England greatly improved this device, by covering the bottle, or jar, outside and in with tinfoil. This piece of electrical apparatus will be easily recognized as the well-known Leyden jar, so called by the Abbot Nollet of Paris, after the place of its discovery. [9]

In 1741, John Ellicott "proposed to measure the strength of electrification by its power to raise a weight in one scale of a balance while the other was held over the electrified body and pulled to it by its attractive power" . In 1749, Sir William Watson conducted numerous experiments to ascertain the velocity of electricity in a wire. These experiments, although perhaps not so intended, also demonstrated the possibility of transmitting signals to a distance by electricity. In these experiments, the signal appeared to travel the 12,276-foot length of the insulated wire instantaneously. Le Monnier in France had previously made somewhat similar experiments, sending shocks through an iron wire 1,319 feet long. [9]

About 1750, first experiments in electrotherapeutics were made. Various experimenters made tests to ascertain the physiological and therapeutical effects of electricity. Demainbray in Edinburgh examined the effects of electricity upon plants and concluded that the growth of two

Benjamin Franklin is frequently confused as the key luminary behind electricity; William Watson and Benjamin Franklin share the discovery of electrical potentials . Benjamin Franklin promoted his investigations of electricity and theories through the famous, though extremely dangerous, experiment of having his son fly a kite through a storm-threatened sky. A key attached to the kite string sparked and charged a Leyden jar, thus establishing the link between lightning and electricity. [39] Following these experiments, he invented a lightning rod. It is either Franklin (more frequently) or Ebenezer Kinnersley of Philadelphia (less frequently) who is considered to have established the convention of positive and negative electricity.

Theories regarding the nature of electricity were quite vague at this period, and those prevalent were more or less conflicting. Franklin considered that electricity was an imponderable fluid pervading everything, and which, in its normal condition, was uniformly distributed in all substances. He assumed that the electrical manifestations obtained by rubbing glass were due to the production of an excess of the electric fluid in that substance and that the manifestations produced by rubbing wax were due to a deficit of the fluid. This theory was opposed by Robert Symmer's "Two-fluid" theory in 1759. By Symmer's theory, the vitreous and resinous electricities were regarded as imponderable fluids, each fluid being composed of mutually repellent particles while the particles of the opposite electricities are mutually attractive. When the two fluids unite as a result of their attraction for one another, their effect upon external objects is neutralized. The act of rubbing a body decomposes the fluids, one of which remains in excess on the body and manifests itself as vitreous or resinous electricity.[9]

Up to the time of Franklin's historic kite experiment,[40] the identity of the electricity developed by rubbing and by electrostatic machines (frictional electricity) with lightning had not been generally established. Dr. Wall,[41] Abbot Nollet, Hauksbee,[42] Stephen Gray[43] and John Henry Winkler[44] had indeed suggested the resemblance between the phenomena of "electricity" and "lightning", Gray having intimated that they only differed in degree. It was doubtless Franklin, however, who first proposed tests to determine the sameness of the phenomena. In a letter to Peter Comlinson of London, on 19 October 1752, Franklin, referring to his kite experiment, wrote,

> "At this key the phial (Leyden jar) may be charged; and from the electric fire thus obtained spirits may be kindled, and all the other electric experiments be formed which are usually done by the help of a rubbed glass globe or tube, and thereby the sameness of the electric matter with that of lightning be completely demonstrated." [45]

On 10 May 1742 Thomas-François Dalibard, at Marley (near Paris), using a vertical iron rod 40 feet long, obtained results corresponding to those recorded by Franklin and somewhat prior to the date of Franklin's experiment. Franklin's important demonstration of the sameness of frictional electricity and lightning doubtless added zest to the efforts of the many experimenters in this field in the last half of the 18th century, to advance the progress of the science.[9]

Franklin's observations aided later scientists such as

Michael Faraday, Luigi Galvani, Alessandro Volta, André-Marie Ampère and Georg Simon Ohm, whose collective work provided the basis for modern electrical technology and for whom fundamental units of electrical measurement are named. Others who would advance the field of knowledge included William Watson, Boze, Smeaton, Louis-Guillaume Le Monnier, Jacques de Romas, Jean Jallabert, Giovanni Battista Beccaria, Tiberius Cavallo, John Canton, Robert Symmer, Abbot Nollet, John Henry Winkler, Richman, Dr. Wilson, Kinnersley, Joseph Priestley, Franz Aepinus, Edward Hussey Délavai, Henry Cavendish and Charles-Augustin de Coulomb. Descriptions of many of the experiments and discoveries of these early electrical scientists may be found in the scientific publications of the time, notably the *Philosophical Transactions*, *Philosophical Magazine*, *Cambridge Mathematical Journal*, *Young's Natural Philosophy*, Priestley's *History of Electricity*, Franklin's *Experiments and Observations on Electricity*, Cavalli's *Treatise on Electricity* and De la Rive's *Treatise on Electricity*.[9]

Henry Elles was one of the first people to suggest links between electricity and magnetism. In 1757 he claimed that he had written to the Royal Society in 1755 about the links between electricity and magnetism, asserting that "there are some things in the power of magnetism very similar to those of electricity" but he did "not by any means think them the same". In 1760 he similarly claimed that in 1750 he had been the first "to think how the electric fire may be the cause of thunder".[46] Among the more important of the electrical research and experiments during this period were those of Franz Aepinus, a noted German scholar (1724–1802) and Henry Cavendish of London, England.[9]

Franz Aepinus is credited as the first to conceive of the view of the reciprocal relationship of electricity and magnetism. In his work *Tentamen Theoria Electricitatis et Magnetism*,[47] published in Saint Petersburg in 1759, he gives the following amplification of Franklin's theory, which in some of its features is measurably in accord with present-day views: "The particles of the electric fluid repel each other, attract and are attracted by the particles of all bodies with a force that decreases in proportion as the distance increases; the electric fluid exists in the pores of bodies; it moves unobstructedly through non-electric (conductors), but moves with difficulty in insulators; the manifestations of electricity are due to the unequal distribution of the fluid in a body, or to the approach of bodies unequally charged with the fluid." Aepinus formulated a corresponding theory of magnetism excepting that, in the case of magnetic phenomena, the fluids only acted on the particles of iron. He also made numerous electrical experiments apparently showing that, in order to manifest electrical effects, tourmaline must be heated to between 37.5°C and 100 °C. In fact, tourmaline remains unelectrified when its temperature is uniform, but manifests electrical properties when its temperature is rising or falling. Crystals that manifest electrical properties in this way are termed pyroelectric; along with tourmaline, these include sulphate of quinine and quartz.[9]

Henry Cavendish independently conceived a theory of electricity nearly akin to that of Aepinus.[48] In 1784, he was perhaps the first to utilize an electric spark to produce an explosion of hydrogen and oxygen in the proper proportions that would create pure water. Cavendish also discovered the inductive capacity of dielectrics (insulators), and, as early as 1778, measured the specific inductive capacity for beeswax and other substances by comparison with an air condenser.

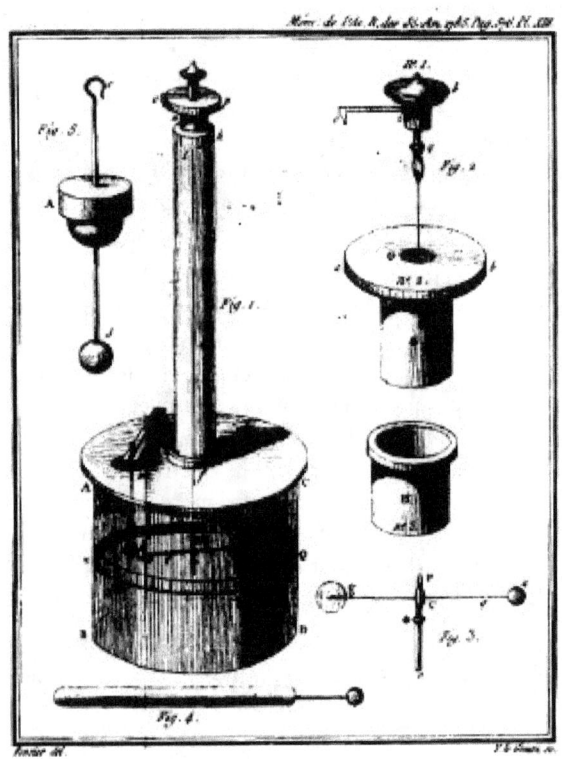

Drawing of Coulomb's torsion balance. From Plate 13 of his 1785 memoir.

Around 1784 C. A. Coulomb devised the torsion balance, discovering what is now known as Coulomb's law: the force exerted between two small electrified bodies varies inversely as the square of the distance, not as Aepinus in his theory of electricity had assumed, merely inversely as the distance. According to the theory advanced by Cavendish, "the particles attract and are attracted inversely as some less power of the distance than the cube."[9] *A large part of the domain of electricity became virtually annexed by Coulomb's discovery of the law of inverse squares.*

Through the experiments of William Watson and others proving that electricity could be transmitted to a distance, the idea of making practical use of this phenomenon began, around 1753, to engross the minds of inquisitive people.

To this end, suggestions as to the employment of electricity in the transmission of intelligence were made. The first of the methods devised for this purpose was probably that of Georges Lesage in 1774.[49][50][51] This method consisted of 24 wires, insulated from one another and each having had a pith ball connected to its distant end. Each wire represented a letter of the alphabet. To send a message, a desired wire was charged momentarily with electricity from an electric machine, whereupon the pith ball connected to that wire would fly out. Other methods of telegraphing in which frictional electricity was employed were also tried, some of which are described in the history on the telegraph.[9]

The era of galvanic or voltaic electricity represented a revolutionary break from the historical focus on frictional electricity. Alessandro Volta discovered that chemical reactions could be used to create positively charged anodes and negatively charged cathodes. When a conductor was attached between these, the difference in the electrical potential (also known as voltage) drove a current between them through the conductor. The potential difference between two points is measured in units of volts in recognition of Volta's work.[9]

The first mention of voltaic electricity, although not recognized as such at the time, was probably made by Johann Georg Sulzer in 1767, who, upon placing a small disc of zinc under his tongue and a small disc of copper over it, observed a peculiar taste when the respective metals touched at their edges. Sulzer assumed that when the metals came together they were set into vibration, acting upon the nerves of the tongue to produce the effects noticed. In 1790, Prof. Luigi Alyisio Galvani of Bologna, while conducting experiments on "animal electricity", noticed the twitching of a frog's legs in the presence of an electric machine. He observed that a frog's muscle, suspended on an iron balustrade by a copper hook passing through its dorsal column, underwent lively convulsions without any extraneous cause, the electric machine being at this time absent.[9]

To account for this phenomenon, Galvani assumed that electricity of opposite kinds existed in the nerves and muscles of the frog, the muscles and nerves constituting the charged coatings of a Leyden jar. Galvani published the results of his discoveries, together with his hypothesis, which engrossed the attention of the physicists of that time. The most prominent of these was Volta, professor of physics at Pavia, who contended that the results observed by Galvani were the result of the two metals, copper and iron, acting as electromotors, and that the muscles of the frog played the part of a conductor, completing the circuit. This precipitated a long discussion between the adherents of the conflicting views. One group agreed with Volta that the electric current was the result of an electromotive force of contact at the two metals; the other adopted a modification of

Galvani's view and asserted that the current was the result of a chemical affinity between the metals and the acids in the pile. Michael Faraday wrote in the preface to his *Experimental Researches*, relative to the question of whether metallic contact is productive of a part of the electricity of the voltaic pile: "I see no reason as yet to alter the opinion I have given; ... but the point itself is of such great importance that I intend at the first opportunity renewing the inquiry, and, if I can, rendering the proofs either on the one side or the other, undeniable to all." *[9]

Even Faraday himself, however, did not settle the controversy, and while the views of the advocates on both sides of the question have undergone modifications, as subsequent investigations and discoveries demanded, up to 1918 diversity of opinion on these points continued to crop out. Volta made numerous experiments in support of his theory and ultimately developed the pile or battery,*[52] which was the precursor of all subsequent chemical batteries, and possessed the distinguishing merit of being the first means by which a prolonged continuous current of electricity was obtainable. Volta communicated a description of his pile to the Royal Society of London and shortly thereafter Nicholson and Cavendish (1780) produced the decomposition of water by means of the electric current, using Volta's pile as the source of electromotive force.*[9]

13.4   19th century

13.4.1   Early 19th century

Alessandro Volta.

In 1800 Alessandro Volta constructed the first device to produce a large electric current, later known as the electric

battery. Napoleon, informed of his works, summoned him in 1801 for a command performance of his experiments. He received many medals and decorations, including the Légion d'honneur.

Davy in 1806, employing a voltaic pile of approximately 250 cells, or couples, decomposed potash and soda, showing that these substances were respectively the oxides of potassium and sodium, which metals previously had been unknown. These experiments were the beginning of electrochemistry, the investigation of which Faraday took up, and concerning which in 1833 he announced his important law of electrochemical equivalents, viz.: "*The same quantity of electricity — that is, the same electric current — decomposes chemically equivalent quantities of all the bodies which it traverses; hence the weights of elements separated in these electrolytes are to each other as their chemical equivalents.*" Employing a battery of 2,000 elements of a voltaic pile Humphry Davy in 1809 gave the first public demonstration of the electric arc light, using for the purpose charcoal enclosed in a vacuum.*[9]

Somewhat important to note, it was not until many years after the discovery of the voltaic pile that the sameness of annual and frictional electricity with voltaic electricity was clearly recognized and demonstrated. Thus as late as January 1833 we find Faraday writing*[53] in a paper on the electricity of the electric ray. "*After an examination of the experiments of Walsh,*[54]*[55] Ingenhousz, Henry Cavendish, Sir H. Davy, and Dr. Davy, no doubt remains on my mind as to the identity of the electricity of the torpedo with common (frictional) and voltaic electricity; and I presume that so little will remain on the mind of others as to justify my refraining from entering at length into the philosophical proof of that identity. The doubts raised by Sir Humphry Davy have been removed by his brother, Dr. Davy; the results of the latter being the reverse of those of the former. ... The general conclusion which must, I think, be drawn from this collection of facts* (a table showing the similarity, of properties of the diversely named electricities) *is, that electricity, whatever may be its source, is identical in its nature.*" *[9]

It is proper to state, however, that prior to Faraday's time the similarity of electricity derived from different sources was more than suspected. Thus, William Hyde Wollaston,*[56] wrote in 1801:*[57] "*This similarity in the means by which both electricity and galvanism (voltaic electricity) appear to be excited in addition to the resemblance that has been traced between their effects shows that they are both essentially the same and confirm an opinion that has already been advanced by others, that all the differences discoverable in the effects of the latter may be owing to its being less intense, but produced in much larger quantity.*" In the same paper Wollaston describes certain experiments in which he uses very fine wire in a solution of sulphate of copper through which he passed electric currents from an electric machine. This is interesting in connection with the later day use of almost similarly arranged fine wires in electrolytic receivers in wireless, or radio-telegraphy.[9]

Hans Christian Ørsted.

In the first half of the 19th century many very important additions were made to the world's knowledge concerning electricity and magnetism. For example, in 1819 Hans Christian Ørsted of Copenhagen discovered the deflecting effect of an electric current traversing a wire upon a suspended magnetic needle.[9]

This discovery gave a clue to the subsequently proved intimate relationship between electricity and magnetism which was promptly followed up by Ampère who shortly thereafter (1821) announced his celebrated theory of electrodynamics, relating to the force that one current exerts upon another, by its electro-magnetic effects, namely[9]

1. Two parallel portions of a circuit attract one another if the currents in them are flowing in the same direction, and repel one another if the currents flow in the opposite direction.

2. Two portions of circuits crossing one another obliquely attract one another if both the currents flow either towards or from the point of crossing, and repel one another if one flows to and the other from that point.

3. When an element of a circuit exerts a force on another element of a circuit, that force always tends to urge the second one in a direction at right angles to its own direction.

Ampere brought a multitude of phenomena into theory by his investigations of the mechanical forces between conductors supporting currents and magnets.

The German physicist Seebeck discovered in 1821 that when heat is applied to the junction of two metals that had been soldered together an electric current is set up. This is termed Thermo-Electricity. Seebeck's device consists of a strip of copper bent at each end and soldered to a plate of bismuth. A magnetic needle is placed parallel with the copper strip. When the heat of a lamp is applied to the junction of the copper and bismuth an electric current is set up which deflects the needle.[9]

Around this time, Siméon Denis Poisson attacked the difficult problem of induced magnetization, and his results, though differently expressed, are still the theory, as a most important first approximation. It was in the application of mathematics to physics that his services to science were performed. Perhaps the most original, and certainly the most permanent in their influence, were his memoirs on the theory of electricity and magnetism, which virtually created a new branch of mathematical physics.

George Green wrote *An Essay on the Application of Mathematical Analysis to the Theories of Electricity and Magnetism* in 1828. The essay introduced several important concepts, among them a theorem similar to the modern Green's theorem, the idea of potential functions as currently used in physics, and the concept of what are now called Green's functions. George Green was the first person to create a mathematical theory of electricity and magnetism and his theory formed the foundation for the work of other scientists such as James Clerk Maxwell, William Thomson, and others.

Peltier in 1834 discovered an effect opposite to Thermo-Electricity, namely, that when a current is passed through a couple of dissimilar metals the temperature is lowered or raised at the junction of the metals, depending on the direction of the current. This is termed the Peltier "effect". The variations of temperature are found to be proportional to the strength of the current and not to the square of the strength of the current as in the case of heat due to the ordinary resistance of a conductor. This second law is the C^2R law,[58] discovered experimentally in 1841 by the English physicist Joule. In other words, this important law is that the heat generated in any part of an electric circuit is directly proportional to the product of the resistance of this part of the circuit and to the square of the strength of current flowing in the circuit.[9]

In 1822 Johann Schweigger devised the first galvanometer. This instrument was subsequently much improved by Wilhelm Weber (1833). In 1825 William Sturgeon of Woolwich, England, invented the horseshoe and straight bar electromagnet, receiving therefor the silver medal of the Society of Arts.[59] In 1837 Carl Friedrich Gauss and Weber (both noted workers of this period) jointly invented a reflecting galvanometer for telegraph purposes. This was the forerunner of the Thomson reflecting and other exceedingly sensitive galvanometers once used in submarine signaling and still widely employed in electrical measurements. Arago in 1824 made the important discovery that when a copper disc is rotated in its own plane, and if a magnetic needle be freely suspended on a pivot over the disc, the needle will rotate with the disc. If on the other hand the needle is fixed it will tend to retard the motion of the disc. This effect was termed Arago's rotations.[9][60][61]

Georg Simon Ohm.

Futile attempts were made by Charles Babbage, Peter Barlow, John Herschel and others to explain this phenomenon. The true explanation was reserved for Faraday, namely, that electric currents are induced in the copper disc by the cutting of the magnetic lines of force of the needle, which currents in turn react on the needle. Georg Simon Ohm did his work on resistance in the years 1825 and 1826, and published his results in 1827 as the book *Die galvanische Kette, mathematisch bearbeitet.*[62][63] He drew considerable inspiration from Fourier's work on heat conduction in the

theoretical explanation of his work. For experiments, he initially used voltaic piles, but later used a thermocouple as this provided a more stable voltage source in terms of internal resistance and constant potential difference. He used a galvanometer to measure current, and knew that the voltage between the thermocouple terminals was proportional to the junction temperature. He then added test wires of varying length, diameter, and material to complete the circuit. He found that his data could be modeled through a simple equation with variable composed of the reading from a galvanometer, the length of the test conductor, thermocouple junction temperature, and a constant of the entire setup. From this, Ohm determined his law of proportionality and published his results. In 1827, he announced the now famous law that bears his name, that is:

$$\text{Electromotive force} = \text{Current} \times \text{Resistance}\,[64]$$

Ohm brought into order a host of puzzling facts connecting electromotive force and electric current in conductors, which all previous electricians had only succeeded in loosely binding together qualitatively under some rather vague statements. Ohm found that the results could be summed up in such a simple law and by Ohm's discovery a large part of the domain of electricity became annexed to theory.

13.4.2   Faraday and Henry

The discovery of electromagnetic induction was made almost simultaneously, although independently, by Michael Faraday, who was first to make the discovery in 1831, and Joseph Henry in 1832.[65][66] Henry's discovery of self-induction and his work on spiral conductors using a copper coil were made public in 1835, just before those of Faraday.[67][68][69]

In 1831 began the epoch-making researches of Michael Faraday, the famous pupil and successor of Humphry Davy at the head of the Royal Institution, London, relating to electric and electromagnetic induction. The remarkable researches of Faraday, the *prince of experimentalists*, on electrostatics and electrodynamics and the induction of currents. These were rather long in being brought from the crude experimental state to a compact system, expressing the real essence. Faraday was not a competent mathematician.[70][71][72] but had he been one, he would have been greatly assisted in his researches, have saved himself much useless speculation, and would have anticipated much later work. He would, for instance, knowing Ampere's theory, by his own results have readily been led to Neumann's theory, and the connected work of Helmholtz and Thomson. Faraday's studies and researches extended from 1831

Joseph Henry.

Michael Faraday.

to 1855 and a detailed description of his experiments, deductions and speculations are to be found in his compiled papers, entitled Experimental Researches in Electricity.' Faraday was by profession a chemist. He was not in the remotest degree a mathematician in the ordinary sense — indeed it is a question if in all his writings there is a single mathematical formula.' [9]

The experiment which led Faraday to the discovery of electromagnetic induction was made as follows: He constructed what is now and was then termed an induction coil, the primary and secondary wires of which were wound on a wooden bobbin, side by side, and insulated from one another. In the circuit of the primary wire he placed a battery of approximately 100 cells. In the secondary wire he inserted a galvanometer. On making his first test he observed no results, the galvanometer remaining quiescent, but on increasing the length of the wires he noticed a deflection of the galvanometer in the secondary wire when the circuit of the primary wire was made and broken. This was the first observed instance of the development of electromotive force by electromagnetic induction.' [9]

He also discovered that induced currents are established in a second closed circuit when the current strength is varied in the first wire, and that the direction of the current in the secondary circuit is opposite to that in the first circuit. Also that a current is induced in a secondary circuit

when another circuit carrying a current is moved to and from the first circuit, and that the approach or withdrawal of a magnet to or from a closed circuit induces momentary currents in the latter. In short, within the space of a few months Faraday discovered by experiment virtually all the laws and facts now known concerning electro-magnetic induction and magneto-electric induction. Upon these discoveries, with scarcely an exception, depends the operation of the telephone, the dynamo machine, and incidental to the dynamo electric machine practically all the gigantic electrical industries of the world, including electric lighting, electric traction, the operation of electric motors for power purposes, and electro-plating, electrotyping, etc.' [9]

In his investigations of the peculiar manner in which iron filings arrange themselves on a cardboard or glass in proximity to the poles of a magnet, Faraday conceived the idea of magnetic "lines of force" extending from pole to pole of the magnet and along which the filings tend to place themselves. On the discovery being made that magnetic effects accompany the passage of an electric current in a wire, it was also assumed that similar magnetic lines of force whirled around the wire. For convenience and to account for induced electricity it was then assumed that when these lines of force are "cut" by a wire in passing across them or when the lines of force in rising and falling cut the wire, a current of electricity is developed, or to be more exact, an electromotive force

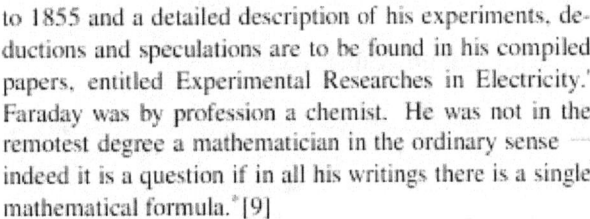

is developed in the wire that sets up a current in a closed circuit. Faraday advanced what has been termed the *molecular theory of electricity*[73] which assumes that electricity is the manifestation of a peculiar condition of the molecule of the body rubbed or the ether surrounding the body. Faraday also, by experiment, discovered paramagnetism and diamagnetism, namely, that all solids and liquids are either attracted or repelled by a magnet. For example, iron, nickel, cobalt, manganese, chromium, etc., are paramagnetic (attracted by magnetism), whilst other substances, such as bismuth, phosphorus, antimony, zinc, etc., are repelled by magnetism or are diamagnetic.[9][74]

Brugans of Leyden in 1778 and Le Baillif and Becquerel in 1827[75] had previously discovered diamagnetism in the case of bismuth and antimony. Faraday also rediscovered specific inductive capacity in 1837, the results of the experiments by Cavendish not having been published at that time. He also predicted[76] the retardation of signals on long submarine cables due to the inductive effect of the insulation of the cable, in other words, the static capacity of the cable.[9]

The 25 years immediately following Faraday's discoveries of electromagnetic induction were fruitful in the promulgation of laws and facts relating to induced currents and to magnetism. In 1834 Heinrich Lenz and Moritz von Jacobi independently demonstrated the now familiar fact that the currents induced in a coil are proportional to the number of turns in the coil. Lenz also announced at that time his important law that, in all cases of electromagnetic induction the induced currents have such a direction that their reaction tends to stop the motion that produces them, a law that was perhaps deducible from Faraday's explanation of Arago's rotations.[9][77]

The induction coil was first designed by Nicholas Callan in 1836. In 1845 Joseph Henry, the American physicist, published an account of his valuable and interesting experiments with induced currents of a high order, showing that currents could be induced from the secondary of an induction coil to the primary of a second coil, thence to its secondary wire, and so on to the primary of a third coil, etc.[78] Heinrich Daniel Ruhmkorff further developed the induction coil, the Ruhmkorff coil was patented in 1851,[79] and he utilized long windings of copper wire to achieve a spark of approximately 2 inches (50 mm) in length. In 1857, after examining a greatly improved version made by an American inventor, Edward Samuel Ritchie,[80][81] Ruhmkorff improved his design (as did other engineers), using glass insulation and other innovations to allow the production of sparks more than 300 millimetres (12 in) long.[82]

13.4.3   Middle 19th century

Up to the middle of the 19th century, indeed up to about 1870, electrical science was, it may be said, a sealed book to the majority of electrical workers. Prior to this time a number of handbooks had been published on electricity and magnetism, notably Auguste de La Rive's exhaustive '*Treatise on Electricity*.'[84] in 1851 (French) and 1853 (English); August Beer's *Einleitung in die Elektrostatik, die Lehre vom Magnetismus und die Elektrodynamik,*[85] Wiedemann's '*Galvanismus,*' and Reiss'[86] '*Reibungsalelektricität.*' But these works consisted in the main in details of experiments with electricity and magnetism, and but little with the laws and facts of those phenomena. Henry d'Abria[87][88] published the results of some researches into the laws of induced currents, but owing to their complexity of the investigation it was not productive of very notable results.[89] Around the mid-19th century, Fleeming Jenkin's work on '*Electricity and Magnetism*[90]' and Clerk Maxwell's '*Treatise on Electricity and Magnetism*' were published.[9]

These books were departures from the beaten path. As Jenkin states in the preface to his work the science of the schools was so dissimilar from that of the practical electrician that it was quite impossible to give students sufficient, or even approximately sufficient, textbooks. A student he said might have mastered de la Rive's large and valuable treatise and yet feel as if in an unknown country and listening to an unknown tongue in the company of practical men. As another writer has said, with the coming of Jenkin's and Maxwell's books all impediments in the way of electrical students were removed, "*the full meaning of Ohm's law becomes clear; electromotive force, difference of potential, resistance, current, capacity, lines of force, magnetization and chemical affinity were measurable, and could be reasoned about, and calculations could be made about them with as much certainty as calculations in dynamics*".[9][91]

About 1850, Kirchhoff published his laws relating to branched or divided circuits. He also showed mathematically that according to the then prevailing electrodynamic theory, electricity would be propagated along a perfectly conducting wire with the velocity of light. Helmholtz investigated mathematically the effects of induction upon the strength of a current and deduced therefrom equations, which experiment confirmed, showing amongst other important points the retarding effect of self-induction under certain conditions of the circuit.[9][92]

In 1853, Sir William Thomson (later Lord Kelvin) predicted as a result of mathematical calculations the oscillatory nature of the electric discharge of a condenser circuit. To Henry, however, belongs the credit of discerning as a result of his experiments in 1842 the oscillatory nature of the Leyden jar discharge. He wrote:[93] *The phenomena*

Sir William Thomson.

require us to admit the existence of a principal discharge in one direction, and then several reflex actions backward and forward, each more feeble than the preceding, until the equilibrium is obtained. These oscillations were subsequently observed by B. W. Feddersen (1857)[94][95] who using a rotating concave mirror projected an image of the electric spark upon a sensitive plate, thereby obtaining a photograph of the spark which plainly indicated the alternating nature of the discharge. Sir William Thomson was also the discoverer of the electric convection of heat (the "Thomson" effect). He designed for electrical measurements of precision his quadrant and absolute electrometers. The reflecting galvanometer and siphon recorder, as applied to submarine cable signaling, are also due to him.[9]

About 1876 the American physicist Henry Augustus Rowland of Baltimore demonstrated the important fact that a static charge carried around produces the same magnetic effects as an electric current.[96][97] The Importance of this discovery consists in that it may afford a plausible theory of magnetism, namely, that magnetism may be the result of directed motion of rows of molecules carrying static charges.[9]

After Faraday's discovery that electric currents could be developed in a wire by causing it to cut across the lines of force of a magnet, it was to be expected that attempts would be made to construct machines to avail of this fact in the development of voltaic currents.[98] The first machine of this

kind was due to Hippolyte Pixii, 1832. It consisted of two bobbins of iron wire, opposite which the poles of a horseshoe magnet were caused to rotate. As this produced in the coils of the wire an alternating current, Pixii arranged a commutating device (commutator) that converted the alternating current of the coils or armature into a direct current in the external circuit. This machine was followed by improved forms of magneto-electric machines due to Ritchie, Saxton, Clarke 1834, Stohrer 1843, Nollet 1849, Shepperd 1856, Van Maldern, Siemens, Wilde and others.[9]

A notable advance in the art of dynamo construction was made by Mr. S. A. Varley in 1866[99] and by Dr. Charles William Siemens and Mr. Charles Wheatstone,[100] who independently discovered that when a coil of wire, or armature, of the dynamo machine is rotated between the poles (or in the "field") of an electromagnet, a weak current is set up in the coil due to residual magnetism in the iron of the electromagnet, and that if the circuit of the armature be connected with the circuit of the electromagnet, the weak current developed in the armature increases the magnetism in the field. This further increases the magnetic lines of force in which the armature rotates, which still further increases the current in the electromagnet, thereby producing a corresponding increase in the field magnetism, and so on, until the maximum electromotive force which the machine is capable of developing is reached. By means of this principle the dynamo machine develops its own magnetic field, thereby much increasing its efficiency and economical operation. Not by any means, however, was the dynamo electric machine perfected at the time mentioned.[9]

In 1860 an important improvement had been made by Dr. Antonio Pacinotti of Pisa who devised the first electric machine with a ring armature. This machine was first used as an electric motor, but afterward as a generator of electricity. The discovery of the principle of the reversibility of the dynamo electric machine (variously attributed to Walenn 1860; Pacinotti 1864 ; Fontaine, Gramme 1873; Deprez 1881, and others) whereby it may be used as an electric motor or as a generator of electricity has been termed one of the greatest discoveries of the 19th century.[9]

In 1872 the drum armature was devised by Hefner-Alteneck. This machine in a modified form was subsequently known as the Siemens dynamo. These machines were presently followed by the Schuckert, Gulcher,[101] Fein,[102][103] Brush, Hochhausen, Edison and the dynamo machines of numerous other inventors. In the early days of dynamo machine construction the machines were mainly arranged as direct current generators, and perhaps the most important application of such machines at that time was in electro-plating, for which purpose machines of low voltage and large current strength were employed.[9][104]

Beginning about 1887 alternating current generators came into extensive operation and the commercial development of the transformer, by means of which currents of low voltage and high current strength are transformed to currents of high voltage and low current strength, and vice versa, in time revolutionized the transmission of electric power to long distances. Likewise the introduction of the rotary converter (in connection with the "step-down" transformer) which converts alternating currents into direct currents (and vice versa) has effected large economies in the operation of electric power systems.[9][105]

Before the introduction of dynamo electric machines, voltaic, or primary, batteries were extensively used for electro-plating and in telegraphy. There are two distinct types of voltaic cells, namely, the "open" and the "closed", or "constant", type. The open type in brief is that type which operated on closed circuit becomes, after a short time, polarized; that is, gases are liberated in the cell which settle on the negative plate and establish a resistance that reduces the current strength. After a brief interval of open circuit these gases are eliminated or absorbed and the cell is again ready for operation. Closed circuit cells are those in which the gases in the cells are absorbed as quickly as liberated and hence the output of the cell is practically uniform. The Leclanché and Daniell cells, respectively, are familiar examples of the "open" and "closed" type of voltaic cell. The "open" cells are used very extensively at present, especially in the dry cell form, and in annunciator and other open circuit signal systems. Batteries of the Daniell or "gravity" type were employed almost generally in the United States and Canada as the source of electromotive force in telegraphy before the dynamo machine became available, and still are largely used for this service or as "local" cells. Batteries of the "gravity" and the Edison-Lalande types are still much used in "closed circuit" systems.[9]

In the late 19th century, the term luminiferous aether, meaning light-bearing aether, was a conjectured medium for the propagation of light.[106] The word *aether* stems via Latin from the Greek αἰθήρ, from a root meaning to kindle, burn, or shine. It signifies the substance which was thought in ancient times to fill the upper regions of space, beyond the clouds.

13.4.4 Maxwell

In 1864 James Clerk Maxwell of Edinburgh announced his electromagnetic theory of light, which was perhaps the greatest single step in the world's knowledge of electricity.[107] Maxwell had studied and commented on the field of electricity and magnetism as early as 1855/6 when *On Faraday's lines of force*[108] was read to the Cambridge Philosophical Society. The paper presented a simplified

James Clerk Maxwell.

model of Faraday's work, and how the two phenomena were related. He reduced all of the current knowledge into a linked set of differential equations with 20 equations in 20 variables. This work was later published as *On Physical Lines of Force* in March 1861.[109] In order to determine the force which is acting on any part of the machine we must find its momentum, and then calculate the rate at which this momentum is being changed. This rate of change will give us the force. The method of calculation which it is necessary to employ was first given by Lagrange, and afterwards developed, with some modifications, by Hamilton's equations. It is usually referred to as Hamilton's principle; when the equations in the original form are used they are known as Lagrange's equations. Now Maxwell logically showed how these methods of calculation could be applied to the electromagnetic field.[110] The energy of a dynamical system is partly kinetic, partly potential. Maxwell supposes that the magnetic energy of the field is kinetic energy, the electric energy potential.[111]

Around 1862, while lecturing at King's College, Maxwell calculated that the speed of propagation of an electromagnetic field is approximately that of the speed of light. He considered this to be more than just a coincidence, and commented "*We can scarcely avoid the conclusion that light consists in the transverse undulations of the same medium which is the cause of electric and magnetic phenomena.*"[112]

Working on the problem further, Maxwell showed that the

equations predict the existence of waves of oscillating electric and magnetic fields that travel through empty space at a speed that could be predicted from simple electrical experiments; using the data available at the time, Maxwell obtained a velocity of 310,740,000 m/s. In his 1864 paper *A Dynamical Theory of the Electromagnetic Field*, Maxwell wrote, *The agreement of the results seems to show that light and magnetism are affections of the same substance, and that light is an electromagnetic disturbance propagated through the field according to electromagnetic laws.*[113]

As already noted herein Faraday, and before him, Ampère and others, had inklings that the luminiferous ether of space was also the medium for electric action. It was known by calculation and experiment that the velocity of electricity was approximately 186,000 miles per second; that is, equal to the velocity of light, which in itself suggests the idea of a relationship between -electricity and "light." A number of the earlier philosophers or mathematicians, as Maxwell terms them, of the 19th century, held the view that electromagnetic phenomena were explainable by action at a distance. Maxwell, following Faraday, contended that the seat of the phenomena was in the medium. The methods of the mathematicians in arriving at their results were synthetical while Faraday's methods were analytical. Faraday in his mind's eye saw lines of force traversing all space where the mathematicians saw centres of force attracting at a distance. Faraday sought the seat of the phenomena in real actions going on in the medium; they were satisfied that they had found it in a power of action at a distance on the electric fluids.[114]

Both of these methods, as Maxwell points out, had succeeded in explaining the propagation of light as an electromagnetic phenomenon while at the same time the fundamental conceptions of what the quantities concerned are, radically differed. The mathematicians assumed that insulators were barriers to electric currents; that, for instance, in a Leyden jar or electric condenser the electricity was accumulated at one plate and that by some occult action at a distance electricity of an opposite kind was attracted to the other plate.

Maxwell, looking further than Faraday, reasoned that if light is an electromagnetic phenomenon and is transmissible through dielectrics such as glass, the phenomenon must be in the nature of electromagnetic currents in the dielectrics. He therefore contended that in the charging of a condenser, for instance, the action did not stop at the insulator, but that some "displacement" currents are set up in the insulating medium, which currents continue until the resisting force of the medium equals that of the charging force. In a closed conductor circuit, an electric current is also a displacement of electricity.

The conductor offers a certain resistance, akin to friction,

to the displacement of electricity, and heat is developed in the conductor, proportional to the square of the current(as already stated herein), which current flows as long as the impelling electric force continues. This resistance may be likened to that met with by a ship as it displaces in the water in its progress. The resistance of the dielectric is of a different nature and has been compared to the compression of multitudes of springs, which, under compression, yield with an increasing back pressure, up to a point where the total back pressure equals the initial pressure. When the initial pressure is withdrawn the energy expended in compressing the "springs" is returned to the circuit, concurrently with the return of the springs to their original condition, this producing a reaction in the opposite direction. Consequently the current due to the displacement of electricity in a conductor may be continuous, while the displacement currents in a dielectric are momentary and, in a circuit or medium which contains but little resistance compared with capacity or inductance reaction, the currents of discharge are of an oscillatory or alternating nature.[115]

Maxwell extended this view of displacement currents in dielectrics to the ether of free space. Assuming light to be the manifestation of alterations of electric currents in the ether, and vibrating at the rate of light vibrations, these vibrations by induction set up corresponding vibrations in adjoining portions of the ether, and in this way the undulations corresponding to those of light are propagated as an electromagnetic effect in the ether. Maxwell's electromagnetic theory of light obviously involved the existence of electric waves in free space, and his followers set themselves the task of experimentally demonstrating the truth of the theory. By 1871, he presented the *Remarks on the mathematical classification of physical quantities.*[116]

13.4.5 End of the 19th century

In 1887, the German physicist Heinrich Hertz in a series of experiments proved the actual existence electromagnetic waves, showing that transverse free space electromagnetic waves can travel over some distance as predicted by Maxwell and Faraday. Hertz published his work in a book titled: *Electric waves: being researches on the propagation of electric action with finite velocity through space.*[117] The discovery of electromagnetic waves in space led to the development in the closing years of the 19th century of radio.

The electron as a unit of charge in electrochemistry was posited by G. Johnstone Stoney in 1874, who also coined the term *electron* in 1894. Plasma was first identified in a Crookes tube, and so described by Sir William Crookes in 1879 (he called it "radiant matter").[118] The place of electricity in leading up to the discovery of those beauti-

Heinrich Hertz.

Oliver Heaviside.

ful phenomena of the Crookes Tube (due to Sir William Crookes), viz., Cathode rays.[119] and later to the discovery of Roentgen or X-rays, must not be overlooked, since without electricity as the excitant of the tube the discovery of the rays might have been postponed indefinitely. It has been noted herein that Dr. William Gilbert was termed the founder of electrical science. This must, however, be regarded as a comparative statement.[9]

Oliver Heaviside was a self-taught scholar who reformulated Maxwell's field equations in terms of electric and magnetic forces and energy flux, and independently co-formulated vector analysis. His series of articles continued the work entitled "*Electromagnetic Induction and its Propagation*", commenced in The Electrician in 1885 to nearly 1887 (ed., the latter part of the work dealing with the propagation of electromagnetic waves along wires through the dielectric surrounding them), when the great pressure on space and the want of readers appeared to necessitate its abrupt discontinuance.[120] (A straggler piece appeared December 31, 1887.) He wrote an interpretation of the transcendental formulae of electromagnetism. Following the real object of true naturalists[121] when they employ mathematics to assist them, he wrote to find out the connections of known phenomena, and by deductive reasoning, to obtain a knowledge of electromagnetic phenomena. Although at odds with the scientific establishment for most of his life, Heaviside changed the face of mathematics and science for years to come.

Of the changes in the field of electromagnetic theory, certain conclusions from *Electro-Magnetic Theory*[122] by Heaviside are, if not drawn, at least indicated in this book. Two of them may be stated as follows:

1. That magnetism is a phenomenon of motion and not a statical phenomenon; also that this motion is more likely to be translational than vortical.

2. That all electric currents are phenomena consequent upon the emission of electro-magnetic wave disturbances in the aether, and that the proper treatment of all the phenomena of currents and magnetic flux should be considered as the consequence, and not as the cause, of electro-magnetic waves.

The ultimate results of his work are twofold. (1) The first ultimate result is purely mathematical, which is important only to those who study mathematical physics. The system of *vectorial algebra*[123] as developed by Mr. Heaviside was used because of ease for physical investigations to the methods of quaternions. (2) The second ultimate result is physical. It consists in more closely uniting the more recondite problems of telegraphy, telephony, *Teslaic phenomena* and *Hertzian phenomena* with the fundamental properties of the aether. In elucidating this connection, the merit of the book appears most prominently as a stepping-stone to the goal in the full view of all physical analysis, namely, the resolution of all physical phenomena to the activities of the aether, and of matter in the aether, under the laws of dynamics.[124]

During the late 1890s a number of physicists proposed that electricity, as observed in studies of electrical conduction in conductors, electrolytes, and cathode ray tubes, consisted of discrete units, which were given a variety of names, but the reality of these units had not been confirmed in a compelling way. However, there were also indications that the cathode rays had wavelike properties.[9]

Faraday, Weber, Helmholtz, Clifford and others had glimpses of this view; and the experimental works of Zeeman, Goldstein, Crookes, J. J. Thomson and others had greatly strengthened this view. Weber predicted that electrical phenomena were due to the existence of electrical atoms, the influence of which on one another depended on their position and relative accelerations and velocities. Helmholtz and others also contended that the existence of electrical atoms followed from Faraday's laws of electrolysis, and Johnstone Stoney, to whom is due the term "electron", showed that each chemical ion of the decomposed electrolyte carries a definite and constant quantity of electricity, and inasmuch as these charged ions are separated on the electrodes as neutral substances there must be an instant, however brief, when the charges must be capable of existing separately as electrical atoms; while in 1887, Clifford wrote: "There is great reason to believe that every material atom carries upon it a small electric current, if it does not wholly consist of this current." [9]

The Serbian American engineer Nikola Tesla learned of Hertz' experiments at the Exposition Universelle in 1889 and launched into his own experiments in high frequency and high potential current developing "high-frequency" alternators (which operated around 15,000 hertz).[125]. He concluded from his observations that Maxwell and Hertz were wrong about the existence of airborne electromagnetic waves (which he attributed it to what he called "electrostatic thrusts")[126] but saw great potential in Maxwell's idea that that electricity and light were part of the same phenomena, seeing it as a way to create a new type of wireless electric lighting.[127] By 1893 he was giving lectures on "On Light and Other High Frequency Phenomena", including a demonstration where he would light a Geissler tubes wirelessly. Tesla worked for many years after that trying to develop a wireless power distribution system.[128]

In 1896, J.J. Thomson performed experiments indicating that cathode rays really were particles, found an accurate value for their charge-to-mass ratio e/m, and found that e/m was independent of cathode material. He made good estimates of both the charge e and the mass m, finding that cathode ray particles, which he called "corpuscles", had

Nikola Tesla, c. 1896.

J.J. Thomson.

perhaps one thousandth of the mass of the least massive ion known (hydrogen). He further showed that the negatively charged particles produced by radioactive materials, by heated materials, and by illuminated materials, were universal. The nature of the Crookes tube "cathode ray" matter was identified by Thomson in 1897.[129]

In the late 19th century, the Michelson–Morley experiment was performed by Albert A. Michelson and Edward W. Morley at what is now Case Western Reserve University. It is generally considered to be the evidence against the theory of a luminiferous aether. The experiment has also been referred to as "the kicking-off point for the theoretical aspects of the Second Scientific Revolution." [130] Primarily for this work, Michelson was awarded the Nobel Prize in 1907. Dayton Miller continued with experiments, conducting thousands of measurements and eventually developing the most accurate interferometer in the world at that time. Miller and others, such as Morley, continue observations and experiments dealing with the concepts.[131] A range of proposed aether-dragging theories could explain the null result but these were more complex, and tended to use arbitrary-looking coefficients and physical assumptions.[9]

By the end of the 19th century electrical engineers had become a distinct profession, separate from physicists and inventors. They created companies that investigated, developed and perfected the techniques of electricity transmission, and gained support from governments all over the world for starting the first worldwide electrical telecommunication network, the telegraph network. Pioneers in this field included Werner von Siemens, founder of Siemens AG in 1847, and John Pender, founder of Cable & Wireless.

The first public demonstration of a "alternator system" took place in 1886. Large two-phase alternating current generators were built by a British electrician, J.E.H. Gordon,[132] in 1882. Lord Kelvin and Sebastian Ferranti also developed early alternators, producing frequencies between 100 and 300 hertz. After 1891, polyphase alternators were introduced to supply currents of multiple differing phases.[133] Later alternators were designed for varying alternating-current frequencies between sixteen and about one hundred hertz, for use with arc lighting, incandescent lighting and electric motors.[134]

The possibility of obtaining the electric current in large quantities, and economically, by means of dynamo electric machines gave impetus to the development of incandescent and arc lighting. Until these machines had attained a commercial basis voltaic batteries were the only available source of current for electric lighting and power. The cost of these batteries, however, and the difficulties of maintaining them in reliable operation were prohibitory of their use for practical lighting purposes. The date of the employment of arc

and incandescent lamps may be set at about 1877.[9]

Even in 1880, however, but little headway had been made toward the general use of these illuminants; the rapid subsequent growth of this industry is a matter of general knowledge.[135] The employment of storage batteries, which were originally termed secondary batteries or accumulators, began about 1879. Such batteries are now utilized on a large scale as auxiliaries to the dynamo machine in electric power-houses and substations, in electric automobiles and in immense numbers in automobile ignition and starting systems, also in fire alarm telegraphy and other signal systems.[9]

World's Fair Tesla presentation.

In 1893, the World's Columbian International Exposition was held in a building which was devoted to electrical exhibits. General Electric Company (backed by Edison and J.P. Morgan) had proposed to power the electric exhibits with direct current at the cost of one million dollars. However, Westinghouse proposed to illuminate the Columbian Exposition in Chicago with alternating current for half that price, and Westinghouse won the bid. It was an historical moment and the beginning of a revolution, as George Westinghouse introduced the public to electrical power by illuminating the Exposition.

13.4.6 Second Industrial Revolution

Main article: Second Industrial Revolution
 Between 1885 and 1890 Galileo Ferraris in Italy, Nikola Tesla in the United States, and Mikhail Dolivo-Dobrovolsky in Germany explored poly-phase currents combined with electromagnetic induction leading to the development of practical AC induction motors.[136] The AC induction motor helped usher in the Second Industrial Revolution. The rapid advance of electrical technology in the latter 19th and early 20th centuries led to commercial rivalries. In the War of Currents in the late 1880s, George Westinghouse and Thomas Edison became adversaries due to Edison's promotion of direct current (DC) for electric power distribution over alternating current (AC) advocated by Westinghouse.

Several inventors helped develop commercial systems.

Thomas Edison.

truthfully written: "The most important and remarkable of the uses which have been made of electricity consists in its application to telegraph purposes" .[139] The statement was, however, quite accurate and perhaps the time could have been carried forward to the year 1876 without material modification of the remarks. In that year the telephone, due to Alexander Graham Bell, was invented, but it was not until several years thereafter that its commercial employment began in earnest. Since that time also the sister branches of electricity just mentioned have advanced and are advancing with such gigantic strides in every direction that it is difficult to place a limit upon their progress. Electrical devices account of the use of electricity in the arts and industries.[9]

Charles Proteus Steinmetz, theoretician of alternating current.

Samuel Morse, inventor of a long-range telegraph; Thomas Edison, inventor of the first commercial electrical energy distribution network; George Westinghouse, inventor of the electric locomotive; Alexander Graham Bell, the inventor of the telephone and founder of a successful telephone business.

In 1871 the electric telegraph had grown to large proportions and was in use in every civilized country in the world, its lines forming a network in all directions over the surface of the land. The system most generally in use was the electromagnetic telegraph due to S. F. B. Morse of New York, or modifications of his system.[137] Submarine cables[138] connecting the Eastern and Western hemispheres were also in successful operation at that time.[9]

When, however, in 1918 one views the vast applications of electricity to electric light, electric railways, electric power and other purposes (all it may be repeated made possible and practicable by the perfection of the dynamo machine), it is difficult to believe that no longer ago than 1871 the author of a book published in that year, in referring to the state of the art of applied electricity at that time, could have

AC replaced DC for central station power generation and power distribution, enormously extending the range and improving the safety and efficiency of power distribution. Edison's low-voltage distribution system using DC ultimately lost to AC devices proposed by others: Westinghouse' AC system, Tesla's AC inventions, and the theoretical work of Charles Proteus Steinmetz. The successful Niagara Falls system was a turning point in the acceptance of alternating current. Eventually, the General Electric company (formed by a merger between Edison's companies and the AC-based rival Thomson-Houston) began manufacture of AC machines. Centralized power generation became possible when it was recognized that alternating current electric power lines can transport electricity at low costs across great distances by taking advantage of the ability to change voltage across the distribution path using power transformers. The voltage is raised at the point of generation (a representative number is a generator voltage in the low kilovolt range) to a much higher voltage (tens of thousands to several hundred thousand volts) for primary transmission, followed to several downward transformations, to as low as that used in residential domestic use.[9]

The International Electro-Technical Exhibition of 1891 featuring the long distance transmission of high-power, three-phase electric current. It was held between 16 May and 19 October on the disused site of the three former "Westbahnhöfe" (Western Railway Stations) in Frankfurt am Main. The exhibition featured the first long distance transmission of high-power, three-phase electric current, which was generated 175 km away at Lauffen am Neckar. As a result of this successful field trial, three-phase current became established for electrical transmission networks throughout the world.[9]

Much was done in the direction in the improvement of railroad terminal facilities, and it is difficult to find one steam railroad engineer who would have denied that all the important steam railroads of this country were not to be operated electrically. In other directions the progress of events as to the utilization of electric power was expected to be equally rapid. In every part of the world the power of falling water, nature's perpetual motion machine, which has been going to waste since the world began, is now being converted into electricity and transmitted by wire hundreds of miles to points where it is usefully and economically employed.[9][140]

The first windmill for electricity production was built in Scotland in July 1887 by the Scottish electrical engineer James Blyth.[141] Across the Atlantic, in Cleveland, Ohio a larger and heavily engineered machine was designed and constructed in 1887–88 by Charles F. Brush,[142] this was built by his engineering company at his home and operated from 1886 until 1900.[143] The Brush wind turbine had a rotor 56 feet (17 m) in diameter and was mounted on a 60-foot (18 m) tower. Although large by today's standards, the machine was only rated at 12 kW; it turned relatively slowly since it had 144 blades. The connected dynamo was used either to charge a bank of batteries or to operate up to 100 incandescent light bulbs, three arc lamps, and various motors in Brush's laboratory. The machine fell into disuse after 1900 when electricity became available from Cleveland's central stations, and was abandoned in 1908.[144]

13.5   20th century

Various units of electricity and magnetism have been adopted and named by representatives of the electrical engineering institutes of the world, which units and names have been confirmed and legalized by the governments of the United States and other countries. Thus the volt, from the Italian Volta, has been adopted as the practical unit of electromotive force, the ohm, from the enunciator of Ohm's law, as the practical unit of resistance; the ampere, after the eminent French scientist of that name, as the practical unit of current strength, the henry as the practical unit of inductance, after Joseph Henry and in recognition of his early and important experimental work in mutual induction.[145]

Dewar and John Ambrose Fleming predicted that at absolute zero, pure metals would become perfect electromagnetic conductors (though, later, Dewar altered his opinion on the disappearance of resistance believing that there would always be some resistance). Walther Hermann Nernst developed the third law of thermodynamics and stated that absolute zero was unattainable. Carl von Linde and William Hampson, both commercial researchers, nearly at the same time filed for patents on the Joule-Thomson effect. Linde's patent was the climax of 20 years of systematic investigation of established facts, using a regenerative counterflow method. Hampson's design was also of a regenerative method. The combined process became known as the Linde-Hampson liquefaction process. Heike Kamerlingh Onnes purchased a Linde machine for his research. Zygmunt Florenty Wroblewski conducted research into electrical properties at low temperatures, though his research ended early due to his accidental death. Around 1864, Karol Olszewski and Wroblewski predicted the electrical phenomena of dropping resistance levels at ultra-cold temperatures. Olszewski and Wroblewski documented evidence of this in the 1880s. A milestone was achieved on 10 July 1908 when Onnes at the Leiden University in Leiden produced, for the first time, liquified helium and achieved superconductivity.

In 1900, William Du Bois Duddell develops the Singing Arc and produced melodic sounds, from a low to a high-tones, from this arc lamp.

13.5.1   Lorentz and Poincaré

Main articles: History of special relativity and Lorentz ether theory
Between 1900 and 1910, many scientists like Wilhelm Wien, Max Abraham, Hermann Minkowski, or Gustav Mie believed that all forces of nature are of electromagnetic origin (the so-called "electromagnetic world view"). This was connected with the electron theory developed between 1892 and 1904 by Hendrik Lorentz. Lorentz introduced a strict separation between matter (electrons) and ether, whereby in his model the ether is completely motionless, and it won't be set in motion in the neighborhood of ponderable matter. Contrary to other electron models before, the electromagnetic field of the ether appears as a mediator between the electrons, and changes in this field can propagate not faster than the speed of light.

Hendrik Lorentz.

is seen by modern historians as being a mathematical transformation from a "real" system resting in the aether into a "fictitious" system in motion.[146][147][148]

Henri Poincaré.

In 1896, three years after submitting his thesis on the Kerr effect, Pieter Zeeman disobeyed the direct orders of his supervisor and used laboratory equipment to measure the splitting of spectral lines by a strong magnetic field. Lorentz theoretically explained the Zeeman effect on the basis of his theory, for which both received the Nobel Prize in Physics in 1902. A fundamental concept of Lorentz's theory in 1895 was the "theorem of corresponding states" for terms of order v/c. This theorem states that a moving observer (relative to the ether) makes the same observations as a resting observer. This theorem was extended for terms of all orders by Lorentz in 1904. Lorentz noticed, that it was necessary to change the space-time variables when changing frames and introduced concepts like physical length contraction (1892) to explain the Michelson–Morley experiment, and the mathematical concept of local time (1895) to explain the aberration of light and the Fizeau experiment. That resulted in the formulation of the so-called Lorentz transformation by Joseph Larmor (1897, 1900) and Lorentz (1899, 1904).[146][147][148] As Lorentz later noted (1921, 1928), he considered the time indicated by clocks resting in the aether as "true" time, while local time was seen by him as a heuristic working hypothesis and a mathematical artifice.[149][150] Therefore, Lorentz's theorem

Continuing the work of Lorentz, Henri Poincaré between 1895 and 1905 formulated on many occasions the Principle of Relativity and tried to harmonize it with electrodynamics. He declared simultaneity only a convenient convention which depends on the speed of light, whereby the constancy of the speed of light would be a useful postulate for making the laws of nature as simple as possible. In 1900 he interpreted Lorentz's local time as the result of clock synchronization by light signals, and introduced the electromagnetic momentum by comparing electromagnetic energy to what he called a "fictitious fluid" of mass $m = E/c^2$. And finally in June and July 1905 he declared the relativity principle a general law of nature, including gravitation. He corrected some mistakes of Lorentz and proved the Lorentz covariance of the electromagnetic equations. Poincaré also suggested that there exist non-electrical forces to stabilize the electron configuration and asserted that gravitation is a non-electrical force as well, contrary to the electromagnetic world view. However, historians pointed out that he still used the notion of an ether and distinguished between "apparent" and "real" time and therefore didn't invent special relativity in its modern understanding.[148][151][152][153][154][155]

13.5.2 Einstein's *Annus Mirabilis*

Main article: Annus Mirabilis Papers
In 1905, while he was working in the patent office, Albert

Albert Einstein, 1905.

Einstein had four papers published in the *Annalen der Physik*, the leading German physics journal. These are the papers that history has come to call the *Annus Mirabilis Papers*:

- His paper on the particulate nature of light put forward the idea that certain experimental results, notably the photoelectric effect, could be simply understood from the postulate that light interacts with matter as discrete "packets" (quanta) of energy, an idea that had been introduced by Max Planck in 1900 as a purely mathematical manipulation, and which seemed to contradict contemporary wave theories of light (Einstein 1905a). This was the only work of Einstein's that he himself called "revolutionary."

- His paper on Brownian motion explained the random movement of very small objects as direct evidence of molecular action, thus supporting the atomic theory. (Einstein 1905b)

- His paper on the electrodynamics of moving bodies introduced the radical theory of special relativity,

which showed that the observed independence of the speed of light on the observer's state of motion required fundamental changes to the notion of simultaneity. Consequences of this include the time-space frame of a moving body slowing down and contracting (in the direction of motion) relative to the frame of the observer. This paper also argued that the idea of a luminiferous aether — one of the leading theoretical entities in physics at the time — was superfluous. (Einstein 1905c)

- In his paper on mass–energy equivalence (previously considered to be distinct concepts), Einstein deduced from his equations of special relativity what later became the well-known expression: $E = mc^2$, suggesting that tiny amounts of mass could be converted into huge amounts of energy. (Einstein 1905d)

All four papers are today recognized as tremendous achievements — and hence 1905 is known as Einstein's "Wonderful Year". At the time, however, they were not noticed by most physicists as being important, and many of those who did notice them rejected them outright. Some of this work — such as the theory of light quanta — remained controversial for years. [156] [157]

13.5.3 Latter half of the 20th Century

The first formulation of a quantum theory describing radiation and matter interaction is due to Paul Adrien Maurice Dirac, who, during 1920, was first able to compute the coefficient of spontaneous emission of an atom. [158] Paul Dirac described the quantization of the electromagnetic field as an ensemble of harmonic oscillators with the introduction of the concept of creation and annihilation operators of particles. In the following years, with contributions from Wolfgang Pauli, Eugene Wigner, Pascual Jordan, Werner Heisenberg and an elegant formulation of quantum electrodynamics due to Enrico Fermi, [159] physicists came to believe that, in principle, it would be possible to perform any computation for any physical process involving photons and charged particles. However, further studies by Felix Bloch with Arnold Nordsieck, [160] and Victor Weisskopf, [161] in 1937 and 1939, revealed that such computations were reliable only at a first order of perturbation theory, a problem already pointed out by Robert Oppenheimer. [162] At higher orders in the series infinities emerged, making such computations meaningless and casting serious doubts on the internal consistency of the theory itself. With no solution for this problem known at the time, it appeared that a fundamental incompatibility existed between special relativity and quantum mechanics.

In December 1938, the German chemists Otto Hahn and

Paul Adrien Maurice Dirac.

Fritz Strassmann sent a manuscript to *Naturwissenschaften* reporting they had detected the element barium after bombarding uranium with neutrons;[163] simultaneously, they communicated these results to Lise Meitner. Meitner, and her nephew Otto Robert Frisch, correctly interpreted these results as being nuclear fission.[164] Frisch confirmed this experimentally on 13 January 1939.[165] In 1944, Hahn received the Nobel Prize for Chemistry for the discovery of nuclear fission. Some historians who have documented the history of the discovery of nuclear fission believe Meitner should have been awarded the Nobel Prize with Hahn.[166][167][168]

Difficulties with the Quantum theory increased through the end of 1940. Improvements in microwave technology made it possible to take more precise measurements of the shift of the levels of a hydrogen atom,[169] now known as the Lamb shift and magnetic moment of the electron.[170] These experiments unequivocally exposed discrepancies which the theory was unable to explain. With the invention of bubble chambers and spark chambers in the 1950s, experimental particle physics discovered a large and ever-growing number of particles called hadrons. It seemed that such a large number of particles could not all

be fundamental.

Shortly after the end of the war in 1945, Bell Labs formed a Solid State Physics Group, led by William Shockley and chemist Stanley Morgan; other personnel including John Bardeen and Walter Brattain, physicist Gerald Pearson, chemist Robert Gibney, electronics expert Hilbert Moore and several technicians. Their assignment was to seek a solid-state alternative to fragile glass vacuum tube amplifiers. Their first attempts were based on Shockley's ideas about using an external electrical field on a semiconductor to affect its conductivity. These experiments failed every time in all sorts of configurations and materials. The group was at a standstill until Bardeen suggested a theory that invoked surface states that prevented the field from penetrating the semiconductor. The group changed its focus to study these surface states and they met almost daily to discuss the work. The rapport of the group was excellent, and ideas were freely exchanged.[171]

As to the problems in the electron experiments, a path to a solution was given by Hans Bethe. In 1947, while he was traveling by train to reach Schenectady from New York,[172] after giving a talk at the conference at Shelter Island on the subject, Bethe completed the first non-relativistic computation of the shift of the lines of the hydrogen atom as measured by Lamb and Retherford.[173] Despite the limitations of the computation, agreement was excellent. The idea was simply to attach infinities to corrections at mass and charge that were actually fixed to a finite value by experiments. In this way, the infinities get absorbed in those constants and yield a finite result in good agreement with experiments. This procedure was named renormalization.

Based on Bethe's intuition and fundamental papers on the subject by Sin-Itiro Tomonaga,[174] Julian Schwinger,[175][176] Richard Feynman[177][178][179] and Freeman Dyson,[180][181] it was finally possible to get fully covariant formulations that were finite at any order in a perturbation series of quantum electrodynamics. Sin-Itiro Tomonaga, Julian Schwinger and Richard Feynman were jointly awarded with a Nobel prize in physics in 1965 for their work in this area.[182] Their contributions, and those of Freeman Dyson, were about covariant and gauge invariant formulations of quantum electrodynamics that allow computations of observables at any order of perturbation theory. Feynman's mathematical technique, based on his diagrams, initially seemed very different from the field-theoretic, operator-based approach of Schwinger and Tomonaga, but Freeman Dyson later showed that the two approaches were equivalent.[180] Renormalization, the need to attach a physical meaning at certain divergences appearing in the theory through integrals, has subsequently become one of the fundamental aspects of quantum field theory and

Richard Feynman.

has come to be seen as a criterion for a theory's general acceptability. Even though renormalization works very well in practice, Feynman was never entirely comfortable with its mathematical validity, even referring to renormalization as a "shell game" and "hocus pocus".[183] QED has served as the model and template for all subsequent quantum field theories. Peter Higgs, Jeffrey Goldstone, and others, Sheldon Glashow, Steven Weinberg and Abdus Salam independently showed how the weak nuclear force and quantum electrodynamics could be merged into a single electroweak force.

Robert Noyce credited Kurt Lehovec for the *principle of p-n junction isolation* caused by the action of a biased p-n junction (the diode) as a key concept behind the integrated circuit.[184] Jack Kilby recorded his initial ideas concerning the integrated circuit in July 1958 and successfully demonstrated the first working integrated circuit on September 12, 1958.[185] In his patent application of February 6, 1959, Kilby described his new device as "a body of semiconductor material ... wherein all the components of the electronic circuit are completely integrated."[186] Kilby won the 2000 Nobel Prize in Physics for his part of the invention of the integrated circuit.[187] Robert Noyce also came up with his own idea of an integrated circuit half a year later than Kilby. Noyce's chip solved many practical problems that Kilby's had not. Noyce's chip, made at Fairchild Semiconductor, was made of silicon, whereas Kilby's chip was made of germanium.

Philo Farnsworth developed the Farnsworth–Hirsch Fusor, or simply fusor, an apparatus designed by Farnsworth to create nuclear fusion. Unlike most controlled fusion systems, which slowly heat a magnetically confined plasma, the fusor injects high temperature ions directly into a reaction chamber, thereby avoiding a considerable amount of complexity. When the Farnsworth-Hirsch Fusor was first introduced to the fusion research world in the late 1960s, the Fusor was the first device that could clearly demonstrate it was producing fusion reactions at all. Hopes at the time were high that it could be quickly developed into a practical power source. However, as with other fusion experiments, development into a power source has proven difficult. Nevertheless, the fusor has since become a practical neutron source and is produced commercially for this role.[188]

The first step towards the Standard Model was Sheldon Glashow's discovery, in 1960, of a way to combine the electromagnetic and weak interactions.[189] In 1967, Steven Weinberg[190] and Abdus Salam[191] incorporated the Higgs mechanism[192][193][194] into Glashow's electroweak theory, giving it its modern form. The Higgs mechanism is believed to give rise to the masses of all the elementary particles in the Standard Model. This includes the masses of the W and Z bosons, and the masses of the fermions - i.e. the quarks and leptons. After the neutral weak currents caused by Z boson exchange were discovered at CERN in 1973,[195][196][197][198] the electroweak theory became widely accepted and Glashow, Salam, and Weinberg shared the 1979 Nobel Prize in Physics for discovering it. The W and Z bosons were discovered experimentally in 1981, and their masses were found to be as the Standard Model predicted. The theory of the strong interaction, to which many contributed, acquired its modern form around 1973–74, when experiments confirmed that the hadrons were composed of fractionally charged quarks. With the establishment of quantum chromodynamics in the 1970s finalized a set of fundamental and exchange particles, which allowed for the establishment of a "standard model" based on the mathematics of gauge invariance, which successfully described all forces except for gravity, and which remains generally accepted within the domain to which it is designed to be applied.

The 'standard model' groups the electroweak interaction theory and quantum chromodynamics into a structure denoted by the gauge group $SU(3) \times SU(2) \times U(1)$. The formulation of the unification of the electromagnetic and weak interactions in the standard model is due to Abdus Salam, Steven Weinberg and, subsequently, Sheldon Glashow. After the discovery, made at CERN, of the existence of neutral weak currents,[199][200][201][202] mediated by the Z boson foreseen in the standard model, the physicists Salam, Glashow and Weinberg received the 1979 Nobel Prize in Physics for their electroweak theory.[203] Since then, discoveries of the bottom quark (1977), the top quark (1995) and the tau neutrino (2000) have given credence to the standard model. Because of its success in explaining a wide variety of experimental results.

13.5.4   Electrodynamic tethers

Main article: Electrodynamic tether

Before the start of the 21st century, the electrodynamic tether[204] being oriented at an angle to the local vertical between the object and a planet with a magnetic field cut the Earth's magnetic field and generated a current; thereby it converted some of the orbiting body's kinetic energy to electrical energy. The tether's far end can be left bare, making electrical contact with the ionosphere, creating a generator. As part of a *tether propulsion* system, crafts can use long, strong conductors[205] to change the orbits of spacecraft. It has the potential to make space travel significantly cheaper. It is a simplified, very low-budget magnetic sail. It can be used either to accelerate or brake an orbiting spacecraft. When direct current is pumped through the tether, it exerts a force against the magnetic field, and the tether accelerates the spacecraft.

13.6   21st century

13.6.1   Electromagnetic technologies

There are a range of emerging energy technologies. By 2007, solid state micrometer-scale electric double-layer capacitors based on advanced superionic conductors had been for low-voltage electronics such as deep-sub-voltage nanoelectronics and related technologies (the 22 nm technological node of CMOS and beyond). Also, the nanowire battery, a lithium-ion battery, was invented by a team led by Dr. Yi Cui in 2007.

Magnetic resonance

Reflecting the fundamental importance and applicability of Magnetic resonance imaging[206] in medicine, Paul Lauterbur of the University of Illinois at Urbana-Champaign and Sir Peter Mansfield of the University of Nottingham were awarded the 2003 Nobel Prize in Physiology or Medicine for their *"discoveries concerning magnetic resonance imaging"*. The Nobel citation acknowledged Lauterbur's insight of using magnetic field gradients to determine spatial localization, a discovery that allowed rapid acquisition of 2D images.

Wireless electricity

Main article: wireless energy transfer

Wireless electricity is a form of wireless energy transfer,[207] the ability to provide electrical energy to remote objects without wires. The term WiTricity was coined in 2005 by Dave Gerding and later used for a project led by Prof. Marin Soljačić in 2007.[208][209] The MIT researchers successfully demonstrated the ability to power a 60 watt light bulb wirelessly, using two 5-turn copper coils of 60 cm (24 in) diameter, that were 2 m (7 ft) away, at roughly 45% efficiency.[210] This technology can potentially be used in a large variety of applications, including consumer, industrial, medical and military. Its aim is to reduce the dependence on batteries. Further applications for this technology include transmission of information—it would not interfere with radio waves and thus could be used as a cheap and efficient communication device without requiring a license or a government permit.

Further information: WiTricity

13.6.2   Unified Theories

Main article: Grand Unified Theory

As of 2010, there is still no hard evidence that nature is described by a Grand Unified Theory (GUT). The Higgs particle has been tentatively verified,.[211] The discovery of neutrino oscillations indicates that the Standard Model is incomplete and has led to renewed interest toward certain GUT such as $SO(10)$. One of the few possible experimental tests of certain GUT is proton decay and also fermion masses. There are a few more special tests for supersymmetric GUT. The gauge coupling strengths of QCD, the weak interaction and hypercharge seem to meet at a common length scale called the GUT scale and equal approximately to 10^{16} GeV, which is slightly suggestive. This interesting numerical observation is called the gauge coupling unification, and it works particularly well if one assumes the existence of superpartners of the Standard Model particles. Still it is possible to achieve the same by postulating, for instance, that ordinary (non supersymmetric) $SO(10)$ models break with an intermediate gauge scale, such as the one of Pati–Salam group.

The Theory of Everything (TOE) is a putative theory of theoretical physics that fully explains and links together all known physical phenomena, and, ideally, has predictive power for the outcome of any experiment that could be carried out in principle. M-Theory is not yet complete, but the underlying structure of the mathematics has been established and is in agreement with not only all the string theories, but with all of our scientific observations of the universe. Furthermore, it has passed many tests of internal mathematical consistency that many other attempts to combine quantum mechanics and gravity had failed. Unfortunately, until we can find some way to observe higher dimensions (impossible with our current level of technology) M-Theory has a very difficult time making predictions which can be tested in a laboratory. Technologically, it may never be possible for it to be "proven". Physicist and author Michio Kaku has remarked that M-Theory may present us with a "Theory of Everything" which is so concise that its underlying formula would fit on a T-shirt.[212] Stephen Hawking originally believed that M-Theory may be the ultimate theory but later suggested that the search for understanding of mathematics and physics will never be complete.[213]

See also: Unified field theory

13.6.3 Open problems

Main article: Open problems in physics

The magnetic monopole[214] in the *quantum* theory of magnetic charge started with a paper by the physicist Paul A.M. Dirac in 1931.[215] The detection of magnetic monopoles is an open problem in experimental physics. In some theoretical models, magnetic monopoles are unlikely to be observed, because they are too massive to be created in particle accelerators, and also too rare in the Universe to enter a particle detector with much probability.

After more than twenty years of intensive research, the origin of high-temperature superconductivity is still not clear, but it seems that instead of *electron-phonon* attraction mechanisms, as in conventional superconductivity, one is dealing with genuine *electronic* mechanisms (e.g. by antiferromagnetic correlations), and instead of s-wave pairing, d-wave pairings[216] are substantial.[217] One goal of all this research is room-temperature superconductivity.[218]

13.7 See also

Histories History of electromagnetic spectrum, History of electrical engineering, History of Maxwell's equations, History of radio, History of optics, History of physics

General Biot–Savart law, Ponderomotive force, Telluric currents, Terrestrial magnetism, ampere-hours, Transverse waves, Longitudinal waves, Plane waves,

Refractive index, torque, Revolutions per minute, Photosphere, Vortex, vortex rings.

Theory permittivity, scalar product, vector product, tensor, divergent series, linear operator, unit vector, parallelepiped, osculating plane, standard candle

Technology Solenoid, electro-magnets, Nicol prisms, rheostat, voltmeter, gutta-percha covered wire, Electrical conductor, ammeters, Gramme machine, binding posts, Induction motor, Lightning arresters, Technological and industrial history of the United States, Western Electric Company,

Lists Outline of energy development

Timelines Timeline of electromagnetism, Timeline of luminiferous aether

13.8 References

Citations and notes

[1] Bruno Kolbe, Francis ed Legge, Joseph Skellon, tr., *"An Introduction to Electricity"*. Kegan Paul, Trench, Trübner, 1908. 429 pages. Page 391. (cf., "[...] *high poles covered with copper plates and with gilded tops were erected 'to break the stones coming from on high'.* J. Dümichen, Baugeschichte des Dendera-Tempels, Strassburg, 1877")

[2] Urbanitzky, A. v., & Wormell, R. (1886). Electricity in the service of man: a popular and practical treatise on the applications of electricity in modern life. London: Cassell &.

[3] Lyons, T. A. (1901). *A treatise on electromagnetic phenomena, and on the compass and its deviations aboard ship. Mathematical, theoretical, and practical.* New York: J. Wiley & Sons.

[4] *Encyclopedia Britannica* (1890). New York: The Henry G. Allen Company.

[5] Whittaker, E. T. (1910). A history of the theories of aether and electricity from the age of Descartes to the close of the 19th century. Dublin University Press series. London: Longmans, Green and Co.; [etc.].

[6] Carlson, John B. (1975) "Lodestone Compass: Chinese or Olmec Primacy?: Multidisciplinary analysis of an Olmec hematite artifact from San Lorenzo, Veracruz, Mexico", *Science*, 189 (4205 : 5 September), p. 753-760, doi:10.1126/science.189.4205.753, p. 753–760

[7] Lodestone Compass: Chinese or Olmec Primacy?: Multidisciplinary analysis of an Olmec hematite artifact from San Lorenzo, Veracruz, Mexico - Carlson 189 (4205): 753 - Science

[8] Li Shu-hua, p. 175

[9] Maver, William, Jr.: "Electricity, its History and Progress", The Encyclopedia Americana; a library of universal knowledge, vol. X, pp. 172ff. (1918). New York: Encyclopedia Americana Corp.

[10] Heinrich Karl Brugsch-Bey and Henry Danby Seymour. "A History of Egypt Under the Pharaohs". J. Murray, 1881. Page 422. (cf., [... the symbol of a] 'serpent' is rather a fish, which still serves, in the Coptic language, to designate the electric fish [...])

[11] Seeman, Bernard and Barry, James E. The Story of Electricity and Magnetism. Harvey House 1967, p. 19

[12] Moller, Peter; Kramer, Bernd (December 1991), "Review: Electric Fish", BioScience (American Institute of Biological Sciences) 41 (11): 794–6 [794], doi:10.2307/1311732, JSTOR 1311732

[13] Bullock, Theodore H. (2005), Electroreception, Springer, pp. 5–7, ISBN 0-387-23192-7

[14] Morris, Simon C. (2003), Life's Solution: Inevitable Humans in a Lonely Universe, Cambridge University Press, pp. 182–185, ISBN 0-521-82704-3

[15] Riddle of 'Baghdad's batteries'. BBC News.

[16] After the Second World War, Willard Gray demonstrated current production by a reconstruction of the inferred battery design when filled with grape juice. W. Jansen experimented with 1,4-Benzoquinone (some beetles produce quinones) and vinegar in a cell and got satisfactory performance.

[17] An alternative, but still electrical explanation was offered by Paul Keyser. It was suggested that a priest or healer, using an iron spatula to compound a vinegar based potion in a copper vessel, may have felt an electrical tingle and used the phenomenon either for electro-acupuncture, or to amaze supplicants by electrifying a metal statue.

[18] Copper and iron form an electrochemical couple, so that in the presence of any electrolyte, an electric potential (voltage) will be produced. König had observed a number of very fine silver objects from ancient Iraq which were plated with very thin layers of gold, and speculated that they were electroplated using batteries of these "cells".

[19] Corder, Gregory. "Using an Unconventional History of the Battery to engage students and explore the importance of evidence", Virginia Journal of Science Education 1

[20] A history of electricity. By Park Benjamin. Pg 33

[21] His Epistola was written in 1269.

[22] Lane, Frederic C. (1963) "The Economic Meaning of the Invention of the Compass", The American Historical Review, 68 (3: April), p. 605–617

[23] Dampier, W. C. D. (1905). The theory of experimental electricity. Cambridge physical series. Cambridge [Eng.: University Press.

[24] consult ' Priestley's 'History of Electricity,' London 1757

[25] Robert Boyle (1675). Experiments and notes about the mechanical origin or production of particular qualities.

[26] Benjamin, P. (1895). A history of electricity: (The intellectual rise in electricity) from antiquity to the days of Benjamin Franklin. New York: J. Wiley & Sons.

[27] Consult Boyle's 'Experiments on the Origin of Electricity,'" and Priestley's 'History of Electricity'.

[28] The Magnet, or Concerning Magnetic Science (Magnes sive de arte magnetica)

[29] From Physico-Mechanical Experiments, 2nd Ed., London 1719

[30] Consult Dr. Carpue's 'Introduction to Electricity and Galvanism.' London 1803.

[31] Krebs, Robert E. (2003), Groundbreaking Scientific Experiments, Inventions, and Discoveries of the 18th Century, Greenwood Publishing Group, p. 82, ISBN 0-313-32015-2

[32] Keithley, Joseph F. (1999), The Story of Electrical and Magnetic Measurements: From 500 B.C. to the 1940s, Wiley, ISBN 0-7803-1193-0

[33] Biography, Pieter (Petrus) van Musschenbroek

[34] According to Priestley ('History of Electricity,' 3d ed., Vol. I, p. 102)

[35] Priestley's 'History of Electricity.' p. 138

[36] Catholic churchmen in science. (Second series) by James Joseph Wals. Pg 172.

[37] The History and Present State of Electricity with Original Experiments By Joseph Priestle. Pg 173.

[38] Cheney Hart: "Part of a letter from Cheney Hart, M.D. to William Watson, F.R.S. giving Account of the Effects of Electricity in the County Hospital at Shrewsbury", Phil. Trans. 1753:48, pp. 786–788. Read on November 14, 1754.

[39] Kite Experiment (2011). IEEE Global History Network.

[40] see atmospheric electricity

[41] Dr. Wall Experiments of the Luminous Qualities of Amber, Diamonds, and Gum Lac, by Dr. Wall, in a Letter to Dr. Sloane, R. S. Secr. Phil. Trans. 1708 26:69-76; doi:10.1098/rstl.1708.0011

[42] Physico-mechanical experiments, on various subjects: with explanations of all the machines engraved on copper

[43] Vail, A. (1845). The American electro magnetic telegraph: With the reports of Congress, and a description of all telegraphs known, employing electricity or galvanism. Philadelphia: Lea & Blanchard

[44] Hutton, C., Shaw, G., Pearson, R., & Royal Society (Great Britain). (1665). Philosophical transactions of the Royal Society of London: From their commencement, in 1665 to the year 1800. London: C. and R. Baldwin. PaGE 345.

[45] Franklin, 'Experiments and Observations on Electricity'

[46] Royal Society Papers, vol. IX (BL. Add MS 4440): Henry Elles, from Lismore, Ireland, to the Royal Society, London, 9 August 1757, f.12b; 9 August 1757, f.166.

[47] *Tr.*, Test Theory of Electricity and Magnetism

[48] Philosophical Transactions 1771

[49] Electric Telegraph, apparatus by wh. signals may be transmitted to a distance by voltaic currents propagated on metallic wires; fnded. on experimts. of Gray 1729, Nollet, Watson 1745, Lesage 1774, Lamond 1787, Reusserl794, Cavallo 1795, Betancourt 1795, Soemmering 1811, Gauss & Weber 1834, &c. Telegraphs constructed by Wheatstone & Independently by Steinheil 1837, improved by Morse, Cooke, Woolaston, &c.

[50] Cassell's miniature cyclopaedia By Sir William Laird Clowes. Page 288.

[51] Die Geschichte Der Physik in Grundzügen: th. In den letzten hundert jahren (1780–1880) 1887-90 (tr. The history of physics in broad terms: th. In the last hundred years (1780–1880) 1887-90) by Ferdinand Rosenberger. F. Vieweg und sohn, 1890. Page 288.

[52] See Voltaic pile

[53] 'Philosophical Transactions,' 1833

[54] Of Torpedos Found on the Coast of England. In a Letter from John Walsh, Esq; F. R. S. to Thomas Pennant, Esq; F. R. S. John Walsh Philosophical Transactions (1683–1775) Vol. 64, (1774), pp. 464-473

[55] The works of Benjamin Franklin: containing several political and historical tracts not included in any former ed., and many letters official and private, not hitherto published; with notes and a life of the author, Volume 6 Page 348.

[56] another noted and careful experimenter in electricity and the discoverer of palladium and rhodium

[57] Philosophical Magazine, Vol. III, p. 211

[58] (coulomb^2) * the molar gas constant = 8.314472 m2 kg A2 K-1 mol-1

[59] 'Trans. Society of Arts,1 1825

[60] Meteorological essays By François Arago, Sir Edward Sabine. Page 290. "On Rotation Magnetism. *Proces verbal*, Academy of Sciences, 22 November 1824."

[61] For more, see Rotating magnetic field.

[62] Tr., 'The galvanic Circuit investigated mathematically".

[63] G. S. Ohm (1827). *Die galvanische Kette, mathematisch bearbeitet* (PDF). Berlin: T. H. Riemann.

[64] The Encyclopedia Americana: a library of universal knowledge, 1918.

[65] "A Brief History of Electromagnetism" (PDF).

[66] "Electromagnetism". *Smithsonian Institution Archives*.

[67] Tsverava, G. K. 1981. "FARADEI, GENRI, I OTKRYTIE INDUKTIROVANNYKH TOKOV." Voprosy Istorii Estestvoznaniia i Tekhniki no. 3: 99-106. Historical Abstracts, EBSCOhost . Retrieved October 17, 2009.

[68] Bowers, Brian. 2004. "Barking Up the Wrong (Electric Motor) Tree." Proceedings of the IEEE 92, no. 2: 388-392. Computers & Applied Sciences Complete, EBSCOhost . Retrieved October 17, 2009.

[69] 1998. "Joseph Henry." Issues in Science & Technology 14, no. 3: 96. Associates Programs Source, EBSCOhost . Retrieved October 17, 2009.

[70] According to Oliver Heaviside

[71] Oliver Heaviside. Electromagnetic theory: Complete and unabridged ed. of v.1, no.2, and: Volume 3. 1950.

[72] Oliver Heaviside. Electromagnetic theory, v.1. "The Electrician" printing and publishing company, limited, 1893.

[73] A treatise on electricity, in theory and practice, Volume 1 By Auguste de La Rive. Page 139.

[74] 'Phil. Trans.,' 1845.

[75] Elementary Lessons in Electricity and Magnetism By Silvanus Phillips Thompson. Page 363.

[76] Phil. Mag.., March 1854

[77] For more, see Counter-electromotive force.

[78] Philosophical Magazine, 1849.

[79] Ruhmkorff's version coil was such a success that in 1858 he was awarded a 50,000-franc prize by Napoleon III for the most important discovery in the application of electricity.

[80] American Academy of Arts and Sciences, *Proceedings of the American Academy of Arts and Sciences*, Vol. XXIII, May 1895 - May 1896, Boston: University Press, John Wilson and Son (1896), pp. 359-360: Ritchie's most powerful version of his induction coil, using staged windings, achieved electrical *bolts* 2 inches (5.1 cm) or longer in length.

[81] Page, Charles G., *History of Induction: The American Claim to the Induction Coil and Its Electrostatic Developments*, Boston: Harvard University, Intelligencer Printing house (1867), pp. 104-106

[82] American Academy, pp. 359-360

[83] Lyons, T. A. (1901). A treatise on electromagnetic phenomena, and on the compass and its deviations aboard ship. Mathematical, theoretical, and practical. New York: J. Wiley & Sons. Page 500.

[84] La, R. A. (1853). A treatise on electricity: In theory and practice. London: Longman, Brown, Green, and Longmans.

[85] tr., Introduction to electrostatics, the study of magnetism and electrodynamics

[86] May be Johann Philipp Reis, of Friedrichsdorf, Germany

[87] "On a permanent Deflection of the Galvanometer-needle under the influence of a rapid series of equal and opposite induced Currents". By Lord Rayleigh, F.R.S., Philosophical magazine, 1877. Page 44.

[88] Annales de chimie et de physique, Page 385. "Sur l'aimantation par les courants" (tr. "On the magnetization by currents").

[89] 'Ann. de Chimie III,' i, 385.

[90] Jenkin, F. (1873). Electricity and magnetism. Text-books of science. London: Longmans, Green, and Co

[91] Introduction to 'Electricity in the Service of Man'.

[92] 'Poggendorf Ann.1 1851.

[93] Proc. Am. Phil. Soc.,Vol. II, pp. 193

[94] Annalen der Physik, Volume 103. *Contributions to the acquaintance with the electric spark*, B. W. Feddersen. Page 69+.

[95] Special information on method and apparatus can be found in Feddersen's Inaugural Dissertation, Kiel 1857th (In the Commission der Schwers'schen Buchhandl Handl. In Kiel.)

[96] Rowland, H. A. (1902). The physical papers of Henry Augustus Rowland: Johns Hopkins University, 1876-1901. Baltimore: The Johns Hopkins Press.

[97] LII. On the electromagnetic effect of convection-currents Henry A. Rowland; Cary T. Hutchinson Philosophical Magazine Series 5, 1941-5990, Volume 27, Issue 169, Pages 445 – 460

[98] See electric machinery, electric direct current, electrical generators.

[99] consult his British patent of that year

[100] consult 'Royal Society Proceedings, 1867 VOL. 10—12

[101] RJ Gulcher, of Biala, near Bielitz, Austria.

[102] The Electrical journal, Volume 7, 1881. Page117+

[103] ETA: Electrical magazine: A. Ed. Volume 1

[104] See electric direct current.

[105] See Electric alternating current machinery.

[106] The 19th century science book A Guide to the Scientific Knowledge of Things Familiar provides a brief summary of scientific thinking in this field at the time.

[107] Consult Maxwell's 'Electricity and Magnetism,1 Vol. II, Chap. xx

[108] On Faraday's Lines of Force' byJames Clerk Maxwell 1855

[109] James Clerk Maxwell, *On Physical Lines of Force*, Philosophical Magazine, 1861

[110] In November 1847, Clerk Maxwell entered the University of Edinburgh, learning mathematics from Kelland, natural philosophy from J. D. Forbes, and logic from Sir W. R. Hamilton.

[111] Glazebrook, R. (1896). James Clerk Maxwell and modern physics. New York: Macmillan.Pg. 190

[112] J J O'Connor and E F Robertson, *James Clerk Maxwell*, School of Mathematics and Statistics, University of St Andrews, Scotland, November 1997

[113] James Clerk Maxwell, *A Dynamical Theory of the Electromagnetic Field*, Philosophical Transactions of the Royal Society of London 155, 459-512 (1865).

[114] Maxwell's 'Electricity and Magnetism,' preface

[115] See oscillating current, telegraphy, wireless.

[116] Proceedings of the London Mathematical Society, Volume 3. London Mathematical Society, 1871. Pg. 224

[117] Heinrich Hertz (1893). *Electric Waves: Being Researches on the Propagation of Electric Action with Finite Velocity Through Space*. Dover Publications.

[118] Crookes presented a lecture to the British Association for the Advancement of Science, in Sheffield, on Friday, 22 August 1879

[119] consult 'Proc. British Association,' 1879

[120] Perhaps there were other reasons than those mentioned for the discontinuance. We do not dwell in the Palace of Truth.

[121] in Sir W. Thomson's meaning of the word

[122] Electro-Magnetic Theory. By Oliver HeaviBide. Vol. I. Electrician Printing: and Publishing Company, Ltd. London, 1893

[123] In mathematics, *vectorial algebra* may mean a linear algebra, specifically the basic algebraic operations of vector addition and scalar multiplication; see vector space. The algebraic operations in vector calculus, namely the specific additional structure of vectors in 3-dimensional Euclidean space $\mathbf{R}^3$ of dot product and especially cross product. In this sense, *vector algebra* is contrasted with geometric algebra, which provides an alternative generalization to higher

dimensions. Original vector algebras of the 19th century like quaternions, tessarines, or coquaternions, each of which has its own product. The vector algebras biquaternions and hyperbolic quaternions can be used in special relativity.

[124] Electrical engineer. Volume 18. Page299

[125] US 447921, Tesla, Nikola, "Alternating Electric Current Generator".

[126] W. Bernard Carlson, Tesla: Inventor of the Electrical Age, page 127

[127] W. Bernard Carlson, Tesla: Inventor of the Electrical Age, page 132

[128] Radio: Brian Regal, The Life Story of a Technology, page 22. Google Books.

[129] Announced in his evening lecture to the Royal Institution on Friday, 30 April 1897, and published in *Philosophical Magazine*, 44, 293

[130] Earl R. Hoover, *Cradle of Greatness: National and World Achievements of Ohio's Western Reserve* (Cleveland: Shaker Savings Association, 1977).

[131] Dayton C. Miller, "Ether-drift Experiments at Mount Wilson Solar Observatory", *Physical Review*, S2, V19, N4, pp. 407-408 (April 1922).

[132] Gordon gave four lectures on static electric induction (S. Low, Marston, Searle, and Rivington, 1879). In 1891, he also published "*A treatise on electricity and magnetism*). Vol 1. Vol 2. (S. Low, Marston, Searle & Rivington, limited).

[133] Thompson, Silvanus P., *Dynamo-Electric Machinery*, pp. 17

[134] Thompson, Silvanus P., *Dynamo-Electric Machinery*, pp. 16

[135] See electric lighting

[136] Giovanni Dosi, David J. Teece, Josef Chytry, Understanding Industrial and Corporate Change, Oxford University Press, 2004, page 336, Google Books.

[137] See telegraph

[138] see transatlantic telegraph cable

[139] Miller's 'Magnetism and Electricity.' p. 460

[140] See Electric transmission of energy.

[141] 'James Blyth - Britain's first modern wind power pioneer', by Trevor Price, 2003, Wind Engineering, vol 29 no. 3, pp 191-200]

[142] [Anon. 1890, 'Mr. Brush's Windmill Dynamo', Scientific American, vol 63 no. 25, 20 December, p. 54]

[143] A Wind Energy Pioneer: Charles F. Brush, Danish Wind Industry Association. Retrieved 2007-05-02.

[144] *History of Wind Energy* in Cutler J. Cleveland,(ed) *Encyclopedia of Energy Vol.6*, Elsevier, ISBN 978-1-60119-433-6, 2007, pp. 421-422

[145] See electrical units, electrical terms.

[146] Miller 1981, Ch. 1

[147] Pais 1982, Ch. 6b

[148] Janssen, 2007

[149] Lorentz, Hendrik Antoon (1921), "Deux Mémoires de Henri Poincaré sur la Physique Mathématique" [Two Papers of Henri Poincaré on Mathematical Physics], *Acta Mathematica* **38** (1): 293–308, doi:10.1007/BF02392073

[150] Lorentz, H.A.; Lorentz, H. A. (1928), "Conference on the Michelson-Morley Experiment", *The Astrophysical Journal* **68**: 345–351, Bibcode:1928ApJ....68..341M, doi:10.1086/143148

[151] Galison 2002

[152] Darrigol 2005

[153] Katzir 2005

[154] Miller 1981, Ch. 1.7 & 1.14

[155] Pais 1982, Ch. 6 & 8

[156] On the reception of relativity theory around the world, and the different controversies it encountered, see the articles in Thomas F. Glick, ed., *The Comparative Reception of Relativity* (Kluwer Academic Publishers, 1987), ISBN 90-277-2498-9.

[157] Pais, Abraham (1982), *Subtle is the Lord. The Science and the Life of Albert Einstein*, Oxford University Press, pp. 382–386, ISBN 0-19-520438-7

[158] P.A.M. Dirac (1927), "The Quantum Theory of the Emission and Absorption of Radiation", *Proceedings of the Royal Society of London A* **114**: 243–265, Bibcode:1927RSPSA.114..243D, doi:10.1098/rspa.1927.0039.

[159] E. Fermi (1932), "Quantum Theory of Radiation", *Reviews of Modern Physics* **4**: 87–132, Bibcode:1932RvMP....4...87F, doi:10.1103/RevModPhys.4.87.

[160] F. Bloch; A. Nordsieck (1937), "Note on the Radiation Field of the Electron", *Physical Review* **52**: 54–59, Bibcode:1937PhRv...52...54B, doi:10.1103/PhysRev.52.54.

[161] V. F. Weisskopf (1939), "On the Self-Energy and the Electromagnetic Field of the Electron", *Physical Review* **56**: 72–85, Bibcode:1939PhRv...56...72W, doi:10.1103/PhysRev.56.72.

[162] R. Oppenheimer (1930).   "Note on the Theory of the Interaction of Field and Matter". *Physical Review* **35**: 461–477. Bibcode:1930PhRv...35..461O. doi:10.1103/PhysRev.35.461.

[163] O. Hahn and F. Strassmann. *Über den Nachweis und das Verhalten der bei der Bestrahlung des Urans mittels Neutronen entstehenden Erdalkalimetalle* ("On the detection and characteristics of the alkaline earth metals formed by irradiation of uranium with neutrons"), *Naturwissenschaften* Volume 27. Number 1, 11–15 (1939). The authors were identified as being at the Kaiser-Wilhelm-Institut für Chemie. Berlin-Dahlem. Received 22 December 1938.

[164] Lise Meitner and O. R. Frisch. "Disintegration of Uranium by Neutrons: a New Type of Nuclear Reaction", *Nature*, Volume 143, Number 3615, 239–240 (11 February 1939). The paper is dated 16 January 1939. Meitner is identified as being at the Physical Institute, Academy of Sciences, Stockholm. Frisch is identified as being at the Institute of Theoretical Physics, University of Copenhagen.

[165] O. R. Frisch. "Physical Evidence for the Division of Heavy Nuclei under Neutron Bombardment", *Nature*, Volume 143, Number 3616, 276–276 (18 February 1939). The paper is dated 17 January 1939. [The experiment for this letter to the editor was conducted on 13 January 1939; see Richard Rhodes *The Making of the Atomic Bomb*. 263 and 268 (Simon and Schuster, 1986).]

[166] Ruth Lewin Sime. *From Exceptional Prominence to Prominent Exception: Lise Meitner at the Kaiser Wilhelm Institute for Chemistry* Ergebnisse 24 Forschungsprogramm Geschichte der Kaiser-Wilhelm-Gesellschaft im Nationalsozialismus (2005).

[167] Ruth Lewin Sime. *Lise Meitner: A Life in Physics* (University of California, 1997).

[168] Elisabeth Crawford, Ruth Lewin Sime, and Mark Walker. "A Nobel Tale of Postwar Injustice", *Physics Today* Volume 50, Issue 9, 26–32 (1997).

[169] W. E. Lamb; R. C. Retherford (1947). "Fine Structure of the Hydrogen Atom by a Microwave Method." . *Physical Review* **72**: 241–243. Bibcode:1947PhRv...72..241L. doi:10.1103/PhysRev.72.241.

[170] P. Kusch; H. M. Foley (1948).   "On the Intrinsic Moment of the Electron" .   *Physical Review* **73**:   412. Bibcode:1948PhRv...73..412F. doi:10.1103/PhysRev.73.412.

[171] Brattain quoted in Michael Riordan and Lillian Hoddeson; *Crystal Fire: The Invention of the Transistor and the Birth of the Information Age*. New York: Norton (1997) ISBN 0-393-31851-6 pbk. p. 127

[172] Schweber, Silvan (1994).   "Chapter 5" . *QED and the Men Who Did it: Dyson, Feynman, Schwinger, and Tomonaga*. Princeton University Press. p. 230. ISBN 978-0-691-03327-3.

[173] H.   Bethe   (1947).   "The   Electromagnetic Shift   of   Energy   Levels" .   *Physical   Review* **72**:   339–341. Bibcode:1947PhRv...72..339B. doi:10.1103/PhysRev.72.339.

[174] S. Tomonaga (1946). "On a Relativistically Invariant Formulation of the Quantum Theory of Wave Fields". *Progress of Theoretical Physics* **1**: 27–42. doi:10.1143/PTP.1.27.

[175] J. Schwinger (1948).   "On Quantum-Electrodynamics and the Magnetic Moment of the Electron" . *Physical Review* **73**: 416–417. Bibcode:1948PhRv...73..416S. doi:10.1103/PhysRev.73.416.

[176] J. Schwinger (1948).   "Quantum Electrodynamics. I. A Covariant Formulation" .   *Physical Review* **74**:   1439–1461. Bibcode:1948PhRv...74.1439S. doi:10.1103/PhysRev.74.1439.

[177] R. P. Feynman (1949).   "Space-Time Approach to Quantum Electrodynamics" .   *Physical Review* **76**:   769–789. Bibcode:1949PhRv...76..769F. doi:10.1103/PhysRev.76.769.

[178] R. P. Feynman (1949). "The Theory of Positrons". *Physical Review* **76**:   749–759. Bibcode:1949PhRv...76..749F. doi:10.1103/PhysRev.76.749.

[179] R. P. Feynman (1950). "Mathematical Formulation of the Quantum Theory of Electromagnetic Interaction" . *Physical Review* **80**:   440–457. Bibcode:1950PhRv...80..440F. doi:10.1103/PhysRev.80.440.

[180] F. Dyson (1949).   "The Radiation Theories of Tomonaga, Schwinger, and Feynman" .   *Physical Review* **75**:   486–502. Bibcode:1949PhRv...75..486D. doi:10.1103/PhysRev.75.486.

[181] F. Dyson (1949).   "The S Matrix in Quantum Electrodynamics" .   *Physical Review* **75**: 1736–1755. Bibcode:1949PhRv...75.1736D. doi:10.1103/PhysRev.75.1736.

[182]   "The Nobel Prize in Physics 1965" . Nobel Foundation. Retrieved 2008-10-09.

[183] Feynman, Richard (1985). *QED: The Strange Theory of Light and Matter*. Princeton University Press. p. 128. ISBN 978-0-691-12575-6.

[184] Kurt Lehovec's patent on the isolation p-n junction; U.S. Patent 3,029,366 granted on April 10, 1962, filed April 22, 1959. Robert Noyce credits Lehovec in his article – "Microelectronics" , *Scientific American*, September 1977, Volume 23, Number 3, pp. 63–9.

[185] *The Chip that Jack Built*, (c. 2008), (HTML), Texas Instruments, accessed May 29, 2008.

[186] Winston, Brian. *Media technology and society: a history: from the telegraph to the Internet*, (1998), Routeledge, London. ISBN 0-415-14230-X ISBN 978-0-415-14230-4, p. 221

[187] Nobel Web AB, (October 10, 2000),(*The Nobel Prize in Physics 2000*, Retrieved on May 29, 2008

[188] Cartlidge, Edwin. *The Secret World of Amateur Fusion*. Physics World. March 2007: IOP Publishing Ltd. pp. 10-11. ISSN: 0953-8585.

[189] S.L. Glashow (1961). "Partial-symmetries of weak interactions". *Nuclear Physics* 22: 579–588. Bibcode:1961NucPh..22..579G. doi:10.1016/0029-5582(61)90469-2.

[190] S. Weinberg (1967). "A Model of Leptons". *Physical Review Letters* 19: 1264–1266. Bibcode:1967PhRvL...19.1264W. doi:10.1103/PhysRevLett.19.1264.

[191] A. Salam (1968). N. Svartholm, ed. *Elementary Particle Physics: Relativistic Groups and Analyticity*. Eighth Nobel Symposium. Stockholm: Almquvist and Wiksell. p. 367.

[192] F. Englert, R. Brout (1964). "Broken Symmetry and the Mass of Gauge Vector Mesons". *Physical Review Letters* 13: 321–323. Bibcode:1964PhRvL..13..321E. doi:10.1103/PhysRevLett.13.321.

[193] P.W. Higgs (1964). "Broken Symmetries and the Masses of Gauge Bosons". *Physical Review Letters* 13: 508–509. Bibcode:1964PhRvL..13..508H. doi:10.1103/PhysRevLett.13.508.

[194] G.S. Guralnik, C.R. Hagen, T.W.B. Kibble (1964). "Global Conservation Laws and Massless Particles". *Physical Review Letters* 13: 585–587. Bibcode:1964PhRvL..13..585G. doi:10.1103/PhysRevLett.13.585.

[195] F.J. Hasert; et al. (1973). "Search for elastic muon-neutrino electron scattering". *Physics Letters B* 46: 121. Bibcode:1973PhLB...46..121H. doi:10.1016/0370-2693(73)90494-2.

[196] F.J. Hasert; et al. (1973). "Observation of neutrino-like interactions without muon or electron in the gargamelle neutrino experiment". *Physics Letters B* 46: 138. Bibcode:1973PhLB...46..138H. doi:10.1016/0370-2693(73)90499-1.

[197] F.J. Hasert; et al. (1974). "Observation of neutrino-like interactions without muon or electron in the Gargamelle neutrino experiment". *Nuclear Physics B* 73: 1. Bibcode:1974NuPhB..73....1H. doi:10.1016/0550-3213(74)90038-8.

[198] D. Haidt (4 October 2004). "The discovery of the weak neutral currents". *CERN Courier*. Retrieved 2008-05-08.

[199] F. J. Hasert *et al. Phys. Lett.* 46B 121 (1973).

[200] F. J. Hasert *et al. Phys. Lett.* 46B 138 (1973).

[201] F. J. Hasert *et al. Nucl. Phys.* B73 1 (1974).

[202] *The discovery of the weak neutral currents*, CERN courier, 2004-10-04, retrieved 2008-05-08

[203] *The Nobel Prize in Physics 1979*, Nobel Foundation, retrieved 2008-09-10

[204] A long conductor attached to an object.

[205] It is noted that though not all space tethers are conductive.

[206] A medical imaging technique used in radiology to visualize detailed internal structures. The good contrast it provides between the different soft tissues of the body make it especially useful in brain, muscles, heart, and cancer compared with other medical imaging techniques such as computed tomography (CT) or X-rays.

[207] Wireless power is the transmission of electrical energy from a power source to an electrical load without interconnecting wires. Wireless transmission is useful in cases where interconnecting wires are inconvenient, hazardous, or impossible.

[208] "Wireless electricity could power consumer, industrial electronics". MIT News. 2006-11-14.

[209] "Goodbye wires...". MIT News. 2007-06-07.

[210] "Wireless Power Demonstrated". Retrieved 2008-12-09.

[211] http://home.web.cern.ch/about/updates/2013/03/new-results-indicate-new-particle-higgs-boson

[212] M-Theory: The Mother of all SuperStrings

[213] Hawking, Stephen. Gödel and the end of physics. July 20, 2002.

[214] A hypothetical particle in particle physics that is a magnet with only one magnetic pole. In more technical terms, a magnetic monopole would have a net "magnetic charge". Modern interest in the concept stems from particle theories, notably the grand unification and superstring theories, which predict their existence. See Particle Data Group summary of magnetic monopole search; Wen, Xiao-Gang; Witten, Edward. *Electric and magnetic charges in superstring models*.Nuclear Physics B. Volume 261, p. 651-677; and Coleman. *The Magnetic Monopole 50 years Later*, reprinted in *Aspects of Symmetry* for more

[215] Paul Dirac. "Quantised Singularities in the Electromagnetic Field". Proc. Roy. Soc. (London) A 133, 60 (1931). Free web link.

[216] d-Wave Pairing. musr.ca.

[217] The Motivation for an Alternative Pairing Mechanism. musr.ca.

[218] A. Mourachkine (2004). *Room-Temperature Superconductivity*. Cambridge International Science Publishing (Cambridge, UK) (also http://xxx.lanl.gov/abs/cond-mat/0606187). ISBN 1-904602-27-4.

Attribution

This article incorporates text from a publication now in the public domain: *"Electricity, its History and Progress"* by William Maver Jr. - article published within *The Encyclopedia Americana; a library of universal knowledge*, vol. X, pp. 172ff. (1918). New York: Encyclopedia Americana Corp.

13.9   Bibliography

- Bakewell, F. C. (1853). Electric science; its history, phenomena, and applications. London: Ingram, Cooke.

- Benjamin, P. (1898). A history of electricity (The intellectual rise in electricity) from antiquity to the days of Benjamin Franklin. New York: J. Wiley & Sons.

- Darrigol, Olivier (2005). "The Genesis of the theory of relativity" (PDF), *Séminaire Poincaré* 1: 1–22, doi:10.1007/3-7643-7436-5_1, retrieved 2009-06-21

- Durgin, W. A. (1912). Electricity, its history and development. Chicago: A.C. McClurg.

- Einstein, Albert: "Ether and the Theory of Relativity" (1920), republished in *Sidelights on Relativity* (Dover, New York, 1922).

- Einstein, Albert, *The Investigation of the State of Aether in Magnetic Fields*, 1895. (PDF format)

- Einstein, Albert (1905a). "On a Heuristic Viewpoint Concerning the Production and Transformation of Light", *Annalen der Physik* 17: 132–148, Bibcode:1905AnP...322..132E, doi:10.1002/andp.19053220607. This annus mirabilis paper on the photoelectric effect was received by Annalen der Physik March 18.

- Einstein, Albert (1905b). "On the Motion —Required by the Molecular Kinetic Theory of Heat —of Small Particles Suspended in a Stationary Liquid", *Annalen der Physik* 17: 549–560, Bibcode:1905AnP...322..549E, doi:10.1002/andp.19053220806. This annus mirabilis paper on Brownian motion was received May 11.

- Einstein, Albert (1905c). "On the Electrodynamics of Moving Bodies", *Annalen der Physik* 17: 891–921, Bibcode:1905AnP...322..891E, doi:10.1002/andp.19053221004. This annus mirabilis paper on special relativity was received June 30.

- Einstein, Albert (1905d). "Does the Inertia of a Body Depend Upon Its Energy Content?", *Annalen der Physik* 18: 639–641, Bibcode:1905AnP...323..639E, doi:10.1002/andp.19053231314. This annus mirabilis paper on mass-energy equivalence was received September 27.

- *"Aether"*. Encyclopædia Britannica, Eleventh Edition (1910–1911). Volume Vol. 1, Page 297.

- The Encyclopedia Americana; a library of universal knowledge; *"Electricity, its history and Progress"*. (1918). New York: Encyclopedia Americana Corp. Page 171

- Galison, Peter (2003), *Einstein's Clocks, Poincaré's Maps: Empires of Time*. New York: W.W. Norton, ISBN 0-393-32604-7

- Gibson, C. R. (1907). Electricity of to-day, its work & mysteries described in non-technical language. London: Seeley and co., limited

- Heaviside, O. (1894). Electromagnetic theory. London: "The Electrician" Print. and Pub.

- Ireland commissioners of nat. educ., (1861). Electricity, galvanism, magnetism, electromagnetism, heat, and the steam engine. Oxford University.

- Janssen, Michel & Mecklenburg, Matthew (2007), V. F. Hendricks; et al., eds., *Interactions: Mathematics, Physics and Philosophy* (Dordrecht: Springer): 65–134 Missing or empty |title= (help); |contribution= ignored (help)

- Jeans, J. H. (1908). The mathematical theory of electricity and magnetism. Cambridge: University Press.

- Katzir, Shaul (2005), "Poincaré's Relativistic Physics: Its Origins and Nature", *Phys. Perspect.* 7: 268–292, Bibcode:2005PhP.....7..268K, doi:10.1007/s00016-004-0234-y

- Lord Kelvin (Sir William Thomson), *"On Vortex Atoms"*. Proceedings of the Royal Society of Edinburgh, Vol. VI, 1867, pp. 94–105. (ed., Reprinted in Phil. Mag. Vol. XXXIV, 1867, pp. 15–24.)

- Kolbe, Bruno; Francis ed Legge, Joseph Skellon, tr., "An Introduction to Electricity". Kegan Paul, Trench, Trübner, 1908.

- Lodge, Oliver, *"Ether"*, Encyclopædia Britannica, Thirteenth Edition (1926).

- Lodge, Oliver. "The Ether of Space". ISBN 1-4021-8302-X (paperback) ISBN 1-4021-1766-3 (hardcover)

- Lodge, Oliver. "Ether and Reality". ISBN 0-7661-7865-X

- Lyons, T. A. (1901). A treatise on electromagnetic phenomena, and on the compass and its deviations aboard ship. Mathematical, theoretical, and practical. New York: J. Wiley & Sons.

- Maxwell, James Clerk, *"Ether"*, Encyclopædia Britannica, Ninth Edition (1875–89).

- Maxwell, J. C., & Thompson, J. J. (1892). A treatise on electricity and magnetism. Clarendon Press series. Oxford: Clarendon.

- Miller, Arthur I. (1981), *Albert Einstein's special theory of relativity. Emergence (1905) and early interpretation (1905–1911)*. Reading: Addison–Wesley. ISBN 0-201-04679-2

- Pais, Abraham (1982), *Subtle is the Lord: The Science and the Life of Albert Einstein*. New York: Oxford University Press. ISBN 0-19-520438-7

- Priestley, J., & Mynde, J. (1775). The history and present state of electricity, with original experiments. London: Printed for C. Bathurst, and T. Lowndes; J. Rivington, and J. Johnson; S. Crowder [and 4 others in London].

- Schaffner, Kenneth F. : 19th-century aether theories. Oxford: Pergamon Press, 1972. (contains several reprints of *original* papers of famous physicists)

- Slingo, M., Brooker, A., Urbanitzky, A., Perry, J., & Dibner, B. (1895). The cyclopædia of electrical engineering: containing a history of the discovery and application of electricity with its practice and achievements from the earliest period to the present time: the whole being a practical guide to artisans, engineers and students interested in the practice and development of electricity, electric lighting, motors, thermopiles, the telegraph, the telephone, magnets and every other branch of electrical application. Philadelphia: The Gebbie Pub. Co., Limited.

- Steinmetz, C. P., "Transient Electric Phenomena". Page 38. (ed., contained in: General Electric Company. General Electric review. Schenectady: General Electric Co.,)

- *A New System of Alternating Current Motors and Transformers*, by Nikola Tesla, 1888

- Thompson, S. P. (1891). The electromagnet, and electromagnetic mechanism. London: E. & F.N. Spon.

- Whittaker, E. T., *"A History of the Theories of Aether and Electricity, from the Age of Descartes to the Close of the 19th century"*. Dublin University Press series. London: Longmans, Green and Co.:

- Urbanitzky, A. v., & Wormell, R. (1886). Electricity in the service of man: a popular and practical treatise on the applications of electricity in modern life. London: Cassell &.

Chapter 14

Neutron

This article is about the subatomic particle. For other uses, see Neutron (disambiguation).

The **neutron** is a subatomic particle, symbol n or n0, with no net electric charge and a mass slightly larger than that of a proton. Protons and neutrons, each with mass approximately one atomic mass unit, constitute the nucleus of an atom, and they are collectively referred to as nucleons.[4] Their properties and interactions are described by nuclear physics.

The nucleus consists of Z protons, where Z is called the atomic number, and N neutrons, where N is the neutron number. The atomic number defines the chemical properties of the atom, and the neutron number determines the isotope or nuclide.[5] The terms isotope and nuclide are often used synonymously, but they refer to chemical and nuclear properties, respectively. The atomic mass number, symbol A, equals Z+N. For example, carbon has atomic number 6, and its abundant carbon-12 isotope has 6 neutrons, whereas its rare carbon-13 isotope has 7 neutrons. Some elements occur in nature with only one stable isotope, such as fluorine (see stable nuclide). Other elements occur as many stable isotopes, such as tin with ten stable isotopes. Even though it is not a chemical element, the neutron is included in the table of nuclides.[6]

Within the nucleus, protons and neutrons are bound together through the nuclear force, and neutrons are required for the stability of nuclei. Neutrons are produced copiously in nuclear fission and fusion. They are a primary contributor to the nucleosynthesis of chemical elements within stars through fission, fusion, and neutron capture processes.

The neutron is essential to the production of nuclear power. In the decade after the neutron was discovered in 1932,[7] neutrons were used to effect many different types of nuclear transmutations. With the discovery of nuclear fission in 1938,[8] it was quickly realized that, if a fission event produced neutrons, each of these neutrons might cause further fission events, etc., in a cascade known as a nuclear chain reaction.[5] These events and findings led to the first self-sustaining nuclear reactor (Chicago Pile-1, 1942) and the first nuclear weapon (Trinity, 1945).

Free neutrons, or individual neutrons free of the nucleus, are effectively a form of ionizing radiation, and as such, are a biological hazard, depending upon dose.[5] A small natural "neutron background" flux of free neutrons exists on Earth, caused by cosmic ray muons, and by the natural radioactivity of spontaneously fissionable elements in the Earth's crust.[9] Dedicated neutron sources like neutron generators, research reactors and spallation sources produce free neutrons for use in irradiation and in neutron scattering experiments.

14.1 Description

Neutrons and protons are both nucleons, which are attracted and bound together by the nuclear force to form atomic nuclei. The nucleus of the most common isotope of the hydrogen atom (with the chemical symbol "H") is a lone proton. The nuclei of the heavy hydrogen isotopes deuterium and tritium contain one proton bound to one and two neutrons, respectively. All other types of atomic nuclei are composed of two or more protons and various numbers of neutrons. The most common nuclide of the common chemical element lead, ^{208}Pb has 82 protons and 126 neutrons, for example.

The free neutron has a mass of about 1.675×10^{-27} kg (equivalent to 939.6 MeV/c^2, or 1.0087 u).[3] The neutron has a mean square radius of about 0.8×10^{-15} m, or 0.8 fm,[10] and it is a spin-½ fermion.[11] The neutron has a magnetic moment with a negative value, because its orientation is opposite to the neutron's spin.[12] The neutron's magnetic moment causes its motion to be influenced by magnetic fields. Although the neutron has no net electric charge, it does have a slight distribution of charge within it. With its positive electric charge, the proton is directly influenced by electric fields, whereas the response of the neutron to this force is much weaker.

A free neutron is unstable, decaying to a proton, electron and antineutrino with a mean lifetime of just under 15 minutes (881.5±1.5 s). This radioactive decay, known as beta decay,[13] is possible since the mass of the neutron is slightly greater than the proton. The free proton is stable. Neutrons or protons bound in a nucleus can be stable or unstable, however, depending on the nuclide. Beta decay, in which neutrons decay to protons, or vice versa, is governed by the weak force, and it requires the emission or absorption of electrons and neutrinos, or their antiparticles.

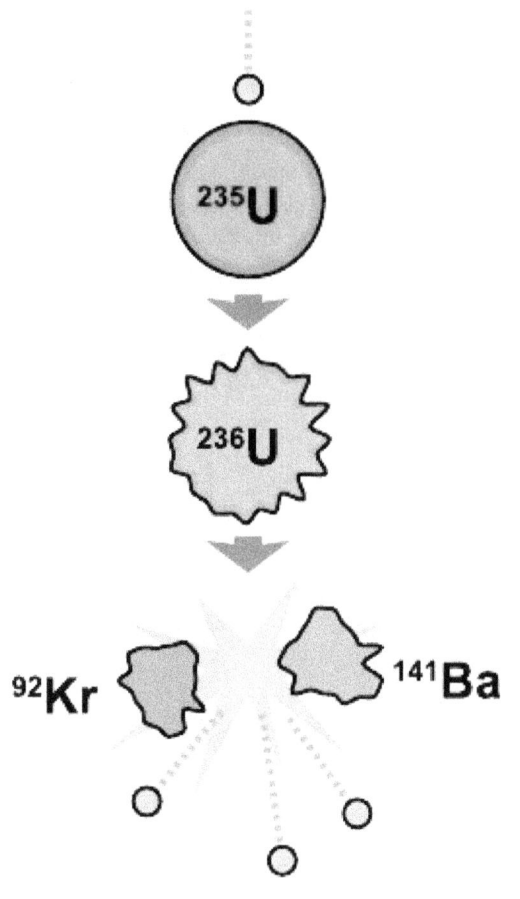

Nuclear fission caused by absorption of a neutron by uranium-235. The heavy nuclide fragments into lighter components and additional neutrons.

Protons and neutrons behave almost identically under the influence of the nuclear force within the nucleus. The concept of isospin, in which the proton and neutron are viewed as two quantum states of the same particle, is used to model the interactions of nucleons by the nuclear or weak forces. Because of the strength of the nuclear force at short distances, the binding energy of nucleons is more than seven orders of magnitude larger than the electromagnetic energy binding electrons in atoms. Nuclear reactions (such as nuclear fission) therefore have an energy density that is more than ten million times that of chemical reactions. Because of the mass–energy equivalence, nuclear binding energies add or subtract from the mass of nuclei. Ultimately, the ability of the nuclear force to store energy arising from the electromagnetic repulsion of nuclear components is the basis for most of the energy that makes nuclear reactors or bombs possible. In nuclear fission, the absorption of a neutron by a heavy nuclide (e.g., uranium-235) causes the nuclide to become unstable and break into light nuclides and additional neutrons. The positively charged light nuclides then repel, releasing electromagnetic potential energy.

The neutron is classified as a hadron, since it is composed of quarks, and as a baryon, since it is composed of three quarks.[14] The finite size of the neutron and its magnetic moment indicate the neutron is a composite, rather than elementary, particle. The neutron consists of two down quarks with charge $-\frac{1}{3}\,e$ and one up quark with charge $+\frac{2}{3}\,e$, although this simple model belies the complexities of the Standard Model for nuclei.[15] The masses of the three quarks sum to only about 12 MeV/c^2, whereas the neutron's mass is about 940 MeV/c^2, for example.[15] Like the proton, the quarks of the neutron are held together by the strong force, mediated by gluons.[16] The nuclear force results from secondary effects of the more fundamental strong force.

14.2 Discovery

Main article: Discovery of the neutron

The story of the discovery of the neutron and its properties is central to the extraordinary developments in atomic physics that occurred in the first half of the 20th century, leading ultimately to the atomic bomb in 1945. In the 1911 Rutherford model, the atom consisted of a small positively charged massive nucleus surrounded by a much larger cloud of negatively charged electrons. In 1920, Rutherford suggested that the nucleus consisted of positive protons and neutrally-charged particles, suggested to be a proton and an electron bound in some way.[17] Electrons were assumed to reside within the nucleus because it was known that beta radiation consisted of electrons emitted from the nucleus.[17] Rutherford called these uncharged particles *neutrons*, by the Latin root for *neutralis* (neuter) and the Greek suffix *-on* (a suffix used in the names of subatomic particles, i.e. *electron* and *proton*).[18][19] References to the word *neutron* in connection with the atom can be found in the literature as early as 1899, however.[20]

Throughout the 1920s, physicists assumed that the atomic nucleus was composed of protons and "nuclear electrons"[21][22] but there were obvious problems. It was difficult to reconcile the proton–electron model for nuclei with the Heisenberg uncertainty relation of quantum mechanics.[23][24] The Klein paradox,[25] discovered by Oskar Klein in 1928, presented further quantum mechanical objections to the notion of an electron confined within a nucleus.[23] Observed properties of atoms and molecules were inconsistent with the nuclear spin expected from proton–electron hypothesis. Since both protons and electrons carry an intrinsic spin of ½ $\hbar$, there is no way to arrange an odd number of spins ±½ $\hbar$ to give a spin integer multiple of $\hbar$. Nuclei with integer spin are common, e.g., ^{14}N.

In 1931, Walther Bothe and Herbert Becker found that if alpha particle radiation from polonium fell on beryllium, boron, or lithium, an unusually penetrating radiation was produced. The radiation was not influenced by an electric field, so Bothe and Becker assumed it was gamma radiation.[26][27] The following year Irène Joliot-Curie and Frédéric Joliot in Paris showed that if this "gamma" radiation fell on paraffin, or any other hydrogen-containing compound, it ejected protons of very high energy.[28] Neither Rutherford nor James Chadwick at the Cavendish Laboratory in Cambridge were convinced by the gamma ray interpretation.[29] Chadwick quickly performed a series of experiments that showed that the new radiation consisted of uncharged particles with about the same mass as the proton.[7][30][31] These particles were neutrons. Chadwick won the Nobel Prize in Physics for this discovery in 1935.[2]

Models for atomic nucleus consisting of protons and neutrons were quickly developed by Werner Heisenberg[32][33][34] and others.[35][36] The proton–neutron model explained the puzzle of nuclear spins. The origins of beta radiation were explained by Enrico Fermi in 1934 by the process of beta decay, in which the neutron decays to a proton by *creating* an electron and a (as yet undiscovered) neutrino.[37] In 1935 Chadwick and his doctoral student Maurice Goldhaber, reported the first accurate measurement of the mass of the neutron.[38][39]

By 1934, Fermi had bombarded heavier elements with neutrons to induce radioactivity in elements of high atomic number. In 1938, Fermi received the Nobel Prize in Physics *"for his demonstrations of the existence of new radioactive elements produced by neutron irradiation, and for his related discovery of nuclear reactions brought about by slow neutrons"*.[40] In 1938 Otto Hahn, Lise Meitner, and Fritz Strassmann discovered nuclear fission, or the fractionation of uranium nuclei into light elements, induced by neutron bombardment.[41][42][43] In 1945 Hahn received the 1944 Nobel Prize in Chemistry *"for his discovery of the fission of heavy atomic nuclei."* [44][45][46] The discovery of nuclear fission would lead to the development of nuclear power and the atomic bomb by the end of World War II.

14.3 Beta decay and the stability of the nucleus

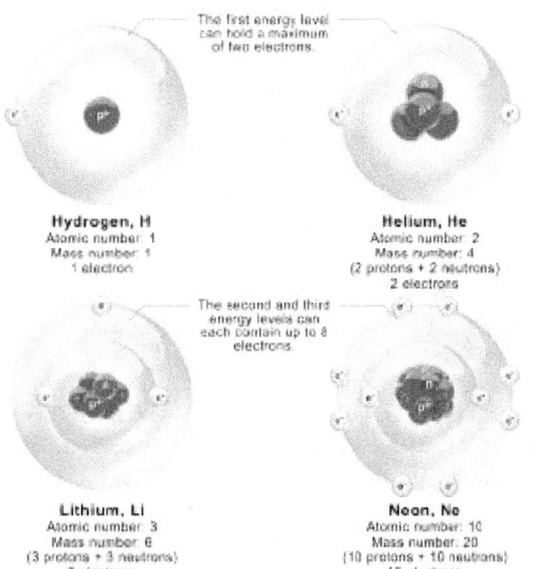

Models depicting the nucleus and electron energy levels in hydrogen, helium, lithium, and neon atoms. In reality, the diameter of the nucleus is about 100,000 times smaller than the diameter of the atom.

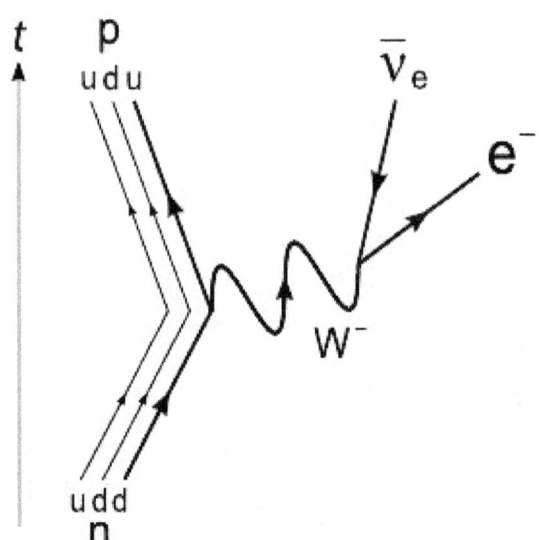

The Feynman diagram for beta decay of a neutron into a proton, electron, and electron antineutrino via an intermediate heavy W boson

Under the Standard Model of particle physics, the only possible decay mode for the neutron that conserves baryon number is for one of the neutron's quarks to change flavour via the weak interaction. The decay of one of the neutron's down quarks into a lighter up quark can be achieved by the emission of a W boson. By this process, the Standard Model description of beta decay, the neutron decays into a proton (which contains one down and two up quarks), an electron, and an electron antineutrino.

Since interacting protons have a mutual electromagnetic repulsion that is stronger than their attractive nuclear interaction, neutrons are a necessary constituent of any atomic nucleus that contains more than one proton (see diproton and neutron–proton ratio).[47] Neutrons bind with protons and one another in the nucleus via the nuclear force, effectively moderating the repulsive forces between the protons and stabilizing the nucleus.

See also: Beta-decay stable isobars and Neutron emission

14.3.1   Free neutron decay

Outside the nucleus, free neutrons are unstable and have a mean lifetime of 881.5 ± 1.5 s (about 14 minutes, 42 seconds); therefore the half-life for this process (which differs from the mean lifetime by a factor of $\ln(2) = 0.693$) is 611.0 ± 1.0 s (about 10 minutes, 11 seconds).[13] Beta decay of the neutron, described above, can be denoted by the radioactive decay:[48]

$$n0 \rightarrow p+ + e- + \nu_e$$

where p+, e−, and ν_e denote the proton, electron and electron antineutrino, respectively. For the free neutron the decay energy for this process (based on the masses of the neutron, proton, and electron) is 0.782343 MeV. The maximal energy of the beta decay electron (in the process wherein the neutrino receives a vanishingly small amount of kinetic energy) has been measured at $0.782 \pm .013$ MeV.[49] The latter number is not well-enough measured to determine the comparatively tiny rest mass of the neutrino (which must in theory be subtracted from the maximal electron kinetic energy) as well as neutrino mass is constrained by many other methods.

A small fraction (about one in 1000) of free neutrons decay with the same products, but add an extra particle in the form of an emitted gamma ray:

$$n0 \rightarrow p+ + e- + \nu_e + \gamma$$

This gamma ray may be thought of as a sort of "internal bremsstrahlung" that arises as the emitted beta particle interacts with the charge of the proton in an electromagnetic way. Internal bremsstrahlung gamma ray production is also a minor feature of beta decays of bound neutrons (as discussed below).

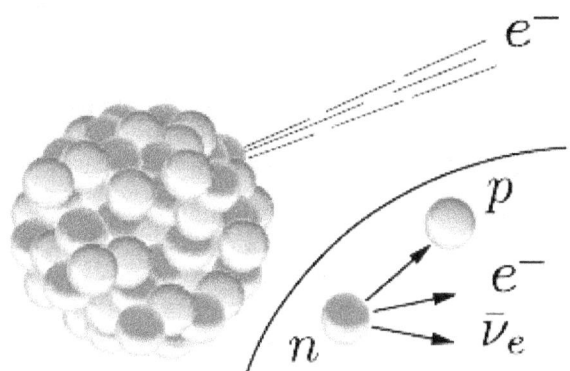

A schematic of the nucleus of an atom indicating β− radiation, the emission of a fast electron from the nucleus (the accompanying antineutrino is omitted). In the Rutherford model for the nucleus, red spheres were protons with positive charge and blue spheres were protons tightly bound to an electron with no net charge.
The **inset** shows beta decay of a free neutron as it is understood today: an electron and antineutrino are created in this process.

A very small minority of neutron decays (about four per million) are so-called "two-body (neutron) decays", in which a proton, electron and antineutrino are produced as usual, but the electron fails to gain the 13.6 eV necessary energy to escape the proton, and therefore simply remains bound to it, as a neutral hydrogen atom (one of the "two bodies"). In this type of free neutron decay, in essence all of the neutron decay energy is carried off by the antineutrino (the other "body").

The transformation of a free proton to a neutron (plus a positron and a neutrino) is energetically impossible, since a free neutron has a greater mass than a free proton.

14.3.2   Bound neutron decay

Main article: Atomic nucleus

While a free neutron has a half life of about 10.2 min, most neutrons within nuclei are stable. According to the nuclear shell model, the protons and neutrons of a nuclide are a quantum mechanical system organized into discrete energy levels with unique quantum numbers. For a neutron to decay, the resulting proton requires an available state at lower energy than the initial neutron state. In stable nuclei the possible lower energy states are all filled, meaning they are each occupied by two protons with spin up and spin down. The Pauli exclusion principle therefore disallows the decay of a neutron to a proton within stable nuclei. The situation is similar to electrons of an atom, where electrons have distinct atomic orbitals and are prevented from decaying to lower energy states, with the emission of a photon, by the exclusion principle.

Neutrons in unstable nuclei can decay by beta decay as described above. In this case, an energetically allowed quantum state is available for the proton resulting from the decay. One example of this decay is carbon-14 (6 protons, 8 neutrons) that decays to nitrogen-14 (7 protons, 7 neutrons) with a half-life of about 5,730 years.

Inside a nucleus, a proton can transform into a neutron via inverse beta decay, if an energetically allowed quantum state is available for the neutron. This transformation occurs by emission of an antielectron (also called positron) and an electron neutrino:

$$p+ \rightarrow n0 + e+ + \nu_e$$

The transformation of a proton to a neutron inside of a nucleus is also possible through electron capture:

$$p+ + e- \rightarrow n0 + \nu_e$$

Positron capture by neutrons in nuclei that contain an excess of neutrons is also possible, but is hindered because positrons are repelled by the positive nucleus, and quickly annihilate when they encounter electrons.

14.3.3   Competition of beta decay types

Three types of beta decay in competition are illustrated by the single isotope copper-64 (29 protons, 35 neutrons), which has a half-life of about 12.7 hours. This isotope has one unpaired proton and one unpaired neutron, so either the proton or the neutron can decay. This particular nuclide (though not all nuclides in this situation) is almost equally likely to decay through proton decay by positron emission (18%) or electron capture (43%), as through neutron decay by electron emission (39%).

14.4   Intrinsic properties

14.4.1   Electric charge

The total electric charge of the neutron is 0 e. This zero value has been tested experimentally, and the present exper-

imental limit for the charge of the neutron is $-2(8) \times 10^{\circ} -22$ e,[50] or $-3(13) \times 10^{\circ} -41$ C. This value is consistent with zero, given the experimental uncertainties (indicated in parentheses). By comparison, the charge of the proton is, of course, +1 e.

14.4.2   Electric dipole moment

Main article: Neutron electric dipole moment

The Standard Model of particle physics predicts a tiny separation of positive and negative charge within the neutron leading to a permanent electric dipole moment.[51] The predicted value is, however, well below the current sensitivity of experiments. From several unsolved puzzles in particle physics, it is clear that the Standard Model is not the final and full description of all particles and their interactions. New theories going beyond the Standard Model generally lead to much larger predictions for the electric dipole moment of the neutron. Currently, there are at least four experiments trying to measure for the first time a finite neutron electric dipole moment, including:

- Cryogenic neutron EDM experiment being set up at the Institut Laue–Langevin[52]

- nEDM experiment under construction at the new UCN source at the Paul Scherrer Institute[53]

- nEDM experiment being envisaged at the Spallation Neutron Source[54]

- nEDM experiment being built at the Institut Laue–Langevin[55]

14.4.3   Magnetic moment

Main article: Neutron magnetic moment

Even though the neutron is a neutral particle, the magnetic moment of a neutron is not zero. Since the neutron is a neutral particle, it is not affected by electric fields, but with its magnetic moment it is affected by magnetic fields. The magnetic moment of the neutron is an indication of its quark substructure and internal charge distribution.[56] The value for the neutron's magnetic moment was first directly measured by Luis Alvarez and Felix Bloch at Berkeley, California in 1940,[57] using an extension of the magnetic resonance methods developed by Rabi. Alvarez and Bloch determined the magnetic moment of the neutron to be $\mu_n = -1.93(2) \mu_N$, where μ_N is the nuclear magneton.

14.4.4 Structure and geometry of charge distribution

An article published in 2007 featuring a model-independent analysis concluded that the neutron has a negatively charged exterior, a positively charged middle, and a negative core.[58] In a simplified classical view, the negative "skin" of the neutron assists it to be attracted to the protons with which it interacts in the nucleus. (However, the main attraction between neutrons and protons is via the nuclear force, which does not involve charge.)

The simplified classical view of the neutron's charge distribution also "explains" the fact that the neutron magnetic dipole points in the opposite direction from its spin angular momentum vector (as compared to the proton). This gives the neutron, in effect, a magnetic moment which resembles a negatively charged particle. This can be reconciled classically with a neutral neutron composed of a charge distribution in which the negative sub-parts of the neutron have a larger average radius of distribution, and therefore contribute more to the particle's magnetic dipole moment, than do the positive parts that are, on average, nearer the core.

14.4.5 Mass

The mass of a neutron cannot be directly determined by mass spectrometry due to lack of electric charge. However, since the mass of protons and deuterons can be measured by mass spectrometry, the mass of a neutron can be deduced by subtracting proton mass from deuteron mass, with the difference being the mass of the neutron plus the binding energy of deuterium (expressed as a positive emitted energy). The latter can be directly measured by measuring the energy (B_d) of the single 0.7822 MeV gamma photon emitted when neutrons are captured by protons (this is exothermic and happens with zero-energy neutrons), plus the small recoil kinetic energy (E_{rd}) of the deuteron (about 0.06% of the total energy).

$$m_n = m_d - m_p + B_d - E_{rd}$$

The energy of the gamma ray can be measured to high precision by X-ray diffraction techniques, as was first done by Bell and Elliot in 1948. The best modern (1986) values for neutron mass by this technique are provided by Greene, et al.[59] These give a neutron mass of:

$$m_{neutron} = 1.008644904(14) \text{ u}$$

The value for the neutron mass in MeV is less accurately known, due to less accuracy in the known conversion of u to MeV:[60]

$$m_{neutron} = 939.56563(28) \text{ MeV}/c^2.$$

Another method to determine the mass of a neutron starts from the beta decay of the neutron, when the momenta of the resulting proton and electron are measured.

14.4.6 Anti-neutron

Main article: Antineutron

The antineutron is the antiparticle of the neutron. It was discovered by Bruce Cork in the year 1956, a year after the antiproton was discovered. CPT-symmetry puts strong constraints on the relative properties of particles and antiparticles, so studying antineutrons yields provide stringent tests on CPT-symmetry. The fractional difference in the masses of the neutron and antineutron is $(9\pm6)\times10^{5}$−5. Since the difference is only about two standard deviations away from zero, this does not give any convincing evidence of CPT-violation.[13]

14.5 Neutron compounds

14.5.1 Dineutrons and tetraneutrons

Main articles: Dineutron and Tetraneutron

The existence of stable clusters of 4 neutrons, or tetraneutrons, has been hypothesised by a team led by Francisco-Miguel Marqués at the CNRS Laboratory for Nuclear Physics based on observations of the disintegration of beryllium−14 nuclei. This is particularly interesting because current theory suggests that these clusters should not be stable.

The dineutron is another hypothetical particle. In 2012, Artemis Spyrou from Michigan State University and coworkers reported that they observed, for the first time, the dineutron emission in the decay of ^{16}Be. The dineutron character is evidenced by a small emission angle between the two neutrons. The authors measured the two-neutron separation energy to be 1.35(10) MeV, in good agreement with shell model calculations, using standard interactions for this mass region.[61]

14.5.2 Neutronium and neutron stars

Main articles: Neutronium and Neutron star

At extremely high pressures and temperatures, nucleons and electrons are believed to collapse into bulk neutronic matter, called neutronium. This is presumed to happen in neutron stars.

The extreme pressure inside a neutron star may deform the neutrons into a cubic symmetry, allowing tighter packing of neutrons.[62]

14.6   Detection

Main article: Neutron detection

The common means of detecting a charged particle by looking for a track of ionization (such as in a cloud chamber) does not work for neutrons directly. Neutrons that elastically scatter off atoms can create an ionization track that is detectable, but the experiments are not as simple to carry out; other means for detecting neutrons, consisting of allowing them to interact with atomic nuclei, are more commonly used. The commonly used methods to detect neutrons can therefore be categorized according to the nuclear processes relied upon, mainly neutron capture or elastic scattering. A good discussion on neutron detection is found in chapter 14 of the book *Radiation Detection and Measurement* by Glenn F. Knoll (John Wiley & Sons, 1979).

14.6.1   Neutron detection by neutron capture

A common method for detecting neutrons involves converting the energy released from neutron capture reactions into electrical signals. Certain nuclides have a high neutron capture cross section, which is the probability of absorbing a neutron. Upon neutron capture, the compound nucleus emits more easily detectable radiation, for example an alpha particle, which is then detected. The nuclides 3He, 6Li, 10B, 233U, 235U, 237Np and 239Pu are useful for this purpose.

14.6.2   Neutron detection by elastic scattering

Neutrons can elastically scatter off nuclei, causing the struck nucleus to recoil. Kinematically, a neutron can transfer more energy to light nuclei such as hydrogen or helium than to heavier nuclei. Detectors relying on elastic scattering are called fast neutron detectors. Recoiling nuclei can ionize and excite further atoms through collisions. Charge and/or scintillation light produced in this way can be collected to produce a detected signal. A major challenge in fast neutron detection is discerning such signals from erroneous signals produced by gamma radiation in the same detector.

Fast neutron detectors have the advantage of not requiring a moderator, and therefore being capable of measuring the neutron's energy, time of arrival, and in certain cases direction of incidence.

14.7   Sources and production

Main articles: Neutron source, neutron generator and research reactor

Free neutrons are unstable, although they have the longest half-life of any unstable sub-atomic particle by several orders of magnitude. Their half-life is still only about 10 minutes, however, so they can be obtained only from sources that produce them freshly.

Natural neutron background.   A small natural background flux of free neutrons exists everywhere on Earth. In the atmosphere and deep into the ocean, the "neutron background" is caused by muons produced by cosmic ray interaction with the atmosphere. These high energy muons are capable of penetration to considerable depths in water and soil. There, in striking atomic nuclei, among other reactions they induce spallation reactions in which a neutron is liberated from the nucleus. Within the Earth's crust a second source is neutrons produced primarily by spontaneous fission of uranium and thorium present in crustal minerals. The neutron background is not strong enough to be a biological hazard, but it is of importance to very high resolution particle detectors that are looking for very rare events, such as (hypothesized) interactions that might be caused by particles of dark matter.[9] Recent research has shown that even thunderstorms can produce neutrons with energies of up to several tens of MeV.[63]

Even stronger neutron background radiation is produced at the surface of Mars, where the atmosphere is thick enough to generate neutrons from cosmic ray muon production and neutron-spallation, but not thick enough to provide significant protection from the neutrons produced. These neutrons not only produce a Martian surface neutron radiation hazard from direct downward-going neutron radiation but may also produce a significant hazard from reflection of neutrons from the Martian surface, which will produce reflected neutron radiation penetrating upward into a Martian craft or habitat from the floor.[64]

Sources of neutrons for research.   These include certain types of radioactive decay (spontaneous fission and neutron emission), and from certain nuclear reactions. Convenient nuclear reactions include tabletop reactions such as natural alpha and gamma bombardment of certain nuclides, often beryllium or deuterium, and induced nuclear fission, such as occurs in nuclear reactors. In addition, high-energy nuclear reactions (such as occur in cosmic radiation showers or accelerator collisions) also produce neutrons from disintigration of target nuclei. Small (tabletop) particle accelerators optimized to produce free neutrons in this way, are called neutron generators.

In practice, the most commonly used small laboratory sources of neutrons use radioactive decay to power neutron production. One noted neutron-producing radioisotope, californium−252 decays (half-life 2.65 years) by spontaneous fission 3% of the time with production of 3.7 neutrons per fission, and is used alone as a neutron source from this process. Nuclear reaction sources (that involve two materials) powered by radioisotopes use an alpha decay source plus a beryllium target, or else a source of high-energy gamma radiation from a source that undergoes beta decay followed by gamma decay, which produces photoneutrons on interaction of the high energy gamma ray with ordinary stable beryllium, or else with the deuterium in heavy water. A popular source of the latter type is radioactive antimony-124 plus beryllium, a system with a half-life of 60.9 days, which can be constructed from natural antimony (which is 42.8% stable antimony-123) by activating it with neutrons in a nuclear reactor, then transported to where the neutron source is needed.°[65]

Institut Laue-Langevin (ILL) in Grenoble, France – a major neutron research facility.

Nuclear fission reactors naturally produce free neutrons; their role is to sustain the energy-producing chain reaction. The intense neutron radiation can also be used to produce various radioisotopes through the process of neutron activation, which is a type of neutron capture.

Experimental nuclear fusion reactors produce free neutrons as a waste product. However, it is these neutrons that possess most of the energy, and converting that energy to a

useful form has proved a difficult engineering challenge. Fusion reactors that generate neutrons are likely to create radioactive waste, but the waste is composed of neutron-activated lighter isotopes, which have relatively short (50–100 years) decay periods as compared to typical half-lives of 10,000 years for fission waste, which is long due primarily to the long half-life of alpha-emitting transuranic actinides.°[66]

14.7.1 Neutron beams and modification of beams after production

Free neutron beams are obtained from neutron sources by neutron transport. For access to intense neutron sources, researchers must go to a specialist neutron facility that operates a research reactor or a spallation source.

The neutron's lack of total electric charge makes it difficult to steer or accelerate them. Charged particles can be accelerated, decelerated, or deflected by electric or magnetic fields. These methods have little effect on neutrons. However, some effects may be attained by use of inhomogeneous magnetic fields because of the neutron's magnetic moment. Neutrons can be controlled by methods that include moderation, reflection, and velocity selection. Thermal neutrons can be polarized by transmission through magnetic materials in a method analogous to the Faraday effect for photons. Cold neutrons of wavelengths of 6–7 angstroms can be produced in beams of a high degree of polarization, by use of magnetic mirrors and magnetized interference filters.°[67]

14.8 Applications

The neutron plays an important role in many nuclear reactions. For example, neutron capture often results in neutron activation, inducing radioactivity. In particular, knowledge of neutrons and their behavior has been important in the development of nuclear reactors and nuclear weapons. The fissioning of elements like uranium-235 and plutonium-239 is caused by their absorption of neutrons.

Cold, thermal and *hot* neutron radiation is commonly employed in neutron scattering facilities, where the radiation is used in a similar way one uses X-rays for the analysis of condensed matter. Neutrons are complementary to the latter in terms of atomic contrasts by different scattering cross sections; sensitivity to magnetism; energy range for inelastic neutron spectroscopy; and deep penetration into matter.

The development of "neutron lenses" based on total internal reflection within hollow glass capillary tubes or by reflection from dimpled aluminum plates has driven ongoing research into neutron microscopy and neutron/gamma ray tomography.[68][69][70]

A major use of neutrons is to excite delayed and prompt gamma rays from elements in materials. This forms the basis of neutron activation analysis (NAA) and prompt gamma neutron activation analysis (PGNAA). NAA is most often used to analyze small samples of materials in a nuclear reactor whilst PGNAA is most often used to analyze subterranean rocks around bore holes and industrial bulk materials on conveyor belts.

Another use of neutron emitters is the detection of light nuclei, in particular the hydrogen found in water molecules. When a fast neutron collides with a light nucleus, it loses a large fraction of its energy. By measuring the rate at which slow neutrons return to the probe after reflecting off of hydrogen nuclei, a neutron probe may determine the water content in soil.

14.9   Medical therapies

Main articles: Fast neutron therapy and Neutron capture therapy of cancer

Because neutron radiation is both penetrating and ionizing, it can be exploited for medical treatments. Neutron radiation can have the unfortunate side-effect of leaving the affected area radioactive, however. Neutron tomography is therefore not a viable medical application.

Fast neutron therapy utilizes high energy neutrons typically greater than 20 MeV to treat cancer. Radiation therapy of cancers is based upon the biological response of cells to ionizing radiation. If radiation is delivered in small sessions to damage cancerous areas, normal tissue will have time to repair itself, while tumor cells often cannot.[71] Neutron radiation can deliver energy to a cancerous region at a rate an order of magnitude larger than gamma radiation[72]

Beams of low energy neutrons are used in boron capture therapy to treat cancer. In boron capture therapy, the patient is given a drug that contains boron and that preferentially accumulates in the tumor to be targeted. The tumor is then bombarded with very low energy neutrons (although often higher than thermal energy) which are captured by the boron-10 isotope in the boron, which produces an excited state of boron-11 that then decays to produce lithium-7 and an alpha particle that have sufficient energy to kill the malignant cell, but insufficient range to damage nearby cells. For such a therapy to be applied to the treatment of cancer, a neutron source having an intensity of the order of billion (10^9) neutrons per second per cm^2 is preferred. Such fluxes require a research nuclear reactor.

14.10   Protection

Exposure to free neutrons can be hazardous, since the interaction of neutrons with molecules in the body can cause disruption to molecules and atoms, and can also cause reactions that give rise to other forms of radiation (such as protons). The normal precautions of radiation protection apply: Avoid exposure, stay as far from the source as possible, and keep exposure time to a minimum. Some particular thought must be given to how to protect from neutron exposure, however. For other types of radiation, e.g., alpha particles, beta particles, or gamma rays, material of a high atomic number and with high density make for good shielding; frequently, lead is used. However, this approach will not work with neutrons, since the absorption of neutrons does not increase straightforwardly with atomic number, as it does with alpha, beta, and gamma radiation. Instead one needs to look at the particular interactions neutrons have with matter (see the section on detection above). For example, hydrogen-rich materials are often used to shield against neutrons, since ordinary hydrogen both scatters and slows neutrons. This often means that simple concrete blocks or even paraffin-loaded plastic blocks afford better protection from neutrons than do far more dense materials. After slowing, neutrons may then be absorbed with an isotope that has high affinity for slow neutrons without causing secondary capture radiation, such as lithium-6.

Hydrogen-rich ordinary water affects neutron absorption in nuclear fission reactors: Usually, neutrons are so strongly absorbed by normal water that fuel enrichment with fissionable isotope is required. The deuterium in heavy water has a very much lower absorption affinity for neutrons than does protium (normal light hydrogen). Deuterium is, therefore, used in CANDU-type reactors, in order to slow (moderate) neutron velocity, to increase the probability of nuclear fission compared to neutron capture.

14.11   Neutron temperature

Main article: Neutron temperature

14.11.1   Thermal neutrons

A *thermal neutron* is a free neutron that is Boltzmann distributed with kT = 0.0253 eV (4.0×10^{-21} J) at room temperature. This gives characteristic (not average, or median) speed of 2.2 km/s. The name 'thermal' comes from their energy being that of the room temperature gas or material they are permeating. (see *kinetic theory* for energies and speeds of molecules). After a number of collisions (often in the range of 10–20) with nuclei, neutrons arrive at this energy level, provided that they are not absorbed.

In many substances, thermal neutron reactions show a much larger effective cross-section than reactions involving faster neutrons, and thermal neutrons can therefore be absorbed more readily (i.e., with higher probability) by any atomic nuclei that they collide with, creating a heavier — and often unstable — isotope of the chemical element as a result.

Most fission reactors use a neutron moderator to slow down, or *thermalize* the neutrons that are emitted by nuclear fission so that they are more easily captured, causing further fission. Others, called fast breeder reactors, use fission energy neutrons directly.

14.11.2 Cold neutrons

Cold neutrons are thermal neutrons that have been equilibrated in a very cold substance such as liquid deuterium. Such a *cold source* is placed in the moderator of a research reactor or spallation source. Cold neutrons are particularly valuable for neutron scattering experiments.

Example of Cold
Neutron Source

Liquid Hydrogen Moderator
Heavy Water Moderator
Hydrogen Vapor
Vacuum
Vacuum

Cold neutron source providing neutrons at about the temperature of liquid hydrogen

14.11.3 Ultracold neutrons

Ultracold neutrons are produced by inelastically scattering cold neutrons in substances with a temperature of a few kelvins, such as solid deuterium or superfluid helium. An alternative production method is the mechanical deceleration of cold neutrons.

14.11.4 Fission energy neutrons

Main article: nuclear fission

A *fast neutron* is a free neutron with a kinetic energy level close to 1 MeV (1.6×10^{-13} J), hence a speed of ~14000 km/s (~ 5% of the speed of light). They are named *fission energy* or *fast* neutrons to distinguish them from lower-energy thermal neutrons, and high-energy neutrons produced in cosmic showers or accelerators. Fast neutrons are produced by nuclear processes such as nuclear fission. Neutrons produced in fission, as noted above, have a Maxwell–Boltzmann distribution of kinetic energies from 0 to ~14 MeV, a mean energy of 2 MeV (for U-235 fission neutrons), and a mode of only 0.75 MeV, which means that more than half of them do not qualify as fast (and thus have almost no chance of initiating fission in fertile materials, such as U-238 and Th-232).

Fast neutrons can be made into thermal neutrons via a process called moderation. This is done with a neutron moderator. In reactors, typically heavy water, light water, or graphite are used to moderate neutrons.

14.11.5 Fusion neutrons

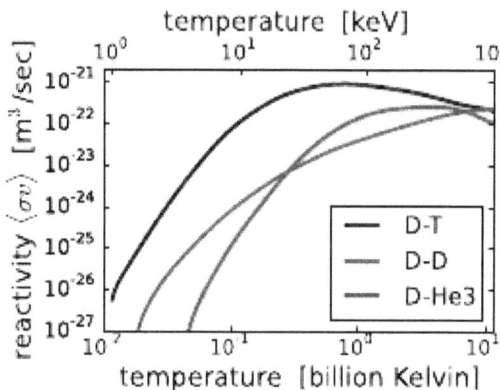

The fusion reaction rate increases rapidly with temperature until it maximizes and then gradually drops off. The DT rate peaks at a lower temperature (about 70 keV, or 800 million kelvins) and at a higher value than other reactions commonly considered for fusion energy.

For more details on this topic, see Nuclear fusion § Criteria and candidates for terrestrial reactions.

D–T (deuterium–tritium) fusion is the fusion reaction that produces the most energetic neutrons, with 14.1 MeV of kinetic energy and traveling at 17% of the speed of light. D–

T fusion is also the easiest fusion reaction to ignite, reaching near-peak rates even when the deuterium and tritium nuclei have only a thousandth as much kinetic energy as the 14.1 MeV that will be produced.

14.1 MeV neutrons have about 10 times as much energy as fission neutrons, and are very effective at fissioning even non-fissile heavy nuclei, and these high-energy fissions produce more neutrons on average than fissions by lower-energy neutrons. This makes D–T fusion neutron sources such as proposed tokamak power reactors useful for transmutation of transuranic waste. 14.1 MeV neutrons can also produce neutrons by knocking them loose from nuclei.

On the other hand, these very high energy neutrons are less likely to simply be captured without causing fission or spallation. For these reasons, nuclear weapon design extensively utilizes D–T fusion 14.1 MeV neutrons to cause more fission. Fusion neutrons are able to cause fission in ordinarily non-fissile materials, such as depleted uranium (uranium-238), and these materials have been used in the jackets of thermonuclear weapons. Fusion neutrons also can cause fission in substances that are unsuitable or difficult to make into primary fission bombs, such as reactor grade plutonium. This physical fact thus causes ordinary non-weapons grade materials to become of concern in certain nuclear proliferation discussions and treaties.

Other fusion reactions produce much less energetic neutrons. D–D fusion produces a 2.45 MeV neutron and helium-3 half of the time, and produces tritium and a proton but no neutron the other half of the time. D–^{3}He fusion produces no neutron.

14.11.6   Intermediate-energy neutrons

A fission energy neutron that has slowed down but not yet reached thermal energies is called an epithermal neutron.

Cross sections for both capture and fission reactions often have multiple resonance peaks at specific energies in the epithermal energy range. These are of less significance in a fast neutron reactor, where most neutrons are absorbed before slowing down to this range, or in a well-moderated thermal reactor, where epithermal neutrons interact mostly with moderator nuclei, not with either fissile or fertile actinide nuclides. However, in a partially moderated reactor with more interactions of epithermal neutrons with heavy metal nuclei, there are greater possibilities for transient changes in reactivity that might make reactor control more difficult.

Ratios of capture reactions to fission reactions are also worse (more captures without fission) in most nuclear fuels such as plutonium-239, making epithermal-spectrum reactors using these fuels less desirable, as captures not only

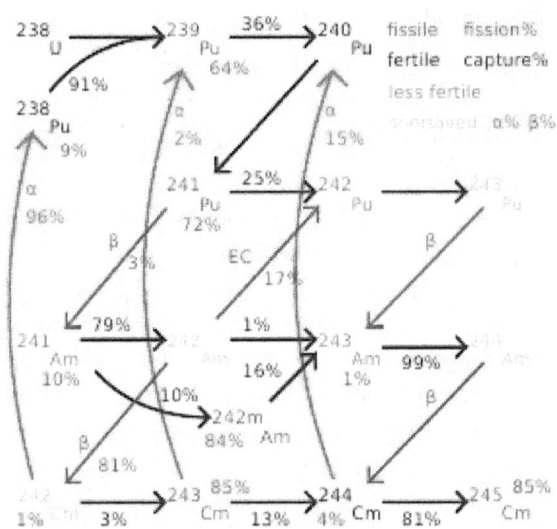

Transmutation flow in light water reactor, which is a thermal-spectrum reactor

waste the one neutron captured but also usually result in a nuclide that is not fissile with thermal or epithermal neutrons, though still fissionable with fast neutrons. The exception is uranium-233 of the thorium cycle, which has good capture-fission ratios at all neutron energies.

14.11.7   High-energy neutrons

These neutrons have much more energy than fission energy neutrons and are generated as secondary particles by particle accelerators or in the atmosphere from cosmic rays. They can have energies as high as tens of joules per neutron. These neutrons are extremely efficient at ionization and far more likely to cause cell death than X-rays or protons.[73][74]

14.12   See also

- Ionizing radiation

- Isotope

- List of particles

- Neutronium

- Neutron magnetic moment

- Neutron radiation and the Sievert radiation scale

- Nuclear reaction

- Thermal reactor

- Nucleosynthesis

    - Neutron capture nucleosynthesis

    - R-process

    - S-process

14.12.1   Neutron sources

- Neutron generator

- Neutron sources

14.12.2   Processes involving neutrons

- Neutron bomb

- Neutron diffraction

- Neutron flux

- Neutron transport

14.13   References

[1] Ernest Rutherford. Chemed.chem.purdue.edu. Retrieved on 2012-08-16.

[2] 1935 Nobel Prize in Physics. Nobelprize.org. Retrieved on 2012-08-16.

[3] Mohr, P.J.; Taylor, B.N. and Newell, D.B. (2014). "The 2014 CODATA Recommended Values of the Fundamental Physical Constants" (Web Version 7.0). The database was developed by J. Baker, M. Douma, and S. Kotochigova. (2014). National Institute of Standards and Technology, Gaithersburg, Maryland 20899.

[4] Thomas, A.W.; Weise, W. (2001). *The Structure of the Nucleon.* Wiley-WCH, Berlin. ISBN 3-527-40297-7

[5] Glasstone, Samuel; Dolan, Philip J., eds. (1977), *The Effects of Nuclear Weapons, Third Edition*, U.S. Dept. of Defense and Energy Research and Development Administration, U.S. Government Printing Office. ISBN 1-60322-016-X

[6] Nudat 2. Nndc.bnl.gov. Retrieved on 2010-12-04.

[7] Chadwick, James (1932). "Possible Existence of a Neutron". *Nature* **129** (3252): 312. Bibcode:1932Natur.129Q.312C. doi:10.1038/129312a0.

[8] Hahn, O. and Strassmann, F. (1939). "Über den Nachweis und das Verhalten der bei der Bestrahlung des Urans mittels Neutronen entstehenden Erdalkalimetalle ("On the detection and characteristics of the alkaline earth metals formed by irradiation of uranium with neutrons")". *Naturwissenschaften* **27** (1): 11–15. Bibcode:1939NW.....27...11H. doi:10.1007/BF01488241.. The authors were identified as being at the Kaiser-Wilhelm-Institut für Chemie, Berlin-Dahlem. Received 22 December 1938.

[9] Carson, M. J.; et al. (2004). "Neutron background in large-scale xenon detectors for dark matter searches". *Astroparticle Physics* **21** (6): 667–687. doi:10.1016/j.astropartphys.2004.05.001.

[10] Povh, B.; Rith, K.; Scholz, C.; Zetsche, F. (2002). *Particles and Nuclei: An Introduction to the Physical Concepts*. Berlin: Springer-Verlag. p. 73. ISBN 978-3-540-43823-6.

[11] J.-L. Basdevant, J. Rich, M. Spiro (2005). *Fundamentals in Nuclear Physics*. Springer. p. 155. ISBN 0-387-01672-4.

[12] Paul Allen Tipler, Ralph A. Llewellyn (2002). *Modern Physics* (4 ed.). Macmillan. p. 310. ISBN 0-7167-4345-0.

[13] Nakamura, K (2010). "Review of Particle Physics". *Journal of Physics G: Nuclear and Particle Physics* **37** (7A): 075021. Bibcode:2010JPhG...37g5021N. doi:10.1088/0954-3899/37/7A/075021. PDF with 2011 partial update for the 2012 edition The exact value of the mean lifetime is still uncertain, due to conflicting results from experiments. The Particle Data Group reports values up to six seconds apart (more than four standard deviations), commenting that "our 2006, 2008, and 2010 Reviews stayed with 885.7±0.8 s; but we noted that in light of SEREBROV 05 our value should be regarded as suspect until further experiments clarified matters. Since our 2010 Review, PICHLMAIER 10 has obtained a mean life of 880.7±1.8 s, closer to the value of SEREBROV 05 than to our average. And SEREBROV 10B[...] claims their values should be lowered by about 6 s, which would bring them into line with the two lower values. However, those reevaluations have not received an enthusiastic response from the experimenters in question; and in any case the Particle Data Group would have to await published changes (by those experimenters) of published values. At this point, we can think of nothing better to do than to average the seven best but discordant measurements, getting 881.5±1.5s. Note that the error includes a scale factor of 2.7. This is a jump of 4.2 old (and 2.8 new) standard deviations. This state of affairs is a particularly unhappy one, because the value is so important. We again call upon the experimenters to clear this up."

[14] R.K. Adair (1989). *The Great Design: Particles, Fields, and Creation.* Oxford University Press. p. 214.

[15] Cho, Adiran (2 April 2010). "Mass of the Common Quark Finally Nailed Down". *http://news.sciencemag.org*. American Association for the Advancement of Science. Retrieved 27 September 2014.

[16] W.N. Cottingham, D.A. Greenwood (1986). *An Introduction to Nuclear Physics*. Cambridge University Press. ISBN 9780521657334.

[17] E. Rutherford (1920). "Nuclear Constitution of Atoms". *Proceedings of the Royal Society A* **97** (686): 374–400. Bibcode:1920RSPSA..97..374R. doi:10.1098/rspa.1920.0040.

[18] "Wolfgang Pauli". *Sources in the History of Mathematics and Physical Sciences*. Sources in the History of Mathematics and Physical Sciences **6**: 105–144. 1985. doi:10.1007/978-3-540-78801-0_3. ISBN 978-3-540-13609-5. |chapter= ignored (help)

[19] Hendry, John, ed. (1984), *Cambridge Physics in the Thirties*, Adam Hilger Ltd, Bristol, ISBN 0852747616

[20] N. Feather (1960). "A history of neutrons and nuclei. Part 1". *Contemporary Physics* **1** (3): 191–203. doi:10.1080/00107516008202611.

[21] Brown, Laurie M. (1978). "The idea of the neutrino". *Physics Today* **31** (9): 23. Bibcode:1978PhT....31i..23B. doi:10.1063/1.2995181.

[22] Friedlander G., Kennedy J.W. and Miller J.M. (1964) *Nuclear and Radiochemistry* (2nd edition), Wiley, pp. 22–23 and 38–39

[23] Stuewer, Roger H. (1985). "Niels Bohr and Nuclear Physics". In French, A. P.; Kennedy, P. J. *Niels Bohr: A Centenary Volume*. Harvard University Press. pp. 197–220. ISBN 0674624165.

[24] Pais, Abraham (1986). *Inward Bound*. Oxford: Oxford University Press. p. 299. ISBN 0198519974.

[25] Klein, O. (1929). "Die Reflexion von Elektronen an einem Potentialsprung nach der relativistischen Dynamik von Dirac". *Zeitschrift für Physik* **53** (3–4): 157–165. Bibcode:1929ZPhy...53..157K. doi:10.1007/BF01339716.

[26] Bothe, W.; Becker, H. (1930). "Künstliche Erregung von Kern-γ-Strahlen" [Artificial excitation of nuclear γ-radiation]. *Zeitschrift für Physik* **66** (5–6): 289–306. Bibcode:1930ZPhy...66..289B. doi:10.1007/BF01390908.

[27] Becker, H.; Bothe, W. (1932). "Die in Bor und Beryllium erregten γ-Strahlen" [Γ-rays excited in boron and beryllium]. *Zeitschrift für Physik* **76** (7–8): 421–438. Bibcode:1932ZPhy...76..421B. doi:10.1007/BF01336726.

[28] Joliot-Curie, Irène and Joliot, Frédéric (1932). "Émission de protons de grande vitesse par les substances hydrogénées sous l'influence des rayons γ très pénétrants" [Emission of high-speed protons by hydrogenated substances under the influence of very penetrating γ-rays]. *Comptes Rendus* **194**: 273.

[29] Brown, Andrew (1997). *The Neutron and the Bomb: A Biography of Sir James Chadwick*. Oxford University Press. ISBN 978-0-19-853992-6.

[30] "Atop the Physics Wave: Rutherford Back in Cambridge, 1919–1937". *Rutherford's Nuclear World*. American Institute of Physics. 2011–2014. Retrieved 19 August 2014.

[31] Chadwick, J. (1933). "Bakerian Lecture. The Neutron". *Proceedings of the Royal Society A: Mathematical, Physical and Engineering Sciences* **142** (846): 1–25. Bibcode:1933RSPSA.142....1C. doi:10.1098/rspa.1933.0152.

[32] Heisenberg, W. (1932). "Über den Bau der Atomkerne. I". *Z. Phys.* **77**: 1–11. doi:10.1007/BF01342433.

[33] Heisenberg, W. (1932). "Über den Bau der Atomkerne. II". *Z. Phys.* **78** (3–4): 156–164. doi:10.1007/BF01337585.

[34] Heisenberg, W. (1933). "Über den Bau der Atomkerne. III". *Z. Phys.* **80** (9–10): 587–596. doi:10.1007/BF01335696.

[35] Iwanenko, D.D., The neutron hypothesis, Nature **129** (1932) 798.

[36] Miller A. I. *Early Quantum Electrodynamics: A Sourcebook*, Cambridge University Press, Cambridge, 1995, ISBN 0521568919, pp. 84–88.

[37] Wilson, Fred L. (1968). "Fermi's Theory of Beta Decay". *Am. J. Phys.* **36** (12): 1150–1160. Bibcode:1968AmJPh..36.1150W. doi:10.1119/1.1974382.

[38] Chadwick, J.; Goldhaber, M. (1934). "A nuclear photo-effect: disintegration of the diplon by gamma rays". *Nature* **134**: 237–238. doi:10.1038/134237a0.

[39] Chadwick, J.; Goldhaber, M. (1935). "A nuclear photo-electric effect" (PDF). *Proc. R. Soc. Lond* **151**: 479–493. doi:10.1098/rspa.1935.0162.

[40] Cooper, Dan (1999). *Enrico Fermi: And the Revolutions in Modern physics*. New York: Oxford University Press. ISBN 0-19-511762-X. OCLC 39508200.

[41] Hahn, O. (1958). "The Discovery of Fission". *Scientific American* **198** (2): 76. doi:10.1038/scientificamerican0258-76.

[42] Rife, Patricia (1999). *Lise Meitner and the dawn of the nuclear age*. Basel, Switzerland: Birkhäuser. ISBN 0-8176-3732-X.

[43] Hahn, O.; Strassmann, F. (10 February 1939). "Proof of the Formation of Active Isotopes of Barium from Uranium and Thorium Irradiated with Neutrons; Proof of the Existence of More Active Fragments Produced by Uranium Fission". *Die Naturwissenschaften* **27**: 89–95.

[44] "The Nobel Prize in Chemistry 1944". Nobel Foundation. Retrieved 2007-12-17.

[45] Bernstein, Jeremy (2001). *Hitler's uranium club: the secret recordings at Farm Hall*. New York: Copernicus. p. 281. ISBN 0-387-95089-3.

[46] "The Nobel Prize in Chemistry 1944: Presentation Speech". Nobel Foundation. Retrieved 2008-01-03.

[47] Sir James Chadwick's Discovery of Neutrons. ANS Nuclear Cafe. Retrieved on 2012-08-16.

[48] Particle Data Group Summary Data Table on Baryons. lbl.gov (2007). Retrieved on 2012-08-16.

[49] Basic Ideas and Concepts in Nuclear Physics: An Introductory Approach, Third Edition K. Heyde Taylor & Francis 2004. Print ISBN 978-0-7503-0980-6. eBook ISBN 978-1-4200-5494-1. DOI: 10.1201/9781420054941.ch5. full text

[50] Olive, K.A.; (Particle Data Group); et al. (2014). "Review of Particle Physics". *Chin. Phys. C* **38**: 090001. doi:10.1088/1674-1137/38/9/090001.

[51] "Pear-shaped particles probe big-bang mystery" (Press release). University of Sussex. 20 February 2006. Retrieved 2009-12-14.

[52] A cryogenic experiment to search for the EDM of the neutron. Hepwww.rl.ac.uk. Retrieved on 2012-08-16.

[53] Search for the neutron electric dipole moment: nEDM. Nedm.web.psi.ch (2001-09-12). Retrieved on 2012-08-16.

[54] SNS Neutron EDM Experiment. P25ext.lanl.gov. Retrieved on 2012-08-16.

[55] Measurement of the Neutron Electric Dipole Moment. Nrd.pnpi.spb.ru. Retrieved on 2012-08-16.

[56] Gell, Y.; Lichtenberg, D. B. (1969). "Quark model and the magnetic moments of proton and neutron". *Il Nuovo Cimento A*. Series 10 **61**: 27–40. Bibcode:1969NCimA..61...27G. doi:10.1007/BF02760010.

[57] Alvarez, L. W; Bloch, F. (1940). "A quantitative determination of the neutron magnetic moment in absolute nuclear magnetons". *Physical Review* **57**: 111–122. doi:10.1103/physrev.57.111.

[58] Miller, G.A. (2007). "Charge Densities of the Neutron and Proton". *Physical Review Letters* **99** (11): 112001. Bibcode:2007PhRvL..99k2001M. doi:10.1103/PhysRevLett.99.112001.

[59] Greene, GL; et al. (1986). "New determination of the deuteron binding energy and the neutron mass". *Phys. Rev. Lett.* **56**: 819–822. Bibcode:1986PhRvL..56..819G. doi:10.1103/PhysRevLett.56.819.

[60] Byrne, J. *Neutrons, Nuclei, and Matter*, Dover Publications, Mineola, New York, 2011, ISBN 0486482383, pp. 18–19

[61] Spyrou, A.; et al. (2012). "First Observation of Ground State Dineutron Decay: 16Be". *Physical Review Letters* **108** (10): 102501. Bibcode:2012PhRvL.108j2501S. doi:10.1103/PhysRevLett.108.102501. PMID 22463404.

[62] Llanes-Estrada, Felipe J.; Moreno Navarro, Gaspar (2011). "Cubic neutrons". arXiv:1108.1859v1 [nucl-th].

[63] Köhn, C., Ebert, U., Calculation of beams of positrons, neutrons and protons associated with terrestrial gamma-ray flashes, J. Geophys. Res. Atmos. (2015), vol. 23, doi: 10.1002/2014JD022229

[64] Clowdsley, MS; Wilson, JW; Kim, MH; Singleterry, RC; Tripathi, RK; Heinbockel, JH; Badavi, FF; Shinn, JL (2001). "Neutron Environments on the Martian Surface" (PDF). *Physica Medica* **17** (Suppl 1): 94–6. PMID 11770546.

[65] Byrne, J. *Neutrons, Nuclei, and Matter*, Dover Publications, Mineola, New York, 2011, ISBN 0486482383, pp. 32–33.

[66] Science/Nature | Q&A: Nuclear fusion reactor. BBC News (2006-02-06). Retrieved on 2010-12-04.

[67] Byrne, J. *Neutrons, Nuclei, and Matter*, Dover Publications, Mineola, New York, 2011, ISBN 0486482383, p. 453.

[68] Kumakhov, M. A.; Sharov, V. A. (1992). "A neutron lens". *Nature* **357** (6377): 390–391. Bibcode:1992Natur.357..390K. doi:10.1038/357390a0.

[69] Physorg.com, "New Way of 'Seeing': A 'Neutron Microscope'". Physorg.com (2004-07-30). Retrieved on 2012-08-16.

[70] "NASA Develops a Nugget to Search for Life in Space". NASA.gov (2007-11-30). Retrieved on 2012-08-16.

[71] Hall EJ (2000). *Radiobiology for the Radiologist*. Lippincott Williams & Wilkins; 5th edition

[72] Johns HE and Cunningham JR (1978). *The Physics of Radiology*. Charles C Thomas 3rd edition

[73] Freeman, Tami (May 23, 2008). "Facing up to secondary neutrons". Medical Physics Web. Retrieved 2011-02-08.

[74] Heilbronn, L.; Nakamura, T; Iwata, Y; Kurosawa, T; Iwase, H; Townsend, LW (2005). "Expand+Overview of secondary neutron production relevant to shielding in space". *Radiation Protection Dosimetry* **116** (1–4): 140–143. doi:10.1093/rpd/nci033. PMID 16604615.

14.14 Further reading

- Annotated bibliography for neutrons from the Alsos Digital Library for Nuclear Issues

- Abraham Pais, *Inward Bound*. Oxford: Oxford University Press, 1986. ISBN 0198519974.

- Sin-Itiro Tomonaga, *The Story of Spin*. The University of Chicago Press, 1997

- Herwig Schopper, *Weak interactions and nuclear beta decay*. Publisher, North-Holland Pub. Co., 1966.

14.15   External links

- neutron properties at Particle Data Group, Lawrence Berkeley National Laboratory in Berkeley, CA. (pdgLive)

Chapter 15

Bohr model

'Rutherford–Bohr model' and 'Bohr–Rutherford diagram' redirect to this page. 'Bohr model' is not to be confused with Bohr equation.

In atomic physics, the Rutherford–Bohr model or **Bohr**

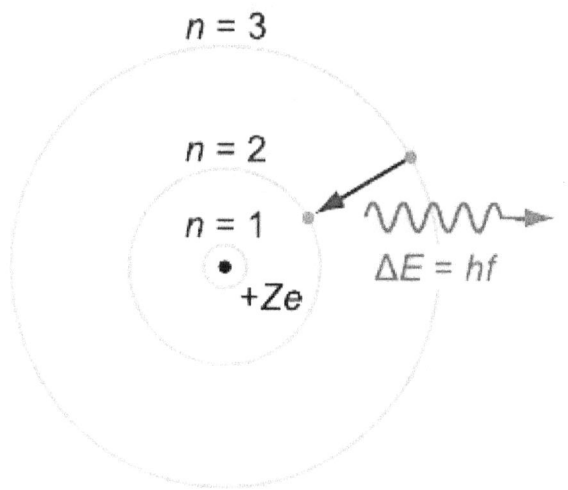

*The **Rutherford–Bohr model** of the hydrogen atom (Z = 1) or a hydrogen-like ion (Z > 1), where the negatively charged electron confined to an atomic shell encircles a small, positively charged atomic nucleus and where an electron jump between orbits is accompanied by an emitted or absorbed amount of electromagnetic energy (hν).[1] The orbits in which the electron may travel are shown as grey circles; their radius increases as n^2, where n is the principal quantum number. The 3 → 2 transition depicted here produces the first line of the Balmer series, and for hydrogen (Z = 1) it results in a photon of wavelength 656 nm (red light).*

model, introduced by Niels Bohr in 1913, depicts the atom as a small, positively charged nucleus surrounded by electrons that travel in circular orbits around the nucleus— similar in structure to the solar system, but with attraction provided by electrostatic forces rather than gravity. After the cubic model (1902), the plum-pudding model (1904), the Saturnian model (1904), and the Rutherford model (1911) came the **Rutherford–Bohr model** or just *Bohr model* for short (1913). The improvement to the Ruther-

ford model is mostly a quantum physical interpretation of it. The Bohr model has been superseded, but the quantum theory remains sound.

The model's key success lay in explaining the Rydberg formula for the spectral emission lines of atomic hydrogen. While the Rydberg formula had been known experimentally, it did not gain a theoretical underpinning until the Bohr model was introduced. Not only did the Bohr model explain the reason for the structure of the Rydberg formula, it also provided a justification for its empirical results in terms of fundamental physical constants.

The Bohr model is a relatively primitive model of the hydrogen atom, compared to the valence shell atom. As a theory, it can be derived as a first-order approximation of the hydrogen atom using the broader and much more accurate quantum mechanics and thus may be considered to be an obsolete scientific theory. However, because of its simplicity, and its correct results for selected systems (see below for application), the Bohr model is still commonly taught to introduce students to quantum mechanics or energy level diagrams before moving on to the more accurate, but more complex, valence shell atom. A related model was originally proposed by Arthur Erich Haas in 1910, but was rejected. The quantum theory of the period between Planck's discovery of the quantum (1900) and the advent of a full-blown quantum mechanics (1925) is often referred to as the old quantum theory.

15.1 Origin

In the early 20th century, experiments by Ernest Rutherford established that atoms consisted of a diffuse cloud of negatively charged electrons surrounding a small, dense, positively charged nucleus.[2] Given this experimental data, Rutherford naturally considered a planetary-model atom, the Rutherford model of 1911 – electrons orbiting a solar nucleus – however, said planetary-model atom has a technical difficulty. The laws of classical mechanics (i.e. the Larmor formula), predict that the electron will release electromagnetic radiation while orbiting a nucleus. Because the electron would lose energy, it would rapidly spiral inwards, collapsing into the nucleus on a timescale of around 16 picoseconds.[3] This atom model is disastrous, because it predicts that all atoms are unstable.[4]

Also, as the electron spirals inward, the emission would rapidly increase in frequency as the orbit got smaller and faster. This would produce a continuous smear, in frequency, of electromagnetic radiation. However, late 19th century experiments with electric discharges have shown that atoms will only emit light (that is, electromagnetic radiation) at certain discrete frequencies.

To overcome this difficulty, Niels Bohr proposed, in 1913, what is now called the *Bohr model of the atom*. He suggested that electrons could only have certain *classical* motions:

1. Electrons in atoms orbit the nucleus.

2. The electrons can only orbit stably, without radiating, in certain orbits (called by Bohr the "stationary orbits"[5]) at a certain discrete set of distances from the nucleus. These orbits are associated with definite energies and are also called energy shells or energy levels. In these orbits, the electron's acceleration does not result in radiation and energy loss as required by classical electromagnetics. The Bohr model of an atom was based upon Planck's quantum theory of radiation.

3. Electrons can only gain and lose energy by jumping from one allowed orbit to another, absorbing or emitting electromagnetic radiation with a frequency ν determined by the energy difference of the levels according to the Planck relation:

$$\Delta E = E_2 - E_1 = h\nu .$$

where h is Planck's constant. The frequency of the radiation emitted at an orbit of period T is as it would be in classical mechanics; it is the reciprocal of the classical orbit period:

$$\nu = \tfrac{1}{T} .$$

The significance of the Bohr model is that the laws of classical mechanics apply to the motion of the electron about the nucleus *only when restricted by a quantum rule*. Although Rule 3 is not completely well defined for small orbits, because the emission process involves two orbits with two different periods, Bohr could determine the energy spacing between levels using Rule 3 and come to an exactly correct quantum rule: the angular momentum L is restricted to be an integer multiple of a fixed unit:

$$L = n\frac{h}{2\pi} = n\hbar$$

where $n = 1, 2, 3, \dots$ is called the principal quantum number, and $\hbar = h/2\pi$. The lowest value of n is 1; this gives a smallest possible orbital radius of 0.0529 nm known as the Bohr radius. Once an electron is in this lowest orbit, it can get no closer to the proton. Starting from the angular momentum quantum rule, Bohr[2] was able to calculate the energies of the allowed orbits of the hydrogen atom and other hydrogen-like atoms and ions.

Other points are:

1. Like Einstein's theory of the Photoelectric effect, Bohr's formula assumes that during a quantum jump a *discrete* amount of energy is radiated. However, unlike Einstein, Bohr stuck to the *classical* Maxwell theory of the electromagnetic field. Quantization of the electromagnetic field was explained by the discreteness of the atomic energy levels; Bohr did not believe in the existence of photons.

2. According to the Maxwell theory the frequency ν of classical radiation is equal to the rotation frequency ν_{rot} of the electron in its orbit, with harmonics at integer multiples of this frequency. This result is obtained from the Bohr model for jumps between energy levels E_n and E_{n-k} when k is much smaller than n. These jumps reproduce the frequency of the k-th harmonic of orbit n. For sufficiently large values of n (so-called Rydberg states), the two orbits involved in the emission process have nearly the same rotation frequency, so that the classical orbital frequency is not ambiguous. But for small n (or large k), the radiation frequency has no unambiguous classical interpretation. This marks the birth of the correspondence principle, requiring quantum theory to agree with the classical theory only in the limit of large quantum numbers.

3. The Bohr-Kramers-Slater theory (BKS theory) is a failed attempt to extend the Bohr model, which violates the conservation of energy and momentum in quantum jumps, with the conservation laws only holding on average.

Bohr's condition, that the angular momentum is an integer multiple of $\hbar$ was later reinterpreted in 1924 by de Broglie as a standing wave condition: the electron is described by a wave and a whole number of wavelengths must fit along the circumference of the electron's orbit:

$$n\lambda = 2\pi r .$$

Substituting de Broglie's wavelength of $\lambda = h/p$ reproduces Bohr's rule. In 1913, however, Bohr justified his rule by appealing to the correspondence principle, without providing any sort of wave interpretation. In 1913, the wave behavior of matter particles such as the electron (i.e., matter waves) was not suspected.

In 1925 a new kind of mechanics was proposed, quantum mechanics, in which Bohr's model of electrons traveling in quantized orbits was extended into a more accurate model of electron motion. The new theory was proposed by Werner Heisenberg. Another form of the same theory, wave mechanics, was discovered by the Austrian physicist Erwin Schrödinger independently, and by different reasoning. Schrödinger employed de Broglie's matter waves, but sought wave solutions of a three-dimensional wave equation describing electrons that were constrained to move about the nucleus of a hydrogen-like atom, by being trapped by the potential of the positive nuclear charge.

15.2 Electron energy levels

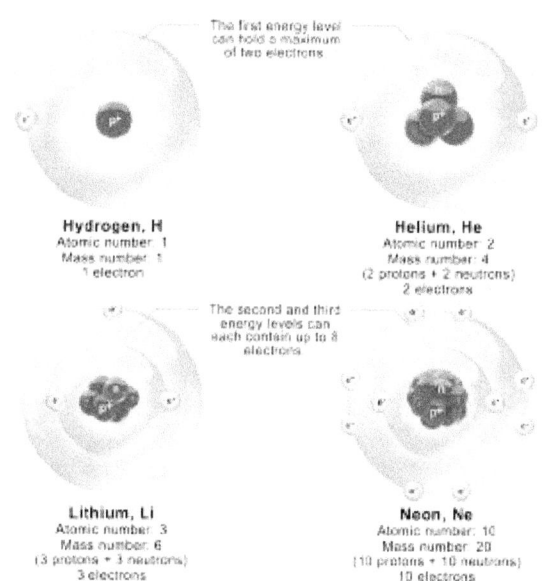

Models depicting electron energy levels in hydrogen, helium, lithium, and neon

The Bohr model gives almost exact results only for a system where two charged points orbit each other at speeds much less than that of light. This not only includes one-electron systems such as the hydrogen atom, singly ionized helium, doubly ionized lithium, but it includes positronium and Rydberg states of any atom where one electron is far away from everything else. It can be used for K-line X-ray transition calculations if other assumptions are added (see

Moseley's law below). In high energy physics, it can be used to calculate the masses of heavy quark mesons.

Calculation of the orbits requires two assumptions.

- **Classical mechanics**

  The electron is held in a circular orbit by electrostatic attraction. The centripetal force is equal to the Coulomb force.

  $$\frac{m_e v^2}{r} = \frac{Z k_e e^2}{r^2}$$

  where m_e is the electron's mass, e is the charge of the electron, k_e is Coulomb's constant and Z is the atom's atomic number. It is assumed here that the mass of the nucleus is much larger than the electron mass (which is a good assumption). This equation determines the electron's speed at any radius:

  $$v = \sqrt{\frac{Z k_e e^2}{m_e r}}.$$

  It also determines the electron's total energy at any radius:

  $$E = \frac{1}{2} m_e v^2 - \frac{Z k_e e^2}{r} = -\frac{Z k_e e^2}{2r}.$$

  The total energy is negative and inversely proportional to r. This means that it takes energy to pull the orbiting electron away from the proton. For infinite values of r, the energy is zero, corresponding to a motionless electron infinitely far from the proton. The total energy is half the potential energy, which is also true for noncircular orbits by the virial theorem.

- **A quantum rule**

  The angular momentum $L = m_e v r$ is an integer multiple of h:

  $$m_e v r = n\hbar$$

  Substituting the expression for the velocity gives an equation for r in terms of n:

  $$m_e \sqrt{\frac{k_e Z e^2}{m_e r}} \, r = n\hbar$$

  so that the allowed orbit radius at any n is:

  $$r_n = \frac{n^2 \hbar^2}{Z k_e e^2 m_e}$$

The smallest possible value of r in the hydrogen atom ($Z=1$) is called the Bohr radius and is equal to:

$$r_1 = \frac{\hbar^2}{k_e e^2 m_e} \approx 5.29 \times 10^{-11}\text{m}$$

The energy of the n-th level for any atom is determined by the radius and quantum number:

$$E = -\frac{Z k_e e^2}{2 r_n} = -\frac{Z^2 (k_e e^2)^2 m_e}{2 \hbar^2 n^2} \approx \frac{-13.6 Z^2}{n^2}\text{eV}$$

An electron in the lowest energy level of hydrogen ($n = 1$) therefore has about 13.6 eV less energy than a motionless electron infinitely far from the nucleus. The next energy level ($n = 2$) is −3.4 eV. The third ($n = 3$) is −1.51 eV, and so on. For larger values of n, these are also the binding energies of a highly excited atom with one electron in a large circular orbit around the rest of the atom.

The combination of natural constants in the energy formula is called the Rydberg energy (R_E):

$$R_E = \frac{(k_e e^2)^2 m_e}{2 \hbar^2}$$

This expression is clarified by interpreting it in combinations that form more natural units:

$m_e c^2$ is the rest mass energy of the electron (511 keV)

$\frac{k_e e^2}{\hbar c} = \alpha \approx \frac{1}{137}$ is the fine structure constant

$R_E = \frac{1}{2}(m_e c^2)\alpha^2$

Since this derivation is with the assumption that the nucleus is orbited by one electron, we can generalize this result by letting the nucleus have a charge $q = Z e$ where Z is the atomic number. This will now give us energy levels for hydrogenic atoms, which can serve as a rough order-of-magnitude approximation of the actual energy levels. So for nuclei with Z protons, the energy levels are (to a rough approximation):

$$E_n = -\frac{Z^2 R_E}{n^2}$$

The actual energy levels cannot be solved analytically for more than one electron (see n-body problem) because the electrons are not only affected by the nucleus but also interact with each other via the Coulomb Force.

When $Z = 1/\alpha$ ($Z \approx 137$), the motion becomes highly relativistic, and Z^2 cancels the α^2 in R; the orbit energy begins

to be comparable to rest energy. Sufficiently large nuclei, if they were stable, would reduce their charge by creating a bound electron from the vacuum, ejecting the positron to infinity. This is the theoretical phenomenon of electromagnetic charge screening which predicts a maximum nuclear charge. Emission of such positrons has been observed in the collisions of heavy ions to create temporary super-heavy nuclei.

The Bohr formula properly uses the reduced mass of electron and proton in all situations, instead of the mass of the electron: $m_{red} = \frac{m_e m_p}{m_e + m_p} = m_e \frac{1}{1 + m_e/m_p}$. However, these numbers are very nearly the same, due to the much larger mass of the proton, about 1836.1 times the mass of the electron, so that the reduced mass in the system is the mass of the electron multiplied by the constant 1836.1/(1+1836.1) = 0.99946. This fact was historically important in convincing Rutherford of the importance of Bohr's model, for it explained the fact that the frequencies of lines in the spectra for singly ionized helium do not differ from those of hydrogen by a factor of exactly 4, but rather by 4 times the ratio of the reduced mass for the hydrogen vs. the helium systems, which was much closer to the experimental ratio than exactly 4.

For positronium, the formula uses the reduced mass also, but in this case, it is exactly the electron mass divided by 2. For any value of the radius, the electron and the positron are each moving at half the speed around their common center of mass, and each has only one fourth the kinetic energy. The total kinetic energy is half what it would be for a single electron moving around a heavy nucleus.

$$E_n = \frac{R_E}{2n^2}$$

15.3 Rydberg formula

The Rydberg formula, which was known empirically before Bohr's formula, is seen in Bohr's theory as describing the energies of transitions or quantum jumps between one orbital energy levels. Bohr's formula gives the numerical value of the already-known and measured Rydberg's constant, but in terms of more fundamental constants of nature, including the electron's charge and Planck's constant.

When the electron gets moved from its original energy level to a higher one, it then jumps back each level until it comes to the original position, which results in a photon being emitted. Using the derived formula for the different energy levels of hydrogen one may determine the wavelengths of light that a hydrogen atom can emit.

The energy of a photon emitted by a hydrogen atom is given by the difference of two hydrogen energy levels:

$$E = E_i - E_f = R_{\mathrm{E}} \left(\frac{1}{n_f^2} - \frac{1}{n_i^2} \right)$$

where n_f is the final energy level, and n_i is the initial energy level.

Since the energy of a photon is

$$E = \frac{hc}{\lambda},$$

the wavelength of the photon given off is given by

$$\frac{1}{\lambda} = R \left(\frac{1}{n_f^2} - \frac{1}{n_i^2} \right).$$

This is known as the Rydberg formula, and the Rydberg constant R is R_{E}/hc, or $R_{\mathrm{E}}/2\pi$ in natural units. This formula was known in the nineteenth century to scientists studying spectroscopy, but there was no theoretical explanation for this form or a theoretical prediction for the value of R, until Bohr. In fact, Bohr's derivation of the Rydberg constant, as well as the concomitant agreement of Bohr's formula with experimentally observed spectral lines of the Lyman ($n_f = 1$), Balmer ($n_f = 2$), and Paschen ($n_f = 3$) series, and successful theoretical prediction of other lines not yet observed, was one reason that his model was immediately accepted.

To apply to atoms with more than one electron, the Rydberg formula can be modified by replacing "Z" with "Z − b" or "n" with "n − b" where b is constant representing a screening effect due to the inner-shell and other electrons (see Electron shell and the later discussion of the "Shell Model of the Atom" below). This was established empirically before Bohr presented his model.

15.4 Shell model of heavier atoms

Bohr extended the model of hydrogen to give an approximate model for heavier atoms. This gave a physical picture that reproduced many known atomic properties for the first time.

Heavier atoms have more protons in the nucleus, and more electrons to cancel the charge. Bohr's idea was that each discrete orbit could only hold a certain number of electrons. After that orbit is full, the next level would have to be used.

This gives the atom a shell structure, in which each shell corresponds to a Bohr orbit.

This model is even more approximate than the model of hydrogen, because it treats the electrons in each shell as non-interacting. But the repulsions of electrons are taken into account somewhat by the phenomenon of screening. The electrons in outer orbits do not only orbit the nucleus, but they also move around the inner electrons, so the effective charge Z that they feel is reduced by the number of the electrons in the inner orbit.

For example, the lithium atom has two electrons in the lowest 1s orbit, and these orbit at Z=2. Each one sees the nuclear charge of Z=3 minus the screening effect of the other, which crudely reduces the nuclear charge by 1 unit. This means that the innermost electrons orbit at approximately 1/4 the Bohr radius. The outermost electron in lithium orbits at roughly Z=1, since the two inner electrons reduce the nuclear charge by 2. This outer electron should be at nearly one Bohr radius from the nucleus. Because the electrons strongly repel each other, the effective charge description is very approximate; the effective charge Z doesn't usually come out to be an integer. But Moseley's law experimentally probes the innermost pair of electrons, and shows that they do see a nuclear charge of approximately Z−1, while the outermost electron in an atom or ion with only one electron in the outermost shell orbits a core with effective charge Z−k where k is the total number of electrons in the inner shells.

The shell model was able to qualitatively explain many of the mysterious properties of atoms which became codified in the late 19th century in the periodic table of the elements. One property was the size of atoms, which could be determined approximately by measuring the viscosity of gases and density of pure crystalline solids. Atoms tend to get smaller toward the right in the periodic table, and become much larger at the next line of the table. Atoms to the right of the table tend to gain electrons, while atoms to the left tend to lose them. Every element on the last column of the table is chemically inert (noble gas).

In the shell model, this phenomenon is explained by shell-filling. Successive atoms become smaller because they are filling orbits of the same size, until the orbit is full, at which point the next atom in the table has a loosely bound outer electron, causing it to expand. The first Bohr orbit is filled when it has two electrons, which explains why helium is inert. The second orbit allows eight electrons, and when it is full the atom is neon, again inert. The third orbital contains eight again, except that in the more correct Sommerfeld treatment (reproduced in modern quantum mechanics) there are extra "d" electrons. The third orbit may hold an extra 10 d electrons, but these positions are not filled until a few more orbitals from the next level are filled (filling the n=3 d orbitals produces the 10 transition elements). The irregular filling pattern is an effect of interactions between electrons, which are not taken into account in either the Bohr or Sommerfeld models and which are difficult to calculate even in the modern treatment.

15.5 Moseley's law and calculation of K-alpha X-ray emission lines

Niels Bohr said in 1962, "You see actually the Rutherford work [the nuclear atom] was not taken seriously. We cannot understand today, but it was not taken seriously at all. There was no mention of it any place. The great change came from Moseley."

In 1913 Henry Moseley found an empirical relationship between the strongest X-ray line emitted by atoms under electron bombardment (then known as the K-alpha line), and their atomic number Z. Moseley's empiric formula was found to be derivable from Rydberg and Bohr's formula (Moseley actually mentions only Ernest Rutherford and Antonius Van den Broek in terms of models). The two additional assumptions that [1] this X-ray line came from a transition between energy levels with quantum numbers 1 and 2, and [2], that the atomic number Z when used in the formula for atoms heavier than hydrogen, should be diminished by 1, to $(Z-1)^2$.

Moseley wrote to Bohr, puzzled about his results, but Bohr was not able to help. At that time, he thought that the postulated innermost "K" shell of electrons should have at least four electrons, not the two which would have neatly explained the result. So Moseley published his results without a theoretical explanation.

Later, people realized that the effect was caused by charge screening, with an inner shell containing only 2 electrons. In the experiment, one of the innermost electrons in the atom is knocked out, leaving a vacancy in the lowest Bohr orbit, which contains a single remaining electron. This vacancy is then filled by an electron from the next orbit, which has n=2. But the n=2 electrons see an effective charge of Z−1, which is the value appropriate for the charge of the nucleus, when a single electron remains in the lowest Bohr orbit to screen the nuclear charge +Z, and lower it by −1 (due to the electron's negative charge screening the nuclear positive charge). The energy gained by an electron dropping from the second shell to the first gives Moseley's law for K-alpha lines:

$$E = h\nu = E_i - E_f = R_E(Z-1)^2 \left(\frac{1}{1^2} - \frac{1}{2^2} \right)$$

or

$$f = \nu = R_v \left(\frac{3}{4} \right) (Z-1)^2 = (2.46 \times 10^{15}\,\text{Hz})(Z-1)^2.$$

Here, $R_v = R_E/h$ is the Rydberg constant, in terms of frequency equal to 3.28×10^{15} Hz. For values of Z between 11 and 31 this latter relationship had been empirically derived by Moseley, in a simple (linear) plot of the square root of X-ray frequency against atomic number (however, for silver, Z = 47, the experimentally obtained screening term should be replaced by 0.4). Notwithstanding its restricted validity,*[6] Moseley's law not only established the objective meaning of atomic number (see Henry Moseley for detail) but, as Bohr noted, it also did more than the Rydberg derivation to establish the validity of the Rutherford/Van den Broek/Bohr nuclear model of the atom, with atomic number (place on the periodic table) standing for whole units of nuclear charge.

The K-alpha line of Moseley's time is now known to be a pair of close lines, written as ($\mathbf{K\alpha_1}$ and $\mathbf{K\alpha_2}$) in Siegbahn notation.

15.6 Shortcomings

The Bohr model gives an incorrect value ι.=ℏ for the ground state orbital angular momentum. The angular momentum in the true ground state is known to be zero from experiment.*[7] Although mental pictures fail somewhat at these levels of scale, an electron in the lowest modern "orbital" with no orbital momentum, may be thought of as not to rotate "around" the nucleus at all, but merely to go tightly around it in an ellipse with zero area (this may be pictured as "back and forth", without striking or interacting with the nucleus). This is only reproduced in a more sophisticated semiclassical treatment like Sommerfeld's. Still, even the most sophisticated semiclassical model fails to explain the fact that the lowest energy state is spherically symmetric - it doesn't point in any particular direction. Nevertheless, in the modern *fully quantum treatment in phase space*, the proper deformation (full extension) of the semi-classical result adjusts the angular momentum value to the correct effective one. As a consequence, the physical ground state expression is obtained through a shift of the vanishing quantum angular momentum expression, which corresponds to spherical symmetry.

In modern quantum mechanics, the electron in hydrogen is a spherical cloud of probability that grows denser near the nucleus. The rate-constant of probability-decay in hydrogen is equal to the inverse of the Bohr radius, but since

Bohr worked with circular orbits, not zero area ellipses, the fact that these two numbers exactly agree is considered a "coincidence". (However, many such coincidental agreements are found between the semiclassical vs. full quantum mechanical treatment of the atom; these include identical energy levels in the hydrogen atom and the derivation of a fine structure constant, which arises from the relativistic Bohr–Sommerfeld model (see below) and which happens to be equal to an entirely different concept, in full modern quantum mechanics).

The Bohr model also has difficulty with, or else fails to explain:

- Much of the spectra of larger atoms. At best, it can make predictions about the K-alpha and some L-alpha X-ray emission spectra for larger atoms, if *two* additional ad hoc assumptions are made (see Moseley's law above). Emission spectra for atoms with a single outer-shell electron (atoms in the lithium group) can also be approximately predicted. Also, if the empiric electron–nuclear screening factors for many atoms are known, many other spectral lines can be deduced from the information, in similar atoms of differing elements, via the Ritz–Rydberg combination principles (see Rydberg formula). All these techniques essentially make use of Bohr's Newtonian energy-potential picture of the atom.

- the relative intensities of spectral lines; although in some simple cases. Bohr's formula or modifications of it, was able to provide reasonable estimates (for example, calculations by Kramers for the Stark effect).

- The existence of fine structure and hyperfine structure in spectral lines, which are known to be due to a variety of relativistic and subtle effects, as well as complications from electron spin.

- The Zeeman effect – changes in spectral lines due to external magnetic fields; these are also due to more complicated quantum principles interacting with electron spin and orbital magnetic fields.

- The model also violates the uncertainty principle in that it considers electrons to have known orbits and locations, two things which can not be measured simultaneously.

- Doublets and Triplets: Appear in the spectra of some atoms: Very close pairs of lines. Bohr's model cannot say why some energy levels should be very close together.

- Multi-electron Atoms: don't have energy levels predicted by the model. It doesn't work for (neutral) helium.

- A rotating charge, such as the electron classically orbiting around the nucleus, would constantly lose energy in form of electromagnetic radiation (via various mechanisms: dipole radiation, Bremsstrahlung,....). But such radiation is not observed.

15.7 Refinements

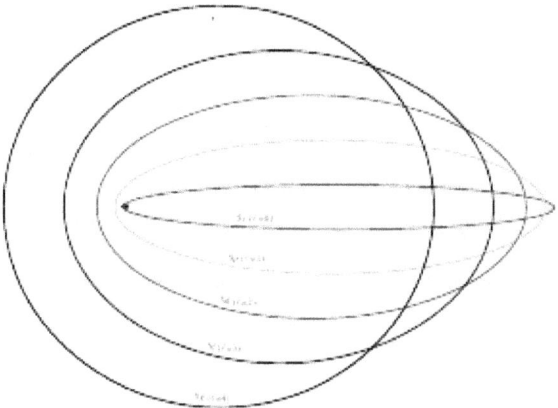

Elliptical orbits with the same energy and quantized angular momentum

Several enhancements to the Bohr model were proposed, most notably the **Sommerfeld model** or **Bohr–Sommerfeld model**, which suggested that electrons travel in elliptical orbits around a nucleus instead of the Bohr model's circular orbits.[1] This model supplemented the quantized angular momentum condition of the Bohr model with an additional radial quantization condition, the **Sommerfeld–Wilson quantization condition**[8][9]

$$\int_0^T p_r \, dq_r = nh$$

where p_r is the radial momentum canonically conjugate to the coordinate q which is the radial position and T is one full orbital period. The integral is the action of action-angle coordinates. This condition, suggested by the correspondence principle, is the only one possible, since the quantum numbers are adiabatic invariants.

The Bohr–Sommerfeld model was fundamentally inconsistent and led to many paradoxes. The magnetic quantum number measured the tilt of the orbital plane relative to the *xy*-plane, and it could only take a few discrete values. This contradicted the obvious fact that an atom could be turned this way and that relative to the coordinates without restriction. The Sommerfeld quantization can be performed in different canonical coordinates and sometimes gives different answers. The incorporation of radiation corrections was difficult, because it required finding action-angle coordinates for a combined radiation/atom system, which is difficult when the radiation is allowed to escape. The whole theory did not extend to non-integrable motions, which meant that many systems could not be treated even in principle. In the end, the model was replaced by the modern quantum mechanical treatment of the hydrogen atom, which was first given by Wolfgang Pauli in 1925, using Heisenberg's matrix mechanics. The current picture of the hydrogen atom is based on the atomic orbitals of wave mechanics which Erwin Schrödinger developed in 1926.

However, this is not to say that the Bohr model was without its successes. Calculations based on the Bohr–Sommerfeld model were able to accurately explain a number of more complex atomic spectral effects. For example, up to first-order perturbations, the Bohr model and quantum mechanics make the same predictions for the spectral line splitting in the Stark effect. At higher-order perturbations, however, the Bohr model and quantum mechanics differ, and measurements of the Stark effect under high field strengths helped confirm the correctness of quantum mechanics over the Bohr model. The prevailing theory behind this difference lies in the shapes of the orbitals of the electrons, which vary according to the energy state of the electron.

The Bohr–Sommerfeld quantization conditions lead to questions in modern mathematics. Consistent semiclassical quantization condition requires a certain type of structure on the phase space, which places topological limitations on the types of symplectic manifolds which can be quantized. In particular, the symplectic form should be the curvature form of a connection of a Hermitian line-bundle, which is called a prequantization.

15.8   See also

15.9   References

15.9.1   Footnotes

[1]   Akhlesh Lakhtakia (Ed.); Salpeter, Edwin E. (1996). "Models and Modelers of Hydrogen". *American Journal of Physics* (World Scientific) **65** (9): 933. Bibcode:1997AmJPh..65..933L. doi:10.1119/1.18691. ISBN 981-02-2302-1.

[2]   Niels Bohr (1913). "On the Constitution of Atoms and Molecules, Part I" (PDF). *Philosophical Magazine* **26** (151): 1–24. doi:10.1080/14786441308634955.

[3]   Olsen and McDonald 2005

[4]   "CK12 – Chemistry Flexbook Second Edition – The Bohr Model of the Atom". Retrieved 30 September 2014.

[5]   Niels Bohr (1913). "On the Constitution of Atoms and Molecules, Part II Systems Containing Only a Single Nucleus" (PDF). *Philosophical Magazine* **26** (153): 476–502. doi:10.1080/14786441308634993.

[6]   M.A.B. Whitaker (1999). "The Bohr–Moseley synthesis and a simple model for atomic x-ray energies". *European Journal of Physics* **20** (3): 213–220. Bibcode:1999EJPh...20..213W. doi:10.1088/0143-0807/20/3/312.

[7]   Smith, Brian. "Quantum Ideas: Week 2" Lecture Notes. p.17. University of Oxford. Retrieved Jan. 23, 2015.

[8]   A. Sommerfeld (1916). "Zur Quantentheorie der Spektrallinien". *Annalen der Physik* **51** (17): 1. Bibcode:1916AnP...356....1S. doi:10.1002/andp.19163561702.

[9]   W. Wilson (1915). "The quantum theory of radiation and line spectra". *Philosophical Magazine* **29** (174): 795–802. doi:10.1080/14786440608635362.

15.9.2   Primary sources

- Niels Bohr (1913). "On the Constitution of Atoms and Molecules, Part I" (PDF). *Philosophical Magazine* **26** (151): 1–24. doi:10.1080/14786441308634955.

- Niels Bohr (1913). "On the Constitution of Atoms and Molecules, Part II Systems Containing Only a Single Nucleus" (PDF). *Philosophical Magazine* **26** (153): 476–502. doi:10.1080/14786441308634993.

- Niels Bohr (1913). "On the Constitution of Atoms and Molecules, Part III Systems containing several nuclei". *Philosophical Magazine* **26**: 857–875. doi:10.1080/14786441308635031.

- Niels Bohr (1914). "The spectra of helium and hydrogen". *Nature* **92** (2295): 231–232. Bibcode:1913Natur..92..231B. doi:10.1038/092231d0.

- Niels Bohr (1921). "Atomic Structure". *Nature* **107** (2682): 104–107. Bibcode:1921Natur.107..104B. doi:10.1038/107104a0.

- A. Einstein (1917). "Zum Quantensatz von Sommerfeld und Epstein". *Verhandlungen der Deutschen Physikalischen Gesellschaft* **19**: 82–92. Reprinted in *The Collected Papers of Albert Einstein*, A. Engel translator, (1997) Princeton University Press, Princeton. **6** p. 434. (provides an elegant reformulation of the Bohr–Sommerfeld quantization conditions, as well as an important insight into the quantization of non-integrable (chaotic) dynamical systems.)

15.10   Further reading

- Linus Carl Pauling (1970). "Chapter 5-1" . *General Chemistry* (3rd ed.). San Francisco: W.H. Freeman & Co.

    - Reprint: Linus Pauling (1988). *General Chemistry*. New York: Dover Publications. ISBN 0-486-65622-5.

- George Gamow (1985). "Chapter 2" . *Thirty Years That Shook Physics*. Dover Publications.

- Walter J. Lehmann (1972). "Chapter 18". *Atomic and Molecular Structure: the development of our concepts*. John Wiley and Sons.

- Paul Tipler and Ralph Llewellyn (2002). *Modern Physics* (4th ed.). W. H. Freeman. ISBN 0-7167-4345-0.

- Klaus Hentschel: Elektronenbahnen, Quantensprünge und Spektren, in: Charlotte Bigg & Jochen Hennig (eds.) Atombilder. Ikonografien des Atoms in Wissenschaft und Öffentlichkeit des 20. Jahrhunderts, Göttingen: Wallstein-Verlag 2009, pp. 51–61

- Steven and Susan Zumdahl (2010). "Chapter 7.4" . *Chemistry* (8th ed.). Brooks/Cole. ISBN 978-0-495-82992-8.

- Helge Kragh (2011). "Conceptual objections to the Bohr atomic theory — do electrons have a "free will"?". *European Physical Journal H* **36** (3): 327. Bibcode:2011EPJH...36..327K. doi:10.1140/epjh/e2011-20031-x.

15.11   External links

- Standing waves in Bohr's atomic model An interactive simulation to intuitively explain the quantization condition of standing waves in Bohr's atomic model

Chapter 16

Effective mass (solid-state physics)

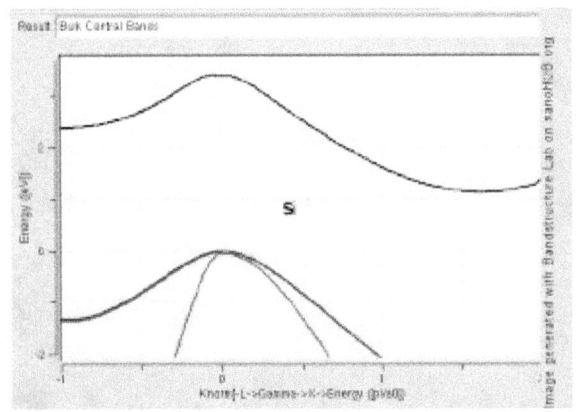

Result Bulk Central Band

Bulk band structure for Si,Ge,GaAs and InAs generated with tight binding model. Note that Si and Ge are indirect with minima at X and L, while GaAs and InAs are direct band gap materials.

In solid state physics, a particle's **effective mass** (often denoted m^*) is the mass that it seems to have when responding to forces, or the mass that it seems to have when *en masse* with other identical particles in a thermal distribution. One of the results from the band theory of solids is that the movement of particles in a periodic potential, over long distances larger than the lattice spacing, can be very different from their motion in a vacuum. The effective mass is a quantity that is used to simplify band structures by constructing an analogy to the behavior of a free particle with that mass. For some purposes and some materials, the effective mass can be considered to be a simple constant of a material. In general, however, the value of effective mass depends on the purpose for which it is used, and can vary depending on a number of factors.

For electrons or electron holes in a solid, the effective mass is usually stated in units of the true mass of the electron m_e (9.11×10^{-31} kg). In these units it is usually in the range 0.01 to 10, but can also be lower or higher, for example reaching 1,000 in exotic heavy fermion materials, or anywhere from zero to infinity (depending on definition) in graphene. As it simplifies the more general band theory, the electronic effective mass can be seen as an important basic

parameter that influences measurable properties of a solid, including everything from the efficiency of a solar cell to the speed of an integrated circuit.

16.1 Simple case: parabolic, isotropic dispersion relation

At the highest energies of the valence band in many semiconductors (Ge, Si, GaAs, ...), and the lowest energies of the conduction band in some semiconductors (GaAs, ...), the band structure $E(\mathbf{k})$ can be locally approximated as

$$E(\mathbf{k}) = E_0 + \frac{\hbar^2 \mathbf{k}^2}{2m^*}$$

where $E(\mathbf{k})$ is the energy of an electron at wavevector $\mathbf{k}$ in that band, E_0 is a constant giving the edge of energy of that band, and m^* is a constant (the effective mass). It can be shown that the electrons placed in these bands behave as free electrons except with a different mass, as long as their energy stays within the range of validity of the approximation above. As a result, the electron mass in models such as the Drude model must be replaced with the effective mass.

One remarkable property is that the effective mass can become *negative*, when the band curves downwards away from a maximum. As a result of the negative mass, the electrons respond to electric and magnetic forces by gaining velocity in the opposite direction compared to normal; even though these electrons have negative charge, they move in trajectories as if they had positive charge (and positive mass). This explains the existence of valence-band holes, the positive-charge, positive-mass quasiparticles that can be found in semiconductors.[1]

In any case, if the band structure has the simple parabolic form described above, then the value of effective mass is unambiguous. Unfortunately, this parabolic form is not valid for describing most materials. In such complex materials there is no single definition of "effective mass" but instead multiple definitions, each suited to a particular purpose. The rest of the article describes these effective masses in detail.

16.2 Intermediate case: parabolic, anisotropic dispersion relation

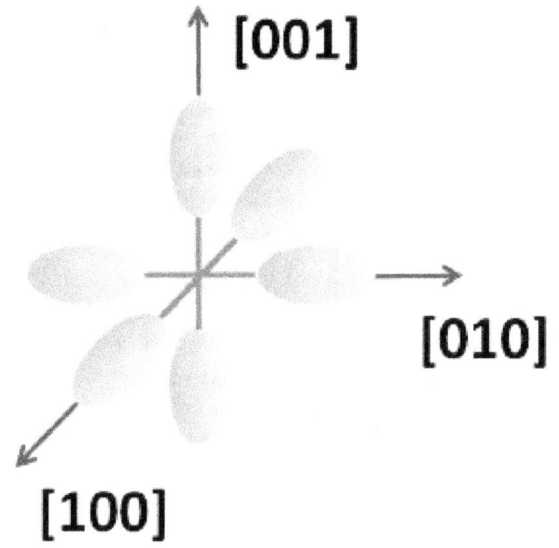

↑ **[001]**

[010] →

[100] ↙

Constant energy ellipsoids in silicon near the six conduction band minima. For each valley (band minimum), the effective masses are $m_l = 0.92m_e$ ("longitudinal"; along one axis) and $m_t = 0.19m_e$ ("transverse"; along two axes). [2]

In some important semiconductors (notably, silicon) the lowest energies of the conduction band are not symmetrical, as the constant-energy surfaces are now ellipsoids, rather than the spheres in the isotropic case. Each conduction band minimum can be approximated only by

$$E(\mathbf{k}) = E_0 + \frac{\hbar^2}{2m_x^*}(k_x - k_{0,x})^2 + \frac{\hbar^2}{2m_y^*}(k_y - k_{0,y})^2 + \frac{\hbar^2}{2m_z^*}(k_z - k_{0,z})^2$$

where x, y, and z axes are aligned to the principal axes of the ellipsoids, and m_x^{**}, m_y^{**} and m_z^{**} are the inertial effective masses along these different axes. The offsets $k_{0,x}$, $k_{0,y}$, and $k_{0,z}$ reflect that the conduction band minimum is no longer centered at zero wavevector. (These effective masses correspond to the principal components of the inertial effective mass tensor, described later.)

In this case, the electron motion is no longer directly comparable to a free electron; the speed of an electron will depend on its direction, and it will accelerate to a different degree depending on the direction of the force. Still, in crystals such as silicon the overall properties such as conductivity appear to be isotropic. This is because there are multiple valleys (conduction-band minima), each with effective masses rearranged along different axes. The valleys collectively act together to give an isotropic conductivity. It is possible to average the different axes' effective masses together in some way, to regain the free electron picture. However, the averaging method turns out to depend on the purpose: [3]

- For calculation of the total density of states and the total carrier density, via the geometric mean combined with a degeneracy factor g which counts the number of valleys (in silicon $g = 6$): [4]

$$m_{\text{density}}^* = \sqrt[3]{g^2 m_x m_y m_z}$$

(This effective mass corresponds to the density of states effective mass, described later.)

For the per-valley density of states and per-valley carrier density, the degeneracy factor is left out.

- For the purposes of calculating conductivity as in the Drude model, via the harmonic mean

$$m_{\text{conductivity}}^* = 3\left[\frac{1}{m_x^*} + \frac{1}{m_y^*} + \frac{1}{m_z^*}\right]^{-1}$$

Since the Drude law also depends on scattering time, which varies greatly, this effective mass is rarely used: conductivity is instead usually expressed in terms of carrier density and an empirically measured parameter, carrier mobility.

16.3 General case

In general the dispersion relation cannot be approximated as parabolic, and in such cases the effective mass should be precisely defined if it is to be used at all. Here a commonly stated definition of effective mass is the *inertial* effective mass tensor defined below; however, in general it is a matrix-valued function of the wavevector, and even more complex than the band structure itself. Other effective masses are more relevant to directly measurable phenomena.

16.3.1 Inertial effective mass tensor

A classical particle under the influence of a force accelerates according to Newton's second law, $\mathbf{a} = m^{*-1}\mathbf{F}$. This intuitive principle appears identically in semiclassical approximations derived from band structure. However, each of the symbols has a slightly modified meaning: acceleration becomes the rate of change in group velocity:

$$\mathbf{a} = \frac{\mathrm{d}}{\mathrm{d}t}\,\mathbf{v}_g = \frac{\mathrm{d}}{\mathrm{d}t}(\nabla_k\,\omega(\mathbf{k})) = \nabla_k\frac{\mathrm{d}\,\omega(\mathbf{k})}{\mathrm{d}t} = \nabla_k\left(\frac{\mathrm{d}\mathbf{k}}{\mathrm{d}t}\cdot\nabla_k\,\omega(\mathbf{k})\right),$$

where ∇_k is the del operator in reciprocal space, and force gives a rate of change in crystal momentum $\mathbf{p}_{\text{crystal}}$:

$$\mathbf{F} = \frac{\mathrm{d}\,\mathbf{p}_{\text{crystal}}}{\mathrm{d}t} = \hbar\frac{\mathrm{d}\mathbf{k}}{\mathrm{d}t}.$$

where $\hbar$ is the *reduced Planck constant*, or $1/2\pi$ times the Planck constant. Combining these two equations yields

$$\mathbf{a} = \nabla_k\left(\frac{\mathbf{F}}{\hbar}\cdot\nabla_k\,\omega(\mathbf{k})\right).$$

Extracting the ith element from both sides gives

$$a_i = \left(\frac{1}{\hbar}\frac{\partial^2\omega(\mathbf{k})}{\partial k_i\partial k_j}\right)F_j = \left(\frac{1}{\hbar^2}\frac{\partial^2 E(\mathbf{k})}{\partial k_i\partial k_j}\right)F_j.$$

where a_i is the ith element of $\mathbf{a}$, F_j is the jth element of $\mathbf{F}$, k_i and k_j are the ith and jth elements of $\mathbf{k}$, respectively, and E is the total energy of the particle according to the Planck–Einstein relation. The index j is contracted by the use of Einstein notation (there is an implicit summation over j). Since Newton's second law uses the inertial mass (not the gravitational mass), we can identify the inverse of this mass in the equation above as the tensor

$$\left[M_{\text{inert}}^{-1}\right]_{ij} = \hbar^{-2}\frac{\partial^2 E}{\partial k_i\partial k_j}.$$

This tensor expresses the change in group velocity due to a change in crystal momentum. Its inverse, M_{inert}, is known as the **effective mass tensor**.

The inertial expression for effective mass is commonly used, but note that its properties can be counter-intuitive:

- The effective mass tensor generally varies depending on $\mathbf{k}$, meaning that the mass of the particle actually changes after it is subject to an impulse. The only cases in which it remains constant are those of parabolic bands, described above.

- The effective mass tensor diverges (becomes infinite) for linear dispersion relations, such as with photons or electrons in graphene.[5] (These particles are sometimes said to be massless, however this refers to their having zero rest mass; rest mass is a distinct concept from effective mass.)

16.3.2   Cyclotron effective mass

Classically, a charged particle in a magnetic field moves in a helix along the magnetic field axis. The period T of its motion depends on its mass m and charge e,

$$T = \left|\frac{2\pi m}{eB}\right|$$

where B is the magnetic flux density.

For particles in asymmetrical band structures, the particle no longer moves exactly in a helix, however its motion transverse to the magnetic field still moves in a closed loop (not necessarily a circle). Moreover, the time to complete one of these loops still varies inversely with magnetic field, and so it is possible to define a *cyclotron effective mass* from the measured period, using the above equation.

The semiclassical motion of the particle can be described by a closed loop in k-space. Throughout this loop, the particle maintains a constant energy, as well as a constant momentum along the magnetic field axis. By defining A to be the k-space area enclosed by this loop (this area depends on the energy E, the direction of the magnetic field, and the on-axis wavevector k_B), then it can be shown that the cyclotron effective mass depends on the band structure via the derivative of this area in energy:

$$m^*(E,\hat{B},k_B) = \frac{\hbar^2}{2\pi}\cdot\frac{\partial}{\partial E}A\left(E,\hat{B},k_B\right)$$

Typically, experiments that measure cyclotron motion (cyclotron resonance, de Haas–van Alphen effect, etc.) are restricted to only probe motion for energies near the Fermi level.

In two-dimensional electron gases, the cyclotron effective mass is defined only for one magnetic field direction (perpendicular) and the out-of-plane wavevector drops out. The cyclotron effective mass therefore is only a function of energy, and it turns out to be exactly related to the density of states at that energy via the relation $g(E) = \frac{g_v m^*}{\pi\hbar^2}$, where g_v is the valley degeneracy. Such a simple relationship does not apply in three-dimensional materials.

16.3.3   Density of states effective masses (lightly doped semiconductors)

In semiconductors with low levels of doping, the electron concentration in the conduction band is in general given by

$$n_e = N_C \exp\left(-\frac{E_C - E_F}{kT}\right)$$

where E_F is the Fermi level, E_C is the minimum energy of the conduction band, and N_C is a concentration coefficient that depends on temperature. The above relationship for n_e can be shown to apply for any conduction band shape (including non-parabolic, asymmetric bands), provided the doping is weak (E_C-E_F >> kT); this is a consequence of Fermi–Dirac statistics limiting towards Maxwell–Boltzmann statistics.

The concept of effective mass is useful to model the temperature dependence of N_C, thereby allowing the above relationship to be used over a range of temperatures. In an idealized three-dimensional material with a parabolic band, the concentration coefficient is given by

$$N_C = 2 \left(\frac{2\pi m_e^* kT}{h^2} \right)^{\frac{3}{2}}$$

In semiconductors with non-simple band structures, this relationship is used to define an effective mass, known as the **density of states effective mass of electrons**. The name "density of states effective mass" is used since the above expression for N_C is derived via the density of states for a parabolic band.

In practice, the effective mass extracted in this way is not quite constant in temperature (N_C does not exactly vary as $T^{3/2}$). In silicon, for example, this effective mass varies by a few percent between absolute zero and room temperature because the band structure itself slightly changes in shape. These band structure distortions are a result of changes in electron-phonon interaction energies, with the lattice's thermal expansion playing a minor role.*[4]

Similarly, the number of holes in the valence band, and the **density of states effective mass of holes** are defined by:

$$n_h = N_V \exp \left(-\frac{E_F - E_V}{kT} \right), \quad N_V = 2 \left(\frac{2\pi m_h^* kT}{h^2} \right)^{\frac{3}{2}}$$

where E_V is the maximum energy of the valence band. Practically, this effective mass tends to vary greatly between absolute zero and room temperature in many materials (e.g., a factor of two in silicon), as there are multiple valence bands with distinct and significantly non-parabolic character, all peaking near the same energy.*[4]

16.4   Experimental determination

Traditionally effective masses were measured using cyclotron resonance, a method in which microwave absorption of a semiconductor immersed in a magnetic field goes through a sharp peak when the microwave frequency

equals the cyclotron frequency $f_c = \frac{eB}{2\pi m^*}$. In recent years effective masses have more commonly been determined through measurement of band structures using techniques such as angle-resolved photo emission (ARPES) or, most directly, the de Haas–van Alphen effect. Effective masses can also be estimated using the coefficient γ of the linear term in the low-temperature electronic specific heat at constant volume c_v. The specific heat depends on the effective mass through the density of states at the Fermi level and as such is a measure of degeneracy as well as band curvature. Very large estimates of carrier mass from specific heat measurements have given rise to the concept of heavy fermion materials. Since carrier mobility depends on the ratio of carrier collision lifetime τ to effective mass, masses can in principle be determined from transport measurements, but this method is not practical since carrier collision probabilities are typically not known a priori.

16.5   Significance

The effective mass is used in transport calculations, such as transport of electrons under the influence of fields or carrier gradients, but also is used to calculate the carrier density and density of states in semiconductors. These masses are related but, as explained in the previous sections, are not the same because the weighting of various directions and wavevectors are different.

Certain III-V compounds such as GaAs and InSb have far smaller effective masses than tetrahedral group IV materials like Si and Ge. In the simplest Drude picture of electronic transport, the maximum obtainable charge carrier velocity is inversely proportional to the effective mass: $v = \|\mu\| \cdot E$ where $\|\mu\| = \frac{e\tau}{\|m^*\|}$ with e being the electronic charge. The ultimate speed of integrated circuits depends on the carrier velocity, so the low effective mass is the fundamental reason that GaAs and its derivatives are used instead of Si in high-bandwidth applications like cellular telephony.*[9]

16.6   Footnotes

[1] Kittel, *Introduction to Solid State Physics* 8th edition, page 194-196

[2] Charles Kittel. *op. cit.* p. 216. ISBN 0-471-11181-3.

[3] http://ecee.colorado.edu/~bart/book/effmass.htm

[4] Green, M. A. (1990). "Intrinsic concentration, effective densities of states, and effective mass in silicon". *Journal of Applied Physics* **67** (6): 2944–2941. Bibcode:1990JAP....67.2944G. doi:10.1063/1.345414.

[5] Viktor Ariel; Amir Natan (2012). "Electron Effective Mass in Graphene". arXiv:1206.6100 [physics.gen-ph].

[6] S.Z. Sze, *Physics of Semiconductor Devices*, ISBN 0-471-05661-8.

[7] W.A. Harrison, *Electronic Structure and the Properties of Solids*, ISBN 0-486-66021-4.

[8] This site gives the effective masses of Silicon at different temperatures.

[9] Silveirinha, M. R. G.; Engheta, N. (2012). "Transformation electronics: Tailoring the effective mass of electrons". *Physical Review B* **86** (16). doi:10.1103/PhysRevB.86.161104.

16.7 References

- Pastori Parravicini, G. (1975). *Electronic States and Optical Transitions in Solids*. Pergamon Press. ISBN 0-08-016846-9. This book contains an exhaustive but accessible discussion of the topic with extensive comparison between calculations and experiment.

- S. Pekar, The method of effective electron mass in crystals, Zh. Eksp. Teor. Fiz. **16**, 933 (1946).

16.8 External links

- NSM archive

Chapter 17

Double-slit experiment

"Slit experiment" redirects here. For other uses, see Diffraction.

The modern **double-slit experiment** is a demonstration

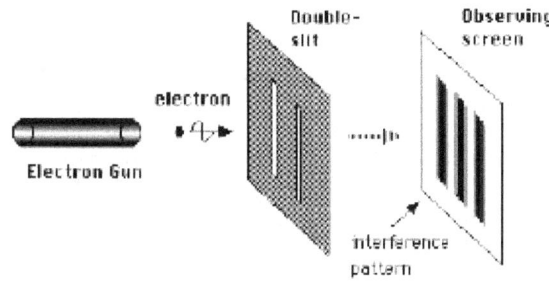

Photons or particles of matter (like an electron) produce a wave pattern when two slits are used

that light and matter can display characteristics of both classically defined waves and particles; moreover, it displays the fundamentally probabilistic nature of quantum mechanical phenomena. A simpler form of the double-slit experiment was performed originally by Thomas Young in 1801 (well before quantum mechanics). He believed it demonstrated that the wave theory of light was correct and his experiment is sometimes referred to as **Young's experiment**[1] or *Young's slits*. The experiment belongs to a general class of "double path" experiments, in which a wave is split into two separate waves that later combine into a single wave. Changes in the path lengths of both waves result in a phase shift, creating an interference pattern. Another version is the Mach–Zehnder interferometer, which splits the beam with a mirror.

In the basic version of this experiment, a coherent light source such as a laser beam illuminates a plate pierced by two parallel slits, and the light passing through the slits is observed on a screen behind the plate.[2][3] The wave nature of light causes the light waves passing through the two slits to interfere, producing bright and dark bands on the screen - a result that would not be expected if light consisted of classical particles.[2][4] However, the light is always found to be absorbed at the screen at discrete

points, as individual particles (not waves), the interference pattern appearing via the varying density of these particle hits on the screen.[5] Furthermore, versions of the experiment that include detectors at the slits find that each detected photon passes through one slit (as would a classical particle), and not through both slits (as would a wave).[6][7][8][9][10] These results demonstrate the principle of wave–particle duality.[11][12]

Other atomic-scale entities such as electrons are found to exhibit the same behavior when fired towards a double slit.[3] Additionally, the detection of individual discrete impacts is observed to be inherently probabilistic, which is inexplicable using classical mechanics.[3]

The experiment can be done with entities much larger than electrons and photons, although it becomes more difficult as size increases. The largest entities for which the double-slit experiment has been performed were molecules that each comprised 810 atoms (whose total mass was over 10,000 atomic mass units).[13][14]

17.1 Overview

If light consisted strictly of ordinary or classical particles, and these particles were fired in a straight line through a slit and allowed to strike a screen on the other side, we would expect to see a pattern corresponding to the size and shape of the slit. However, when this "single-slit experiment" is actually performed, the pattern on the screen is a diffraction pattern in which the light is spread out. The smaller the slit, the greater the angle of spread. The top portion of the image on the right shows the central portion of the pattern formed when a red laser illuminates a slit and, if one looks carefully, two faint side bands. More bands can be seen with a more highly refined apparatus. Diffraction explains the pattern as being the result of the interference of light waves from the slit.

If one illuminates two parallel slits with a more intense red laser, the light from the two slits again interferes. Here

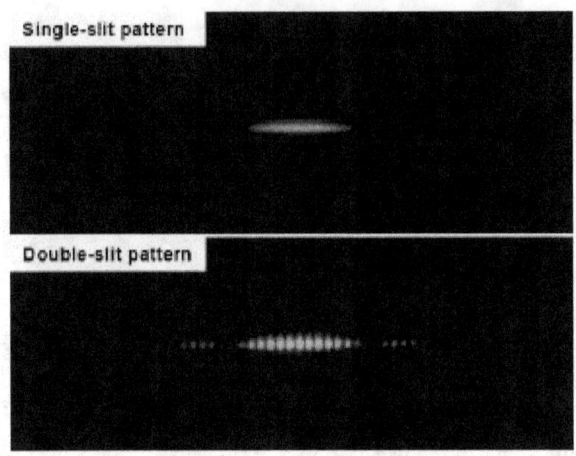

Same double-slit assembly (0.7 mm between slits); in top image, one slit is closed. In the single-slit image, a diffraction pattern (the faint spots on either side of the main band) forms due to the nonzero width of the slit. A diffraction pattern is also seen in the double-slit image, but at twice the intensity and with the addition of many smaller interference fringes.

the interference is a more pronounced pattern with a series of light and dark bands. The width of the bands is a property of the frequency of the illuminating light.[15] (See the bottom photograph to the right.) When Thomas Young (1773–1829) first demonstrated this phenomenon, it indicated that light consists of waves, as the distribution of brightness can be explained by the alternately additive and subtractive interference of wavefronts.[3] Young's experiment, performed in the early 1800s, played a vital part in the acceptance of the wave theory of light, vanquishing the corpuscular theory of light proposed by Isaac Newton, which had been the accepted model of light propagation in the 17th and 18th centuries. However, the later discovery of the photoelectric effect demonstrated that under different circumstances, light can behave as if it is composed of discrete particles. These seemingly contradictory discoveries made it necessary to go beyond classical physics and take the quantum nature of light into account.

The double-slit experiment (and its variations) has become a classic thought experiment, for its clarity in expressing the central puzzles of quantum mechanics. Because it demonstrates the fundamental limitation of the ability of the observer to predict experimental results, Richard Feynman called it "a phenomenon which is impossible [⋯] to explain in any classical way, and which has in it the heart of quantum mechanics. In reality, it contains the *only* mystery [of quantum mechanics]." [3] Feynman was fond of saying that all of quantum mechanics can be gleaned from carefully thinking through the implications of this single experiment.[16] Richard Feynman also proposed (as a thought experiment) that if detectors were placed before each slit, the interference pattern would disappear.[17]

The Englert–Greenberger duality relation provides a detailed treatment of the mathematics of double-slit interference in the context of quantum mechanics.

A low-intensity double-slit experiment was first performed by G. I. Taylor in 1909,[18] by reducing the level of incident light until photon emission/absorption events were mostly nonoverlapping. A double-slit experiment was not performed with anything other than light until 1961, when Claus Jönsson of the University of Tübingen performed it with electrons.[19][20] In 1974 the Italian physicists Pier Giorgio Merli, Gian Franco Missiroli, and Giulio Pozzi repeated the experiment using single electrons, showing that each electron interferes with itself as predicted by quantum theory.[21][22] In 2002, the single-electron version of the experiment was voted "the most beautiful experiment" by readers of *Physics World*.[23]

17.2   Variations of the experiment

17.2.1   Interference of individual particles

An important version of this experiment involves single particles (or waves—for consistency, they are called particles here). Sending particles through a double-slit apparatus one at a time results in single particles appearing on the screen, as expected. Remarkably, however, an interference pattern emerges when these particles are allowed to build up one by one (see the image to the right). This demonstrates the wave-particle duality, which states that all matter exhibits both wave and particle properties: the particle is measured as a single pulse at a single position, while the wave describes the probability of absorbing the particle at a specific place of the detector.[24] This phenomenon has been shown to occur with photons, electrons, atoms and even some molecules, including buckyballs.[25][26][27][28][29] So experiments with electrons add confirmatory evidence to the view that electrons, protons, neutrons, and even larger entities that are ordinarily called particles nevertheless have their own wave nature and even their own specific frequencies.

The probability of detection is the square of the amplitude of the wave and can be calculated with classical waves (see below). The particles do not arrive at the screen in a predictable order, so knowing where all the previous particles appeared on the screen and in what order tells nothing about where a future particle will be detected.[30] If there is a cancellation of waves at some point, that does not mean that a particle disappears; it will appear somewhere else. Ever since the origination of quantum mechanics, some theorists have searched for ways to incorporate additional determinants or "hidden variables" that, were they to become known, would account for the location of each individual impact with the target.*[31]

More complicated systems that involve two or more particles in superposition are not amenable to the above explanation.*[32]

17.2.2 "Which-way"experiments and the principle of complementarity

A well-known thought experiment predicts that if particle detectors are positioned at the slits, showing through which slit a photon goes, the interference pattern will disappear.*[3] This *which-way* experiment illustrates the complementarity principle that photons can behave as either particles or waves, but cannot be observed as both at the same time.*[33]*[34]*[35] Despite the importance of this *gedanken* in the history of quantum mechanics (for example, see the discussion on Einstein's version of this experiment), technically feasible realizations of this experiment were not proposed until the 1970s.*[36] (Naive implementations of the textbook *gedanken* are not possible because photons cannot be detected without absorbing the photon.) Currently, multiple experiments have been performed illustrating various aspects of complementarity.*[37]

An experiment performed in 1987 *[38]*[39] produced results that demonstrated that information could be obtained regarding which path a particle had taken without destroying the interference altogether. This showed the effect of measurements that disturbed the particles in transit to a lesser degree and thereby influenced the interference pattern only to a comparable extent. In other words, if one does not insist that the method used to determine which slit each photon passes through be completely reliable, one can still detect a (degraded) interference pattern.*[40]

17.2.3 Delayed choice and quantum eraser variations

Main article: Delayed choice quantum eraser

Wheeler's delayed choice experiments demonstrate that extracting "which path" information *after* a particle passes through the slits can seem to retroactively alter its previous behavior at the slits.

Quantum eraser experiments demonstrate that wave behavior can be restored by erasing or otherwise making permanently unavailable the "which path" information.

A simple do-it-at-home demonstration of the quantum eraser phenomenon was given in an article in *Scientific American*.*[41] If one sets polarizers before each slit with

their axes orthogonal to each other, the interference pattern will be eliminated. The polarizers can be considered as introducing which-path information to each beam. Introducing a third polarizer in front of the detector with an axis of 45° relative to the other polarizers "erases" this information, allowing the interference pattern to reappear. This can also be accounted for by considering the light to be a classical wave.*[41]*:91 and also when using circular polarizers and single photons.*[42]*:6 Implementations of the polarizers using entangled photon pairs have no classical explanation.*[42]

17.2.4 Weak measurement

Main article: Weak measurement

In a highly publicized experiment in 2012, researchers claimed to have identified the path each particle had taken without any adverse effects at all on the interference pattern generated by the particles.*[43] In order to do this, they used a setup such that particles coming to the screen were not from a point-like source, but from a source with two intensity maxima. However, commentators such as Motl*[44] and Svensson*[45] have pointed out that there is in fact no conflict between the weak measurements performed in this variant of the double-slit experiment and the Heisenberg uncertainty principle. Weak measurement followed by post-selection did not allow simultaneous position and momentum measurements for each individual particle, but rather allowed measurement of the average trajectory of the particles that arrived at different positions. In other words, the experimenters were creating a statistical map of the full trajectory landscape.*[45]

17.2.5 Other variations

In 1967, Pfleegor and Mandel demonstrated two-source interference using two separate lasers as light sources.*[46]*[47]

It was shown experimentally in 1972 that in a double-slit system where only one slit was open at any time, interference was nonetheless observed provided the path difference was such that the detected photon could have come from either slit.*[48]*[49] The experimental conditions were such that the photon density in the system was much less than unity.

In 1999, the double-slit experiment was successfully performed with buckyball molecules (each of which comprises 60 carbon atoms).*[26]*[50] A buckyball is large enough (diameter about 0.7 nm, nearly half a million times larger than a proton) to be seen under an electron microscope.

In 2005, E. R. Eliel presented an experimental and theoretical study of the optical transmission of a thin metal screen perforated by two subwavelength slits, separated by many optical wavelengths. The total intensity of the far-field double-slit pattern is shown to be reduced or enhanced as a function of the wavelength of the incident light beam.[51]

In 2012, researchers at the University of Nebraska–Lincoln performed the double-slit experiment with electrons as described by Richard Feynman, using new instruments that allowed control of the transmission of the two slits and the monitoring of single-electron detection events. Electrons were fired by an electron gun and passed through one or two slits of 62 nm wide × 4 μm tall.[52]

In 2013, the double-slit experiment was successfully performed with molecules that each comprised 810 atoms (whose total mass was over 10,000 atomic mass units).[13][14]

Hydrodynamic pilot wave analogs

Hydrodynamic analogs have been developed that can recreate various aspects of quantum mechanical systems, including single-particle interference through a double-slit.[53] A silicone oil droplet, bouncing along the surface of a liquid, self-propels via resonant interactions with its own wave field. The droplet gently sloshes the liquid with every bounce. At the same time, ripples from past bounces affect its course. The droplet's interaction with its own ripples, which form what is known as a pilot wave, causes it to exhibit behaviors previously thought to be peculiar to elementary particles — including behaviors customarily taken as evidence that elementary particles are spread through space like waves, without any specific location, until they are measured.[54][55]

Behaviors mimicked via this hydrodynamic pilot-wave system include quantum single particle diffraction,[56] tunneling, quantized orbits, orbital level splitting, spin, and multimodal statistics. It is also possible to infer uncertainty relations and exclusion principles. On the other hand, no hydrodynamic analog of entanglement has yet been developed.[53] Videos are available illustrating various features of this system. (See the External links.)

17.3 Classical wave-optics formulation

Much of the behaviour of light can be modelled using classical wave theory. The Huygens–Fresnel principle is one such model; it states that each point on a wavefront generates a secondary wavelet, and that the disturbance at any subsequent point can be found by summing the contributions of the individual wavelets at that point. This summation needs to take into account the phase as well as the amplitude of the individual wavelets. It should be noted that only the intensity of a light field can be measured — this is proportional to the square of the amplitude.

In the double-slit experiment, the two slits are illuminated by a single laser beam. If the width of the slits is small enough (less than the wavelength of the laser light), the slits diffract the light into cylindrical waves. These two cylindrical wavefronts are superimposed, and the amplitude, and therefore the intensity, at any point in the combined wavefronts depends on both the magnitude and the phase of the two wavefronts. The difference in phase between the two waves is determined by the difference in the distance travelled by the two waves.

If the viewing distance is large compared with the separation of the slits (the far field), the phase difference can be found using the geometry shown in the figure below right. The path difference between two waves travelling at an angle θ is given by:

$$d \sin \theta \approx d\theta$$

Where d is the distance between the two slits. When the two waves are in phase, i.e. the path difference is equal to an integral number of wavelengths, the summed amplitude, and therefore the summed intensity is maximum, and when they are in anti-phase, i.e. the path difference is equal to half a wavelength, one and a half wavelengths, etc., then the two waves cancel and the summed intensity is zero. This effect is known as interference. The interference fringe maxima occur at angles

$$d\theta_n = n\lambda, \ n = 0, 1, 2, \ldots$$

where λ is the wavelength of the light. The angular spacing of the fringes, θ_f, is given by

$$\theta_f \approx \lambda/d$$

The spacing of the fringes at a distance z from the slits is given by

$$w = z\theta_f = z\lambda/d$$

For example, if two slits are separated by 0.5 mm (d), and are illuminated with a 0.6μm wavelength laser (λ), then at a distance of 1m (z), the spacing of the fringes will be 1.2 mm.

If the width of the slits b is greater than the wavelength, the Fraunhofer diffraction equation gives the intensity of the diffracted light as: [57]

$$I(\theta) \propto \cos^2 \left[\frac{\pi d \sin \theta}{\lambda} \right] \; \text{sinc}^2 \left[\frac{\pi b \sin \theta}{\lambda} \right]$$

Where the sinc function is defined as $\text{sinc}(x) = \sin(x)/(x)$ for $x \neq 0$, and $\text{sinc}(0) = 1$.

This is illustrated in the figure above, where the first pattern is the diffraction pattern of a single slit, given by the sinc function in this equation, and the second figure shows the combined intensity of the light diffracted from the two slits, where the cos function represent the fine structure, and the coarser structure represents diffraction by the individual slits as described by the sinc function.

Similar calculations for the near field can be done using the Fresnel diffraction equation. As the plane of observation gets closer to the plane in which the slits are located, the diffraction patterns associated with each slit decrease in size, so that the area in which interference occurs is reduced, and may vanish altogether when there is no overlap in the two diffracted patterns. [58]

17.4 Interpretations of the experiment

Like the Schrödinger's cat thought experiment, the double-slit experiment is often used to highlight the differences and similarities between the various interpretations of quantum mechanics.

17.4.1 Copenhagen interpretation

The Copenhagen interpretation, put forth by some of the pioneers in the field of quantum mechanics, asserts that it is undesirable to posit anything that goes beyond the mathematical formulae and the kinds of physical apparatus and reactions that enable us to gain some knowledge of what goes on at the atomic scale. One of the mathematical constructs that enables experimenters to predict very accurately certain experimental results is sometimes called a probability wave. In its mathematical form it is analogous to the description of a physical wave, but its "crests" and "troughs" indicate levels of probability for the occurrence of certain phenomena (e.g., a spark of light at a certain point on a detector screen) that can be observed in the macro world of ordinary human experience.

The probability "wave" can be said to "pass through space" because the probability values that one can com-

pute from its mathematical representation are dependent on time. One cannot speak of the location of any particle such as a photon between the time it is emitted and the time it is detected simply because in order to say that something is located somewhere at a certain time one has to detect it. The requirement for the eventual appearance of an interference pattern is that particles be emitted, and that there be a screen with at least two distinct paths for the particle to take from the emitter to the detection screen. Experiments observe nothing whatsoever between the time of emission of the particle and its arrival at the detection screen. If a ray tracing is next made as if a light wave (as understood in classical physics) is wide enough to take both paths, then that ray tracing will accurately predict the appearance of maxima and minima on the detector screen when many particles pass through the apparatus and gradually "paint" the expected interference pattern.

17.4.2 Path-integral formulation

The Copenhagen interpretation is similar to the path integral formulation of quantum mechanics provided by Feynman. The path integral formulation replaces the classical notion of a single, unique trajectory for a system, with a sum over all possible trajectories. The trajectories are added together by using functional integration.

Each path is considered equally likely, and thus contributes the same amount. However, the phase of this contribution at any given point along the path is determined by the action along the path:

$$A_{\text{path}}(x, y, z, t) = e^{iS(x,y,z,t)}$$

All these contributions are then added together, and the magnitude of the final result is squared, to get the probability distribution for the position of a particle:

$$p(x, y, z, t) \propto \left| \int_{\text{all paths}} e^{iS(x,y,z,t)} \right|^2$$

As is always the case when calculating probability, the results must then be normalized by imposing:

$$\iiint_{\text{all space}} p(x, y, z, t) \, dV = 1$$

To summarize, the probability distribution of the outcome is the normalized square of the norm of the superposition, over all paths from the point of origin to the final point, of waves propagating proportionally to the action along each path. The differences in the cumulative action along the different paths (and thus the relative phases of the contributions) produces the interference pattern observed by the double-slit experiment. Feynman stressed that his formulation is merely a mathematical description, not an attempt to describe a real process that we can measure.

17.4.3   Relational interpretation

According to the relational interpretation of quantum mechanics, first proposed by Carlo Rovelli,[59] observations such as those in the double-slit experiment result specifically from the interaction between the observer (measuring device) and the object being observed (physically interacted with), not any absolute property possessed by the object. In the case of an electron, if it is initially "observed" at a particular slit, then the observer–particle (photon–electron) interaction includes information about the electron's position. This partially constrains the particle's eventual location at the screen. If it is "observed" (measured with a photon) not at a particular slit but rather at the screen, then there is no "which path" information as part of the interaction, so the electron's "observed" position on the screen is determined strictly by its probability function. This makes the resulting pattern on the screen the same as if each individual electron had passed through both slits. It has also been suggested that space and distance themselves are relational, and that an electron can appear to be in "two places at once"—for example, at both slits—because its spatial relations to particular points on the screen remain identical from both slit locations.[60]

17.4.4   Many-worlds interpretation

Physicist David Deutsch argues in his book *The Fabric of Reality* that the double-slit experiment is evidence for the many-worlds interpretation.

17.5   See also

- Complementarity (physics)
- Delayed choice quantum eraser
- Dual-polarization interferometry
- Elitzur–Vaidman bomb tester
- N-slit interferometer
- Photon polarization
- Quantum coherence
- Schrödinger's cat
- Young's interference experiment
- Measurement problem

17.6   References

[1] . While there is no doubt that Young's demonstration of optical interference, using sunlight, pinholes and cards, played a vital part in the acceptance of the wave theory of light, there is some question as to whether he ever actually performed a double-slit interference experiment.

  - Robinson, Andrew (2006). *The Last Man Who Knew Everything*. New York, NY: Pi Press. pp. 123–124. ISBN 0-13-134304-1.

[2] Lederman, Leon M.; Christopher T. Hill (2011). *Quantum Physics for Poets*. US: Prometheus Books. pp. 102–111. ISBN 1616142812.

[3] Feynman, Richard P.; Robert B. Leighton; Matthew Sands (1965). *The Feynman Lectures on Physics, Vol. 3*. US: Addison-Wesley. pp. 1.1–1.8. ISBN 0201021188.

[4] Feynman, 1965, p. 1.5

[5] Darling, David (2007). "Wave–Particle Duality". *The Internet Encyclopedia of Science*. The Worlds of David Darling. Retrieved 2008-10-18.

[6] Feynman, 1965, p. 1.7

[7] Lederman, 2011, p. 109

[8] "...*if in a double-slit experiment, the detectors which register outcoming photons are placed immediately behind the diaphragm with two slits: A photon is registered in one detector, not in both...*" Müller-Kirsten, H. J. W. (2006). *Introduction to Quantum Mechanics: Schrödinger Equation and Path Integral*. US: World Scientific. p. 14. ISBN 9812566910.

[9] Plotnitsky, Arkady (2012). *Niels Bohr and Complementarity: An Introduction*. US: Springer. pp. 75–76. ISBN 1461445175.

[10] "*It seems that light passes through one slit or the other in the form of photons if we set up an experiment to detect which slit the photon passes, but passes through both slits in the form of a wave if we perform an interference experiment.*" Rae, Alastair I. M. (2004). *Quantum Physics: Illusion Or Reality?*. UK: Cambridge University Press. pp. 9–10. ISBN 1139455273.

[11] Feynman, *Lectures on Physics* **3**:Quantum Mechanics p.1-1 "There is one lucky break, however—electrons behave just like light." .

[12] See: Davisson–Germer experiment "The diffraction of electrons by a crystal of nickel". *BSTJ* 7: 90–105, 1928.

[13] "Physicists Smash Record For Wave-Particle Duality"

[14] Eibenberger, Sandra; et al. (2013). "Matter-wave interference with particles selected from a molecular library with masses exceeding 10000 amu". *Physical Chemistry Chemical Physics* 15: pp. 14696–14700. arXiv:1310.8343. Bibcode:2013PCCP...1514696E. doi:10.1039/C3CP51500A.

[15] Charles Sanders Peirce first proposed the use of this effect as an artifact-independent reference standard for length

- C.S. Peirce (July 1879) "Note on the Progress of Experiments for Comparing a Wave-length with a Metre" *American Journal of Science*, as referenced by Crease, Robert P. (2011). *World in the Balance: the historic quest for an absolute system of measurement*. New York: W.W. Norton. p. 317. ISBN 978-0-393-07298-3. p. 203.

[16] Greene, Brian (1999). *The Elegant Universe: Superstrings, Hidden Dimensions, and the Quest for the Ultimate Theory*. New York: W.W. Norton. pp. 97–109. ISBN 0-393-04688-5.

[17] Feynman, 1965, chapter 3

[18] Sir Geoffrey, Ingram Taylor (1909). "Interference Fringes with Feeble Light". *Proc. Cam. Phil. Soc.* 15: 114.

[19] Jönsson C. (1961) *Zeitschrift für Physik*, 161:454–474 doi:10.1007/BF01342460

[20] Jönsson, C (1974). "Electron diffraction at multiple slits". *American Journal of Physics* 42: 4–11. Bibcode:1974AmJPh..42....4J. doi:10.1119/1.1987592.

[21] Merli, P G; Missiroli, G F; Pozzi, G (1976). "On the statistical aspect of electron interference phenomena". *American Journal of Physics* 44: 306–307. Bibcode:1976AmJPh..44..306M. doi:10.1119/1.10184.

[22] Rosa, R (2012). "The Merli–Missiroli–Pozzi Two-Slit Electron-Interference Experiment". *Physics in Perspective* 14: 178–195. Bibcode:2012PhP....14..178R. doi:10.1007/s00016-011-0079-0.

[23] "The most beautiful experiment". Physics World 2002.

[24] Greene, Brian (2007). *The Fabric of the Cosmos: Space, Time, and the Texture of Reality*. Random House LLC. p. 90. ISBN 0-307-42853-2.. Extract of page 90

[25] Donati, O; Missiroli, G F; Pozzi, G (1973). "An Experiment on Electron Interference". *American Journal of Physics* 41: 639–644. Bibcode:1973AmJPh..41..639D. doi:10.1119/1.1987321.

[26] New Scientist: Quantum wonders: Corpuscles and buckyballs, 2010 (Introduction, subscription needed for full text, quoted in full in)

[27] Wave Particle Duality of C60

[28] Nairz, Olaf; Brezger, Björn; Arndt, Markus; Anton Zeilinger, Abstract (2001). "Diffraction of Complex Molecules by Structures Made of Light". *Phys. Rev. Lett.* 87: 160401. arXiv:quant-ph/0110012. Bibcode:2001PhRvL..87p0401N. doi:10.1103/physrevlett.87.160401.

[29] Nairz, O; Arndt, M; Zeilinger, A (2003). "Quantum interference experiments with large molecules" (PDF). *American Journal of Physics* 71: 319–325. Bibcode:2003AmJPh..71..319N. doi:10.1119/1.1531580.

[30] Brian Greene, *The Elegant Universe*, p. 104, pp. 109–114

[31] Greene, Brian (2004). *The Fabric of the Cosmos: Space, Time, and the Texture of Reality*. Knopf. pp. 204–213. ISBN 0-375-41288-3.

[32] Baggott, Jim (2011). *The Quantum Story: A History in 40 Moments*. New York: Oxford University Press. pp. 76. ("The wavefunction of a system containing N particles depends on $3N$ position coordinates and is a function in a $3N$-dimensional configuration space or 'phase space'. It is difficult to visualize a reality comprising imaginary functions in an abstract, multi-dimensional space. No difficulty arises, however, if the imaginary functions are not to be given a real interpretation.")

[33] Harrison, David (2002). "Complementarity and the Copenhagen Interpretation of Quantum Mechanics". *UPSCALE*. Dept. of Physics, U. of Toronto. Retrieved 2008-06-21.

[34] Cassidy, David (2008). "Quantum Mechanics 1925–1927: Triumph of the Copenhagen Interpretation". *Werner Heisenberg*. American Institute of Physics. Retrieved 2008-06-21.

[35] Boscá Díaz-Pintado, Maria C. (29–31 March 2007). "Updating the wave-particle duality". *15th UK and European Meeting on the Foundations of Physics*. Leeds, UK. Retrieved 2008-06-21.

[36] Bartell, L. (1980). "Complementarity in the double-slit experiment: On simple realizable systems for observing intermediate particle-wave behavior". *Physical Review D* 21 (6): 1698. Bibcode:1980PhRvD..21.1698B. doi:10.1103/PhysRevD.21.1698.

[37] Zeilinger, A. (1999). "Experiment and the foundations of quantum physics". *Reviews of Modern Physics* 71 (2): S288. Bibcode:1999RvMPS..71..288Z. doi:10.1103/RevModPhys.71.S288.

[38] P. Mittelstaedt; A. Prieur; R. Schieder (1987). "Unsharp particle-wave duality in a photon split-beam experiment". *Foundations of Physics* 17 (9): 891–903. Bibcode:1987FoPh...17..891M. doi:10.1007/BF00734319.

[39] D.M. Greenberger and A. Yasin. "Simultaneous wave and particle knowledge in a neutron interferometer". *Physics Letters A* 128, 391–4 (1988).

[40] Wootters, W. K.; Zurek, W. H. (1979). "Complementarity in the double-slit experiment: Quantum nonseparability and a quantitative statement of Bohr's principle" (PDF). *Phys. Rev. D* **19** (473–484). Bibcode:1979PhRvD..19..473W. doi:10.1103/PhysRevD.19.473. Retrieved 5 February 2014.

[41] Hillmer, R.; Kwiat, P. (2007). "A do-it-yourself quantum eraser" (PDF). *Scientific American Magazine* **296** (5): 90–95. doi:10.1038/scientificamerican0507-90. Retrieved 5 February 2014.

[42] Chiao, R. Y.; P. G. Kwiat; Steinberg, A. M. (1995). "Quantum non-locality in two-photon experiments at Berkeley". *Quantum and Semiclassical Optics: Journal of the European Optical Society Part B* **7** (3): 259. arXiv:quant-ph/9501016. Bibcode:1995QuSOp...7..259C. doi:10.1088/1355-5111/7/3/006.

[43] Francis, Matthew. "Disentangling the wave-particle duality in the double-slit experiment". Ars Technica.

[44] Motl, Luboš. "Pseudoscience hiding behind "weak measurements"". Retrieved 8 November 2013.

[45] Svensson, Bengt E. Y. "Pedagogical Review of Quantum Measurement Theory with an Emphasis on Weak Measurements". *Quanta* **2** (1): 18–49. doi:10.12743/quanta.v2i1.12. Retrieved 4 February 2014.

[46] Pfleegor, R. L. and Mandel, L. (July 1967). "Interference of Independent Photon Beams". *Physical Review* **159** (5): 1084–1088. Bibcode:1967PhRv..159.1084P. doi:10.1103/PhysRev.159.1084.

[47] http://scienceblogs.com/principles/2010/11/interference_of_independent_ph.php>

[48] Sillitto, R.M. and Wykes, Catherine (1972). "An interference experiment with light beams modulated in anti-phase by an electro-optic shutter". *Physics Letters A* **39** (4): 333–334. Bibcode:1972PhLA...39..333S. doi:10.1016/0375-9601(72)91015-8.

[49] "To a light particle"

[50] Nature: Wave–particle duality of C60 molecules, 14 October 1999. Abstract, subscription needed for full text

[51] Schouten, H.F.; Kuzmin, N.; Dubois, G.; Visser, T.D.; Gbur, G.; Alkemade, P.F.A.; Blok, H.; Hooft, G.W.; Lenstra, D.; Eliel, E.R. (7 February 2005). "Plasmon-Assisted Two-Slit Transmission: Young's Experiment Revisited". *Phys. Rev. Lett.* **94**: 053901. Bibcode:2005PhRvL..94e3901S. doi:10.1103/physrevlett.94.053901.

[52] Bach, Roger; et al. (March 2013). "Controlled double-slit electron diffraction". *New Journal of Physics* **15** (3): 033018. arXiv:1210.6243. Bibcode:2013NJPh...15c3018B. doi:10.1088/1367-2630/15/3/033018.

[53] Bush, John WM (2015). "Pilot-wave hydrodynamics" (PDF). *Annual Review of Fluid Mechanics* **47**: 269–292. Bibcode:2015AnRFM..47..269B. doi:10.1146/annurev-fluid-010814-014506. Retrieved 21 June 2015.

[54] Bush, John W. M. "Quantum mechanics writ large". *PNAS* **107** (41): 17455–17456. doi:10.1073/pnas.1012399107. Retrieved 23 June 2015.

[55] Natalie Wolchover, Quanta Magazine. Science. 06.30.14. "Have We Been Interpreting Quantum Mechanics Wrong This Whole Time?".

[56] Couder, Y.; Fort, E. (2012). "Probabilities and trajectories in a classical wave-particle duality" (PDF). *Journal of Physics: Conference Series* **361**: 012001. Bibcode:2012JPhCS.361a2001C. doi:10.1088/1742-6596/361/1/012001. Retrieved 23 June 2015.

[57] Jenkins FA and White HE, Fundamentals of Optics, 1967, McGraw Hill, New York

[58] Longhurst RS, Physical and Geometrical Optics, 1967, 2nd Edition, Longmans

[59] Rovelli, Carlo (1996). "Relational Quantum Mechanics". *International Journal of Theoretical Physics* **35** (8): 1637–1678. arXiv:quant-ph/9609002. Bibcode:1996IJTP...35.1637R. doi:10.1007/BF02302261.

[60] Filk, Thomas (2006). "Relational Interpretation of the Wave Function and a Possible Way Around Bell's Theorem". *International Journal of Theoretical Physics* **45**: 1205–1219. arXiv:quant-ph/0602060. Bibcode:2006IJTP...45.1166F. doi:10.1007/s10773-006-9125-0.

17.6.1 Further reading

- Al-Khalili, Jim (2003). *Quantum: A Guide for the Perplexed*. London: Weidenfeld and Nicholson. ISBN 0-297-84305-2.
- Feynman, Richard P. (1988). *QED: The Strange Theory of Light and Matter*. Princeton University Press. ISBN 0-691-02417-0.
- Frank, Philipp (1957). *Philosophy of Science*. Prentice-Hall.
- French, A.P.; Taylor, Edwin F. (1978). *An Introduction to Quantum Physics*. Norton. ISBN 0-393-09106-6.
- Quznetsov, Gunn (2011). *Final Book on Fundamental Theoretical Physics*. American Research Press. ISBN 978-1-59973-172-8.
- Greene, Brian (2000). *The Elegant Universe*. Vintage. ISBN 0-375-70811-1.

- Greene, Brian (2005). *The Fabric of the Cosmos*. Vintage. ISBN 0-375-72720-5.

- Gribbin, John (1999). *Q is for Quantum: Particle Physics from A to Z*. Weidenfeld & Nicolson. ISBN 0-7538-0685-1.

- Hey, Tony (2003). *The New Quantum Universe*. Cambridge University Press. ISBN 0-521-56457-3.

- Sears, Francis Weston (1949). *Optics*. Addison Wesley.

- Tipler, Paul (2004). *Physics for Scientists and Engineers: Electricity, Magnetism, Light, and Elementary Modern Physics* (5th ed.). W. H. Freeman. ISBN 0-7167-0810-8.

17.7 External links

17.7.1 Interactive animations

- Huygens and interference

17.7.2 Single particle experiments

- Website with the movie and other information from the first single electron experiment by Merli, Missiroli, and Pozzi.

- Movie showing single electron events build up to form an interference pattern in double-slit experiments. Several versions with and without narration (File size = 3.6 to 10.4 MB) (Movie Length = 1m 8s)

- Freeview video 'Electron Waves Unveil the Microcosmos' A Royal Institution Discourse by Akira Tonomura provided by the Vega Science Trust

- Hitachi website that provides background on Tonomura video and link to the video

17.7.3 Hydrodynamic analog

- "Single-particle interference observed for macroscopic objects"

- Pilot-Wave Hydrodynamics: Supplemental Video

- *Through the Wormhole*: Yves Couder . Explains Wave/Particle Duality via Silicon Droplets

17.7.4 Computer simulations

- Java demonstration of double slit experiment, animated

- Java demonstration of Young's double slit interference

- A simulation that runs in Mathematica Player, in which the number of quantum particles, the frequency of the particles, and the slit separation can be independently varied

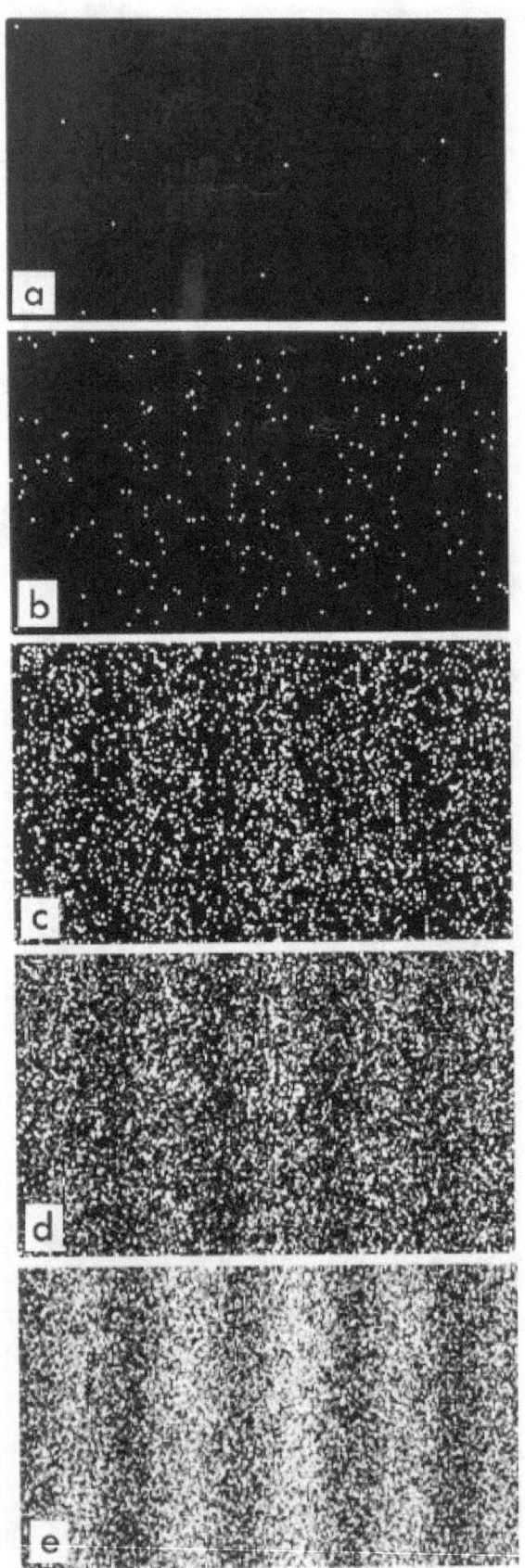

A laboratory double-slit assembly; distance between top posts approximately 2.5 cm (one inch).

Near-field intensity distribution patterns for plasmonic slits with equal widths (A) and non-equal widths (B).

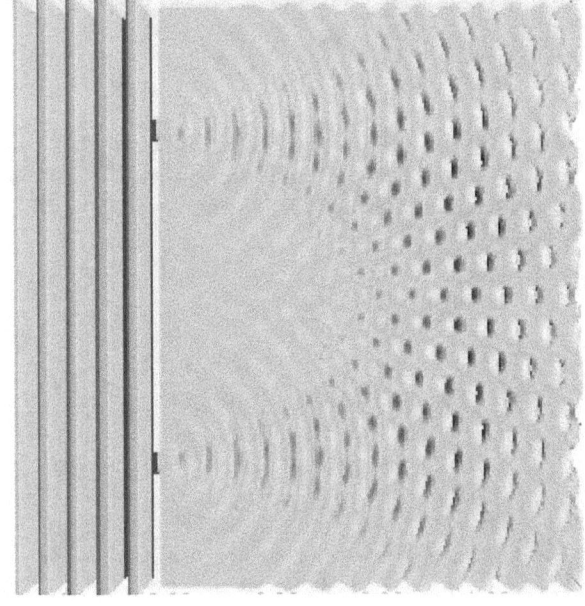

Two-slit diffraction pattern by a plane wave

Electron buildup over time

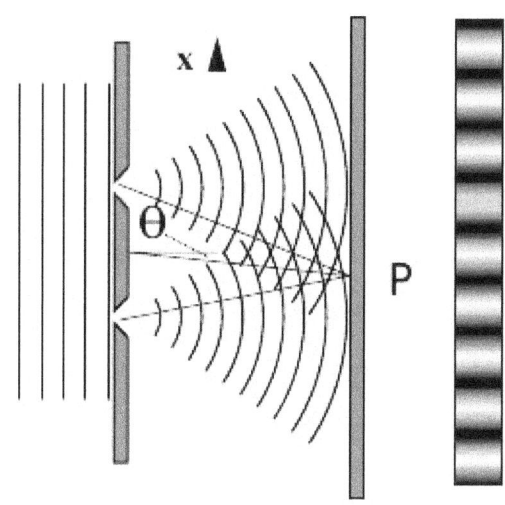

Two slits are illuminated by a plane wave.

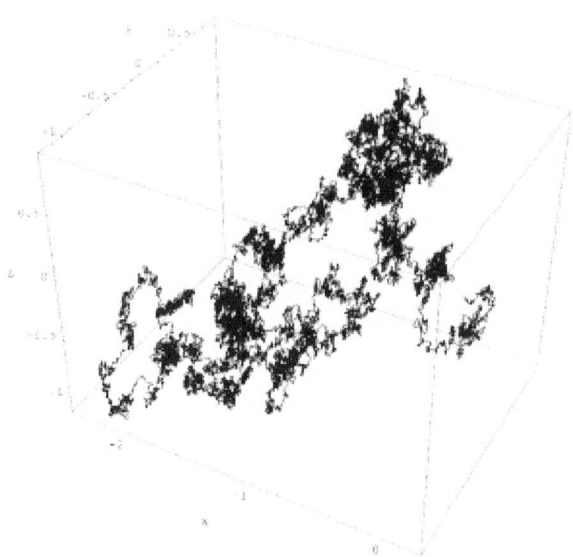

One of an infinite number of equally likely paths used in the Feynman path integral. (see also: Wiener process.)

Chapter 18

Wave function

Not to be confused with Wave equation.

A **wave function** in quantum mechanics describes the

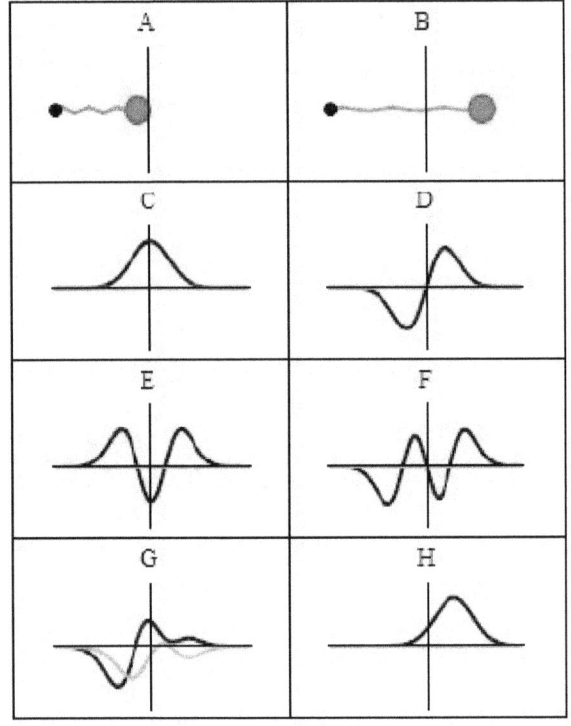

Comparison of classical and quantum harmonic oscillator conceptions for a single spinless particle. The two processes differ greatly. The classical process (A–B) is represented as the motion of a particle along a trajectory. The quantum process (C–H) has no such trajectory. Rather, it is represented as a wave. Panels (C–F) show four different standing wave solutions of the Schrödinger equation. Panels (G–H) further show two different wave functions that are solutions of the Schrödinger equation but not standing waves.

quantum state of an isolated system of one or more particles. There is *one* wave function containing all the information about the entire system, not a separate wave function for each particle in the system. Its interpretation is that of a probability amplitude. Quantities associated with measurements, such as the average momentum of a particle, can

be derived from the wave function. It is a central entity in quantum mechanics and is important in all modern theories, like quantum field theory incorporating quantum mechanics, while its interpretation may differ. The most common symbols for a wave function are the Greek letters ψ or Ψ (lower-case and capital psi).

For a given system, once a representation corresponding to a maximal set of commuting observables and a suitable coordinate system is chosen, the wave function is a complex-valued function of the system's degrees of freedom corresponding to the chosen representation and coordinate system, continuous as well as discrete. Such a set of observables, by a postulate of quantum mechanics, are Hermitian linear operators on the space of states representing a set of **physical observables**, like position, momentum and spin that can, in principle, be simultaneously measured with arbitrary precision. Wave functions can be added together and multiplied by complex numbers to form new wave functions, and hence are elements of a vector space. This is the superposition principle of quantum mechanics. This vector space is endowed with an inner product such that it is a complete metric topological space with respect to the metric induced by the inner product. In this way the set of wave functions for a system form a function space that is a Hilbert space. The inner product is a measure of the overlap between physical states and is used in the foundational probabilistic interpretation of quantum mechanics, the Born rule, relating transition probabilities to inner products. The actual space depends on the system's degrees of freedom (hence on the chosen representation and coordinate system) and the exact form of the Hamiltonian entering the equation governing the dynamical behavior. In the non-relativistic case, disregarding spin, this is the Schrödinger equation.

The Schrödinger equation determines the allowed wave functions for the system and how they evolve over time. A wave function behaves qualitatively like other waves, such as water waves or waves on a string, because the Schrödinger equation is mathematically a type of wave equation. This explains the name "wave function", and gives rise to wave–particle duality. The wave of the wave function, however, is not a wave in physical space: it is a wave in an abstract mathematical "space", and in this respect it differs fundamentally from water waves or waves on a string.[1][2][3][4][5][6][7]

For a given system, the choice of which relevant degrees of freedom to use are not unique, and correspondingly the domain of the wave function is not unique. It may be taken to be a function of all the position coordinates of the particles over *position space*, or the momenta of all the particles over *momentum space*, the two are related by a Fourier transform. These descriptions are the most important, but they are not the only possibilities. Just like in classical mechanics, canonical transformations may be used in the description of a quantum system. Some particles, like electrons and photons, have nonzero spin, and the wave function must include this fundamental property as an intrinsic discrete degree of freedom. In general, for a particle with *half-integer* spin the wave function is a spinor, for a particle with *integer* spin the wave function is a tensor. Particles with spin zero are called scalar particles, those with spin 1 vector particles, and more generally for higher integer spin, tensor particles. The terminology derives from how the wave functions transform under a rotation of the coordinate system. No *elementary* particle with spin $\frac{3}{2}$ or higher is known, except for the hypothesized spin 2 graviton. Other discrete variables can be included, such as isospin. When a system has internal degrees of freedom, the wave function at each point in the continuous degrees of freedom (e.g. a point in space) assigns a complex number for *each* possible value of the discrete degrees of freedom (e.g. z-component of spin). These values are often displayed in a column matrix (e.g. a 2×1 column vector for a non-relativistic electron with spin $\frac{1}{2}$).

In the Copenhagen interpretation, an interpretation of quantum mechanics, the squared modulus of the wave function, $|\psi|^2$, is a real number interpreted as the probability density of measuring a particle as being at a given place at a given time or having a definite momentum, and possibly having definite values for discrete degrees of freedom. The integral of this quantity, over all the system's degrees of freedom, must be 1 in accordance with the probability interpretation, this general requirement a wave function must satisfy is called the *normalization condition*. Since the wave function is complex valued, only its relative phase and relative magnitude can be measured. Its value does not in isolation tell anything about the magnitudes or directions of measurable observables; one has to apply quantum operators, whose eigenvalues correspond to sets of possible results of measurements, to the wave function ψ and calculate the statistical distributions for measurable quantities.

The unit of measurement for ψ depends on the system, and can be found by dimensional analysis of the normalization condition for the system. For one particle in three dimensions, its units are $[length]^{-3/2}$, because an integral of $|\psi|^2$ over a region of three-dimensional space is a dimensionless probability.[8]

18.1   Historical background

In 1905 Einstein postulated the proportionality between the frequency of a photon and its energy, $E = hf$,[9] and in 1916 the corresponding relation between photon momentum and wavelength, $\lambda = h/p$.[10] In 1923, De Broglie was the first to suggest that the relation $\lambda = h/p$, now called the De Broglie relation, holds for *massive* particles, the chief clue being Lorentz invariance,[11] and this can be viewed as the starting point for the modern development of quantum mechanics. The equations represent wave–particle duality for both massless and massive particles.

In the 1920s and 1930s, quantum mechanics was developed using calculus and linear algebra. Those who used the techniques of calculus included Louis de Broglie, Erwin Schrödinger, and others, developing "wave mechanics". Those who applied the methods of linear algebra included Werner Heisenberg, Max Born, and others, developing "matrix mechanics". Schrödinger subsequently showed that the two approaches were equivalent.[12]

In 1926, Schrödinger published the famous wave equation now named after him, indeed the Schrödinger equation, based on classical energy conservation using quantum operators and the de Broglie relations such that the solutions of the equation are the wave functions for the quantum system.[13] However, no one was clear on how to *interpret it*.[14] At first, Schrödinger and others thought that wave functions represent particles that are spread out with most of the particle being where the wave function is large.[15] This was shown to be incompatible with how elastic scattering of a wave packet representing a particle off a target appears; it spreads out in all directions.[16] While a scattered particle may scatter in any direction, it does not break up and take off in all directions. In 1926, Born provided the perspective of probability amplitude.[16][17][18] This relates calculations of quantum mechanics directly to probabilistic experimental observations. It is accepted as part of the Copenhagen interpretation of quantum mechanics. There are many other interpretations of quantum mechanics. In 1927, Hartree and Fock made the first step in an attempt to solve the N-body wave function, and developed the *self-consistency cycle*: an iterative algorithm to approximate the solution. Now it is also known as the Hartree–Fock method.[19] The Slater determinant and permanent (of a matrix) was part of the method, provided by John C. Slater.

Schrödinger did encounter an equation for the wave function that satisfied relativistic energy conservation *before* he published the non-relativistic one, but discarded it as it predicted negative probabilities and negative energies. In 1927, Klein, Gordon and Fock also found it, but incorporated the electromagnetic interaction and proved that it was Lorentz invariant. De Broglie also arrived at the same equation in 1928. This relativistic wave equation is now most commonly known as the Klein–Gordon equation.[20]

In 1927, Pauli phenomenologically found a non-relativistic equation to describe spin-1/2 particles in electromagnetic fields, now called the Pauli equation.[21] Pauli found the wave function was not described by a single complex function of space and time, but needed two complex numbers, which respectively correspond to the spin +1/2 and −1/2 states of the fermion. Soon after in 1928, Dirac found an equation from the first successful unification of special relativity and quantum mechanics applied to the electron, now called the Dirac equation. In this, the wave function is a *spinor* represented by four complex-valued components:[19] two for the electron and two for the electron's antiparticle, the positron. In the non-relativistic limit, the Dirac wave function resembles the Pauli wave function for the electron. Later, other relativistic wave equations were found.

18.1.1   Wave functions and wave equations in modern theories

All these wave equations are of enduring importance. The Schrödinger equation and the Pauli equation are under many circumstances excellent approximations of the relativistic variants. They are considerably easier to solve in practical problems than the relativistic equations. The Klein-Gordon equation and the Dirac equation, while being relativistic, do not represent full reconciliation of quantum mechanics and special relativity. The branch of quantum mechanics where these equations are studied the same way as the Schrödinger equation, often called relativistic quantum mechanics, while very successful, has its limitations (see e.g. Lamb shift) and conceptual problems (see e.g. Dirac sea).

Relativity makes it inevitable that the number of particles in a system is not constant. For full reconciliation, quantum field theory is needed.[22] In this theory, the wave equations and the wave functions have their place, but in a somewhat different guise. The main objects of interest are not the wave functions, but rather operators, so called *field operators* (or just fields where "operator" is understood) on the Hilbert space of states (to be described next section). It turns out that the original relativistic wave equations and their solutions are still needed to build the Hilbert space.

Moreover, the *free fields operators*, i.e. when interactions are assumed not to exist, turn out to (formally) satisfy the same equation as do the fields (wave functions) in many cases.

Thus the Klein-Gordon equation (spin 0) and the Dirac equation (spin 1/2) in this guise remain in the theory. Higher spin analogues include the Proca equation (spin 1), Rarita–Schwinger equation (spin 3/2), and, more generally, the Bargmann–Wigner equations. For *massless* free fields two examples are the free field Maxwell equation (spin 1) and the free field Einstein equation (spin 2) for the field operators.[23] All of them are essentially a direct consequence of the requirement of Lorentz invariance. Their solutions must transform under Lorentz transformation in a prescribed way, i.e. under a particular representation of the Lorentz group and that together with few other reasonable demands, e.g. the *cluster decomposition principle*,[24] with implications for causality is enough to fix the equations.

It should be emphasized that this applies to free field equations; interactions are not included. It should also be noted that the equations and their solutions, though needed for the theories, are not the central objects of study.

18.2   Wave functions and function spaces

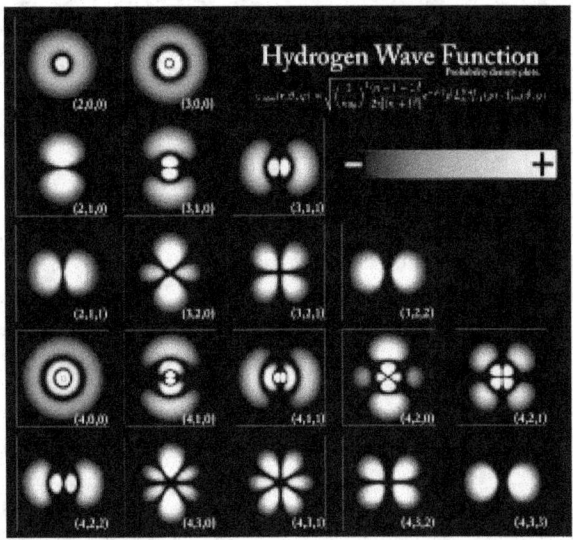

The electron probability density for the first few hydrogen atom electron orbitals shown as cross-sections. These orbitals form an orthonormal basis for the wave function of the electron. Different orbitals are depicted with different scale.

The concept of Function spaces enters naturally in the discussion about wave functions. A function space is a set of functions, usually with some defining requirements on the functions, together with a topology on that set. The latter will sparsely be used here, it is only needed to obtain a precise definition of what it means for a subset of a function space to be closed. A wave function is an element of a function space partly characterized by the following concrete and abstract descriptions.

- The Schrödinger equation is linear. This means that the solutions to it, wave functions, can be added and multiplied by scalars to form a new solution.

- The superposition principle of quantum mechanics. If Ψ and Φ are two states in the abstract space of **states** of a quantum mechanical system, then $a\Psi + b\Phi$ is a valid state as well.

The first item says that the set of solutions to the Schrödinger equation is a vector space. The second item says that the set of allowable states is a vector space. This similarity is of course not accidental. Not all properties of the respective spaces have been given so far. There are also a distinctions between the spaces to keep in mind.

- Basic states are characterized by a set of quantum numbers. This is a set of eigenvalues of a maximal set of commuting observables. A choice of such a set may be called a choice of **representation**. It is a postulate of quantum mechanics that a physically observable quantity of a system, like position, momentum and spin, is represented by a linear Hermitian operator on the state space. The possible outcomes of measurement of the quantity are the eigenvalues of the operator.[15] Maximality refers to that no more algebraically independent linear Hermitian operator can be added to the set that commutes with the ones already present. The physical interpretation is that such a set represents what can – in theory – be simultaneously be measured with arbitrary position at a given time. The set is non-unique. It may for a one-particle system, for example, be position and spin z-projection, (x, S_z), or it may be momentum and spin y-projection, (p, S_y). At a deeper level, most observables, perhaps all, arise as generators of symmetries.[15][25][nb 1]

- Once a representation is chosen, there is still arbitrariness. It remains to choose a coordinate system. This may, for example, correspond to a choice of x, y- and z-axis, or a choice of **curvlinear coordinates** as exemplified by the spherical coordinates used for the atomic wave functions illustrated below. This final choice also fixes a basis in abstract Hilbert space. The basic states are labeled by the quantum numbers

corresponding to the maximal set of commuting observables and an appropriate coordinate system.[nb 2]

- Wave functions corresponding to a state are accordingly not unique. This has been exemplified already with momentum and position space wave functions describing the same abstract state. This non-uniqueness reflects the non-uniqueness in the choice of a maximal set of commuting observables.

- The abstract states are "abstract" only in that an arbitrary choice necessary for a particular *explicit* description of it is not given. This is the same as saying that no choice of maximal set of commuting observables has been given. This is analogous to a vector space without a specified basis.

- The wave functions of position and momenta, respectively, can be seen as a choice of representation yielding two different, but entirely equivalent, explicit descriptions of the same state for a system with no discrete degrees of freedom.

- Corresponding to the two examples in the first item, to a particular state there corresponds two wave functions, $\Psi(x, S_z)$ and $\Psi(p, S_y)$, both describing the same state. For each choice of maximal commuting sets of observables for the abstract state space, there is a corresponding representation that is associated to a function space of wave functions.

- Each choice of representation should be thought of as specifying a unique function space in which wave functions corresponding to that choice of representation lives. This distinction is best kept, even if one could argue that two such function spaces are mathematically equal, e.g. being the set of square integrable functions. One can then think of the function spaces as two distinct copies of that set.

- Between all these different function spaces and the abstract state space, there are one-to-one correspondences (here disregarding normalization and unobservable phase factors), the common denominator here being a particular abstract state. The relationship between the momentum and position space wave functions, for instance, describing the same state is the Fourier transform.

To make this concrete, in the figure to the right, the 19 subimages are images of wave functions in position space (their norm squared). The wave functions each represent the abstract state characterized by the triple of quantum numbers (n, l, m), in the lower right of each image. These are the principal quantum number, the orbital angular momentum quantum number and the magnetic quantum number. Together with one spin-projection quantum number of the electron, this is a complete set of observables.

The figure can serve to illustrate some further properties of the function spaces of wave functions.

- In this case, the wave functions are square integrable. One can initially take the function space as the space of square integrable functions, usually denoted L^2.

- The displayed functions are solutions to the Schrödinger equation. Obviously, not every function in L^2 satisfies the Schrödinger equation for the hydrogen atom. The function space is thus a subspace of L^2.

- The displayed functions form part of a basis for the function space. To each triple (n, l, m), there corresponds a basis wave function. If spin is taken into account, there are two basis functions for each triple. The function space thus has a countable basis.

- The basis functions are mutually orthonormal. For this concept to have a meaning, there must exist an inner product. The function space is thus an inner product space. The inner product between two states intuitively measures the "overlap" between the states. The physical interpretation is that the norm squared is proportional to the transition probability between the states. That is,

$$P(\Psi \to \Phi_i) = |(\Psi, \Phi_i)|^2$$

where the i is an index composed of quantum numbers corresponding to a representation and the probabilities are the probabilities of finding the state Ψ in the definite state represented by Φ_i upon measurement of the physical observables corresponding to the representation, for instance, i could be the quadruple (n, l, m, S_z). This is the Born rule,[16] and is one of the fundamental postulates of quantum mechanics.

These observations encapsulate the essence of the function spaces of which wave functions are elements. Mathematically, this is expressed (in one spatial dimension, disregarding here unimportant issues of normalization) for a particle with no internal degrees of freedom as

$$\Psi = I\Psi = \int \Phi_x(\Phi_x, \Psi) dx = \int \Psi(x)\Phi_x dx = \int \Phi_p(\Phi_p, \Psi) dp = \int \Psi(p)\Phi_p dp$$

where Ψ is any "abstract" state, Φ_x is an eigenfunction of the position operator representing a particle localized at x, $(\cdot, \cdot)$ represents the inner product, Φ_p is an eigenfunction of the momentum operator representing a particle with precise momentum p, I is the **identity operator** and the integrals (first and third) represent the completeness of momentum and position eigenstates, $\Psi(x)$ is the coordinate space wave function and $\Psi(p)$ is the wave function in momentum space. In Dirac notation, the above equation reads

$$|\Psi\rangle = I|\Psi\rangle = \int |x\rangle\langle x|\Psi\rangle dx = \int \Psi(x)|x\rangle dx = \int |p\rangle\langle p|\Psi\rangle dp =$$

The description is not yet complete. There is a further technical requirement on the function space, that of completeness, that allows one to take limits of sequences in the function space, and be ensured that, if the limit exists, it is an element of the function space. A complete inner product space is called a Hilbert space. The property of completeness is crucial in advanced treatments and applications of quantum mechanics. It will not be very important in the subsequent discussion of wave functions, and technical details and links may be found in footnotes like the one that follows.[nb 3] The space L^2 is a Hilbert space, with inner product presented later. The function space of the example of the figure is a subspace of L^2. A subspace of a Hilbert space is a Hilbert space if it is closed. It is here that the topology of the function space enters into its description.

It is also important to note, in order to avoid confusion, that not all functions to be discussed are elements of some Hilbert space, say L^2. The most glaring example is the set of functions $e^{2\pi i p x/h}$. These are solutions of the Schrödinger equation for a free particle, but are not normalizable, hence not in L^2. But they are nonetheless fundamental for the description. One can, using them, express functions that *are* normalizable using wave packets. They are, in a sense to be made precise later, a basis (but not a Hilbert space basis) in which wave functions of interest can be expressed. There is also the artifact "normalization to a delta function" that is frequently employed for notational convenience, see further down. The delta functions themselves aren't square integrable either.

18.2.1   Physical requirements

The above description of the function space containing the wave functions is mostly mathematically motivated. The function spaces are, due to completeness, very *large* in a certain sense. Not all functions are realistic descriptions of any physical system. For instance, in the function space L^2 one can find the function that takes on the value 0 for all rational numbers and 1 for the irrationals in the interval [0,

Continuously differentiable

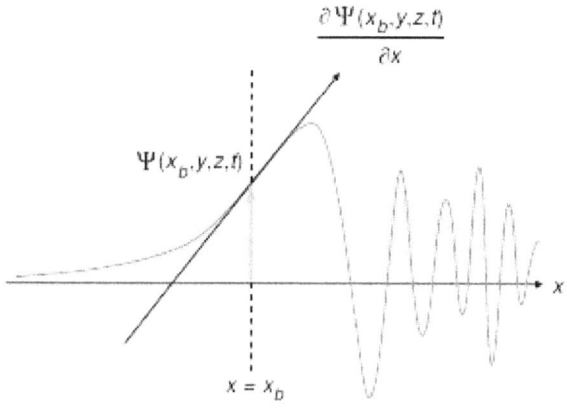

Discontinuous

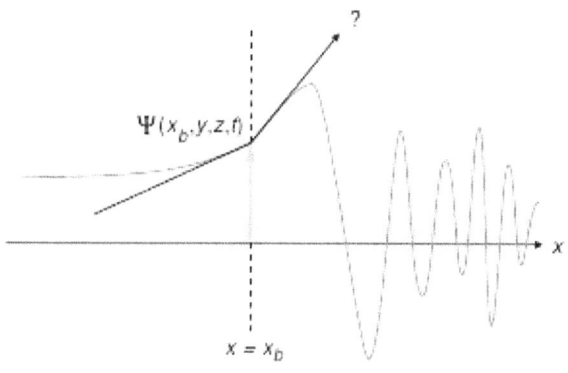

Continuity of the wave function and its first spatial derivative (in the x direction, y and z coordinates not shown), at some time t.

1]. This *is* square integrable,[nb 4] but can hardly represent a physical state.

The following constraints on the wave function are sometimes explicitly formulated for the calculations and physical interpretation to make sense:[26][27]

- The wave function must be square integrable. This is motivated by the Copenhagen interpretation of the wave function as a probability amplitude.

- It must everywhere be everywhere continuous and everywhere continuously differentiable. This is motivated by the appearance of the Schrödinger equation.

It is possible to relax these conditions somewhat for special purposes.[nb 5] If these requirements are not met, it is not possible to interpret the wave function as a probability amplitude.[28]

This does not alter the structure of the Hilbert space that these particular wave functions inhabit, but it should be pointed out that the subspace of the square-integrable functions L^2, which is a Hilbert space, satisfying the second requirement *is not closed* in L^2, hence not a Hilbert space in itself.[nb 6] The functions that does not meet the requirements are still needed for both technical and practical reasons.[nb 7][nb 8]

18.3 Definition (one spinless particle in 1d)

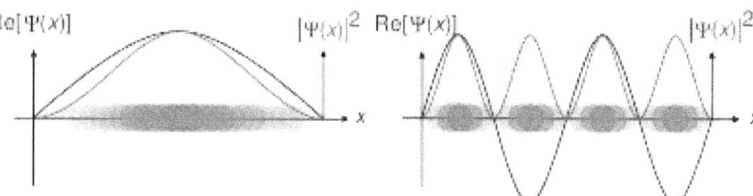

Standing waves for a particle in a box, examples of stationary states.

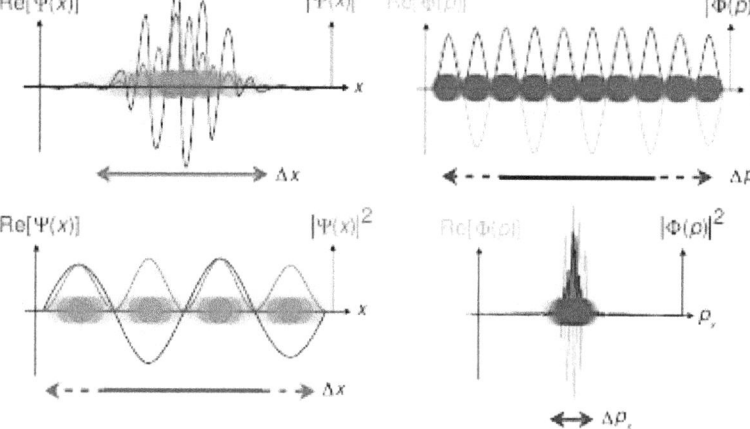

Travelling waves of a free particle.
The real parts of position wave function $\Psi(x)$ and momentum wave function $\Phi(p)$, and corresponding probability densities $|\Psi(x)|^2$ and $|\Phi(p)|^2$, for one spin-0 particle in one x or p dimension. The colour opacity of the particles corresponds to the probability density (*not* the wave function) of finding the particle at position x or momentum p.

For now, consider the simple case of a single particle, without spin, in one spatial dimension. More general cases are discussed below.

18.3.1   Position-space wave function

The state of such a particle is completely described by its wave function,

$$\Psi(x, t),$$

where x is position and t is time. This is a complex-valued function of two real variables x and t.

If interpreted as a probability amplitude, the square modulus of the wave function, the positive real number

$$|\Psi(x, t)|^2 = \Psi(x, t)^* \Psi(x, t) = \rho(x, t),$$

is interpreted as the probability density that the particle is at x. The asterisk indicates the complex conjugate. If the particle's position is measured, its location cannot be determined from the wave function, but is described by a probability distribution. The probability that its position x will be in the interval $a \leq x \leq b$ is the integral of the density over this interval:

$$P_{a \leq x \leq b}(t) = \int_a^b dx\, |\Psi(x, t)|^2$$

where t is the time at which the particle was measured. This leads to the **normalization condition**:

$$\int_{-\infty}^{\infty} dx\, |\Psi(x, t)|^2 = 1.$$

because if the particle is measured, there is 100% probability that it will be *somewhere*.

Since the Schrödinger equation is linear, if any number of wave functions Ψ_n for $n = 1, 2, \ldots$ are solutions of the equation, then so is their sum, and their scalar multiples by complex numbers a_n. Taking scalar multiplication and addition together is known as a linear combination:

$$\sum_n a_n \Psi_n(x, t) = a_1 \Psi_1(x, t) + a_2 \Psi_2(x, t) + \cdots$$

This is the superposition principle. Multiplying a wave function Ψ by any nonzero constant complex number c to obtain $c\Psi$ does not change any information about the quantum system, because c cancels in the Schrödinger equation for $c\Psi$.

18.3.2   Momentum-space wave function

The particle also has a wave function in momentum space:

$$\Phi(p, t)$$

where p is the momentum in one dimension, which can be any value from $-\infty$ to $+\infty$, and t is time.

All the previous remarks on superposition, normalization, etc. apply similarly. In particular, if the particle's momentum is measured, the result is not deterministic, but is described by a probability distribution:

$$P_{a \leq p \leq b}(t) = \int_a^b dp\, |\Phi(p, t)|^2.$$

and the normalization condition is:

$$\int_{-\infty}^{\infty} dp\, |\Phi(p, t)|^2 = 1.$$

18.3.3   Relation between wave functions

The position-space and momentum-space wave functions are Fourier transforms of each other, therefore both contain the same information, and either one alone is sufficient to calculate any property of the particle. As elements of **abstract physical Hilbert space**, whose elements are the possible states of the system under consideration, they represent the same object, but they are not equal when viewed as square-integrable functions. (A function and its Fourier transform are not equal.) For one dimension,[29]

$$\Phi(p, t) = \frac{1}{\sqrt{2\pi\hbar}} \int_{-\infty}^{\infty} dx\, e^{-ipx/\hbar} \Psi(x, t) \quad \rightleftharpoons \quad \Psi(x, t) = \frac{1}{\sqrt{2\pi\hbar}}$$

In practice, the position-space wave function is used much more often than the momentum-space wave function. The potential entering the Schrödinger equation determines in which basis the description is easiest. For the harmonic oscillator, x and p enter symmetrically, so there it doesn't matter which description one uses.

18.4   Definitions (other cases)

Following are the general forms of the wave function for systems in higher dimensions and more particles, as well as

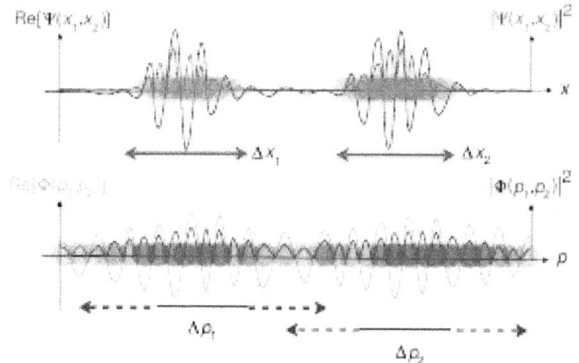

Traveling waves of two free particles, with two of three dimensions suppressed. Top is position space wave function, bottom is momentum space wave function, with corresponding probability densities.

including other degrees of freedom than position coordinates or momentum components.

The position-space wave function of a single particle in three spatial dimensions is similar to the case of one spatial dimension above:

$$\Psi(\mathbf{r}, t)$$

where $\mathbf{r}$ is the position vector in three-dimensional space, and t is time. As always $\Psi(\mathbf{r}, t)$ is a complex number, for this case a complex-valued function of four real variables.

If there are many particles, in general there is only one wave function, not a separate wave function for each particle. The fact that *one* wave function describes *many* particles is what makes quantum entanglement and the EPR paradox possible. The position-space wave function for N particles is written:[19]

$$\Psi(\mathbf{r}_1, \mathbf{r}_2 \cdots \mathbf{r}_N, t)$$

where $\mathbf{r}_i$ is the position of the ith particle in three-dimensional space, and t is time. Altogether, this is a complex-valued function of $3N + 1$ real variables.

In quantum mechanics there is a fundamental distinction between *identical particles* and *distinguishable* particles. For example, any two electrons are identical and fundamentally indistinguishable from each other; the laws of physics make it impossible to "stamp an identification number" on a certain electron to keep track of it.[29] This translates to a requirement on the wave function for a system of identical particles:

$$\Psi(\ldots \mathbf{r}_a, \ldots, \mathbf{r}_b, \ldots) = \pm\Psi(\ldots \mathbf{r}_b, \ldots, \mathbf{r}_a, \ldots)$$

where the + sign occurs if the particles are *all bosons* and − sign if they are *all fermions*. In other words, the wave function is either totally symmetric in the positions of bosons, or totally antisymmetric in the positions of fermions.[30] The physical interchange of particles corresponds to mathematically switching arguments in the wave function. The antisymmetry of fermionic wave functions leads to the Pauli principle. Generally, bosonic and fermionic symmetry requirements are the manifestation of particle statistics and are present in other quantum state formalisms.

For N *distinguishable* particles (no two being identical, i.e. no two having the same set of quantum numbers), there is no requirement for the wave function to be either symmetric or antisymmetric.

For a collection of particles, some identical with coordinates $\mathbf{r}_1, \mathbf{r}_2, \ldots$ and others distinguishable $\mathbf{x}_1, \mathbf{x}_2, \ldots$ (not identical with each other, and not identical to the aforementioned identical particles), the wave function is symmetric or antisymmetric in the identical particle coordinates $\mathbf{r}_i$ only:

$$\Psi(\ldots \mathbf{r}_a, \ldots, \mathbf{r}_b, \ldots, \mathbf{x}_1, \mathbf{x}_2, \ldots) = \pm\Psi(\ldots \mathbf{r}_b, \ldots, \mathbf{r}_a, \ldots, \mathbf{x}_1, \mathbf{x}_2, \ldots)$$

Again, there is no symmetry requirement for the distinguishable particle coordinates $\mathbf{x}_i$.

For a particle with spin, the wave function can be written in "position–spin space" as:

$$\Psi(\mathbf{r}, t, s_z)$$

which is a complex-valued function of position $\mathbf{r}$ in three-dimensional space, time t, and s_z, the spin projection quantum number along the z axis. (The z axis is an arbitrary choice; other axes can be used instead if the wave function is transformed appropriately, see below.) The s_z parameter, unlike $\mathbf{r}$ and t, is a *discrete variable*. For example, for a spin-1/2 particle, s_z can only be $+1/2$ or $-1/2$, and not any other value. (In general, for spin s, s_z can be $s, s - 1, \ldots, -s + 1, -s$.)

Often, the complex values of the wave function for all the spin numbers are arranged into a column vector, in which there are as many entries in the column vector as there are allowed values of s_z. In this case, the spin dependence is placed in indexing the entries and the wave function is a complex vector-valued function of space and time only:

$$\Psi(\mathbf{r}, t) = \begin{bmatrix} \Psi(\mathbf{r}, t, s) \\ \Psi(\mathbf{r}, t, s - 1) \\ \vdots \\ \Psi(\mathbf{r}, t, -(s - 1)) \\ \Psi(\mathbf{r}, t, -s) \end{bmatrix}$$

The wave function for N particles each with spin is the complex-valued function:

$$\Psi(\mathbf{r}_1, \mathbf{r}_2 \cdots \mathbf{r}_N, s_{z1}, s_{z2} \cdots s_{zN}, t)$$

Concerning the general case of N particles with spin in 3d, if Ψ is interpreted as a probability amplitude, the probability density is:

$$\rho(\mathbf{r}_1 \cdots \mathbf{r}_N, s_{z1} \cdots s_{zN}, t) = |\Psi(\mathbf{r}_1 \cdots \mathbf{r}_N, s_{z1} \cdots s_{zN}, t)|^2$$

and the probability that particle 1 is in region R_1 with spin $s_{z1} = m_1$ and particle 2 is in region R_2 with spin $s_{z2} = m_2$ etc. at time t is the integral of the probability density over these regions and spins:

$$P_{\mathbf{r}_1 \in R_1, s_{z1} = m_1, \ldots, \mathbf{r}_N \in R_N, s_{zN} = m_N}(t) = \int_{R_1} d^3\mathbf{r}_1 \int_{R_2} d^3\mathbf{r}_2 \cdots \int d^3\mathbf{r}_N |\Psi(\mathbf{r}_1 \cdots \mathbf{r}_N, m_1 \cdots m_N, t)|^2$$

The multidimensional Fourier transforms of the position or position–spin space wave functions yields momentum or momentum–spin space wave functions.

18.4.1 Decompositions into products

For systems in time-independent potentials, the wave function can always be written as a function of the degrees of freedom multiplied by a time-dependent phase factor, the form of which is given by the Schrödinger equation. For the case of N particles position-spin space,

$$\Psi(\mathbf{r}_1, \mathbf{r}_2, \ldots, \mathbf{r}_N, t, s_{z1}, s_{z2}, \ldots, s_{zN}) = e^{-iEt/\hbar}\psi(\mathbf{r}_1, \mathbf{r}_2, \ldots, \mathbf{r}_N, s_{z1}, s_{z2}, \ldots, s_{zN})$$

where E is the energy eigenvalue of the system corresponding to the eigenstate Ψ. Wave functions of this form are called stationary states.

In some situations, the wave function for a particle with spin factors into a product of a space function ψ and a spin function ξ, where each are complex-valued functions, and the time dependence can be placed in either function:

$$\Psi(\mathbf{r}, t, s_z) = \psi(\mathbf{r}, t)\xi(s_z) = \phi(\mathbf{r})\zeta(s_z, t).$$

The dynamics of each factor can be studied in isolation. This factorization is always possible when the orbital and spin angular momenta of the particle are separable in the Hamiltonian operator, that is, the Hamiltonian can be split into an orbital term and a spin term.[31] It is not possible

for those interactions where an external field or any space-dependent quantity couples to the spin; examples include a particle in a magnetic field, and spin-orbit coupling. For the time-independent case this reduces to

$$\Psi(\mathbf{r}, t, s_z) = e^{-iEt/\hbar}\psi(\mathbf{r})\xi(s_z).$$

where again E is the energy eigenvalue of the system corresponding to the eigenstate Ψ. This extends to the case of N particles:

$$\Psi(\mathbf{r}, t, s_z) = \psi(\mathbf{r}_1, \mathbf{r}_2, \ldots, \mathbf{r}_N, t)\xi(s_{z1}, s_{z2}, \ldots, s_{zN}) = \phi(\mathbf{r}_1, \mathbf{r}_2, \ldots$$

and for the case of identical particles, each factor has to have the correct antisymmetry or symmetry, to make the overall wave function antisymmetric for fermions or symmetric for bosons.

18.5 Inner product

18.5.1 Position-space inner products

The **inner product** of two wave functions Ψ_1 and Ψ_2 is useful and important for a number of reasons given below. For the case of one spinless particle in 1d, it can be defined as the complex number (at time t)[nb 9]

$$\langle \Psi_1, \Psi_2 \rangle = \int_{-\infty}^{\infty} dx\, \Psi_1^*(x, t)\Psi_2(x, t).$$

More generally, the formulae for the inner products are integrals over all coordinates or momenta and sums over all spin quantum numbers. That is, for one spinless particle in 3d the inner product of two wave functions can be defined as the complex number:

$$\langle \Psi_1, \Psi_2 \rangle = \int_{\text{all space}} d^3\mathbf{r}\, \Psi_1^*(\mathbf{r}, t)\Psi_2(\mathbf{r}, t).$$

while for many spinless particles in 3d:

$$\langle \Psi_1, \Psi_2 \rangle = \int_{\text{all space}} d^3\mathbf{r}_1 \int_{\text{all space}} d^3\mathbf{r}_2 \cdots \int_{\text{all space}} d^3\mathbf{r}_N\, \Psi_1^*(\mathbf{r}_1 \cdots \mathbf{r}_N, t)\Psi$$

(altogether, this is N three-dimensional volume integrals with differential volume elements $d^3\mathbf{r}_i$, also written "dV_i" or "$dx_i\, dy_i\, dz_i$"). For one particle with spin in 3d:

$$\langle \Psi_1, \Psi_2 \rangle = \sum_{\text{all } s_z} \int_{\text{all space}} d^3\mathbf{r} \Psi_1^*(\mathbf{r}, t, s_z) \Psi_2(\mathbf{r}, t, s_z).$$

and for the general case of N particles with spin in 3d:

$$\langle \Psi_1, \Psi_2 \rangle = \sum_{s_z N} \cdots \sum_{s_z 2} \sum_{s_z 1} \int_{\text{all space}} d^3\mathbf{r}_1 \int_{\text{all space}} d^3\mathbf{r}_2 \cdots \int_{\text{all space}}$$

(altogether, N three-dimensional volume integrals followed by N sums over the spins).

In the Copenhagen interpretation, the modulus squared of the inner product (a complex number) gives a real number

$$|\langle \Psi_1, \Psi_2 \rangle|^2 = P(\Psi_2 \to \Psi_1).$$

which is interpreted as the probability of the wave function Ψ_2 "collapsing" to the new wave function Ψ_1 upon measurement of an observable, whose eigenvalues are the possible results of the measurement, with Ψ_1 being an eigenvector of the resulting eigenvalue.

Although the inner product of two wave functions is a complex number, the inner product of a wave function Ψ with itself,

$$\langle \Psi, \Psi \rangle = \| \Psi \|^2.$$

is *always* a positive real number. The number $\|\Psi\|$ (not $\|\Psi\|^2$) is called the **norm** of the wave function Ψ, and is not the same as the modulus $|\Psi|$.

A wave function is normalized if:

$$\langle \Psi, \Psi \rangle = 1.$$

If Ψ is not normalized, then dividing by its norm gives the normalized function $\Psi/\|\Psi\|$.

Two wave functions Ψ_1 and Ψ_2 are orthogonal if their inner product is zero:

$$\langle \Psi_1, \Psi_2 \rangle = 0.$$

A set of wave functions $\Psi_1, \Psi_2, \ldots$ are orthonormal if they are each normalized and are all orthogonal to each other:

$$\langle \Psi_m, \Psi_n \rangle = \delta_{mn}.$$

where m and n each take values 1, 2, ..., and δ_{mn} is the Kronecker delta ($+1$ for $m = n$ and 0 for $m \neq n$). Orthonormality of wave functions is instructive to consider since this guarantees linear independence of the functions. (However, the wave functions do not have to be orthonormal and can still be linearly independent, but the inner product of Ψ_m and Ψ_n is more complicated than the mere δ_{mn}).

Returning to the superposition above:

$$d^3\mathbf{r}_N \Psi_1^*(\mathbf{r}_1 \cdots \mathbf{r}_N, s_{z1} \cdots s_{zN}, t) \Psi_2(\mathbf{r}_1 \cdots \mathbf{r}_N, s_{z1} \cdots s_{zN}, t)$$

$$\Psi = \sum_n a_n \psi_n$$

if the basis wave functions ψ_n are orthonormal, then the coefficients have a particularly simple form:

$$a_n = \langle \psi_n, \Psi \rangle$$

If the basis wave functions were not orthonormal, then the coefficients would be different.

18.5.2 Momentum-space inner products

Analogous to the position case, the inner product of two wave functions $\Phi_1(p, t)$ and $\Phi_2(p, t)$ can be defined as:

$$\langle \Phi_1, \Phi_2 \rangle = \int_{-\infty}^{\infty} dp \, \Phi_1^*(p, t) \Phi_2(p, t).$$

and similarly for more particles in higher dimensions.

One particular solution to the time-independent Schrödinger equation is

$$\Psi_p(x) = e^{ipx/\hbar}.$$

a plane wave, which can be used in the description of a particle with momentum exactly p, since it is an eigenfunction of the momentum operator. These functions are not normalizable to unity (they aren't square-integrable), so they are not really elements of physical Hilbert space. The set

$$\{\Psi_p(x, t), -\infty \leq p \leq \infty\}$$

forms what is called the **momentum basis**. This "basis" is not a basis in the usual mathematical sense. For one thing, since the functions aren't normalizable, they are instead **normalized to a delta function**.

$$\langle \Psi_p, \Psi_{p'} \rangle = \delta(p - p').$$

For another thing, though they are linearly independent, there are too many of them (they form an uncountable set) for a basis for physical Hilbert space. They can still be used to express all functions in it using Fourier transforms as described above.

18.6   Units of the wave function

Although wave functions are complex numbers, both the real and imaginary parts each have the same units (the imaginary unit i is a pure number without physical units). The units of ψ depend on the number of particles N the wave function describes, and the number of spatial or momentum dimensions n of the system.

When integrating $|\psi|^2$ over all the coordinates, the volume element $d^n\mathbf{r}_1 d^n\mathbf{r}_2 ... d^n\mathbf{r}_N$ has units of [length]^{Nn}. Since the normalization conditions require the integral to be the unitless number 1, $|\psi|^2$ must have units of [length]$^{-Nn}$, thus the units of $|\psi|$ and hence ψ are [length]$^{-Nn/2}$. Likewise, in momentum space, length is replaced by momentum, and the units are [momentum]$^{-Nn/2}$. These results are true for particles of any spin, since for particles with spin, the summations are over dimensionless spin quantum numbers.

18.7   More on wave functions and abstract state space

Main article: Quantum state

As has been demonstrated, the set of all possible normalizable wave functions for a system with a particular choice of basis constitute a Hilbert space. This vector space is in general infinite-dimensional. Due to the multiple possible choices of basis, these Hilbert spaces are not unique. One therefore talks about an abstract Hilbert space, **state space**, where the choice of basis is left undetermined. The choice of basis corresponds to a choice of a maximal set of quantum numbers, each quantum number corresponding to an observable. Two observables corresponding to quantum numbers in the maximal set must commute, therefore, the basis isn't entirely arbitrary, but nonetheless, there are always several choices.

Specifically, each state is represented as an abstract vector in state space[32]

$$|\Psi\rangle$$

where $|\Psi\rangle$ is a "ket" (a vector) written in Dirac's bra–ket notation.[33] Kets that differ by multiplication by a scalar represent the same state. A **ray** in Hilbert space is a set of normalized vectors differing by a complex number of modulus 1. If $|\psi\rangle$ and $|\phi\rangle$ are two states in the vector space, and a and b are two complex numbers, then the linear combination

$$|\Psi\rangle = a|\psi\rangle + b|\phi\rangle$$

is also in the same vector space. The state space is postulated to have an inner product, denoted by

$$\langle \Psi_1 | \Psi_2 \rangle.$$

that is (usually, this differs) linear in the first argument and antilinear in the second argument. The dual vectors are denoted as "bras", $\langle \Psi |$. These are linear functionals, elements of the dual space to the state space. The inner product, once chosen, can be used to define a unique map from state space to its dual, see Riesz representation theorem. this map is antilinear. One has

$$\langle \Psi | = a^*\langle \psi | + b^*\langle \phi | \leftrightarrow a|\psi\rangle + b|\phi\rangle = |\Psi\rangle,$$

where the asterisk denotes the complex conjugate. For this reason one has under this map

$$\langle \Phi | \Psi \rangle = \langle \Phi | (|\Psi\rangle),$$

and one may, as a practical consequence, at least notation-wise in this formalism, ignore that bra's are dual vectors.

The state vector for the system evolves in time according to the Schrödinger equation, or other dynamical pictures of quantum mechanics- In bra-ket notation this reads,

$$i\hbar \frac{d}{dt}|\Psi\rangle = \hat{H}|\Psi\rangle$$

Abstract state space is also, by definition, required to be a Hilbert space. The only requirement missing for this in the description so far is completeness. See the quantum state article for more explanation of the Hilbert space formalism and its consequences to quantum physics.

The connection to the Hilbert spaces of wave functions is made as follows. If (a, b, ···, l, m, ···) is a maximal set of quantum numbers, denote the state corresponding to *fixed choices* of these quantum numbers by

$$|a, b, \ldots, l, m, \ldots\rangle.$$

The wave function corresponding to an arbitrary state $|\Psi\rangle$ is denoted

$$\langle a, b, \ldots, l, m, \ldots |\Psi\rangle,$$

for a concrete example,

$$\Psi(x) = \langle x|\Psi\rangle.$$

There are several advantages to understanding wave functions as representing elements of an abstract vector space:

- All the powerful tools of linear algebra can be used to manipulate and understand wave functions. For example:

    - Linear algebra explains how a vector space can be given a basis, and then any vector in the vector space can be expressed in this basis. This explains the relationship between a wave function in position space and a wave function in momentum space, and suggests that there are other possibilities too.

    - Bra–ket notation can be used to manipulate wave functions.

- The idea that quantum states are vectors in an abstract vector space (technically, a complex projective Hilbert space) is completely general in all aspects of quantum mechanics and quantum field theory, whereas the idea that quantum states are complex-valued "wave" functions of space is only true in certain situations.

Following is a summary of the bra–ket formalism applied to wave functions, with general discrete or continuous bases.

18.7.1 Discrete and continuous bases

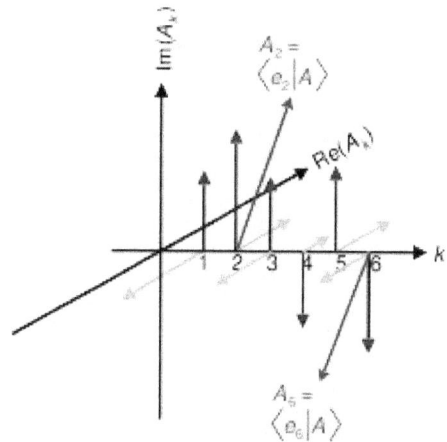

Discrete components A_k of a complex vector $|A\rangle = \sum_k A_k|e_k\rangle$, which belongs to a *countably infinite*-dimensional Hilbert space; there are countably infinitely many k values and basis vectors $|e_k\rangle$.

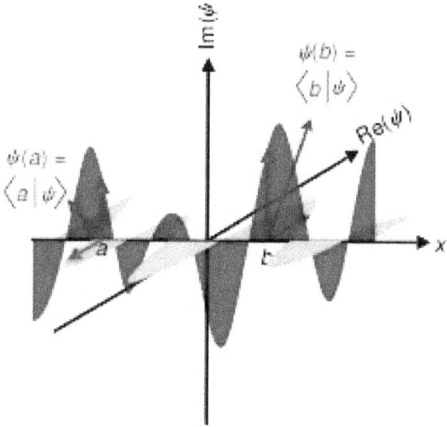

Continuous components $\psi(x)$ of a complex vector $|\psi\rangle = \int dx\ \psi(x)|x\rangle$, which belongs to an *uncountably infinite*-dimensional Hilbert space; there are uncountably infinitely many x values and basis vectors $|x\rangle$.

Components of complex vectors plotted against index number: discrete k and continuous x. Two probability amplitudes out of infinitely many are highlighted.

A Hilbert space with a discrete basis $|e_i\rangle$ for $i = 1, 2\ldots n$ is orthonormal if the inner product of all pairs of basis kets are given by the Kronecker delta:

$$\langle e_i|e_j\rangle = \delta_{ij}.$$

Orthonormal bases are convenient to work with because the inner product of two vectors have simple expressions. A wave function $|\Psi\rangle$ expressed in this discrete basis of the Hilbert space, and the corresponding bra in the dual space, are respectively given by:

$$|\Psi\rangle = \sum_{i=1}^{n} c_i|\varepsilon_i\rangle = \begin{bmatrix} c_1 \\ \vdots \\ c_n \end{bmatrix} \quad \langle\Psi| = |\Psi\rangle^{\dagger} = \sum_{i=1}^{n} c_i^*\langle\varepsilon_i| = \begin{bmatrix} c_1^* & \cdots & c_n^* \end{bmatrix} \qquad \sum_{i=1}^{n} |\varepsilon_i\rangle\langle\varepsilon_i| = 1 \,, \quad \int d\varepsilon\, |\varepsilon\rangle\langle\varepsilon| = 1$$

where the complex numbers

$$c_i = \langle\varepsilon_i|\Psi\rangle$$

are the components of the vector. The column vector is a useful way to list the numbers, and operations on the entire vector can be done according to matrix addition and multiplication. The entire vector $|\Psi\rangle$ is independent of the basis, but the components depend on the basis. If a change of basis is made, the components of the vector must also change to compensate.

A Hilbert space with a continuous basis $\{\,|\varepsilon\rangle\,\}$ is orthonormal if the inner product of all pairs of basis kets are given by the Dirac delta function:

$$\langle\varepsilon|\varepsilon'\rangle = \delta(\varepsilon - \varepsilon') \,.$$

As with the discrete bases, a symbol ε is used in the basis states, two common notations are $|\varepsilon\rangle$ and sometimes $|\Psi_\varepsilon\rangle$. A particular basis ket may be subscripted $|\varepsilon_0\rangle \equiv |\Psi_{\varepsilon_0}\rangle$ or primed $|\varepsilon'\rangle \equiv |\Psi_{\varepsilon'}\rangle$, or simply given another symbol in place of ε.

While discrete basis vectors are summed over a discrete index, continuous basis vectors are integrated over a continuous index (a variable of a function). In what follows, all integrals are with respect to the real-valued basis variable ε (not complex-valued), over the required range. Usually this is just the real line or subsets of it. The state $|\Psi\rangle$ in the continuous basis of the Hilbert space, with the corresponding bra in the dual space, are respectively given by:[34]

$$|\Psi\rangle = \int d\varepsilon\, |\varepsilon\rangle \Psi(\varepsilon) \,, \quad \langle\Psi| = \int d\varepsilon\, \langle\varepsilon| \Psi(\varepsilon)^* \,.$$

where the components are the complex-valued functions

$$\Psi(\varepsilon) = \langle\varepsilon|\Psi\rangle$$

of a real variable ε.

18.7.2 Completeness conditions

The **completeness conditions** (also called **closure relations**) are

for discrete and continuous orthonormal bases, respectively. An orthonormal set of kets form bases if and only if they satisfy these relations.[34] In each case, the equality to unity means this is an identity operator; its action on any state leaves it unchanged. Multiplying any state on the right of these gives the representation of the state $|\Psi\rangle$ in the basis. The inner product of a first state $|\Psi_1\rangle$ with a second $|\Psi_2\rangle$ can also be obtained by multiplying $|\Psi_1\rangle$ on the left and $|\Psi_2\rangle$ on the right of the relevant completeness condition.

18.7.3 Inner product

Physically, the nature of the inner product is dependent on the basis in use, because the basis is chosen to reflect the quantum state of the system.

If $|\Psi_1\rangle$ is a state in the above basis with components c_1, c_2, ..., c_n and $|\Psi_2\rangle$ is another state in the same basis with components z_1, z_2, ..., z_n, the inner product is the complex number:

$$\langle\Psi_1|\Psi_2\rangle = \left(\sum_i z_i^*\langle\varepsilon_i|\right)\left(\sum_j c_j|\varepsilon_j\rangle\right) = \sum_{ij} z_i^* c_j \langle\varepsilon_i|\varepsilon_j\rangle = \sum_i$$

If $|\Psi_1\rangle$ is a state in the above continuous basis with components $\Psi_1(\varepsilon')$, and $|\Psi_2\rangle$ is another state in the same basis with components $\Psi_2(\varepsilon)$, the inner product is the complex number:

$$\langle\Psi_1|\Psi_2\rangle = \left(\int d\varepsilon'\, \Psi_1(\varepsilon')^*\langle\varepsilon'|\right)\left(\int d\varepsilon\, \Psi_2(\varepsilon)|\varepsilon\rangle\right) = \int d\varepsilon' \int d\varepsilon^*$$

where the integrals are taken over all ε and ε'.

The square of the **norm (magnitude)** of the state vector $|\Psi\rangle$ is given by the inner product of $|\Psi\rangle$ with itself, a real number:

$$\|\Psi\|^2 = \langle\Psi|\Psi\rangle = \sum_{j=1}^{n} |c_j|^2 \,, \quad \|\Psi\|^2 = \langle\Psi|\Psi\rangle = \int d\varepsilon\, |\Psi(\varepsilon)|^2$$

for the discrete and continuous bases, respectively. Each say the projection of a complex probability amplitude onto itself is real. If $|\Psi\rangle$ is normalized, these expressions would be each separately equal to 1. If the state is not normalized, then dividing by its magnitude normalizes the state:

$$|\Psi_N\rangle = \frac{1}{\|\Psi\|}|\Psi\rangle$$

18.7.4 Normalized components and probabilities

In the literature, the following results are often presented with normalized wavefunctions. Here, we keep the normalization factors to show where they appear if the wavefunction is not already normalized.

For the discrete basis, projecting the normalized state $|\Psi_N\rangle$ onto a particular state the system may collapse to, $|\varepsilon_q\rangle$, gives the complex number;

$$\langle \varepsilon_q|\Psi_N\rangle = \langle \varepsilon_q| \frac{1}{\|\Psi\|}\left(\sum_{i=1}^n c_i|\varepsilon_i\rangle\right) = \frac{c_q}{\|\Psi\|}.$$

so the modulus squared of this gives a real number;

$$P(\varepsilon_q) = |\langle \varepsilon_q|\Psi_N\rangle|^2 = \frac{|c_q|^2}{\|\Psi\|^2}.$$

In the Copenhagen interpretation, this is the probability of state $|\varepsilon_q\rangle$ occurring.

In the continuous basis, the projection of the normalized state onto some particular basis $|\varepsilon'\rangle$ is a complex-valued function;

$$\langle \varepsilon'|\Psi_N\rangle = \langle \varepsilon'| \left(\frac{1}{\|\Psi\|}\int d\varepsilon|\varepsilon\rangle\Psi(\varepsilon)\right) = \frac{1}{\|\Psi\|}\int d\varepsilon\langle\varepsilon'|\varepsilon\rangle\Psi(\varepsilon)$$

so the squared modulus is a real-valued function

$$\rho(\varepsilon') = |\langle \varepsilon'|\Psi_N\rangle|^2 = \frac{|\Psi(\varepsilon')|^2}{\|\Psi\|^2}$$

In the Copenhagen interpretation, this function is the *probability density function* of measuring the observable ε', so integrating this with respect to ε' between $a \leq \varepsilon' \leq b$ gives:

$$P_{a\leq\varepsilon\leq b} = \frac{1}{\|\Psi\|^2}\int_a^b d\varepsilon'|\Psi(\varepsilon')|^2 = \frac{1}{\|\Psi\|^2}\int_a^b d\varepsilon'|\langle\varepsilon'|\Psi\rangle|^2$$

the probability of finding the system with ε' between $\varepsilon' = a$ and $\varepsilon' = b$.

18.7.5 Wave function collapse

The physical meaning of the components of $|\Psi\rangle$ is given by the *wave function collapse postulate*, also known as wave function collapse. If the observable(s) ε (momentum and/or spin, position and/or spin, etc.) corresponding to states $|\varepsilon_i\rangle$ has distinct and definite values, λ_i, and a measurement of that variable is performed on a system in the state $|\Psi\rangle$ then the probability of measuring λ_i is $|\langle\varepsilon_i|\Psi\rangle|^2$. If the measurement yields λ_i, the system "collapses" to the state $|\varepsilon_i\rangle$ irreversibly and instantaneously.

18.7.6 Time dependence

Main article: Dynamical pictures (quantum mechanics)

In the Schrödinger picture, the states evolve in time, so the time dependence is placed in $|\Psi\rangle$ according to [35]

$$|\Psi(t)\rangle = \sum_i |\varepsilon_i\rangle\langle\varepsilon_i|\Psi(t)\rangle = \sum_i c_i(t)|\varepsilon_i\rangle$$

for discrete bases, or

$$|\Psi(t)\rangle = \int d\varepsilon\, |\varepsilon\rangle\langle\varepsilon|\Psi(t)\rangle = \int d\varepsilon\, \Psi(\varepsilon,t)|\varepsilon\rangle$$

for continuous bases. However, in the Heisenberg picture the states $|\Psi\rangle$ are constant in time and time dependence is placed in the Heisenberg operators, so $|\Psi\rangle$ is not written as $|\Psi(t)\rangle$. The Heisenberg picture wave function is a snapshot of a Schrödinger picture wave function, representing the whole spacetime history of the system. In the interaction picture (also called Dirac picture), the time dependence is placed in both the states and operators, the subdivision depending on the interaction term in the Hamiltonian, and can be viewed as intermediate between the Heisenberg and Schrödinger pictures. It is useful primarily in computing S-matrix elements. [36]

18.7.7 Tensor product

Further information: Bra-ket notation § Composite bras and kets

It is useful to introduce another operation with the physical interpretation of forming composite states from a collection of other states. This is the tensor product. Given two systems described by states $|\Psi\rangle$ and $|\Phi\rangle$, the tensor product of the states forms the composite state denoted by $|\Psi\rangle\otimes|\Phi\rangle$ or simply without any operation symbol $|\Psi\rangle|\Phi\rangle$, and the new system includes both of the original systems together. The tensor product state $|\Psi\rangle|\Phi\rangle$ lives in a new space: the tensor product of the original Hilbert spaces. The bases spanning this space are the tensor products of the original bases. The product is not commutative in general, so $|\Psi\rangle|\Phi\rangle \neq |\Phi\rangle|\Psi\rangle$. If $|\Psi\rangle$ has components c_i and $|\Phi\rangle$ has components z_j, each in a discrete orthonormal basis $|\varepsilon_k\rangle$, then:

$$|\Psi\rangle|\Phi\rangle = \left(\sum_i c_i|\varepsilon_i\rangle\right)\left(\sum_j z_j|\varepsilon_j\rangle\right) = \sum_{i,j} c_i z_j|\varepsilon_i\rangle|\varepsilon_j\rangle$$

and the notation can be simplified by abbreviating $|A\rangle = |\Psi\rangle|\Phi\rangle$, $A_{ij} = c_i z_j$, and $|E_{ij}\rangle = |\varepsilon_i\rangle|\varepsilon_j\rangle$, so that

$$|A\rangle = \sum_{i,j} A_{ij}|E_{ij}\rangle$$

The same procedure follows for continuous bases using integration. This can also be extended to any number of states, however taking tensor products for fermions and bosons is complicated by the symmetry requirements, see identical particles for general results.

18.8   Position representations

This section applies mostly to non-relativistic quantum mechanics. In relativistic quantum mechanics, eigenstates of the position operator are problematic due to a relativistic extension of Heisenberg's uncertainty principle. In relativistic quantum field theory, they are not used at all to label physical states. Associated to a particle perfectly localized to a point in space is an infinite uncertainty in energy. This leads to pair production in the relativistic regime. Thus such a particle automatically has companions, leading to a breakdown of the description.

18.8.1   State space for one spin-0 particle in 1d

For a spinless particle in one spatial dimension (the x-axis or real line), the state $|\Psi\rangle$ can be expanded in terms of a continuum of basis states; $|x\rangle$, also written $|\Psi_x\rangle$, corresponding to the set of all position coordinates x. The completeness condition for this basis is

$$1 = \int_{-\infty}^{\infty} dx\,|x\rangle\langle x|$$

and the orthogonality relation is

$$\langle x'|x\rangle = \delta(x' - x)$$

The state $|\Psi\rangle$ is expressed by:

$$|\Psi\rangle = \left(\int_{-\infty}^{\infty} dx\,|x\rangle\langle x|\right)|\Psi\rangle = \int_{-\infty}^{\infty} dx\,|x\rangle\langle x|\Psi\rangle = \int_{-\infty}^{\infty} dx\,\Psi(x)|x\rangle$$

in which the "wave function" described as a function is a component of the complex state vector.

$$\Psi(x) = \langle x|\Psi\rangle$$

The inner product as stated at the beginning of this article is:

$$\langle\Psi_1|\Psi_2\rangle = \langle\Psi_1|\left(\int_{-\infty}^{\infty} dx\,|x\rangle\langle x|\right)|\Psi_2\rangle = \int_{-\infty}^{\infty} dx\,\langle\Psi_1|x\rangle\langle x|\Psi_2\rangle =$$

If the particle is confined to a region R (a subset of the x-axis), the integrals in the inner product and completeness condition would be integrals over R.

18.8.2   State space (other cases)

The previous example can be extended to more particles in higher dimensions, and include spin.

For one spinless particle in 3d, the basis states are $|\mathbf{r}\rangle$ and any state vector $|\Psi\rangle$ in this space is expressed in terms of the basis vectors as $|\mathbf{r}\rangle$:

$$|\Psi\rangle = \int_{\text{all space}} d^3\mathbf{r}\,|\mathbf{r}\rangle\langle\mathbf{r}|\Psi\rangle$$

with components:

$$\langle\mathbf{r}|\Psi\rangle = \Psi(\mathbf{r})$$

For N spinless particles in 3d, the basis states are $|\mathbf{r}_1, ..., \mathbf{r}_N\rangle$. This is the tensor product of the one-particle position bases $|\mathbf{r}_1\rangle$, $|\mathbf{r}_2\rangle$, ..., $|\mathbf{r}_N\rangle$, each of which spans the separate one-particle Hilbert spaces, so $|\mathbf{r}_1, ..., \mathbf{r}_N\rangle$ are the basis states for the tensor product of the one-particle Hilbert spaces (the

Hilbert space for the composite many particle system). Any state vector $|\Psi\rangle$ in this space is

$$|\Psi\rangle = \int\limits_{\text{all space}} d^3\mathbf{r}_N \cdots \int\limits_{\text{all space}} d^3\mathbf{r}_2 \int\limits_{\text{all space}} d^3\mathbf{r}_1 |\mathbf{r}_1, \mathbf{r}_2, \ldots, \mathbf{r}_N\rangle\langle\mathbf{r}_1, \mathbf{r}_2, \ldots, \mathbf{r}_N |\Psi\rangle$$

with components:

$$\langle\mathbf{r}_1, \mathbf{r}_2, \ldots, \mathbf{r}_N |\Psi\rangle = \Psi(\mathbf{r}_1, \mathbf{r}_2, \ldots, \mathbf{r}_N)$$

For one particle with spin in 3d, the basis states are $|\mathbf{r}, s_z\rangle$, the tensor product of the position basis $|\mathbf{r}\rangle$ and spin basis $|s_z\rangle$, which exists in a new space from the spin space and position space alone. Any state $|\Psi\rangle$ in this space is:

$$|\Psi\rangle = \sum_{s_z} \int\limits_{\text{all space}} d^3\mathbf{r}|\mathbf{r}, s_z\rangle\langle\mathbf{r}, s_z|\Psi\rangle$$

with components:

$$\langle\mathbf{r}, s_z|\Psi\rangle = \Psi(\mathbf{r}, s_z)$$

For N particles with spin in 3d, the basis states are $|\mathbf{r}_1, \ldots, \mathbf{r}_N, s_{z1}, \ldots, s_{zN}\rangle$, the tensor product of the position basis $|\mathbf{r}_1, \ldots, \mathbf{r}_N\rangle$ and spin basis $|s_{z1}, \ldots, s_{zN}\rangle$, which exists in a new space from the spin space and position space alone. Any state in this space is:

$$|\Psi\rangle = \sum_{s_{z1}, \ldots, s_{zN}} \int\limits_{\text{all space}} d^3\mathbf{r}_N \cdots \int\limits_{\text{all space}} d^3\mathbf{r}_1 |\mathbf{r}_1, \ldots, \mathbf{r}_N, s_{z1}, \ldots, s_{zN}\rangle\langle\mathbf{r}_1, \ldots, \mathbf{r}_N, s_{z1}, \ldots, s_{zN}|\Psi\rangle$$

with components:

$$\langle\mathbf{r}_1, \ldots, \mathbf{r}_N, s_{z1}, \ldots, s_{zN}|\Psi\rangle = \Psi(\mathbf{r}_1, \ldots, \mathbf{r}_N, s_{z1}, \ldots, s_{zN})$$

If the particles are restricted to regions of position space, then the integrals in the completeness relations are taken over those regions, rather than the entire coordinate space. For the general case of many particles with spin in 3d, if particle 1 is in region R_1, particle 2 is in region R_2, and so on, the state in this position–spin representation is:

$$|\Psi\rangle = \sum_{s_{z1}, \ldots, s_{zN}} \int\limits_{R_N} d^3\mathbf{r}_N \cdots \int\limits_{R_1} d^3\mathbf{r}_1 \, \Psi(\mathbf{r}_1, \ldots, \mathbf{r}_N, s_{z1}, \ldots, s_{zN})|\mathbf{r}_1, \ldots, \mathbf{r}_N, s_{z1}, \ldots, s_{zN}\rangle$$

The orthogonality relation for this basis is:

$$\langle\mathbf{x}_1, \ldots, \mathbf{x}_N, m_1, \ldots, m_N |\mathbf{r}_1, \ldots, \mathbf{r}_N, s_{z1}, \ldots, s_{zN}\rangle = \delta_{m_1 s_{z1}} \cdots \delta_{m_N s_{zN}}$$

and the inner product of $|\Psi_1\rangle$ and $|\Psi_2\rangle$ is:

$$\langle\Psi_1|\Psi_2\rangle = \sum_{s_{z1}, \ldots, s_{zN}} \int\limits_{R_N} d^3\mathbf{r}_N \cdots \int\limits_{R_1} d^3\mathbf{r}_1 \Psi_1(\mathbf{r}_1, \ldots, \mathbf{r}_N, s_{z1}, \ldots, s_{zN})^*$$

Momentum space wave functions are similar, using the momentum vectors of the particles as continuous bases, namely $|\mathbf{p}\rangle$, $|\mathbf{p}_1, \mathbf{p}_2, \ldots, \mathbf{p}_N\rangle$, etc.

18.9 Ontology

Main article: Interpretations of quantum mechanics

Whether the wave function really exists, and what it represents, are major questions in the interpretation of quantum mechanics. Many famous physicists of a previous generation puzzled over this problem, such as Schrödinger, Einstein and Bohr. Some advocate formulations or variants of the Copenhagen interpretation (e.g. Bohr, Wigner and von Neumann) while others, such as Wheeler or Jaynes, take the more classical approach [37] and regard the wave function as representing information in the mind of the observer, i.e. a measure of our knowledge of reality. Some, including Schrödinger, Bohm and Everett and others, argued that the wave function must have an objective, physical existence. Einstein thought that a complete description of physical reality should refer directly to physical space and time, as distinct from the wave function, which refers to an abstract mathematical space. [38]

18.10 Examples

18.10.1 Free particle

Main article: Free particle

A free particle in 3d with wave vector $\mathbf{k}$ and angular frequency ω has a wave function

$$\Psi(\mathbf{r}, t) = Ae^{i(\mathbf{k}\cdot\mathbf{r}-\omega t)}.$$

18.10.2 Particle in a box

Main article: Particle in a box

A particle is restricted to a 1D region between $x = 0$ and $x = L$; its wave function is:

$$\Psi(x,t) = Ae^{i(kx-\omega t)}, \qquad 0 \leq x \leq L$$
$$\Psi(x,t) = 0, \qquad\qquad x < 0, x > L$$

To normalize the wave function we need to find the value of the arbitrary constant A; solved from

$$\int_{-\infty}^{\infty} dx\, |\Psi|^2 = 1.$$

From Ψ, we have $|\Psi|^2 = A^2$, so the integral becomes:

$$\int_{-\infty}^{0} dx \cdot 0 + \int_{0}^{L} dx\, A^2 + \int_{L}^{\infty} dx \cdot 0 = 1.$$

Solving this equation gives $A = 1/\sqrt{L}$, so the normalized wave function in the box is:

$$\Psi(x,t) = \frac{1}{\sqrt{L}} e^{i(kx-\omega t)}, \quad 0 \leq x \leq L.$$

18.10.3   One-dimensional quantum tunnelling

Main articles: Finite potential barrier and Quantum tunnelling

One of most prominent features of the wave mechanics is a possibility for a particle to reach a location with a prohibitive (in classical mechanics) force potential. In the one-dimensional case of particles with energy less than V_0 in the square potential

$$V(x) = \begin{cases} V_0 & |x| < a \\ 0 & \text{otherwise,} \end{cases}$$

the steady-state solutions to the wave equation have the form (for some constants k, κ)

$$\psi(x) = \begin{cases} A_r \exp(ikx) + A_l \exp(-ikx) & x < -a, \\ B_r \exp(\kappa x) + B_l \exp(-\kappa x) & |x| \leq a, \\ C_r \exp(ikx) + C_l \exp(-ikx) & x > a. \end{cases}$$

Note that these wave functions are not normalized; see scattering theory for discussion.

The standard interpretation of this is as a stream of particles being fired at the step from the left (the direction of negative x): setting $A_r = 1$ corresponds to firing particles singly; the terms containing A_r and C_r signify motion to the right, while A_l and C_l – to the left. Under this beam interpretation, put $C_l = 0$ since no particles are coming from the right. By applying the continuity of wave functions and their derivatives at the boundaries, it is hence possible to determine the constants above.

18.10.4   Quantum Dots

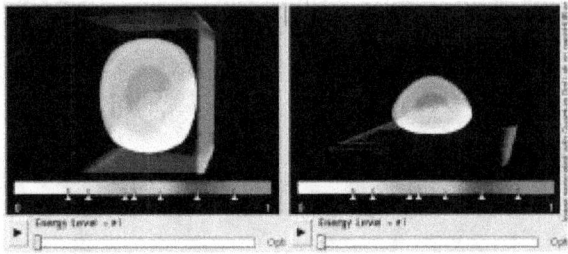

3D confined electron wave functions in a quantum dot. Here, rectangular and triangular-shaped quantum dots are shown. Energy states in rectangular dots are more s-type and p-type. However, in a triangular dot the wave functions are mixed due to confinement symmetry. (Click for animation)

In a semiconductor crystallite whose radius is smaller than the size of its exciton Bohr radius, the excitons are squeezed, leading to quantum confinement. The energy levels can then be modeled using the particle in a box model in which the energy of different states is dependent on the length of the box.

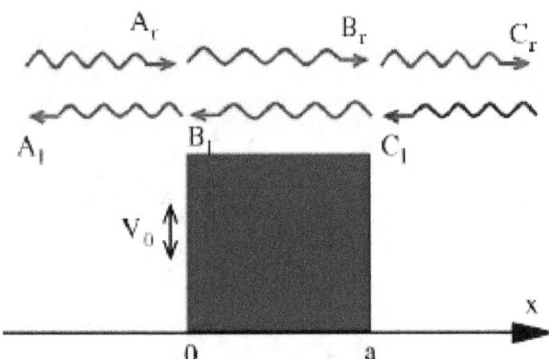

Scattering at a finite potential barrier of height V_0 The amplitudes and direction of left and right moving waves are indicated. In red, those waves used for the derivation of the reflection and transmission amplitude. $E > V_0$ for this illustration.

18.10.5 Other

Some examples of wave functions for specific applications include:

- Finite square well

- Delta potential

- Quantum harmonic oscillator

- Hydrogen atom and Hydrogen-like atom

18.11 See also

- Boson

- de Broglie–Bohm theory

- Double-slit experiment

- Faraday wave

- Fermion

- Schrödinger equation

- Wave function collapse

- Wave packet

- Phase space formulation of quantum mechanics, wave functions are replaced by quasi-probability distributions that place the position and momenta variables on equal footing.

18.12 Remarks

[1] For this statement to make sense, the observables need to be elements of a maximal commuting set. To see this, it is a simple matter to note that, for example, the momentum operator of the i'th particle in an n-particle system is *not* a generator of any symmetry in nature. On the other hand, the *total* angular momentum *is* a generator of a symmetry in nature; the translational symmetry.

[2] The resulting basis may or may not technically be a basis in the mathematical sense of Hilbert spaces. For instance, states of definite position and definite momentum are not square integrable. This may be overcome with the use of wave packets or by enclosing the system in a "box". See further remarks below.

[3] In technical terms, this is formulated the following way. The inner product yields a norm. This norm in turn induces a metric. If this metric is complete, then the aforementioned limits will be in the function space. The inner product space is then called complete. A complete inner product space is a Hilbert space. The abstract state space is always taken as a Hilbert space. The matching requirement for the function spaces is a natural one. The Hilbert space property of the abstract state space was originally extracted from the observation that the function spaces forming normalizable solutions to the Schrödinger equation are Hilbert spaces.

[4] As is explained in a later footnote, the integral must be taken to be the Lebesgue integral, the Riemann integral is not sufficient.

[5] One such relaxation is that the wave function must belong to the Sobolev space $W^{1,2}$. It means that it is differentiable in the sense of distributions, and its gradient is square-integrable. This relaxation is necessary for potentials that are not functions but are distributions, such as the Dirac delta function.

[6] It is easy to visualize a sequence of functions meeting the requirement that converges to a *discontinuous* function. For this, modify an example given in Inner product space#Examples. This element though *is* an element of L^2.

[7] For instance, in perturbation theory one may construct a sequence of functions approximating the true wave function. This sequence will be guaranteed to converge in a larger space, but without the assumption of a full-fledged Hilbert space, it will not be guaranteed that the convergence is to a function in the relevant space and hence solving the original problem.

[8] Some functions not being square-integrable, like the plane-wave free particle solutions are necessary for the description as outlined in a previous note and also further below.

[9] The functions are here assumed to be elements of L^2, the space of square integrable functions. The elements of this space are more precisely equivalence classes of square integrable functions, two functions declared equivalent if they differ on a set of Lebesgue measure 0. This is necessary to obtain an inner product (that is, $(\Psi, \Psi) = 0 \Rightarrow \Psi \equiv 0$) as opposed to a **semi-inner product**. The integral is taken to be the Lebesgue integral. This is essential for completeness of the space, thus yielding a complete inner product space = Hilbert space.

18.13 Notes

[1] Born 1927, pp. 354–357

[2] Heisenberg 1958, p. 143

[3] Heisenberg, W. (1927/1985/2009). Heisenberg is translated by Camilleri 2009, p. 71, (from Bohr 1985, p. 142).

[4] Murdoch 1987, p. 43

[5] de Broglie 1960, p. 48

[6] Landau Lifshitz, p. 6

[7] Newton 2002, pp. 19–21

[8] Lerner & Trigg 1991, pp. 1223–1229

[9] Einstein 1905, pp. 132–148 (in German), Arons & Peppard 1965, p. 367 (in English)

[10] Einstein 1916, pp. 47–62 and a nearly identical version Einstein 1917, pp. 121–128 translated in ter Haar 1967, pp. 167–183.

[11] de Broglie 1923, pp. 507–510,548,630

[12] Hanle 1977, pp. 606–609

[13] Schrödinger 1926, pp. 1049–1070

[14] Tipler, Mosca & Freeman 2008

[15] Weinberg 2013

[16] Born 1926a, translated in Wheeler & Zurek 1983 at pages 52–55.

[17] Born 1926b, translated in Ludwig 1968, pp. 206–225. Also here.

[18] Young & Freedman 2008, p. 1333

[19] Atkins 1974

[20] Martin & Shaw 2008

[21] Pauli 1927, pp. 601–623.

[22] Weinberg (2002) takes the standpoint that quantum field theory appears the way it does because it is the *only* way to reconcile quantum mechanics with special relativity.

[23] Weinberg (2002) See especially chapter 5, where some of these results are derived.

[24] Weinberg 2002 Chapter 4.

[25] Weinberg 2002

[26] Eisberg & Resnick 1985

[27] Rae 2008

[28] Atkins 1974, p. 258

[29] Griffiths 2004

[30] Zettili 2009, p. 463

[31] Shankar 1994, p. 378–379

[32] Dirac 1982

[33] Dirac 1939

[34] (Peleg et al. 2010) pp. 64–65.

[35] (Peleg et al. 2010, pp. 68–69)

[36] Weinberg 2002 Chapter 3, Scattering matrix.

[37] Jaynes 2003

[38] Einstein 1998, p. 682

18.14   References

• Atkins, P. W. (1974). *Quanta: A Handbook of Concepts*. ISBN 0-19-855494-X.

• Arons, A. B.; Peppard, M. B. (1965). "Einstein's proposal of the photon concept: A translation of the *Annalen der Physik* paper of 1905" (PDF). *American Journal of Physics* **33** (5): 367. Bibcode:1965AmJPh..33..367A. doi:10.1119/1.1971542.

• Bohr, N. (1985). J. Kalckar, ed. *Niels Bohr - Collected Works: Foundations of Quantum Physics I (1926 - 1932)* **6**. Amsterdam: North Holland. ISBN 9780444532893.

• Born, M. (1926a). "Zur Quantenmechanik der Stossvorgange". *Z. f. Physik* **37**: 863–867. Bibcode:1926ZPhy...37..863B. doi:10.1007/bf01397477.

• Born, M. (1926b). "Quantenmechanik der Stossvorgange". *Z. f. Physik* **38**: 803–827. Bibcode:1926ZPhy...38..803B. doi:10.1007/bf01397184.

• Born, M. (1927). "Physical aspects of quantum mechanics". *Nature* **119**: 354–357. Bibcode:1927Natur.119..354B. doi:10.1038/119354a0.

• de Broglie, L. (1923). "Radiations — Ondes et quanta" [Radiation — Waves and quanta]. *Comptes Rendus* (in French) **177**: 507–510, 548, 630. Online copy (French) Online copy (English)

• de Broglie, L. (1960). *Non-linear Wave Mechanics: a Causal Interpretation*. Amsterdam: Elsevier.

• Camilleri, K. (2009). *Heisenberg and the Interpretation of Quantum Mechanics: the Physicist as Philosopher*. Cambridge UK: Cambridge University Press. ISBN 978-0-521-88484-6.

- Dirac, P. A. M. (1982). *The principles of quantum mechanics*. The international series on monographs on physics (4th ed.). Oxford University Press. ISBN 0 19 852011 5.

- Dirac, P. A. M. (1939). "A new notation for quantum mechanics". *Mathematical Proceedings of the Cambridge Philosophical Society* **35** (3): 416–418. Bibcode:1939PCPS...35..416D. doi:10.1017/S0305004100021162.

- Einstein, A. (1905). "Über einen die Erzeugung und Verwandlung des Lichtes betreffenden heuristischen Gesichtspunkt". *Annalen der Physik* (in German) **17** (6): 132–148. Bibcode:1905AnP...322..132E. doi:10.1002/andp.19053220607.

- Einstein, A. (1916). "Zur Quantentheorie der Strahlung". *Mitteilungen der Physikalischen Gesellschaft Zürich* **18**: 47–62.

- Einstein, A. (1917). "Zur Quantentheorie der Strahlung". *Physikalische Zeitschrift* (in German) **18**: 121–128. Bibcode:1917PhyZ...18..121E.

- Einstein, A. (1998). P. A. Schlipp, ed. *Albert Einstein: Philosopher-Scientist*. The Library of Living Philosophers **VII** (3rd ed.). La Salle Publishing Company. Illinois: Open Court. ISBN 0-87548-133-7.

- Eisberg, R.; Resnick, R. (1985). *Quantum Physics of Atoms, Molecules, Solids, Nuclei and Particles* (2nd ed.). John Wiley & Sons. ISBN 978-0-471-87373-0.

- Griffiths, D. J. (2004). *Introduction to Quantum Mechanics* (2nd ed.). Essex England: Pearson Education Ltd. ISBN 978-0131118928.

- Heisenberg, W. (1958). *Physics and Philosophy: the Revolution in Modern Science*. New York: Harper & Row.

- Hanle, P.A. (1977). "Erwin Schrodinger's Reaction to Louis de Broglie's Thesis on the Quantum Theory." , *Isis* **68** (4). doi:10.1086/351880

- Jaynes, E. T. (2003). G. Larry Bretthorst, ed. *Probability Theory: The Logic of Science*. Cambridge University Press. ISBN 978-0-521 59271-0.

- Landau, L.D.; Lifshitz, E. M. (1977). *Quantum Mechanics: Non-Relativistic Theory*. Vol. 3 (3rd ed.). Pergamon Press. ISBN 978-0-08-020940-1. Online copy

- Lerner, R.G.; Trigg, G.L. (1991). *Encyclopaedia of Physics* (2nd ed.). VHC Publishers. ISBN 0-89573-752-3.

- Ludwig, G. (1968). *Wave Mechanics*. Oxford UK: Pergamon Press. ISBN 0-08-203204-1. LCCN 66-30631.

- Murdoch, D. (1987). *Niels Bohr's Philosophy of Physics*. Cambridge UK: Cambridge University Press. ISBN 0-521-33320-2.

- Newton, R.G. (2002). *Quantum Physics: a Text for Graduate Student*. New York: Springer. ISBN 0-387-95473-2.

- Pauli, Wolfgang (1927). "Zur Quantenmechanik des magnetischen Elektrons". *Zeitschrift für Physik* (in German) **43**. Bibcode:1927ZPhy...43..601P. doi:10.1007/bf01397326.

- Peleg, Y.; Pnini, R.; Zaarur, E.; Hecht, E. (2010). *Quantum mechanics*. Schaum's outlines (2nd ed.). McGraw Hill. ISBN 978-0-07-162358-2.

- Rae, A.I.M. (2008). *Quantum Mechanics* **2** (5th ed.). Taylor & Francis Group. ISBN 1-5848-89705.

- Schrödinger, E. (1926). "An Undulatory Theory of the Mechanics of Atoms and Molecules" (PDF). *Physical Review* **28** (6): 1049–1070. Bibcode:1926PhRv...28.1049S. doi:10.1103/PhysRev.28.1049. Archived from the original (PDF) on 17 December 2008.

- Shankar, R. (1994). *Principles of Quantum Mechanics* (2nd ed.). ISBN 0306447908.

- Martin, B.R.; Shaw, G. (2008). *Particle Physics*. Manchester Physics Series (3rd ed.). John Wiley & Sons. ISBN 978-0-470-03294-7.

- ter Haar, D. (1967). *The Old Quantum Theory*. Pergamon Press. pp. 167–183. LCCN 66029628.

- Tipler, P. A.; Mosca, G.; Freeman (2008). *Physics for Scientists and Engineers – with Modern Physics* (6th ed.). ISBN 0-7167-8964-7.

- Weinberg, S. (2013). *Lectures in Quantum Mechanics*. Cambridge University Press. ISBN 978-1-107-02872-2

- Weinberg, S. (2002), *The Quantum Theory of Fields 1*, Cambridge University Press, ISBN 0-521-55001-7

- Young, H. D.; Freedman, R. A. (2008). Pearson, ed. *Sears' and Zemansky's University Physics* (12th ed.). Addison-Wesley. ISBN 978-0-321-50130-1.

- Wheeler, J.A.; Zurek, W.H. (1983). *Quantum Theory and Measurement*. Princeton NJ: Princeton University Press.

- Zettili, N. (2009). *Quantum Mechanics: Concepts and Applications* (2nd ed.). ISBN 978-0-470-02679-3.

18.15   Further reading

- Yong-Ki Kim (September 2, 2000).   "Practical Atomic Physics" (PDF). *National Institute of Standards and Technology* (Maryland): 1 (55 pages). Retrieved 2010-08-17.

- Polkinghorne, John (2002). *Quantum Theory, A Very Short Introduction*. Oxford University Press. ISBN 0-19-280252-6.

18.16   External links

- . . .

- Normalization.

- Quantum Mechanics and Quantum Computation at BerkeleyX

- Einstein, *The quantum theory of radiation*

Chapter 19

Virtual particle

In physics, a **virtual particle** is an explanatory conceptual entity that is found in mathematical calculations about quantum field theory. It refers to mathematical terms that have some appearance of representing particles inside a subatomic process such as a collision. Virtual particles, however, do not appear directly amongst the observable and detectable input and output quantities of those calculations, which refer only to actual, as distinct from virtual, particles. Virtual particle terms represent "particles" that are said to be 'off mass shell'. For example, they can progress backwards in time, can have apparent mass very different from their regular particle namesake's, and can travel faster than light. That is to say, when looked at individually, they appear to be able to violate basic laws of physics. Regular particles of course never do so. On the other hand, any particle that is actually observed never precisely satisfies the conditions theoretically imposed on regular particles. Virtual particles occur in combinations that mutually more or less nearly cancel from the actual output quantities, so that no actual violation of the laws of physics occurs in completed processes. Often the virtual-particle virtual "events" appear to occur close to one another in time, for example within the time scale of a collision, so that they are virtually and apparently "short-lived". If the mathematical terms that are interpreted as representing virtual particles are omitted from the calculations, the result is an approximation that may or may not be near the correct and accurate answer obtained from the proper full calculation.[1][2][3]

Quantum theory is different from classical theory. The difference is in accounting for the inner workings of subatomic processes. Classical physics cannot account for such. It was pointed out by Heisenberg that what "actually" or "really" occurs inside such subatomic processes as collisions is not directly observable and no unique and physically definite visualization is available for it. Quantum mechanics has the specific merit of by-passing speculation about such inner workings. It restricts itself to what is actually observable and detectable. Virtual particles are conceptual devices that in a sense try to by-pass Heisenberg's insight, by offering putative or virtual explanatory visualizations for the inner workings of subatomic processes.

A virtual particle does not necessarily appear to carry the same mass as the corresponding real particle. This is because it appears as "short-lived" and "transient", so that the uncertainty principle allows it to appear not to conserve energy and momentum. The longer a virtual particle appears to "live", the closer its characteristics come to those of an actual particle.

Virtual particles appear in many processes, including particle scattering and Casimir forces. In quantum field theory, even classical forces — such as the electromagnetic repulsion or attraction between two charges — can be thought of as due to the exchange of many virtual photons between the charges.

Virtual particles appear in calculations of subatomic interactions, but never as asymptotic states or indices to the scattering matrix. A subatomic process involving virtual particles is schematically representable by a Feynman diagram in which they are represented by internal lines.

Antiparticles and quasiparticles should not be confused with virtual particles or virtual antiparticles.

Many physicists believe that, because of its intrinsically perturbative character, the concept of virtual particles is often confusing and misleading, and is thus best avoided.[4][5]

19.1 Properties

The concept of virtual particles arises in the perturbation theory of quantum field theory, an approximation scheme in which interactions (in essence, forces) between actual particles are calculated in terms of exchanges of virtual particles. Such calculations are often performed using schematic representations known as Feynman diagrams, in which virtual particles appear as internal lines. By expressing the interaction in terms of the exchange of a virtual particle with four-momentum q, where q is given by the difference between the four-momenta of the particles entering and leaving the interaction vertex, both momentum and energy are conserved at the interaction vertices of the Feynman diagram.[6]:119

A virtual particle does not precisely obey the energy–momentum relation $m^2c^4 = E^2 - p^2c^2$. Its kinetic energy may not have the usual relationship to velocity–indeed, it can be negative.[7]:110 This is expressed by the phrase *off mass shell*.[6]:119 The probability amplitude for a virtual particle to exist tends to be canceled out by destructive interference over longer distances and times. Quantum tunnelling may be considered a manifestation of virtual particle exchanges.[8]:235 The range of forces carried by virtual particles is limited by the uncertainty principle, which regards energy and time as conjugate variables; thus, virtual particles of larger mass have more limited range.[9]

Written in the usual mathematical notations, in the equations of physics, there is no mark of the distinction between virtual and actual particles. The amplitude that a virtual particle exists interferes with the amplitude for its non-existence, whereas for an actual particle the cases of existence and non-existence cease to be coherent with each other and do not interfere any more. In the quantum field theory view, actual particles are viewed as being detectable excitations of underlying quantum fields. Virtual particles are also viewed as excitations of the underlying fields, but appear only as forces, not as detectable particles. They are "temporary" in the sense that they appear in calculations, but are not detected as single particles. Thus, in mathematical terms, they never appear as indices to the scattering matrix, which is to say, they never appear as the observable inputs and outputs of the physical process being modelled.

There are two principal ways in which the notion of virtual particles appears in modern physics. They appear as intermediate terms in Feynman diagrams; that is, as terms in a perturbative calculation. They also appear as an infinite set of states to be summed or integrated over in the calculation of a semi-non-perturbative effect. In the latter case, it is sometimes said that virtual particles contribute to a mechanism that mediates the effect, or that the effect occurs through the virtual particles.[6]:118

19.2   Manifestations

There are many observable physical phenomena that arise in interactions involving virtual particles. For bosonic particles that exhibit rest mass when they are free and actual, virtual interactions are characterized by the relatively short range of the force interaction produced by particle exchange. Examples of such short-range interactions are the strong and weak forces, and their associated field bosons. For the gravitational and electromagnetic forces,

the zero rest-mass of the associated boson particle permits long-range forces to be mediated by virtual particles. However, in the case of photons, power and information transfer by virtual particles is a relatively short-range phenomenon (existing only within a few wavelengths of the field-disturbance, which carries information or transferred power), as for example seen in the characteristically short range of inductive and capacitive effects in the near field zone of coils and antennas.

Some field interactions which may be seen in terms of virtual particles are:

- The Coulomb force (static electric force) between electric charges. It is caused by the exchange of virtual photons. In symmetric 3-dimensional space this exchange results in the inverse square law for electric force. Since the photon has no mass, the coulomb potential has an infinite range.

- The magnetic field between magnetic dipoles. It is caused by the exchange of virtual photons. In symmetric 3-dimensional space this exchange results in the inverse cube law for magnetic force. Since the photon has no mass, the magnetic potential has an infinite range.

- Electromagnetic induction. This phenomenon transfers energy to and from a magnetic coil via a changing (electro)magnetic field.

- The strong nuclear force between quarks is the result of interaction of virtual gluons. The residual of this force outside of quark triplets (neutron and proton) holds neutrons and protons together in nuclei, and is due to virtual mesons such as the pi meson and rho meson.

- The weak nuclear force - it is the result of exchange by virtual W and Z bosons.

- The spontaneous emission of a photon during the decay of an excited atom or excited nucleus; such a decay is prohibited by ordinary quantum mechanics and requires the quantization of the electromagnetic field for its explanation.

- The Casimir effect, where the ground state of the quantized electromagnetic field causes attraction between a pair of electrically neutral metal plates.

- The van der Waals force, which is partly due to the Casimir effect between two atoms.

- Vacuum polarization, which involves pair production or the decay of the vacuum, which is the spontaneous production of particle-antiparticle pairs (such as electron-positron).

- Lamb shift of positions of atomic levels.

- Hawking radiation, where the gravitational field is so strong that it causes the spontaneous production of photon pairs (with black body energy distribution) and even of particle pairs.

- Much of the so-called near-field of radio antennas, where the magnetic and electric effects of the changing current in the antenna wire and the charge effects of the wire's capacitive charge may be (and usually are) important contributors to the total EM field close to the source, but both of which effects are dipole effects that decay with increasing distance from the antenna much more quickly than do the influence of "conventional" electromagnetic waves that are "far" from the source. ["Far" in terms of ratio of antenna length or diameter, to wavelength]. These far-field waves, for which E is (in the limit of long distance) equal to cB, are composed of actual photons. It should be noted that actual and virtual photons are mixed near an antenna, with the virtual photons responsible only for the "extra" magnetic-inductive and transient electric-dipole effects, which cause any imbalance between E and cB. As distance from the antenna grows, the near-field effects (as dipole fields) die out more quickly, and only the "radiative" effects that are due to actual photons remain as important effects. Although virtual effects extend to infinity, they drop off in field strength as $1/r^2$ rather than the field of EM waves composed of actual photons, which drop $1/r$ (the powers, respectively, decrease as $1/r^4$ and $1/r^2$). See near and far field for a more detailed discussion. See near field communication for practical communications applications of near fields.

Most of these have analogous effects in solid-state physics; indeed, one can often gain a better intuitive understanding by examining these cases. In semiconductors, the roles of electrons, positrons and photons in field theory are replaced by electrons in the conduction band, holes in the valence band, and phonons or vibrations of the crystal lattice. A virtual particle is in a virtual state where the probability amplitude is not conserved. Examples of macroscopic virtual phonons, photons, and electrons in the case of the tunneling process were presented by Günter Nimtz [10] and Alfons A. Stahlhofen.[11]

19.3   History

Paul Dirac was the first to propose that empty space (a vacuum) can be visualized as consisting of a sea of electrons with negative energy, known as the Dirac sea. The Dirac

sea has a direct analog to the electronic band structure in crystalline solids as described in solid state physics. Here, particles correspond to conduction electrons, and antiparticles to holes. A variety of interesting phenomena can be attributed to this structure. The development of quantum field theory (QFT) in the 1930s made it possible to reformulate the Dirac equation in a way that treats the positron as a "real" particle rather than the absence of a particle, and makes the vacuum the state in which no particles exist instead of an infinite sea of particles.

19.4   Feynman diagrams

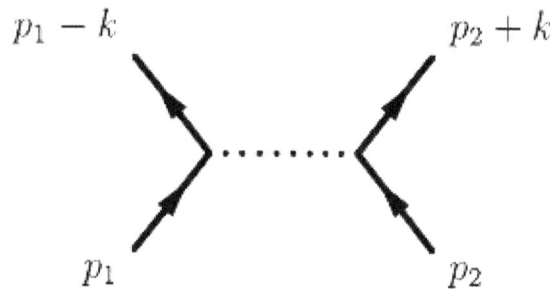

One particle exchange scattering diagram

The calculation of scattering amplitudes in theoretical particle physics requires the use of some rather large and complicated integrals over a large number of variables. These integrals do, however, have a regular structure, and may be represented as Feynman diagrams. The appeal of the Feynman diagrams is strong, as it allows for a simple visual presentation of what would otherwise be a rather arcane and abstract formula. In particular, part of the appeal is that the outgoing legs of a Feynman diagram can be associated with actual, on-shell particles. Thus, it is natural to associate the other lines in the diagram with particles as well, called the "virtual particles". In mathematical terms, they correspond to the propagators appearing in the diagram.

In the image to the right, the solid lines correspond to actual particles (of momentum p_1 and so on), while the dotted line corresponds to a virtual particle carrying momentum k. For example, if the solid lines were to correspond to electrons interacting by means of the electromagnetic interaction, the dotted line would correspond to the exchange of a virtual photon. In the case of interacting nucleons, the dotted line would be a virtual pion. In the case of quarks interacting by means of the strong force, the dotted line would be a virtual gluon, and so on.

Virtual particles may be mesons or vector bosons, as in the example above; they may also be fermions. However, in or-

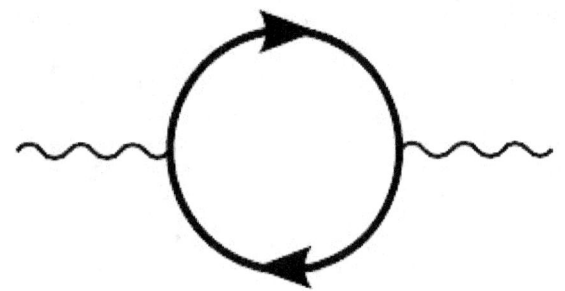

One-loop diagram with fermion propagator

der to preserve quantum numbers, most simple diagrams involving fermion exchange are prohibited. The image to the right shows an allowed diagram, a one-loop diagram. The solid lines correspond to a fermion propagator, the wavy lines to bosons.

19.5   Vacuums

Main article: Quantum fluctuation

In formal terms, a particle is considered to be an eigenstate of the particle number operator $a^* \dagger a$, where a is the particle annihilation operator and $a^* \dagger$ the particle creation operator (sometimes collectively called ladder operators). In many cases, the particle number operator does not commute with the Hamiltonian for the system. This implies the number of particles in an area of space is not a well-defined quantity but, like other quantum observables, is represented by a probability distribution. Since these particles do not have a permanent existence, they are called *virtual particles* or *vacuum fluctuations* of vacuum energy. In a certain sense, they can be understood to be a manifestation of the time-energy uncertainty principle in a vacuum.[12]

An important example of the "presence" of virtual particles in a vacuum is the Casimir effect.[13] Here, the explanation of the effect requires that the total energy of all of the virtual particles in a vacuum can be added together. Thus, although the virtual particles themselves are not directly observable in the laboratory, they do leave an observable effect: Their zero-point energy results in forces acting on suitably arranged metal plates or dielectrics.[14] On the other hand, the Casimir effect can be interpreted as the relativistic van der Waals force.[15]

19.6   Pair production

Main article: Pair production

Virtual particles are often popularly described as coming in pairs, a particle and antiparticle, which can be of any kind. These pairs exist for an extremely short time, and then mutually annihilate. In some cases, however, it is possible to boost the pair apart using external energy so that they avoid annihilation and become actual particles.

This may occur in one of two ways. In an accelerating frame of reference, the virtual particles may appear to be actual to the accelerating observer; this is known as the Unruh effect. In short, the vacuum of a stationary frame appears, to the accelerated observer, to be a warm gas of actual particles in thermodynamic equilibrium.

Another example is pair production in very strong electric fields, sometimes called vacuum decay. If, for example, a pair of atomic nuclei are merged to very briefly form a nucleus with a charge greater than about 140, (that is, larger than about the inverse of the fine structure constant, which is a dimensionless quantity), the strength of the electric field will be such that it will be energetically favorable to create positron-electron pairs out of the vacuum or Dirac sea, with the electron attracted to the nucleus to annihilate the positive charge. This pair-creation amplitude was first calculated by Julian Schwinger in 1951.

19.7   Actual and virtual particles compared

As a consequence of quantum mechanical uncertainty, any object or process that exists for a limited time or in a limited volume cannot have a precisely defined energy or momentum. This is the reason that virtual particles — which exist only temporarily as they are exchanged between ordinary particles — do not necessarily obey the mass-shell relation. However, the longer a virtual particle exists, the more closely it adheres to the mass-shell relation. A "virtual" particle that exists for an arbitrarily long time is simply an ordinary particle.

However, all particles have a finite lifetime, as they are created and eventually destroyed by some processes. As such, there is no absolute distinction between "real" and "virtual" particles. In practice, the lifetime of "ordinary" particles is far longer than the lifetime of the virtual particles that contribute to processes in particle physics, and as such the distinction is useful to make.

19.8 See also

- Force carrier

- QCD vacuum

- QED vacuum

- Static forces and virtual-particle exchange

- Vacuum genesis

- Vacuum Rabi oscillation

- Vacuum state

- Virtual State

19.9 References

[1] Peskin, M.E., Schroeder, D.V. (1995). *An Introduction to Quantum Field Theory*, Westview Press, ISBN 0-201-50397-2, p. 80.

[2] Mandl, F., Shaw, G. (1984/2002). *Quantum Field Theory*, John Wiley & Sons, Chichester UK, revised edition, ISBN 0-471-94186-7, pp. 56, 176.

[3] Bayfield, James E. (1999). *Quantum evolution : an introduction to time-dependent quantum mechanics*. New York: John Wiley. p. 62. ISBN 9780471181743.

[4] Unruh, William (23 August 2013). *The View from GR* (flash tv). KITP Rapid Response Workshop: Black Holes: Complementarity, Fuzz, or Fire?. Santa Barbara, CA: The Kavli Institute for Theoretical Physics. Retrieved 2 August 2015.

[5] Anderson, Philip W. (February 2000). "Brainwashed by Feynman?". *Physics Today* **53** (2): 11–12. doi:10.1063/1.882955.

[6] Cambridge, Mark Thomson, University of (2013). *Modern particle physics*. Cambridge: Cambridge University Press. ISBN 978-1107034266.

[7] Hawking, Stephen (1998). *A brief history of time* (Updated and expanded tenth anniversary ed.). New York: Bantam Books. ISBN 9780553896923.

[8] Walters, Tony Hey ; Patrick (2004). *The new quantum universe* (Reprint. ed.). Cambridge [u.a.]: Cambridge Univ. Press. ISBN 9780521564571.

[9] Calle, Carlos I. (2010). *Superstrings and other things : a guide to physics* (2nd ed.). Boca Raton: CRC Press/Taylor & Francis. pp. 443–444. ISBN 9781439810743.

[10] G. Nimtz, On Virtual Phonons, Photons and Electrons, Found. Phys. 39, 1346-1355 (2009)

[11] A.Stahlhofen and G. Nimtz, Evanescent Modes are Virtual Photons, Europhys. Lett. 76, 198 (2006)

[12] Raymond, David J. (2012). *A radically modern approach to introductory physics: volume 2: four forces*. Socorro, NM: New Mexico Tech Press. pp. 252–254. ISBN 978-0-98303-946-4.

[13] Choi, Charles Q. (13 February 2013). "A vacuum can yield flashes of light". *Nature*. doi:10.1038/nature.2013.12430. Retrieved 2 August 2015.

[14] Lambrecht, Astrid (September 2002). "The Casimir effect: a force from nothing". *Physics world* **15** (9): 29–32.

[15] Jaffe, R. L. (12 July 2005). "Casimir effect and the quantum vacuum". *Physical Review D* **72** (2). doi:10.1103/PhysRevD.72.021301.

19.10 External links

- Are virtual particles really constantly popping in and out of existence? — Gordon Kane, director of the Michigan Center for Theoretical Physics at the University of Michigan at Ann Arbor, proposes an answer at the *Scientific American* website.

- Virtual Particles: What are they?

Chapter 20

Cathode ray

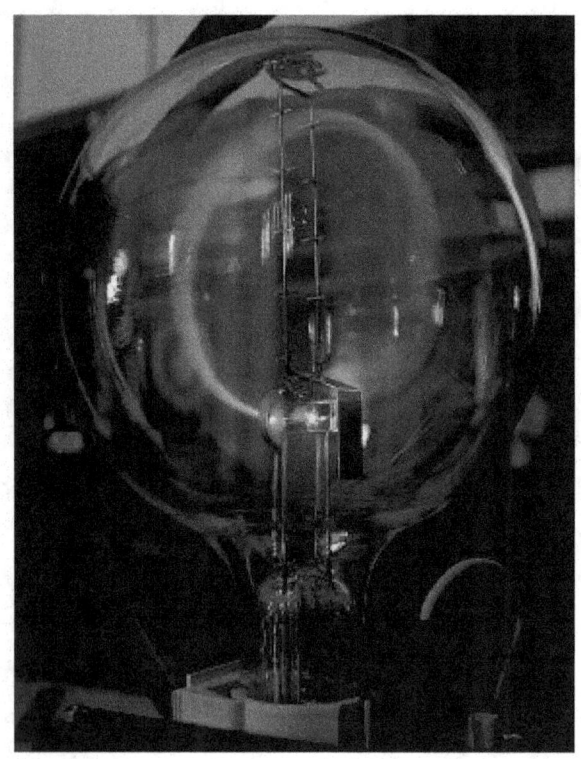

A beam of cathode rays bent into a circle by a magnetic field generated by a Helmholtz coil. Cathode rays are normally invisible; in this tube enough residual gas has been left that the gas atoms glow from fluorescence when struck by the fast moving electrons.

Cathode rays (also called an **electron beam** or **e-beam**) are streams of electrons observed in vacuum tubes. If an evacuated glass tube is equipped with two electrodes and a voltage is applied, the glass opposite of the negative electrode is observed to glow, due to electrons emitted from and travelling perpendicular to the cathode (the electrode connected to the negative terminal of the voltage supply). They were first observed in 1869 by German physicist Johann Hittorf, and were named in 1876 by Eugen Goldstein *Kathodenstrahlen*, or cathode rays.[1] [2]

Electrons were first discovered as the constituents of cath-

ode rays. In 1897 British physicist J. J. Thomson showed the rays were composed of a previously unknown negatively charged particle, which was later named the *electron*. Cathode ray tubes (CRTs) use a focused beam of electrons deflected by electric or magnetic fields to create the image in a classic television set.

20.1 Description

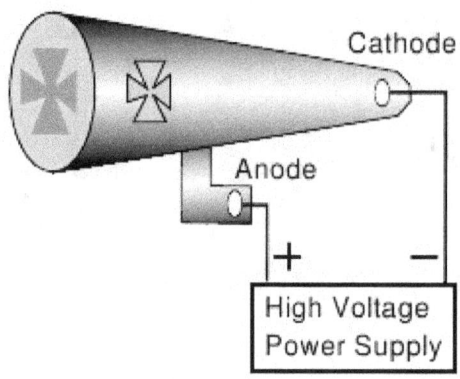

A diagramatic Crookes tube showing the connections for the high voltage supply. The Maltese cross has no external electrical connection.

Cathode rays are so named because they are emitted by the negative electrode, or cathode, in a vacuum tube. To release electrons into the tube, they first must be detached from the atoms of the cathode. In the early cold cathode vacuum tubes, called Crookes tubes, this was done by using a high electrical potential between the anode and the cathode to ionize the residual gas in the tube; the ions were accelerated by the electric field and released electrons when they collided with the cathode. Modern vacuum tubes use thermionic emission, in which the cathode is made of a thin wire filament which is heated by a separate electric current passing through it. The increased random heat motion of the filament atoms knocks electrons out of the atoms at the surface of the filament, into the evacuated space of the tube.

Since the electrons have a negative charge, they are repelled by the cathode and attracted to the anode. They travel in straight lines through the empty tube. The voltage applied between the electrodes accelerates these low mass particles to high velocities. Cathode rays are invisible, but their presence was first detected in early vacuum tubes when they struck the glass wall of the tube, exciting the atoms of the glass and causing them to emit light, a glow called fluorescence. Researchers noticed that objects placed in the tube in front of the cathode could cast a shadow on the glowing wall, and realized that something must be travelling in straight lines from the cathode. After the electrons reach the anode, they travel through the anode wire to the power supply and back to the cathode, so cathode rays carry electric current through the tube.

The current in a beam of cathode rays through a tube can be controlled by passing it through a metal screen of wires (a grid) to which a small voltage is applied. The electric field of the wires deflects some of the electrons, preventing them from reaching the anode. Thus a small voltage on the grid can be made to control a much larger voltage on the anode. This is the principle used in vacuum tubes to amplify electrical signals. High speed beams of cathode rays can also be steered and manipulated by electric fields created by additional metal plates in the tube to which voltage is applied, or magnetic fields created by coils of wire (electromagnets). These are used in cathode ray tubes, found in televisions and computer monitors, and in electron microscopes.

- Crookes tube

- Cathode rays travel from the cathode at the rear of the tube, striking the glass front, making it glow green by fluorescence. A metal cross in the tube casts a shadow, demonstrating that the rays travel in straight lines.

- A magnet creates a horizontal magnetic field through the neck of the tube, bending the rays up, so the shadow of the cross is higher.

- When the magnet is reversed, it bends the rays down, so the shadow is lower. The pink glow is caused by cathode rays striking residual gas atoms in the tube.

20.2 History

After the 1654 invention of the vacuum pump by Otto von Guericke, physicists began to experiment with passing high voltage electricity through rarefied air. In 1705, it was noted

that electrostatic generator sparks travel a longer distance through low pressure air than through atmospheric pressure air.

20.2.1 Gas discharge tubes

In 1838, Michael Faraday passed a current through a rarefied air filled glass tube and noticed a strange light arc with its beginning at the cathode (negative electrode) and its end are at the anode (positive electrode).[3] In 1857, German physicist and glassblower Heinrich Geissler sucked even more air out with an improved pump, to a pressure of around 10^-3 atm and found that, instead of an arc, a glow filled the tube. The voltage applied between the two electrodes of the tubes, generated by an induction coil, was anywhere between a few kilovolts and 100 kV. These were called Geissler tubes, similar to today's neon signs.

The explanation of these effects was that the high voltage accelerated electrically charged atoms (ions) naturally present in the air of the tube. At low pressure, there was enough space between the gas atoms that the ions could accelerate to high enough speeds that when they struck another atom they knocked electrons off of it, creating more positive ions and free electrons in a chain reaction. The positive ions were all attracted to the cathode. When they struck it they knocked many electrons out of the metal. The free electrons were all attracted to the anode.

Geissler tubes had enough air in them that the electrons could only travel a tiny distance before colliding with an atom. The electrons in these tubes moved in a slow diffusion process, never gaining much speed, so these tubes didn't produce cathode rays. Instead they produced a colorful glow discharge (as in a modern neon light), caused when the electrons or ions struck gas atoms, exciting their orbital electrons to higher energy levels. The electrons released this energy as light. This process is called fluorescence.

20.2.2 Cathode rays

By the 1870s, British physicist William Crookes and others were able to evacuate tubes to a lower pressure, below 10^-6 atm. These were called Crookes tubes. Faraday had been the first to notice a dark space just in front of the cathode, where there was no luminescence. This came to be called the "cathode dark space", "Faraday dark space" or "Crookes dark space". Crookes found that as he pumped more air out of the tubes, the Faraday dark space spread down the tube from the cathode toward the anode, until the tube was totally dark. But at the anode (positive) end of the tube, the glass of the tube itself began to glow.

What was happening was that as more air was pumped from

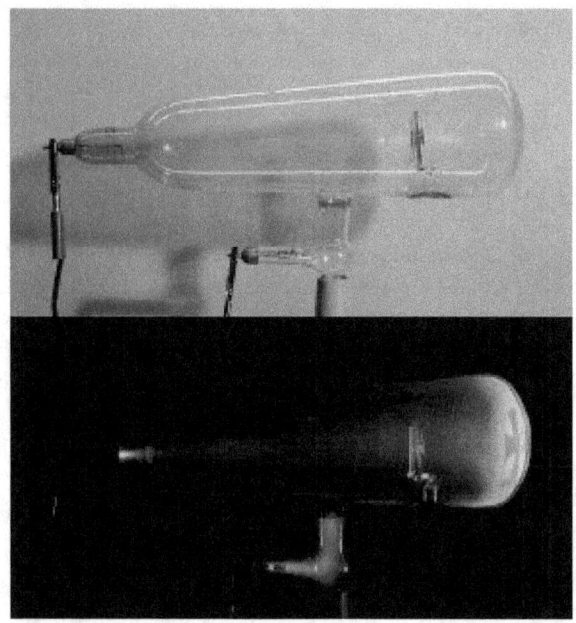

A Crookes tube. The cathode rays travel in straight lines from the cathode (left) and strike the right wall of the tube, making it glow by fluorescence.

the tube, the electrons could travel farther, on average, before they struck a gas atom. By the time the tube was dark, most of the electrons could travel in straight lines from the cathode to the anode end of the tube without a collision. With no obstructions, these low mass particles were accelerated to high velocities by the voltage between the electrodes.These were the cathode rays.

When they reached the anode end of the tube, they were traveling so fast that, although they were attracted to it, they often flew past the anode and struck the back wall of the tube. When they struck atoms in the glass wall, they excited their orbital electrons to higher energy levels, causing them to fluoresce. Later researchers painted the inside back wall with fluorescent chemicals such as zinc sulfide, to make the glow more visible.

Cathode rays themselves are invisible, but this accidental fluorescence allowed researchers to notice that objects in the tube in front of the cathode, such as the anode, cast sharp-edged shadows on the glowing back wall. In 1869, German physicist Johann Hittorf was first to realize that something must be traveling in straight lines from the cathode to cast the shadows. Eugen Goldstein named them *cathode rays*.

20.2.3   Discovery of the electron

At this time, atoms were the smallest particles known, and were believed to be indivisible. What carried electric currents was a mystery. During the last quarter of the 19th

century many experiments were done to determine what cathode rays were. There were two theories. Crookes and Arthur Schuster believed they were particles of "radiant matter", that is, electrically charged atoms. German scientists Eilhard Wiedemann, Heinrich Hertz and Goldstein believed they were "aether waves", some new form of electromagnetic radiation, and were separate from what carried the electric current through the tube.

The debate was resolved in 1897 when J. J. Thomson measured the mass of cathode rays, showing they were made of particles, but were around 1800 times lighter than the lightest atom, hydrogen. Therefore they were not atoms, but a new particle, the first *subatomic* particle to be discovered, which he originally called "*corpuscle*" but was later named *electron*, after particles postulated by George Johnstone Stoney in 1874. He also showed they were identical with particles given off by photoelectric and radioactive materials.[4] It was quickly recognized that they are the particles that carry electric currents in metal wires, and carry the negative electric charge of the atom.

Thomson was given the 1906 Nobel prize for physics for this work. Philipp Lenard also contributed a great deal to cathode ray theory, winning the Nobel prize for physics in 1905 for his research on cathode rays and their properties.

20.2.4   Vacuum tubes

The gas ionization (or cold cathode) method of producing cathode rays used in Crookes tubes was unreliable, because it depended on the pressure of the residual air in the tube. Over time, the air was adsorbed by the walls of the tube, and it stopped working.

A more reliable and controllable method of producing cathode rays was investigated by Hittorf and Goldstein, and rediscovered by Thomas Edison in 1880. A cathode made of a wire filament heated red hot by a separate current passing through it would release electrons into the tube by a process called thermionic emission. The first true electronic vacuum tubes, invented in 1904, used this hot cathode technique, and they superseded Crookes tubes. These tubes didn't need gas in them to work, so they were evacuated to a lower pressure, around 10^{-9} atm (10^{-4} P). The ionization method of creating cathode rays used in Crookes tubes is today only used in a few specialized gas discharge tubes such as krytrons.

Lee De Forest in 1906 found that a small voltage on a grid of metal wires could control a much larger current in a beam of cathode rays passing through a vacuum tube. His invention, called the triode, was the first device that could amplify electric signals, and founded the field of *electronics*. Vacuum tubes made radio and television broadcasting possible, as well as radar, talking movies, audio recording, and long distance telephone service, and were the foundation of consumer electronic devices until the 1960s when the transistor brought the era of vacuum tubes to a close.

Cathode rays are now usually called electron beams. The technology of manipulating electron beams pioneered in these early tubes was applied practically in the design of vacuum tubes, particularly in the invention of the cathode ray tube by Ferdinand Braun in 1897 and is today employed in sophisticated devices such as electron microscopes, electron beam lithography and particle accelerators.

20.3   Properties of cathode rays

Like a wave, cathode rays travel in straight lines, and produce a shadow when obstructed by objects. Ernest Rutherford demonstrated that rays could pass through thin metal foils, behavior expected of a particle. These conflicting properties caused disruptions when trying to classify it as a wave or particle. Crookes insisted it was a particle, while Hertz maintained it was a wave. The debate was resolved when an electric field was used to deflect the rays by J. J. Thomson. This was evidence that the beams were composed of particles because scientists knew it was impossible to deflect electromagnetic waves with an electric field. These can also create mechanical effects, fluorescence, etc.

Louis de Broglie later (1924) showed in his doctoral dissertation that electrons are in fact much like photons in the respect that they act both as waves and as particles in a dual manner as Einstein had shown earlier for light. The wave-like behaviour of cathode rays was later directly demonstrated using a crystal lattice by Davisson and Germer in 1927.

20.4   See also

- α (alpha) particles
- β (beta) particles
- Electron beam processing
- Electron microscope
- Electron beam melting
- Electron beam welding
- Electron gun
- Electron irradiation
- Ionizing radiation
- Particle accelerator
- Rays:
  - γ (gamma) rays
  - n (neutron) rays
  - δ (delta) rays
  - ε (epsilon) rays
- Sterilisation (microbiology)
- Electron beam technology
- phosphorescent screen

20.5   References

[1] E. Goldstein (May 4, 1876) "Vorläufige Mittheilungen über elektrische Entladungen in verdünnten Gasen" (Preliminary communications on electric discharges in rarefied gases). *Monatsberichte der Königlich Preussischen Akademie der Wissenschaften zu Berlin* (Monthly Reports of the Royal Prussian Academy of Science in Berlin), 279-295. From page 286: *"13. Das durch die Kathodenstrahlen in der Wand hervorgerufene Phosphorescenzlicht ist höchst selten von gleichförmiger Intensität auf der von ihm bedeckten Fläche, und zeigt oft sehr barocke Muster."* (13. The phosphorescent light that's produced in the wall by the cathode rays is very rarely of uniform intensity on the surface that it covers, and [it] often shows very baroque patterns.)

[2] Joseph F. Keithley *The story of electrical and magnetic measurements: from 500 B.C. to the 1940s* John Wiley and Sons, 1999 ISBN 0-7803-1193-0, page 205

[3] Michael Faraday (1838) "VIII. Experimental researches in electricity. —Thirteenth series.," *Philosophical Transactions of the Royal Society of London*, **128** : 125-168.

[4] Thomson, J. J. (August 1901). "On bodies smaller than atoms". *The Popular Science Monthly* (Bonnier Corp.): 323–335. Retrieved 2009-06-21.

5. General Chemistry (structure and properties of matter) by Aruna Bandara (2010)

20.6   External links

- The Cathode Ray Tube site
- Crookes tube with maltese cross operating

20.6.1   Animations and Simulations

- The simulation show electrons in crossed fields made by BIGS

Chapter 21

Electron-beam lithography

An example of 'Electron beam lithograph' setup

Electron-beam lithography (often abbreviated as **e-beam lithography**) is the practice of scanning a focused beam of electrons to draw custom shapes on a surface covered with an electron-sensitive film called a resist ("exposing").[1] The electron beam changes the solubility of the resist, enabling selective removal of either the exposed or non-exposed regions of the resist by immersing it in a solvent ("developing"). The purpose, as with photolithography, is to create very small structures in the resist that can subsequently be transferred to the substrate material, often by etching.

The primary advantage of electron-beam lithography is that it can draw custom patterns (direct-write) with sub-10 nm resolution. This form of maskless lithography has high resolution and low throughput, limiting its usage to photomask fabrication, low-volume production of semiconductor devices, and research & development.

21.1 Electron-beam lithography systems

Electron-beam lithography systems used in commercial applications are dedicated e-beam writing systems that are very expensive (> US$1M). For research applications, it is very common to convert an electron microscope into an electron beam lithography system using a relatively low cost accessories (< US$100K). Such converted systems have produced linewidths of ~20 nm since at least 1990, while current dedicated systems have produced linewidths on the order of 10 nm or smaller.

Electron-beam lithography systems can be classified according to both beam shape and beam deflection strategy. Older systems used Gaussian-shaped beams and scanned these beams in a raster fashion. Newer systems use shaped beams, which may be deflected to various positions in the writing field (this is also known as **vector scan**).

21.1.1 Electron sources

Lower-resolution systems can use thermionic sources, which are usually formed from lanthanum hexaboride. However, systems with higher-resolution requirements need to use field electron emission sources, such as heated W/ZrO_2 for lower energy spread and enhanced brightness. Thermal field emission sources are preferred over cold emission sources, in spite of the former's slightly larger beam size, because they offer better stability over typical writing times of several hours.

21.1.2 Lenses

Both electrostatic and magnetic lenses may be used. However, electrostatic lenses have more aberrations and so are not used for fine focusing. There is no current mechanism to make achromatic electron beam lenses, so extremely narrow dispersions of the electron beam energy are needed for finest focusing.

21.1.3   Stage, stitching and alignment

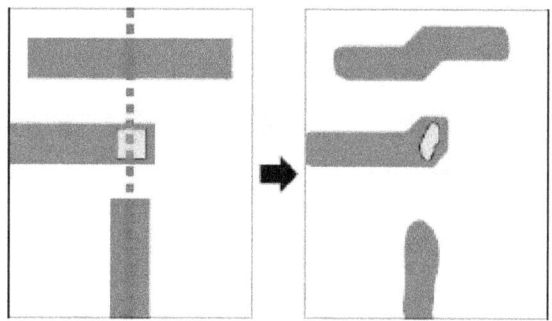

Field stitching. Stitching is a concern for critical features crossing a field boundary (red dotted line).

Typically, for very small beam deflections electrostatic deflection "lenses" are used, larger beam deflections require electromagnetic scanning. Because of the inaccuracy and because of the finite number of steps in the exposure grid the writing field is of the order of 100 micrometre – 1 mm. Larger patterns require stage moves. An accurate stage is critical for stitching (tiling writing fields exactly against each other) and pattern overlay (aligning a pattern to a previously made one).

21.1.4   Electron beam write time

The minimum time to expose a given area for a given dose is given by the following formula:[2]

$$D \cdot A = T \cdot I$$

where T is the time to expose the object (can be divided into exposure time/step size), I is the beam current, D is the dose and A is the area exposed.

For example, assuming an exposure area of 1 cm^2, a dose of 10^{-3} coulombs/cm^2, and a beam current of 10^{-9} amperes, the resulting minimum write time would be 10^6 seconds (about 12 days). This minimum write time does not include time for the stage to move back and forth, as well as time for the beam to be blanked (blocked from the wafer during deflection), as well as time for other possible beam corrections and adjustments in the middle of writing. To cover the 700 cm^2 surface area of a 300 mm silicon wafer, the minimum write time would extend to $7*10^8$ seconds, about 22 years. This is a factor of about 10 million times slower than current optical lithography tools. It is clear that

throughput is a serious limitation for electron beam lithography, especially when writing dense patterns over a large area.

E-beam lithography is not suitable for high-volume manufacturing because of its limited throughput. The smaller field of electron beam writing makes for very slow pattern generation compared with photolithography (the current standard) because more exposure fields must be scanned to form the final pattern area (≤mm^2 for electron beam vs. ≥40 mm^2 for an optical mask projection scanner). The stage moves in between field scans. The electron beam field is small enough that a rastering or serpentine stage motion is needed to pattern a 26 mm X 33 mm area for example, whereas in a photolithography scanner only a one-dimensional motion of a 26 mm X 2 mm slit field would be required.

Currently an optical maskless lithography tool[3] is much faster than an electron beam tool used at the same resolution for photomask patterning.

21.1.5   Shot noise

As features sizes shrink, the number of incident electrons at fixed dose also shrinks. As soon as the number reaches ~10000, shot noise effects become predominant, leading to substantial natural dose variation within a large feature population. With each successive process node, as the feature area is halved, the minimum dose must double to maintain the same noise level. Consequently, the tool throughput would be halved with each successive process node.

Note: 1 ppm of population is about 5 standard deviations away from the mean dose.

Ref.: SPIE Proc. 8683-36 (2013)

Shot noise is a significant consideration even for mask fabrication. For example, a commercial mask e-beam resist like FEP-171 would use doses less than 10 μC/cm^2,[4][5] whereas this leads to noticeable shot noise for a target CD even on the order of ~200 nm on the mask.[6][7]

21.1.6   Defects in electron-beam lithography

Despite the high resolution of electron-beam lithography, the generation of defects during electron-beam lithography is often not considered by users. Defects may be classified into two categories: data-related defects, and physical defects.

Data-related defects may be classified further into two sub-categories. **Blanking** or **deflection errors** occur when the electron beam is not deflected properly when it is supposed to, while **shaping errors** occur in variable-shaped beam systems when the wrong shape is projected onto the sample. These errors can originate either from the electron optical control hardware or the input data that was taped out. As might be expected, larger data files are more susceptible to data-related defects.

Physical defects are more varied, and can include sample charging (either negative or positive), backscattering calculation errors, dose errors, fogging (long-range reflection of backscattered electrons), outgassing, contamination, beam drift and particles. Since the write time for electron beam lithography can easily exceed a day, "randomly occurring" defects are more likely to occur. Here again, larger data files can present more opportunities for defects.

Photomask defects largely originate during the electron beam lithography used for pattern definition.

21.2 Electron energy deposition in matter

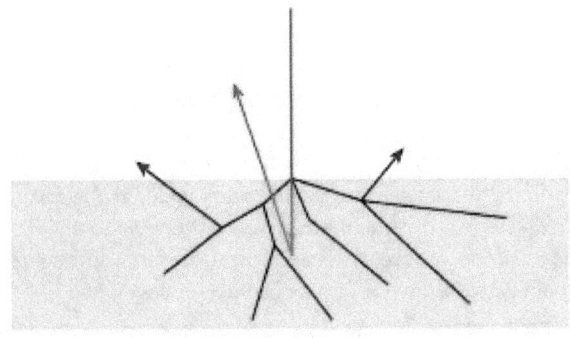

Electron trajectories in resist: *An incident electron (red) produces secondary electrons (blue). Sometimes, the incident electron may itself be backscattered as shown here and leave the surface of the resist (amber).*

The primary electrons in the incident beam lose energy upon entering a material through inelastic scattering or collisions with other electrons. In such a collision the momentum transfer from the incident electron to an atomic electron can be expressed as [8] $dp = 2e^2/bv$, where b is the distance of closest approach between the electrons, and v is the incident electron velocity. The energy transferred by the collision is given by $T = (dp)^2/2m = e^4/Eb^2$, where m is the electron mass and E is the incident electron energy, given by $E = (1/2)mv^2$. By integrating over all values of T between the lowest binding energy, E_0 and the incident energy, one obtains the result that the total cross section for collision is inversely proportional to the incident energy E, and proportional to $1/E_0 - 1/E$. Generally, $E \gg E_0$, so the result is essentially inversely proportional to the binding energy.

By using the same integration approach, but over the range $2E_0$ to E, one obtains by comparing cross-sections that half of the inelastic collisions of the incident electrons produce electrons with kinetic energy greater than E_0. These secondary electrons are capable of breaking bonds (with binding energy E_0) at some distance away from the original collision. Additionally, they can generate additional, lower energy electrons, resulting in an electron cascade. Hence, it is important to recognize the significant contribution of secondary electrons to the spread of the energy deposition.

In general, for a molecule AB: [9]

$$e^* - + AB \rightarrow AB^* - \rightarrow A + B^* -$$

This reaction, also known as "electron attachment" or "dissociative electron attachment" is most likely to occur after the electron has essentially slowed to a halt, since it is easiest to capture at that point. The cross-section for electron attachment is inversely proportional to electron energy at high energies, but approaches a maximum limiting value at zero energy. [10] On the other hand, it is already known that the mean free path at the lowest energies (few to several eV or less, where dissociative attachment is significant) is well over 10 nm, [11] [12] thus limiting the ability to consistently achieve resolution at this scale.

21.2.1 Resolution capability

With today's electron optics, electron beam widths can routinely go down to a few nm. This is limited mainly by aberrations and space charge. However, the feature resolution limit is determined not by the beam size but by forward scattering (or effective beam broadening) in the resist while the pitch resolution limit is determined by secondary electron travel in the resist. [13] [14] This point is driven home by the 2007 demonstration of double patterning using electron beam lithography in the fabrication of 15 nm half-pitch zone plates. [15] Although a 15 nm feature was resolved, a 30 nm pitch was still difficult to do, due to secondary electrons scattering from the adjacent feature. The use of double patterning allowed the spacing between features to be wide enough for the secondary electron scattering to be significantly reduced. The forward scattering can be decreased by using higher energy electrons or thinner resist, but the generation of secondary electrons is inevitable. It is now recognized that for insulating materials like PMMA, low energy electrons can travel quite a far distance (several nm is possible). This is due to the fact that below the ionization potential the only energy loss mechanism is mainly through phonons and polarons, although the latter is basically an ionic lattice effect. [16] Polaron hopping could extend as far as 20 nm. [17] The travel distance of secondary electrons is not a fundamentally derived physical value, but a statistical parameter often determined from many experiments or Monte Carlo simulations down to < 1 eV. This is necessary since the energy distribution of secondary electrons peaks well below 10 eV.[18] Hence, the resolution limit is not usually cited as a well-fixed number as with an optical diffraction-limited system.[13] Repeatability and control at the practical resolution limit often require considerations not related to image formation, e.g., resist development and intermolecular forces.

A study by the College of Nanoscale Science and Engineering (CNSE) presented at the 2013 EUVL Workshop indicated that, as a measure of electron blur, 50-100 eV electrons easily penetrated beyond 10 nm of resist thickness (PMMA or commercial resist); furthermore dielectric breakdown discharge is possible.[19]

21.2.2 Scattering

In addition to producing secondary electrons, primary electrons from the incident beam with sufficient energy to penetrate the resist can be multiply scattered over large distances from underlying films and/or the substrate. This leads to exposure of areas at a significant distance from the desired exposure location. For thicker resist, as the primary electrons move forward, they have an increasing opportunity to scatter laterally from the beam-defined location. This scattering is called **forward scattering**. Sometimes the primary electrons are scattered at angles exceeding 90 degrees, i.e., they no longer advance further into the resist. These electrons are called **backscattered electrons** and have the same effect as long-range flare in optical projection systems. A large enough dose of backscattered electrons can lead to complete exposure of resist over an area much larger than defined by the beam spot.

21.2.3 Proximity effect

The smallest features produced by electron-beam lithography have generally been isolated features, as nested features exacerbate the proximity effect, whereby electrons from exposure of an adjacent region spill over into the exposure of the currently written feature, effectively enlarging its image, and reducing its contrast, i.e., difference between maximum and minimum intensity. Hence, nested feature resolution is harder to control. For most resists, it is difficult to go below 25 nm lines and spaces, and a limit of 20 nm lines and spaces has been found.[20] In actuality, though, the range of secondary electron scattering is quite far, sometimes exceeding 100 nm,[21] but becoming very significant below 30 nm.[22]

The proximity effect is also manifest by secondary electrons leaving the top surface of the resist and then returning some tens of nanometers distance away.[23]

Proximity effects (due to electron scattering) can be addressed by solving the inverse problem and calculating the exposure function $E(x, y)$ that leads to a dose distribution as close as possible to the desired dose $D(x, y)$ when convolved by the scattering distribution point spread function $PSF(x, y)$. However, it must be remembered that an error in the applied dose (e.g., from shot noise) would cause the proximity effect correction to fail.

21.3 Charging

Since electrons are charged particles, they tend to charge the substrate negatively unless they can quickly gain access to a path to ground. For a high-energy beam incident on a silicon wafer, virtually all the electrons stop in the wafer where they can follow a path to ground. However, for a quartz substrate such as a photomask, the embedded electrons will take a much longer time to move to ground. Often the negative charge acquired by a substrate can be compensated or even exceeded by a positive charge on the surface due to secondary electron emission into the vacuum. The presence of a thin conducting layer above or below the resist is generally of limited use for high energy (50 keV or more) electron beams, since most electrons pass through the layer into the substrate. The charge dissipation layer is generally useful only around or below 10 keV, since the resist is thinner and most of the electrons either stop in the resist or close to the conducting layer. However, they are of limited use due to their high sheet resistance, which can lead to ineffective grounding.

The range of low-energy secondary electrons (the largest component of the free electron population in the resist-substrate system) which can contribute to charging is not a fixed number but can vary from 0 to as high as 50 nm (see section New frontiers in electron beam lithography and extreme ultraviolet lithography). Hence, resist-substrate charging is not repeatable and is difficult to compensate consistently. Negative charging deflects the electron beam away from the charged area while positive charging deflects the electron beam toward the charged area.

21.4 Electron-beam resist performance

Due to the scission efficiency generally being an order of magnitude higher than the crosslinking efficiency, most polymers used for positive-tone electron-beam lithography will crosslink (and therefore become negative tone) at doses an order of magnitude than doses used for positive tone exposure.[24] Such large dose increases may be required to avoid shot noise effects.[25][26][27]

A study performed at the Naval Research Laboratory[28] indicated that low-energy (10–50 eV) electrons were able to damage ~30 nm thick PMMA films. The damage was manifest as a loss of material.

- For the popular electron-beam resist ZEP-520, a pitch resolution limit of 60 nm (30 nm lines and spaces), independent of thickness and beam energy, was found.[29]

- A 20 nm resolution had also been demonstrated using a 3 nm 100 keV electron beam and PMMA resist.[30] 20 nm unexposed gaps between exposed lines showed inadvertent exposure by secondary electrons.

- Hydrogen silsesquioxane (HSQ) is a negative tone resist that is capable of forming isolated 2-nm-wide lines and 10 nm periodic dot arrays (10 nm pitch) in very thin layers.[31] HSQ itself is similar to porous, hydrogenated SiO_2. It may be used to etch silicon but not silicon dioxide or other similar dielectrics.

21.5 New frontiers in electron-beam lithography

To get around the secondary electron generation, it will be imperative to use low-energy electrons as the primary radiation to expose resist. Ideally, these electrons should have energies on the order of not much more than several eV in order to expose the resist without generating any secondary electrons, since they will not have sufficient excess energy. Such exposure has been demonstrated using a scanning tunneling microscope as the electron beam source.[32] The data suggest that electrons with energies as low as 12 eV can penetrate 50 nm thick polymer resist. The drawback to using low energy electrons is that it is hard to prevent spreading of the electron beam in the resist.[33] Low energy electron optical systems are also hard to design for high resolution.[34] Coulomb inter-electron repulsion always becomes more severe for lower electron energy.

Another alternative in electron-beam lithography is to use extremely high electron energies (at least 100 keV) to essentially "drill" or sputter the material. This phenomenon has been observed frequently in transmission electron microscopy.[35] However, this is a very inefficient process, due to the inefficient transfer of momentum from the electron beam to the material. As a result it is a slow process, requiring much longer exposure times than conven-

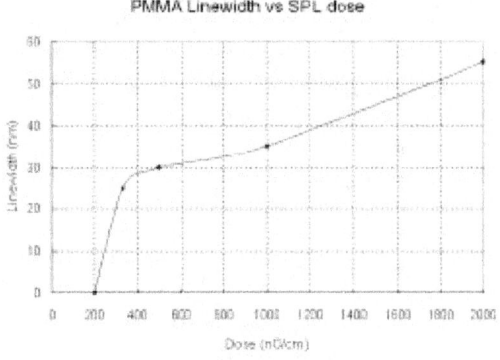

Scanning probe lithography. A scanning probe can be used for low-energy electron beam lithography, offering sub-100 nm resolution, determined by the dose of low-energy electrons.

tional electron beam lithography. Also high energy beams always bring up the concern of substrate damage.

Interference lithography using electron beams is another possible path for patterning arrays with nanometer-scale periods. A key advantage of using electrons over photons in interferometry is the much shorter wavelength for the same energy.

Despite the various intricacies and subtleties of electron beam lithography at different energies, it remains the most practical way to concentrate the most energy into the smallest area.

There has been significant interest in the development of multiple electron beam approaches to lithography in order to increase throughput. This work has been supported by SEMATECH and start-up companies such as Multibeam Corporation.[36] Mapper[37] and IMS.[38] However, the degree of parallelism required to be competitive would need to be very high (at least 10 million, as estimated above); this is far in excess of most scheduled demonstrations. A key difficulty is that the total supplied beam current needs to be multiplied by the number of parallel beams(e.g., 10 million), which dramatically increases cost of ownership. Also, the field size does not change, which means increasing the number of beams increases the strength of Coulomb interaction between beams.[39]

21.6 See also

- Photolithography

- Maskless lithography

- Ion beam lithography

- Electron beam technology

21.7   References

[1] McCord, M. A.; M. J. Rooks (2000). "2". *SPIE Handbook of Microlithography, Micromachining and Microfabrication*.

[2] Parker, N. W.; et al. (2000). "High-throughput NGL electron-beam direct-write lithography system". *Proc. SPIE* **3997**: 713. doi:10.1117/12.390042.

[3] Faster and lower cost for 65 nm and 45 nm photomask patterning

[4] M. L. Kempsell *et al.*, J. Microlith/Nanolith. MEMS MOEMS, vol. 8, 043001(2009).

[5] H. Sunaoshi *et al.*, Prof. SPIE vol. 6283, 628306 (2006).

[6] K. Ugajin *et al.*, Proc. SPIE vol. 6607, 66070A (2007).

[7] F. T. Chen *et al.*, Proc. SPIE vol. 8683, 868311 (2013).

[8] L. Feldman and J. Mayer (1986). *Fundamentals of Surface and Thin Film Analysis* **54**. pp. 130–133. ISBN 0-444-00989-2.

[9] Euronanochem. None. Retrieved on 2011-08-27.

[10] Stoffels, E; Stoffels, W W; Kroesen, G M W (2001). "Plasma chemistry and surface processes of negative ions". *Plasma Sources Science and Technology* **10** (2): 311. Bibcode:2001PSST...10..311S. doi:10.1088/0963-0252/10/2/321.

[11] Seah, M. P.; Dench. W. A. (1979). "Quantitative electron spectroscopy of surfaces: A standard data base for electron inelastic mean free paths in solids". *Surface and Interface Analysis* **1**: 2. doi:10.1002/sia.740010103.

[12] Tanuma, S.; Powell, C. J.; Penn, D. R. (1994). "Calculations of electron inelastic mean free paths. V. Data for 14 organic compounds over the 50–2000 eV range". *Surface and Interface Analysis* **21** (3): 165. doi:10.1002/sia.740210302.

[13] Broers, A. N.; et al. (1996). "Electron beam lithography—Resolution limits". *Microelectronic Engineering* **32**: 131–142. doi:10.1016/0167-9317(95)00368-1.

[14] K. W. Lee (2009). "Secondary electron generation in electron-beam-irradiated solids:resolution limits to nanolithography". *J. Kor. Phys. Soc.* **55** (4): 1720. Bibcode:2009JKPS...55.1720L. doi:10.3938/jkps.55.1720.

[15] SPIE Newsroom: Double exposure makes dense high-resolution diffractive optics. Spie.org (2009-11-03). Retrieved on 2011-08-27.

[16] Dapor, M.; et al. (2010). "Monte Carlo modeling in the low-energy domain of the secondary electron emission of polymethylmethacrylate for critical-dimension scanning electron microscopy". *J. Micro/Nanolith. MEMS MOEMS* **9**: 023001. doi:10.1117/1.3373517.

[17] P. T. Henderson; et al. (1999). "Long-distance charge transport in duplex DNA: The phonon-assisted polaron-like hopping mechanism". *Proc. Natl. Acad. Sci. U.S.A.* **96** (15): 8353–8358. Bibcode:1999PNAS...96.8353H. doi:10.1073/pnas.96.15.8353. PMC 17521. PMID 10411879.

[18] H. Seiler (1983). "Secondary electron emission in the scanning electron microscope". *J. Appl. Phys.* **54** (11): R1–R18. Bibcode:1983JAP....54R...1S. doi:10.1063/1.332840.

[19] G. Denbeaux *et al.*, 2013 International Workshop on EUV Lithography.

[20] J. A. Liddle; et al. (2003). "Resist Requirements and Limitations for Nanoscale Electron-Beam Patterning". *Mat. Res. Soc. Symp. Proc.* **739** (19): 19–30.

[21] Ivin, V (2002). "The inclusion of secondary electrons and Bremsstrahlung X-rays in an electron beam resist model". *Microelectronic Engineering*. **61–62**: 343. doi:10.1016/S0167-9317(02)00531-2.

[22] Yamazaki, Kenji; Kurihara, Kenji; Yamaguchi, Toru; Namatsu, Hideo; Nagase, Masao (1997). "Novel Proximity Effect Including Pattern-Dependent Resist Development in Electron Beam Nanolithography". *Japanese Journal of Applied Physics* **36**: 7552. Bibcode:1997JaJAP..36.7552Y. doi:10.1143/JJAP.36.7552.

[23] Renoud, R; Attard, C; Ganachaud, J-P; Bartholome, S; Dubus, A (1998). "Influence on the secondary electron yield of the space charge induced in an insulating target by an electron beam". *Journal of Physics: Condensed Matter* **10** (26): 5821. Bibcode:1998JPCM...10.5821R. doi:10.1088/0953-8984/10/26/010.

[24] J. N. Helbert et al., *Macromolecules*, vol. 11, 1104 (1978).

[25] M. J. Wieland *et al.*, Proc. SPIE vol. 7271, 72710O (2009)

[26] F. T. Chen *et al.*, Proc. SPIE vol. 8326, 83262L (2012)

[27] P. Kruit *et al.*, J. Vac. Sci. Tech. B 22, 2948 (2004).

[28] Bermudez, V. M. (1999). "Low-energy electron-beam effects on poly(methyl methacrylate) resist films". *Journal of Vacuum Science and Technology B* **17** (6): 2512. Bibcode:1999JVSTB..17.2512B. doi:10.1116/1.591134.

[29] H. Yang *et al.*, Proceedings of the 1st IEEE Intl. Conf. on Nano/Micro Engineered and Molecular Systems, pp. 391–394 (2006).

[30] Cumming, D. R. S.; Thoms, S.; Beaumont, S. P.; Weaver, J. M. R. (1996). "Fabrication of 3 nm wires using 100 keV electron beam lithography and poly(methyl methacrylate) resist". *Applied Physics Letters* **68** (3.) James Watt Nanofabrication Centre): 322. Bibcode:1996ApPhL..68..322C. doi:10.1063/1.116073.

[31] Manfrinato, Vitor R.; Zhang, Lihua; Su, Dong; Duan, Huigao; Hobbs, Richard G.; Stach, Eric A.; Berggren, Karl K. (2013). "Resolution limits of electron-beam lithography toward the atomic scale". *Nano Lett.* **13** (4): 1555–1558. doi:10.1021/nl304715p.

[32] C. R. K. Marrian (1992). "Electron-beam lithography with the scanning tunneling microscope". *Journal of Vacuum Science and Technology* **10** (B): 2877–2881. Bibcode:1992JVSTB..10.2877M. doi:10.1116/1.585978.

[33] T. M. Mayer; et al. (1996). "Field emission characteristics of the scanning tunneling microscope for nanolithography". *Journal of Vacuum Science and Technology* **14** (B): 2438–2444. Bibcode:1996JVSTB..14.2438M. doi:10.1116/1.588751.

[34] L. S. Hordon; et al. (1993). "Limits of low-energy electron optics". *Journal of Vacuum Science and Technology* **11** (B): 2299–2303. Bibcode:1993JVSTB..11.2299H. doi:10.1116/1.586894.

[35] Egerton, R. F.; et al. (2004). "Radiation damage in the TEM and SEM". *Micron* **35** (6): 399–409. doi:10.1016/j.micron.2004.02.003. PMID 15120123.

[36] Multibeam Corporation. Multibeamcorp.com (2011-03-04). Retrieved on 2011-08-27.

[37] Mapper Lithography. Mapper Lithography (2010-01-18). Retrieved on 2011-08-27.

[38] IMS Nanofabrications AG. IMS Nanofabrication AG (2011-12-07). Retrieved on 2012-01-15.

[39] M. L. Yu *et al.*, JVST B 23, 2589 (2005).

Chapter 22

Electron beam processing

For Electron beam, see Cathode ray.

Electron beam processing or **electron irradiation** is

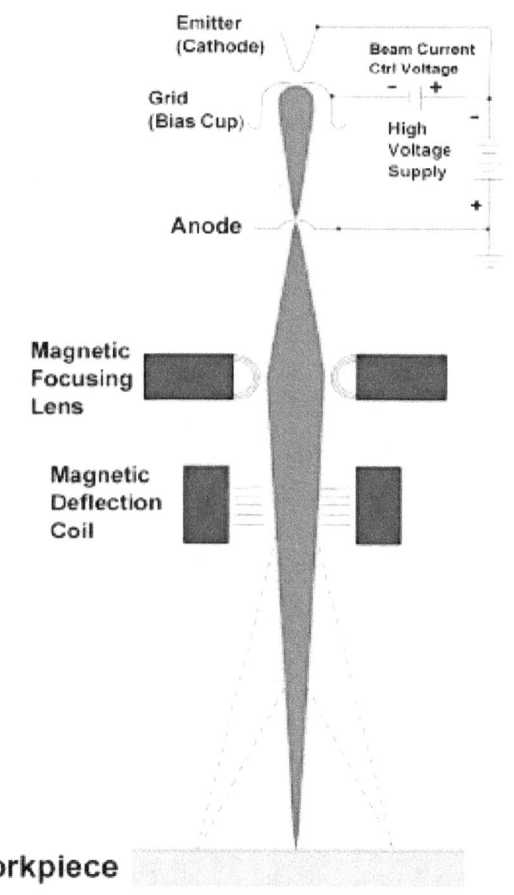

Schematic showing basic components and operation of electron beam materials processing

a process which involves using electrons, usually of high energy, to treat an object for a variety of purposes. This may take place under elevated temperatures and nitrogen atmosphere. Possible uses for electron irradiation include sterilization and to cross-link polymers.

Electron energies typically varies from the keV to MeV

range, depending on the depth of penetration required. The irradiation dose is usually measured in Gray but also in Mrads. Where 1 Gy is equivalent to 100 rad.

The basic components of a typical electron beam processing device are illustrated in the figure.[1] An electron gun (consisting of a cathode, grid, and anode) is used to generate and accelerate the primary beam. A magnetic optical (focusing and deflection) system is used for controlling the way in which the electron beam impinges on the material being processed (the "workpiece"). In operation, the gun cathode is the source of thermally-emitted electrons that are both accelerated and shaped into a collimated beam by the electrostatic field geometry established by the gun electrode (grid and anode) configuration used. The electron beam then emerges from the gun assembly through an exit hole in the ground-plane anode with an energy equal to the value of the negative high voltage (gun operating voltage) being applied to the cathode. This use of a direct high voltage to produce a high energy electron beam allows the conversion of input ac power to beam power at greater than 95% efficiency, making electron beam material processing a highly energy-efficient technique. After exiting the gun, the beam passes through an electromagnetic lens and deflection coil system. The lens is used for producing either a focused or defocused beam spot on the workpiece, while the deflection coil is used to either position the beam spot on a stationary location or provide some form of oscillatory motion.

In polymers, an electron beam may be used on the material to induce effects such as chain scission (which makes the polymer chain shorter) and cross linking. The result is a change in the properties of the polymer which is intended to extend the range of applications for the material. The effects of irradiation may also include changes in crystallinity as well as microstructure. Usually, the irradiation process degrades the polymer. The irradiated polymers may sometimes be characterized using DSC, XRD, FTIR, or SEM.

In poly(vinylidene fluoride-trifluoroethylene) copolymers, high-energy electron irradiation lowers the energy barrier for the ferroelectric-paraelectric phase transition and re-

217

duces polarization hysteresis losses in the material.[2]

Electron beam processing involves irradiation (treatment) of products using a high-energy electron beam accelerator. Electron beam accelerators utilize an on-off technology, with a common design being similar to that of a cathode ray television.

Electron beam processing is used in industry primarily for three product modifications:

- Crosslinking of polymer-based products to improve mechanical, thermal, chemical and other properties,

- Material degradation often used in the recycling of materials, and

- Sterilization of medical and pharmaceutical goods.[3]

Nanotechnology is one of the fastest growing new areas in science and engineering. Radiation is early applied tool in this area; arrangement of atoms and ions has been performed using ion or electron beams for many years. New applications concern nanocluster and nanocomposites synthesis.[4]

22.1   Crosslinking

The cross-linking of polymers through electron beam processing changes a thermoplastic material into a thermoset.[5] When polymers are crosslinked, the molecular movement is severely impeded, making the polymer stable against heat. This locking together of molecules is the origin of all of the benefits of crosslinking, including the improvement of the following properties:[6]

- Thermal: resistance to temperature, aging, low temperature impact, etc.

- Mechanical: tensile strength, modulus, abrasion resistance, pressure rating, creep resistance, etc.

- Chemical: stress crack resistance, etc.

- Other: heat shrink memory properties, positive temperature coefficient, etc.

Cross-Linking is the interconnection of adjacent long molecules with networks of bonds induced by chemical treatment or Electron Beam treatment. Electron Beam processing of thermoplastic material results in an array of enhancements, such as an increase in tensile strength, and resistance to abrasions, stress cracking and solvents. Joint replacements such as knees and hips are being manufactured

from Cross-Linked Ultra High Molecular Weight Polyethylene because of the excellent wear characteristics due to extensive research by the Harris Orthopaedics Lab.[7]

Polymers which are commonly crosslinked using the electron beam irradiation process include polyvinyl chloride (PVC), thermoplastic polyurethanes and elastomers (TPUs), polybutylene terephthalate (PBT), polyamides / nylon (PA66, PA6, PA11, PA12), polyvinylidene fluoride (PVDF), polymethylpentene (PMP), polyethylenes (LLDPE, LDPE, MDPE, HDPE, UHMWPE), and ethylene copolymers such as ethylene-vinyl acetate (EVA) and ethylene tetrafluoroethylene (ETFE). Some of the polymers utilize additives to make the polymer more readily irradiation crosslinkable.[8]

An example of an electron-beam crosslinked part is connector made from polyamide, designed to withstand the higher temperatures needed for soldering with the lead-free solder required by the RoHS initiative.[9]

Cross-linked polyethylene piping called PEX is commonly used as an alternative to copper piping for water lines in newer home construction. PEX piping will outlast copper and has performance characteristics that are superior to copper in many ways. [10]

Foam is also produced using electron beam processing to produce high quality, fine-celled, aesthetically pleasing product.[11][12]

22.2   Long-chain branching

The resin pellets used to produce the foam and thermoformed parts can be electron beam processed to a lower dose level than when crosslinking and gels occur. These resin pellets, such as polypropylene and polyethylene can be used to create lower density foams and other parts as the "melt strength" of the polymer is increased.[13]

22.3   Chain-scissioning

Chain scissioning or polymer degradation can also be achieved through electron beam processing. The effect of the electron beam can cause the degradation of polymers, breaking chains and therefore reducing the molecular weight. The chain scissioning effects observed in polytetrafluoroethylene (PTFE) have been used to created fine micropowders from scrap or off-grade materials.[3]

Chain Scission is the breaking apart of molecular chains to produce required molecular sub-units from the chain. Electron Beam processing provides Chain Scission without the use of harsh chemicals usually utilized to initiate Chain

Scission.

An example of this process is the breaking down of cellulose fibers extracted from wood in order to shorten the molecules, thereby producing a raw material that can then be used to produce biodegradable detergents and diet-food substitutes.

Teflon (PTFE) is also Electron Beam processed, allowing it to be ground to a fine powder for use in inks and as coatings for the automotive industry.*[14]

22.4 Microbiologal sterilization

Electron beam processing has the ability to break the chains of DNA in living organisms, such as bacteria, resulting in microbial death and rendering the space they inhabit sterile. E-beam processing has been used for the sterilization of medical products and aseptic packaging materials for foods as well as disinfestation, the elimination of live insects from grain, tobacco, and other unprocessed bulk crops.*[15]

Sterilization with electrons has significant advantages over other methods of sterilization currently in use. The process is quick, reliable, and compatible with most materials, and does not require any quarantine following the processing.*[16] For some materials and products that are sensitive to oxidative effects, radiation tolerance levels for electron beam irradiation may be slightly higher than for gamma exposure. This is due to the higher dose rates and shorter exposure times of e-beam irradiation which have been shown to reduce the degradative effects of oxygen.*[17]

22.5 Pest and pathogen control

Electron Beam processing as a disinfestation method replaces antiquated environmentally unfriendly methods such as fumigation and chemical dipping. A significant area for this technology is the herb and spice industry. These commodities are valued for their distinctive flavors, aromas and colors. They can be processed by this technology to reduce bacterial contamination without compromise to their sensory properties.

Fruits, vegetables, grains and other food items can be processed by Electron Beam to control fruit flies and other insects that use these commodities as a host for propagation. Suitable as a quarantine measure, several countries rely on this technology to treat food commodities prior to exporting.*[16]

22.6 Notes

[1] Hamm, Robert W.; Hamm, Marianne E. (2012). *Industrial Accelerators and Their Applications*. World Scientific. ISBN 978-981-4307-04-8.

[2] Cheng, Zhoung-Yang; V. Bharti, T. Mai, T.-B. Xu, Q. M. Zhang, et al. (Nov 2000). "Effect of High Energy Electron Irradiation on the Electromechanical Properties of Poly(vinylidene Fluoride-Trifluoroethylene) 50/50 and 65/35 Copolymers". *IEEE Transactions on Ultrasonics, Ferroelectrics and Frequency Control* (IEEE Ultrasonics, Ferroelectrics, and Frequency Control Society) **47** (6): 1296–1307. doi:10.1109/58.883518.

[3] Bly, J.H.: Electron Beam Processing. Yardley, PA: International Information Associates, 1988.

[4] Chmielewski, Andrzej G. (2006). "Worldwide developments in the field of radiation processing of materials in the down of 21st century" (PDF). *NUKLEONIKA* (Institute of Nuclear Chemistry and Technology) **51** (Supplement 1): S3–S9.

[5] Berejka, Anthony J.; Daniel Montoney; Marshall R. Cleland; Loïc Loiseau (2010). "Radiation curing: coatings and composites" (PDF). *NUKLEONIKA* (Institute of Nuclear Chemistry and Technology) **55** (1): 97–106.

[6] "Technology". E-BEAM.

[7] http://www.massgeneral.org/research/researchlab.aspx?id=1018

[8] "Fluorinated Polymers". BGS.

[9]

[10] "Cross-Linking". Iotron Industries: Electron Beam Sterilization Processing Services.

[11] (http://www.toraytpa.com/polyolefin-foams/technology)

[12] http://www.ebeamservices.com/release0407_c.htm

[13] http://www.ebeamservices.com/pdf/E-BEAM-Foam-Applications.pdf

[14] "Chain Scission". Iotron Industries: Electron Beam Sterilization Processing Services.

[15] Singh, A., Silverman, J., eds. *Radiation Processing of Polymers*. New York, NY: Oxford University Press, 1992.

[16] "Iotron Industries". Iotron Industries: Electron Beam Sterilization Processing Services.

[17] "Material Considerations: Irradiation Processing" (PDF). Sterigenics.

Chapter 23

Linear particle accelerator

"Linac" redirects here. For the commune in France, see Linac, Lot.

A **linear particle accelerator** (often shortened to **linac**)

The linac within the Australian Synchrotron uses radio waves from a series of RF cavities at the start of the linac to accelerate the electron beam in bunches to energies of 100 MeV.

is a type of particle accelerator that greatly increases the kinetic energy of charged subatomic particles or ions by subjecting the charged particles to a series of oscillating electric potentials along a linear beamline; this method of particle acceleration was invented by Leó Szilárd. It was

patented in 1928 by Rolf Wideröe,[1] who also built the first operational device and was influenced by a publication of Gustav Ising.[2]

Linacs have many applications: they generate X-rays and high energy electrons for medicinal purposes in radiation therapy, serve as particle injectors for higher-energy accelerators, and are used directly to achieve the highest kinetic energy for light particles (electrons and positrons) for particle physics.

The design of a linac depends on the type of particle that is being accelerated: electrons, protons or ions. Linacs range in size from a cathode ray tube (which is a type of linac) to the 3.2-kilometre-long (2.0 mi) linac at the SLAC National Accelerator Laboratory in Menlo Park, California.

23.1 Construction and operation

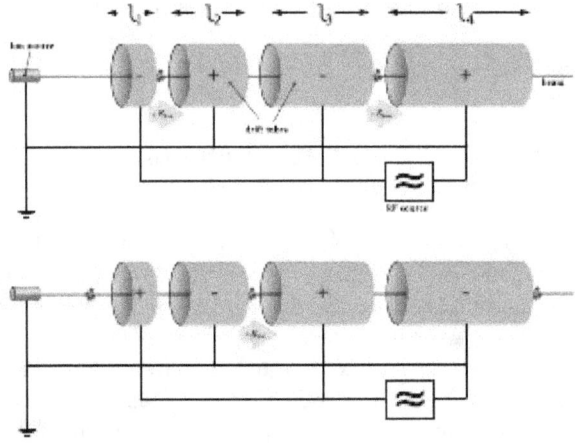

Schema of a linear accelerator

A linear particle accelerator consists of the following elements:

- The particle source. The design of the source depends on the particle that is being moved. Electrons are generated by a cold cathode, a hot cathode, a photocathode, or radio frequency (RF) ion sources. Protons are generated in an ion source, which can have many different designs. If heavier particles are to be accelerated, (e.g., uranium ions), a specialized ion source is needed.

- A high voltage source for the initial injection of particles.

- A hollow pipe vacuum chamber. The length will vary with the application. If the device is used for the production of X-rays for inspection or therapy the pipe may be only 0.5 to 1.5 meters long. If the device is to be an injector for a synchrotron it may be about ten meters long. If the device is used as the primary accelerator for nuclear particle investigations, it may be several thousand meters long.

- Within the chamber, electrically isolated cylindrical electrodes are placed, whose length varies with the distance along the pipe. The length of each electrode is determined by the frequency and power of the driving power source and the nature of the particle to be accelerated, with shorter segments near the source and longer segments near the target. The mass of the particle has a large effect on the length of the cylindrical electrodes; for example an electron is considerably lighter than a proton and so will generally require a much smaller section of cylindrical electrodes as it accelerates very quickly. Likewise, because its mass is so small, electrons have much less kinetic energy than protons at the same speed. Because of the possibility of electron emissions from highly charged surfaces, the voltages used in the accelerator have an upper limit, so this can't be as simple as just increasing voltage to match increased mass.

- One or more sources of radio frequency energy, used to energize the cylindrical electrodes. A very high power accelerator will use one source for each electrode. The sources must operate at precise power, frequency and phase appropriate to the particle type to be accelerated to obtain maximum device power.

- An appropriate target. If electrons are accelerated to produce X-rays then a water cooled tungsten target is used. Various target materials are used when protons or other nuclei are accelerated, depending upon the specific investigation. For particle-to-particle collision investigations the beam may be directed to a pair of storage rings, with the particles kept within the ring by magnetic fields. The beams may then be extracted from the storage rings to create head on particle collisions.

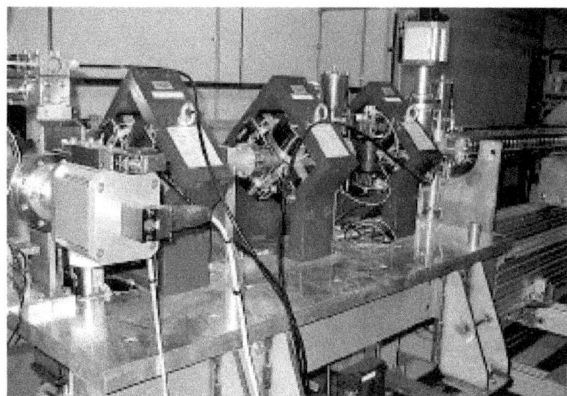

Quadrupole magnets surrounding the linac of the Australian Synchrotron are used to help focus the electron beam

As the particle bunch passes through the tube it is unaffected (the tube acts as a Faraday cage), while the frequency of the driving signal and the spacing of the gaps between electrodes are designed so that the maximum voltage differential appears as the particle crosses the gap. This accelerates the particle, imparting energy to it in the form of increased velocity. At speeds near the speed of light, the incremental velocity increase will be small, with the energy appearing as an increase in the mass of the particles. In portions of the accelerator where this occurs, the tubular electrode lengths will be almost constant.

- Additional magnetic or electrostatic lens elements may be included to ensure that the beam remains in the center of the pipe and its electrodes.

- Very long accelerators may maintain a precise alignment of their components through the use of servo systems guided by a laser beam.

23.2 Advantages

Linacs of appropriate design are capable of accelerating heavy ions to energies exceeding those available in ring-type accelerators, which are limited by the strength of the magnetic fields required to maintain the ions on a curved path. High power linacs are also being developed for production of electrons at relativistic speeds, required since fast electrons traveling in an arc will lose energy through synchrotron radiation; this limits the maximum power that can be imparted to electrons in a synchrotron of given size. Linacs are also capable of prodigious output, producing a nearly continuous stream of particles, whereas a synchrotron will only periodically raise the particles to sufficient energy to merit a "shot" at the target. (The burst can be held or stored in the ring at energy to give the experimen-

The Stanford University superconducting linear accelerator, housed on campus below the Hansen Labs until 2007. This facility is separate from SLAC

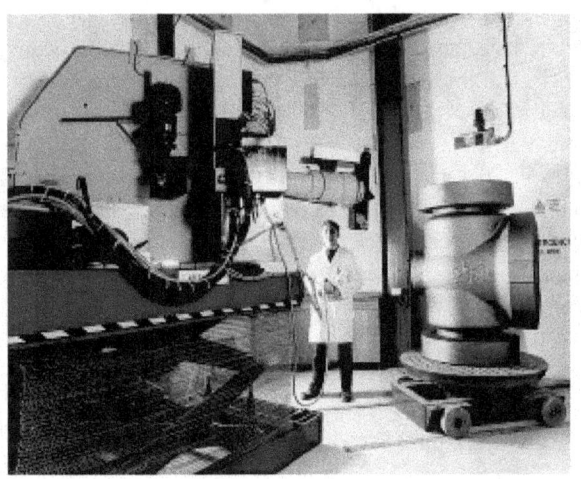

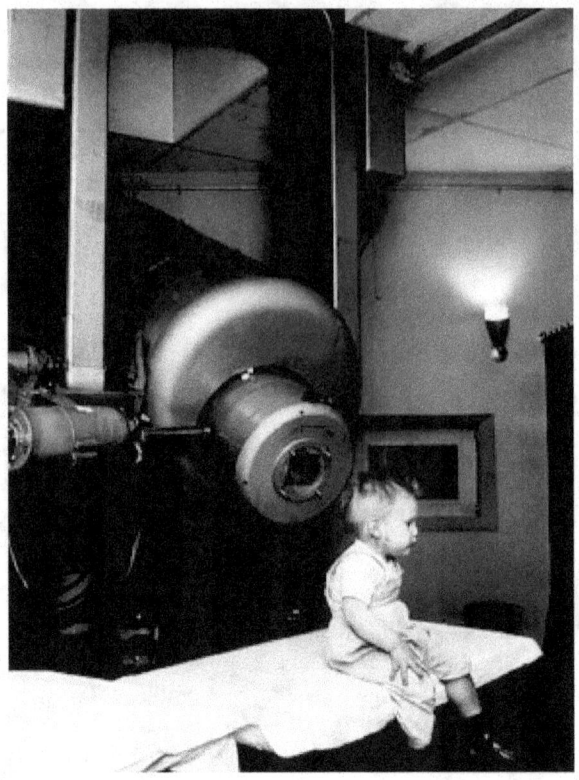

Historical image showing Gordon Isaacs, the first patient treated for retinoblastoma with linear accelerator radiation therapy (in this case an electron beam), in 1957, in the U.S. Other patients had been treated by linac for other diseases since 1953. Gordon's right eye was removed on January 11, 1957 because cancer had spread there. His left eye, however, had only a localized tumor that prompted Henry Kaplan to treat it with the electron beam.

Steel casting undergoing x-ray using the linear accelerator at Goodwin Steel Castings Ltd

tal electronics time to work, but the average output current is still limited.) The high density of the output makes the linac particularly attractive for use in loading storage ring facilities with particles in preparation for particle to particle collisions. The high mass output also makes the device practical for the production of antimatter particles, which are generally difficult to obtain, being only a small fraction of a target's collision products. These may then be stored and further used to study matter-antimatter annihilation.

23.2.1   Medical linacs

Linac-based radiation therapy for cancer therapy began with treatment of the first patient in 1953 in London at Hammersmith Hospital, with an 8 MV machine built by Metropolitan-Vickers, as the first dedicated medical linac.[3] A short while later in 1955, 6 MV linac ther-

apy from a different machine was being used in the United States.

Medical grade linacs accelerate electrons using a tuned-cavity waveguide, in which the RF power creates a standing wave. Some linacs have short, vertically mounted waveguides, while higher energy machines tend to have a horizontal, longer waveguide and a bending magnet to turn the beam vertically towards the patient. Medical linacs use monoenergetic electron beams between 4 and 25 MeV, giving an X-ray output with a spectrum of energies up to and including the electron energy when the electrons are directed at a high-density (such as tungsten) target. The electrons or X-rays can be used to treat both benign and malignant disease. The LINAC produces a reliable, flexible and accurate radiation beam. The versatility of LINAC is a potential advantage over cobalt therapy as a treatment tool. In addition, the device can simply be powered off when not in use; there is no source requiring heavy shielding – although the treatment room itself requires considerable shielding of the walls, doors, ceiling etc. to prevent escape of scattered radiation. Prolonged use of high powered (>18 MeV) machines can induce a significant amount of radiation within the metal parts of the head of the machine after power to the machine has been removed (i.e. they become an active source and the necessary precautions must be observed).

23.3 Application for Medical Isotope Development

The expected shortages with regard to Mo-99, and the technetium-99m medical isotope obtained from it, has also shed light onto linear accelerator technology to produce Mo-99 from non-enriched Uranium-235 through neutron bombardment. This would enable the medical isotope industry to manufacture this crucial isotope by a sub-critical process. The aging facilities, for example the Chalk River Laboratories in Ontario Canada, which still now produce most Mo-99 from highly enriched Uranium-235 could be replaced by this new process. In this way, the sub-critical loading of soluble uranium salts in heavy water with subsequent photo neutron bombardment and extraction of the target product, Mo-99, will be achieved.[4]

23.4 Disadvantages

- The device length limits the locations where one may be placed.

- A great number of driver devices and their associated power supplies are required, increasing the construction and maintenance expense of this portion.

- If the walls of the accelerating cavities are made of normally conducting material and the accelerating fields are large, the wall resistivity converts electric energy into heat quickly. On the other hand superconductors also need constant cooling to keep them below their critical temperature, and the accelerating fields are limited by quenches. Therefore, high energy accelerators such as SLAC, still the longest in the world (in its various generations), are run in short pulses, limiting the average current output and forcing the experimental detectors to handle data coming in short bursts.

23.5 See also

- Accelerator physics
- Beamline
- CERN

- Compact Linear Collider
- Dielectric wall accelerator
- Duoplasmatron
- Electromagnetism
- International Linear Collider
- KEK
- Los Alamos Neutron Science Center
- List of particles
- Particle accelerator
- Particle beam
- Particle physics
- Quadrupole magnet
- SLAC National Accelerator Laboratory
- Soreq Applied Research Accelerator Facility
- Superconducting radio frequency

23.6 References

[1] Wideroe, R. (17 December 1928). "Ueber Ein Neues Prinzip Zur Herstellung Hoher Spannungen". *Archiv fuer Elektronik und Uebertragungstechnik* **21** (4): 387.

[2] Ising, Gustav (1928). "Prinzip Einer Methode Zur Herstellung Von Kanalstrahlen Hoher Voltzahl". *Arkiv Fuer Matematik, Astronomi Och Fysik* **18** (4).

[3] LINAC-3. Advances in Medical Linear Accelerator Technology. ampi-nc.org

[4] Gahl and Flagg (2009).Solution Target Radioisotope Generator Technical Review. Subcritical Fission Mo99 Production. Retrieved 6 January 2013.

23.7 External links

- Linear Particle Accelerator (LINAC) Animation by Ionactive
- 2MV Tandetron linear particle accelerator in Ljubljana, Slovenia

Chapter 24

Particle accelerator

"Atom smasher" and "Supercollider" redirect here. For other uses, see Atom smasher (disambiguation) and Supercollider (disambiguation).

A **particle accelerator** is a device that uses

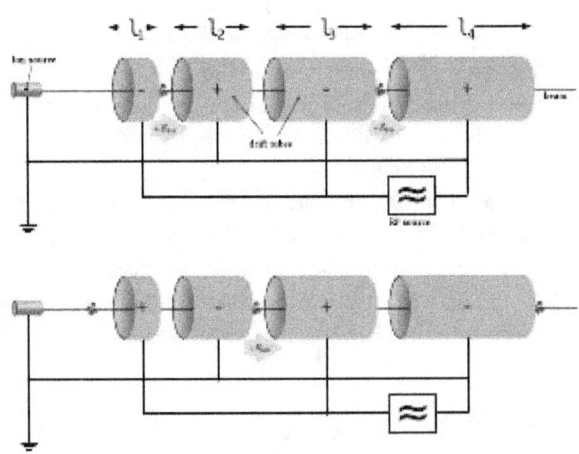

Sketch of the Ising/Widerøe linear accelerator concept, employing oscillating fields (1928)

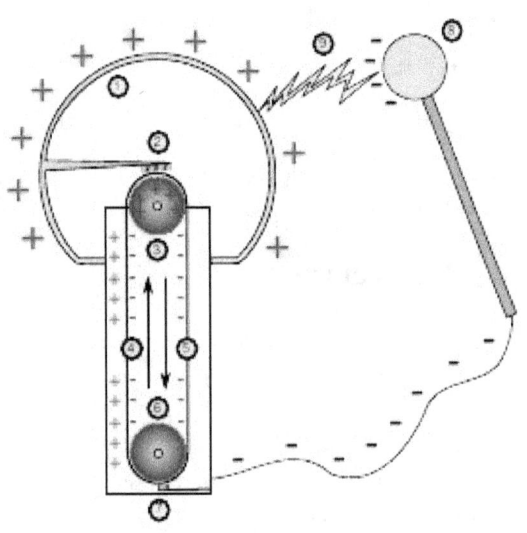

Sketch of an electrostatic Van de Graaff accelerator

electromagnetic fields to propel charged particles to high speeds and to contain them in well-defined beams.[1] Large accelerators are best known for their use in particle physics as colliders (e.g. the LHC at CERN, RHIC at Brookhaven National Laboratory, and Tevatron at Fermilab). Other kinds of particle accelerators are used in a large variety of applications, including particle therapy for oncological purposes, and as synchrotron light sources for the study of condensed matter physics. There are currently more than 30,000 accelerators in operation around the world.[2]

There are two basic classes of accelerators: electrostatic and oscillating field accelerators. *Electrostatic* accelerators use static electric fields to accelerate particles. A small-scale example of this class is the cathode ray tube in an ordinary old

television set. Other examples are the Cockcroft–Walton generator and the Van de Graaff generator. The achievable kinetic energy for particles in these devices is limited by electrical breakdown. *Oscillating field* accelerators, on the other hand, use radio frequency electromagnetic fields to accelerate particles, and circumvent the breakdown problem. This class, which was first developed in the 1920s, is the basis for all modern accelerator concepts and large-scale facilities.

Rolf Wideröe, Gustav Ising, Leó Szilárd, Donald Kerst, and Ernest Lawrence are considered pioneers of this field, conceiving and building the first operational linear particle accelerator,[3] the betatron, and the cyclotron.

Because colliders can give evidence of the structure of the subatomic world, accelerators were commonly referred to as **atom smashers** in the 20th century.[4] Despite the fact that most accelerators (but not ion facilities) actually propel subatomic particles, the term persists in popular usage when referring to particle accelerators in general.[5][6][7]

24.1   Uses

Beamlines leading from the Van de Graaff accelerator to various experiments, in the basement of the Jussieu Campus in Paris.

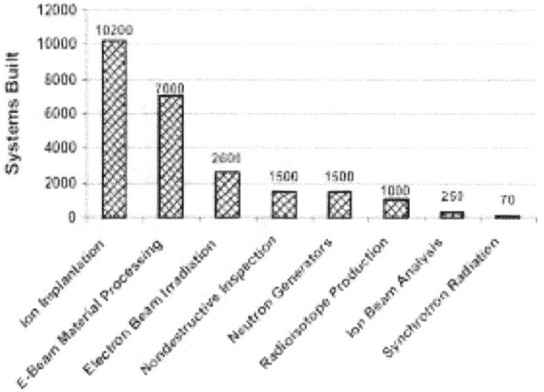

Breakdown of the cumulative number of industrial particle accelerators according to their applications.

The now disused Koffler particle accelerator at the Weizmann Institute, Rehovot, Israel.

Beams of high-energy particles are useful for both funda-

mental and applied research in the sciences, and also in many technical and industrial fields unrelated to fundamental research. It has been estimated that there are approximately 30,000 accelerators worldwide. Of these, only about 1% are research machines with energies above 1 GeV, while about 44% are for radiotherapy, 41% for ion implantation, 9% for industrial processing and research, and 4% for biomedical and other low-energy research.[8] The bar graph shows the breakdown of the number of industrial accelerators according to their applications. The numbers are based on 2012 statistics available from various sources, including production and sales data published in presentations or market surveys, and data provided by a number of manufacturers.[9]

24.1.1   High-energy physics

The largest particle accelerators with the highest particle energies are the Relativistic Heavy Ion Collider (RHIC) at Brookhaven National Laboratory and the Large Hadron Collider (LHC) at CERN (which came on-line in mid-November 2009).[10][11][12] These accelerators are used for experimental particle physics.

For the most basic inquiries into the dynamics and structure of matter, space, and time, physicists seek the simplest kinds of interactions at the highest possible energies. These typically entail particle energies of many GeV, and the interactions of the simplest kinds of particles: leptons (e.g. electrons and positrons) and quarks for the matter, or photons and gluons for the field quanta. Since isolated quarks are experimentally unavailable due to color confinement, the simplest available experiments involve the interactions of, first, leptons with each other, and second, of leptons with nucleons, which are composed of quarks and gluons. To study the collisions of quarks with each other, scientists resort to collisions of nucleons, which at high energy may be usefully considered as essentially 2-body interactions of the quarks and gluons of which they are composed. Thus elementary particle physicists tend to use machines creating beams of electrons, positrons, protons, and antiprotons, interacting with each other or with the simplest nuclei (e.g., hydrogen or deuterium) at the highest possible energies, generally hundreds of GeV or more. Nuclear physicists and cosmologists may use beams of bare atomic nuclei, stripped of electrons, to investigate the structure, interactions, and properties of the nuclei themselves, and of condensed matter at extremely high temperatures and densities, such as might have occurred in the first moments of the Big Bang. These investigations often involve collisions of heavy nuclei – of atoms like iron or gold – at energies of several GeV per nucleon.

Particle accelerators can also produce proton beams, which

can produce proton-rich medical or research isotopes as opposed to the neutron-rich ones made in fission reactors; however, recent work has shown how to make ^{99}Mo, usually made in reactors, by accelerating isotopes of hydrogen,[13] although this method still requires a reactor to produce tritium. An example of this type of machine is LANSCE at Los Alamos.

24.1.2   Synchrotron radiation

Besides being of fundamental interest, high energy electrons may be coaxed into emitting extremely bright and coherent beams of high energy photons via synchrotron radiation, which have numerous uses in the study of atomic structure, chemistry, condensed matter physics, biology, and technology. Examples include the ESRF in Grenoble, France, which has recently been used to extract detailed 3-dimensional images of insects trapped in amber.[14] Thus there is a great demand for electron accelerators of moderate (GeV) energy and high intensity.

24.1.3   Low-energy machines and particle therapy

Everyday examples of particle accelerators are cathode ray tubes found in television sets and X-ray generators. These low-energy accelerators use a single pair of electrodes with a DC voltage of a few thousand volts between them. In an X-ray generator, the target itself is one of the electrodes. A low-energy particle accelerator called an ion implanter is used in the manufacture of integrated circuits.

At lower energies, beams of accelerated nuclei are also used in medicine as particle therapy, for the treatment of cancer.

DC accelerator types capable of accelerating particles to speeds sufficient to cause nuclear reactions are Cockcroft-Walton generators or voltage multipliers, which convert AC to high voltage DC, or Van de Graaff generators that use static electricity carried by belts.

24.2   Electrostatic particle accelerators

Main article: Electrostatic nuclear accelerator

Historically, the first accelerators used simple technology of a single static high voltage to accelerate charged particles. The charged particle was accelerated through an evacuated tube with an electrode at either end, with the static potential across it. Since the particle passed only once through the potential difference, the output energy was limited to

A Cockcroft-Walton generator (Philips, 1937), residing in Science Museum (London).

A 1960s single stage 2 MeV linear Van de Graaff accelerator, here opened for maintenance

the accelerating voltage of the machine. While this method is still extremely popular today, with the electrostatic accelerators greatly out-numbering any other type, they are more suited to lower energy studies owing to the practical voltage limit of about 1 MV for air insulated machines, or 30 MV when the accelerator is operated in a tank of pressurized gas with high dielectric strength, such as sulfur hexafluoride. In a *tandem accelerator* the potential is used twice to accelerate the particles, by reversing the charge of the particles while they are inside the terminal. This is possible with the acceleration of atomic nuclei by using anions (negatively charged ions), and then passing the beam through a thin foil to strip electrons off the anions inside the high voltage terminal, converting them to cations (positively charged ions), which are accelerated again as they leave the terminal.

The two main types of electrostatic accelerator are the Cockcroft-Walton accelerator, which uses a diode-capacitor voltage multiplier to produce high voltage, and the Van de Graaff accelerator, which uses a moving fabric belt to carry charge to the high voltage electrode. Although electrostatic accelerators accelerate particles along a straight line, the term linear accelerator is more often used for accelerators that employ oscillating rather than static electric fields.

24.3 Oscillating field particle accelerators

Due to the high voltage ceiling imposed by electrical discharge, in order to accelerate particles to higher energies, techniques involving more than one lower, but oscillating, high voltage sources are used. The electrodes can either be arranged to accelerate particles in a line or circle, depending on whether the particles are subject to a magnetic field while they are accelerated, causing their trajectories to arc.

24.3.1 Linear particle accelerators

Main article: Linear particle accelerator
In a linear particle accelerator (linac), particles are accelerated in a straight line with a target of interest at one end. They are often used to provide an initial low-energy kick to particles before they are injected into circular accelerators. The longest linac in the world is the Stanford Linear Accelerator, SLAC, which is 3 km (1.9 mi) long. SLAC is an electron-positron collider.

Linear high-energy accelerators use a linear array of plates (or drift tubes) to which an alternating high-energy field is applied. As the particles approach a plate they are accelerated towards it by an opposite polarity charge applied to the plate. As they pass through a hole in the plate, the polarity is switched so that the plate now repels them and they are now

Modern superconducting radio frequency, multicell linear accelerator component.

accelerated by it towards the next plate. Normally a stream of "bunches" of particles are accelerated, so a carefully controlled AC voltage is applied to each plate to continuously repeat this process for each bunch.

As the particles approach the speed of light the switching rate of the electric fields becomes so high that they operate at radio frequencies, and so microwave cavities are used in higher energy machines instead of simple plates.

Linear accelerators are also widely used in medicine, for radiotherapy and radiosurgery. Medical grade linacs accelerate electrons using a klystron and a complex bending magnet arrangement which produces a beam of 6-30 MeV energy. The electrons can be used directly or they can be collided with a target to produce a beam of X-rays. The reliability, flexibility and accuracy of the radiation beam produced has largely supplanted the older use of cobalt-60 therapy as a treatment tool.

24.3.2   Circular or cyclic accelerators

In the circular accelerator, particles move in a circle until they reach sufficient energy. The particle track is typically bent into a circle using electromagnets. The advantage of circular accelerators over linear accelerators (*linacs*) is that the ring topology allows continuous acceleration, as the particle can transit indefinitely. Another advantage is that a circular accelerator is smaller than a linear accelerator of comparable power (i.e. a linac would have to be extremely long to have the equivalent power of a circular accelerator).

Depending on the energy and the particle being accelerated, circular accelerators suffer a disadvantage in that the particles emit synchrotron radiation. When any charged particle is accelerated, it emits electromagnetic radiation and secondary emissions. As a particle traveling in a circle is always accelerating towards the center of the circle, it continuously radiates towards the tangent of the circle. This radiation is called synchrotron light and depends highly on the mass of the accelerating particle. For this reason, many high energy electron accelerators are linacs. Certain accelerators (synchrotrons) are however built specially for producing synchrotron light (X-rays).

Since the special theory of relativity requires that matter always travels slower than the speed of light in a vacuum, in high-energy accelerators, as the energy increases the particle speed approaches the speed of light as a limit, but never attains it. Therefore, particle physicists do not generally think in terms of speed, but rather in terms of a particle's energy or momentum, usually measured in electron volts (eV). An important principle for circular accelerators, and particle beams in general, is that the curvature of the particle trajectory is proportional to the particle charge and to the magnetic field, but inversely proportional to the (typically relativistic) momentum.

Cyclotrons

Main article: Cyclotron

The earliest operational circular accelerators were cyclotrons, invented in 1929 by Ernest O. Lawrence at the University of California, Berkeley. Cyclotrons have a single pair of hollow 'D'-shaped plates to accelerate the particles and a single large dipole magnet to bend their path into a circular orbit. It is a characteristic property of charged particles in a uniform and constant magnetic field B that they orbit with a constant period, at a frequency called the cyclotron frequency, so long as their speed is small compared to the speed of light c. This means that the accelerating D's of a cyclotron can be driven at a constant frequency by a radio frequency (RF) accelerating power

Lawrence's 60 inch cyclotron, with magnet poles 60 inches (5 feet, 1.5 meters) in diameter, at the University of California Lawrence Radiation Laboratory, Berkeley, in August, 1939, the most powerful accelerator in the world at the time. Glenn T. Seaborg and Edwin M. McMillan (right) used it to discover plutonium, neptunium and many other transuranic elements and isotopes, for which they received the 1951 Nobel Prize in chemistry.

source, as the beam spirals outwards continuously. The particles are injected in the centre of the magnet and are extracted at the outer edge at their maximum energy.

Cyclotrons reach an energy limit because of relativistic effects whereby the particles effectively become more massive, so that their cyclotron frequency drops out of synch with the accelerating RF. Therefore, simple cyclotrons can accelerate protons only to an energy of around 15 million electron volts (15 MeV, corresponding to a speed of roughly 10% of c), because the protons get out of phase with the driving electric field. If accelerated further, the beam would continue to spiral outward to a larger radius but the particles would no longer gain enough speed to complete the larger circle in step with the accelerating RF. To accommodate relativistic effects the magnetic field needs to be increased to higher radii like it is done in isochronous cyclotrons.for An example of an isochronous cyclotron is the PSI Ring cyclotron in Switzerland, which provides protons at the energy of 590 MeV which corresponds to roughly 80% of the speed of light. The advantage of such a cyclotron is the maximum achievable extracted proton current which is currently 2.2 mA. The energy and current correspond to 1.3 MW beam power which is the highest of any accelerator currently existing.

Synchrocyclotrons and isochronous cyclotrons

Main articles: Synchrocyclotron and Isochronous cyclotron

A classic cyclotron can be modified to increase its

A magnet in the synchrocyclotron at the Orsay proton therapy center

energy limit.   The historically first approach was the synchrocyclotron, which accelerates the particles in bunches. It uses a constant magnetic field B, but reduces the accelerating field's frequency so as to keep the particles in step as they spiral outward, matching their mass-dependent cyclotron resonance frequency. This approach suffers from low average beam intensity due to the bunching, and again from the need for a huge magnet of large radius and constant field over the larger orbit demanded by high energy.

The second approach to the problem of accelerating relativistic particles is the isochronous cyclotron. In such a structure, the accelerating field's frequency (and the cyclotron resonance frequency) is kept constant for all energies by shaping the magnet poles so to increase magnetic field with radius. Thus, all particles get accelerated in isochronous time intervals. Higher energy particles travel a shorter distance in each orbit than they would in a classical cyclotron, thus remaining in phase with the accelerating field. The advantage of the isochronous cyclotron is that it can deliver continuous beams of higher average intensity, which is useful for some applications. The main disadvantages are the size and cost of the large magnet needed, and the difficulty in achieving the high magnetic field values required at the outer edge of the structure.

Synchrocyclotrons have not been built since the isochronous cyclotron was developed.

Betatrons

Main article: Betatron

Another type of circular accelerator, invented in 1940 for accelerating electrons, is the Betatron, a concept which originates ultimately from Norwegian-German scientist Rolf

Widerøe. These machines, like synchrotrons, use a donut-shaped ring magnet (see below) with a cyclically increasing B field, but accelerate the particles by induction from the increasing magnetic field, as if they were the secondary winding in a transformer, due to the changing magnetic flux through the orbit.[15]

Achieving constant orbital radius while supplying the proper accelerating electric field requires that the magnetic flux linking the orbit be somewhat independent of the magnetic field on the orbit, bending the particles into a constant radius curve. These machines have in practice been limited by the large radiative losses suffered by the electrons moving at nearly the speed of light in a relatively small radius orbit.

Synchrotrons

Main article: Synchrotron
To reach still higher energies, with relativistic mass ap-

Aerial photo of the Tevatron at Fermilab, which resembles a figure eight. The main accelerator is the ring above; the one below (about half the diameter, despite appearances) is for preliminary acceleration, beam cooling and storage, etc.

proaching or exceeding the rest mass of the particles (for protons, billions of electron volts or GeV), it is necessary to use a synchrotron. This is an accelerator in which the particles are accelerated in a ring of constant radius. An immediate advantage over cyclotrons is that the magnetic field need only be present over the actual region of the particle orbits, which is much narrower than that of the ring. (The largest cyclotron built in the US had a 184-inch-diameter (4.7 m) magnet pole, whereas the diameter of synchrotrons such as the LEP and LHC is nearly 10 km. The aperture of the two beams of the LHC is of the order of a millimeter.)

However, since the particle momentum increases during acceleration, it is necessary to turn up the magnetic field B in proportion to maintain constant curvature of the orbit. In consequence, synchrotrons cannot accelerate particles continuously, as cyclotrons can, but must operate cyclically, supplying particles in bunches, which are delivered to a target or an external beam in beam "spills" typically every few seconds.

Since high energy synchrotrons do most of their work on particles that are already traveling at nearly the speed of light c, the time to complete one orbit of the ring is nearly constant, as is the frequency of the RF cavity resonators used to drive the acceleration.

In modern synchrotrons, the beam aperture is small and the magnetic field does not cover the entire area of the particle orbit as it does for a cyclotron, so several necessary functions can be separated. Instead of one huge magnet, one has a line of hundreds of bending magnets, enclosing (or enclosed by) vacuum connecting pipes. The design of synchrotrons was revolutionized in the early 1950s with the discovery of the strong focusing concept.[16][17][18] The focusing of the beam is handled independently by specialized quadrupole magnets, while the acceleration itself is accomplished in separate RF sections, rather similar to short linear accelerators. Also, there is no necessity that cyclic machines be circular, but rather the beam pipe may have straight sections between magnets where beams may collide, be cooled, etc. This has developed into an entire separate subject, called "beam physics" or "beam optics".[19]

More complex modern synchrotrons such as the Tevatron, LEP, and LHC may deliver the particle bunches into storage rings of magnets with constant B, where they can continue to orbit for long periods for experimentation or further acceleration. The highest-energy machines such as the Tevatron and LHC are actually accelerator complexes, with a cascade of specialized elements in series, including linear accelerators for initial beam creation, one or more low energy synchrotrons to reach intermediate energy, storage rings where beams can be accumulated or "cooled" (reducing the magnet aperture required and permitting tighter focusing; see beam cooling), and a last large ring for final acceleration and experimentation.

Electron synchrotrons   See also: Synchrotron light source

Circular electron accelerators fell somewhat out of favor for particle physics around the time that SLAC's linear particle accelerator was constructed, because their synchrotron losses were considered economically prohibitive and because their beam intensity was lower than for the unpulsed linear machines. The Cornell Electron Synchrotron, built at low cost in the late 1970s, was the first in a series of high-energy circular electron accelerators built for fundamental particle physics, the last being LEP, built at CERN, which

Segment of an electron synchrotron at DESY

was used from 1989 until 2000.

A large number of electron synchrotrons have been built in the past two decades, as part of synchrotron light sources that emit ultraviolet light and X rays; see below.

Storage rings

Main article: Storage ring

For some applications, it is useful to store beams of high energy particles for some time (with modern high vacuum technology, up to many hours) without further acceleration. This is especially true for colliding beam accelerators, in which two beams moving in opposite directions are made to collide with each other, with a large gain in effective collision energy. Because relatively few collisions occur at each pass through the intersection point of the two beams, it is customary to first accelerate the beams to the desired energy, and then store them in storage rings, which are essentially synchrotron rings of magnets, with no significant RF power for acceleration.

Synchrotron radiation sources

Main article: Synchrotron light sources

Some circular accelerators have been built to deliberately generate radiation (called synchrotron light) as X-rays also called synchrotron radiation, for example the Diamond Light Source which has been built at the Rutherford Appleton Laboratory in England or the Advanced Photon Source at Argonne National Laboratory in Illinois, USA. High-energy X-rays are useful for X-ray spectroscopy of proteins or X-ray absorption fine structure (XAFS), for example.

Synchrotron radiation is more powerfully emitted by lighter

particles, so these accelerators are invariably electron accelerators. Synchrotron radiation allows for better imaging as researched and developed at SLAC's SPEAR.

FFAG accelerators

Main article: FFAG accelerator

Fixed-Field Alternating Gradient accelerators (FFAG)s, in which a very strong radial field gradient, combined with strong focusing, allows the beam to be confined to a narrow ring, are an extension of the isochronous cyclotron idea that is lately under development.[20] They use RF accelerating sections between the magnets, and so are isochronous for relativistic particles like electrons (which achieve essentially the speed of light at only a few MeV), but only over a limited energy range for protons and heavier particles at sub-relativistic energies. Like the isochronous cyclotrons, they achieve continuous beam operation, but without the need for a huge dipole bending magnet covering the entire radius of the orbits.

History

Main article: List of accelerators in particle physics

Alex Arango's first cyclotron was a mere 4 inches (100 mm) in diameter. Later, in 1939, he built a machine with a 60-inch diameter pole face, and planned one with a 184-inch diameter in 1942, which was, however, taken over for World War II-related work connected with uranium isotope separation; after the war it continued in service for research and medicine over many years.

The first large proton synchrotron was the Cosmotron at Brookhaven National Laboratory, which accelerated protons to about 3 GeV (1953–1968). The Bevatron at Berkeley, completed in 1954, was specifically designed to accelerate protons to sufficient energy to create antiprotons, and verify the particle-antiparticle symmetry of nature, then only theorized. The Alternating Gradient Synchrotron (AGS) at Brookhaven (1960–) was the first large synchrotron with alternating gradient, "strong focusing" magnets, which greatly reduced the required aperture of the beam, and correspondingly the size and cost of the bending magnets. The Proton Synchrotron, built at CERN (1959–), was the first major European particle accelerator and generally similar to the AGS.

The Stanford Linear Accelerator, SLAC, became operational in 1966, accelerating electrons to 30 GeV in a 3 km long waveguide, buried in a tunnel and powered by hundreds of large klystrons. It is still the largest linear accelerator in existence, and has been upgraded with the addition of storage rings and an electron-positron collider facility. It is also an X-ray and UV synchrotron photon source.

The Fermilab Tevatron has a ring with a beam path of 4 miles (6.4 km). It has received several upgrades, and has functioned as a proton-antiproton collider until it was shut down due to budget cuts on September 30, 2011. The largest circular accelerator ever built was the LEP synchrotron at CERN with a circumference 26.6 kilometers, which was an electron/positron collider. It achieved an energy of 209 GeV before it was dismantled in 2000 so that the underground tunnel could be used for the Large Hadron Collider (LHC). The LHC is a proton collider, and currently the world's largest and highest-energy accelerator, expected to achieve 14 TeV energy per beam, and currently operating at half that.

The aborted Superconducting Super Collider (SSC) in Texas would have had a circumference of 87 km. Construction was started in 1991, but abandoned in 1993. Very large circular accelerators are invariably built in underground tunnels a few metres wide to minimize the disruption and cost of building such a structure on the surface, and to provide shielding against intense secondary radiations that occur, which are extremely penetrating at high energies.

Current accelerators such as the Spallation Neutron Source, incorporate superconducting cryomodules. The Relativistic Heavy Ion Collider, and Large Hadron Collider also make use of superconducting magnets and RF cavity resonators to accelerate particles.

24.4   Targets and detectors

The output of a particle accelerator can generally be directed towards multiple lines of experiments, one at a given time, by means of a deviating electromagnet. This makes it possible to operate multiple experiments without needing to move things around or shutting down the entire accelerator beam. Except for synchrotron radiation sources, the purpose of an accelerator is to generate high-energy particles for interaction with matter.

This is usually a fixed target, such as the phosphor coating on the back of the screen in the case of a television tube; a piece of uranium in an accelerator designed as a neutron source; or a tungsten target for an X-ray generator. In a linac, the target is simply fitted to the end of the accelerator. The particle track in a cyclotron is a spiral outwards from the centre of the circular machine, so the accelerated particles emerge from a fixed point as for a linear accelerator.

For synchrotrons, the situation is more complex. Particles are accelerated to the desired energy. Then, a fast acting dipole magnet is used to switch the particles out of the circular synchrotron tube and towards the target.

A variation commonly used for particle physics research is a collider, also called a *storage ring collider*. Two circular synchrotrons are built in close proximity – usually on top of each other and using the same magnets (which are then of more complicated design to accommodate both beam tubes). Bunches of particles travel in opposite directions around the two accelerators and collide at intersections between them. This can increase the energy enormously; whereas in a fixed-target experiment the energy available to produce new particles is proportional to the square root of the beam energy, in a collider the available energy is linear.

24.5   Higher energies

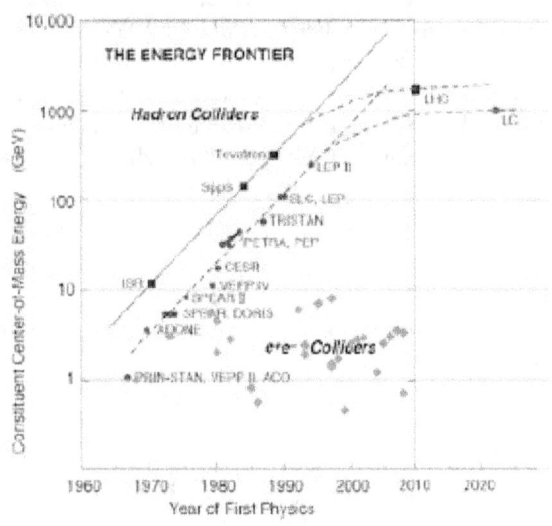

A Livingston chart depicting progress in collision energy through 2010. The LHC is the largest collision energy to date, but also represents the first break in the log-linear trend.

At present the highest energy accelerators are all circular colliders, but both hadron accelerators and electron accelerators are running into limits. Higher energy hadron and ion cyclic accelerators will require accelerator tunnels of larger physical size due to the increased beam rigidity.

For cyclic electron accelerators, a limit on practical bend radius is placed by synchrotron radiation losses and the next generation will probably be linear accelerators 10 times the current length. An example of such a next generation electron accelerator is the 40 km long International Linear Collider, due to be constructed between 2015–2020.

It is believed that plasma wakefield acceleration in the form of electron-beam 'afterburners' and standalone laser pulsers might be able to provide dramatic increases in ef-

ficiency over RF accelerators within two to three decades. In plasma wakefield accelerators, the beam cavity is filled with a plasma (rather than vacuum). A short pulse of electrons or laser light either constitutes or immediately trails the particles that are being accelerated. The pulse disrupts the plasma, causing the charged particles in the plasma to integrate into and move toward the rear of the bunch of particles that are being accelerated. This process transfers energy to the particle bunch, accelerating it further, and continues as long as the pulse is coherent.[21]

Energy gradients as steep as 200 GeV/m have been achieved over millimeter-scale distances using laser pulsers[22] and gradients approaching 1 GeV/m are being produced on the multi-centimeter-scale with electron-beam systems, in contrast to a limit of about 0.1 GeV/m for radio-frequency acceleration alone. Existing electron accelerators such as SLAC could use electron-beam afterburners to greatly increase the energy of their particle beams, at the cost of beam intensity. Electron systems in general can provide tightly collimated, reliable beams; laser systems may offer more power and compactness. Thus, plasma wakefield accelerators could be used – if technical issues can be resolved – to both increase the maximum energy of the largest accelerators and to bring high energies into university laboratories and medical centres.

Higher than 0.25 GeV/m gradients have been achieved by a dielectric laser accelerator, which may present another viable approach to building compact high-energy accelerators.[23]

24.5.1   Black hole production and public safety concerns

See also: Safety of high energy particle collision experiments

In the future, the possibility of black hole production at the highest energy accelerators may arise if certain predictions of superstring theory are accurate.[24][25] This and other exotic possibilities have led to public safety concerns that have been widely reported in connection with the LHC, which began operation in 2008. The various possible dangerous scenarios have been assessed as presenting "no conceivable danger" in the latest risk assessment produced by the LHC Safety Assessment Group.[26] If black holes are produced, it is theoretically predicted that such small black holes should evaporate extremely quickly via Bekenstein-Hawking radiation, but which is as yet experimentally unconfirmed. If colliders can produce black holes, cosmic rays (and particularly ultra-high-energy cosmic rays, UHECRs) must have been producing them for eons, but they have yet to harm anybody.[27] It has been argued that to conserve energy and momentum, any black holes created in a collision between an UHECR and local matter would necessarily be produced moving at relativistic speed with respect to the Earth, and should escape into space, as their accretion and growth rate should be very slow, while black holes produced in colliders (with components of equal mass) would have some chance of having a velocity less than Earth escape velocity, 11.2 km per sec, and would be liable to capture and subsequent growth. Yet even on such scenarios the collisions of UHECRs with white dwarfs and neutron stars would lead to their rapid destruction, but these bodies are observed to be common astronomical objects. Thus if stable micro black holes should be produced, they must grow far too slowly to cause any noticeable macroscopic effects within the natural lifetime of the solar system." [26]

24.6 See also

- Accelerator physics
- Atom smasher (disambiguation)
- Dielectric wall accelerator
- Nuclear transmutation
- List of accelerators in particle physics
- Rolf Widerøe

The idea of a particle accelerator has also been used in television shows such as *The Flash*.

24.7 References

[1] Livingston, M. S.; Blewett, J. (1969). *Particle Accelerators*. New York: McGraw-Hill. ISBN 1-114-44384-0.

[2] Witman, Sarah. "Ten things you might not know about particle accelerators". *Symmetry Magazine*. Fermi National Accelerator Laboratory. Retrieved 21 April 2014.

[3] Pedro Waloschek (ed.): *The Infancy of Particle Accelerators: Life and Work of Rolf Widerøe*, Vieweg, 1994

[4] "Six Million Volt Atom Smasher Creates New Elements". *Popular Mechanics*: 580. April 1935.

[5] Higgins, A. G. (December 18, 2009). "Atom Smasher Preparing 2010 New Science Restart". U.S. News & World Report.

[6] Cho, A. (June 2, 2006). "Aging Atom Smasher Runs All Out in Race for Most Coveted Particle". *Science* 312 (5778): 1302. doi:10.1126/science.312.5778.1302.

[7] "Atom smasher". *American Heritage Science Dictionary*. Houghton Mifflin Harcourt. 2005. p. 49. ISBN 978-0-618-45504-1.

[8] Feder, T. (2010). "Accelerator school travels university circuit" (PDF). *Physics Today* 63 (2): 20. Bibcode:2010PhT....63b..20F. doi:10.1063/1.3326981.

[9] Hamm, Robert W.; Hamm, Marianne E. (2012). *Industrial Accelerators and Their Applications*. World Scientific. ISBN 978-981-4307-04-8.

[10] "CERN management confirms new LHC restart schedule" (Press release). CERN Press Office. February 9, 2009. Retrieved 2009-02-10.

[11] "CERN reports on progress towards LHC restart" (Press release). CERN Press Office. June 19, 2009. Retrieved 2009-07-21.

[12] "Two circulating beams bring first collisions in the LHC" (Press release). CERN Press Office. November 23, 2009. Retrieved 2009-11-23.

[13] Nagai, Y.; Hatsukawa, Y. (2009). "Production of ^{99}Mo for Nuclear Medicine by ^{100}Mo$(n,2n)^{99}$Mo". *Journal of the Physical Society of Japan* 78 (3): 033201. Bibcode:2009JPSJ...78c3201N. doi:10.1143/JPSJ.78.033201.

[14] Amos, J. (April 1, 2008). "Secret 'dino bugs' revealed". BBC News. Retrieved 2008-09-11.

[15] Chao, A. W.; Mess, K. H.; Tigner, M.; et al., eds. (2013). *Handbook of Accelerator Physics and Engineering* (2nd ed.). World Scientific. ISBN 978-981-4417-17-4.

[16] Courant, E. D.; Livingston, M. S.; Snyder, H. S. (1952). "The Strong-Focusing Synchrotron — A New High Energy Accelerator". *Physical Review* 88 (5): 1190–1196. Bibcode:1952PhRv...88.1190C. doi:10.1103/PhysRev.88.1190.

[17] Blewett, J. P. (1952). "Radial Focusing in the Linear Accelerator". *Physical Review* 88 (5): 1197–1199. Bibcode:1952PhRv...88.1197B. doi:10.1103/PhysRev.88.1197.

[18] "The Alternating Gradient Concept". Brookhaven National Laboratory.

[19] "World of Beams Homepage". Lawrence Berkeley National Laboratory.

[20] Clery, D. (2010). "The Next Big Beam?". *Science* 327 (5962): 142–144. Bibcode:2010Sci...327..142C. doi:10.1126/science.327.5962.142.

[21] Wright, M. E. (April 2005). "Riding the Plasma Wave of the Future". *Symmetry Magazine* 2 (3): 12.

[22] Briezman, B. N.; et al. "Self-Focused Particle Beam Drivers for Plasma Wakefield Accelerators" (PDF). Retrieved 2005-05-13.

[23] Peralta, E. A.; et al. "Demonstration of electron acceleration in a laser-driven dielectric microstructure". Retrieved 2014-05-01.

[24] "An Interview with Dr. Steve Giddings". *ESI Special Topics*. Thomson Reuters. July 2004.

[25] Chamblin, A.; Nayak, G. C. (2002). "Black hole production at the CERN LHC: String balls and black holes from pp and lead-lead collisions" . *Physical Review D* **66** (9): 091901. arXiv:hep-ph/0206060. Bibcode:2002PhRvD..66i1901C. doi:10.1103/PhysRevD.66.091901.

[26] Ellis, J. LHC Safety Assessment Group; et al. (5 September 2008). "Review of the Safety of LHC Collisions" (PDF). *Journal of Physics G* **35** (11): 115004. arXiv:0806.3414. Bibcode:2008JPhG...35k5004E. doi:10.1088/0954-3899/35/11/115004. CERN record.

[27] Jaffe, R.; Busza, W.; Sandweiss, J.; Wilczek, F. (2000). "Review of Speculative "Disaster Scenarios" at RHIC" . *Reviews of Modern Physics* **72** (4): 1125–1140. arXiv:hep-ph/9910333. Bibcode:2000RvMP...72.1125J. doi:10.1103/RevModPhys.72.1125.

24.8   External links

- What are particle accelerators used for?

- Stanley Humphries (1999) Principles of Charged Particle Acceleration

- Particle Accelerators around the world

- Wolfgang K. H. Panofsky: The Evolution of Particle Accelerators & Colliders, (PDF), Stanford, 1997

- P.J. Bryant, A Brief History and Review of Accelerators (PDF), CERN, 1994.

- Heilbron, J.L.; Robert W. Seidel (1989). Lawrence and His Laboratory: A History of the Lawrence Berkeley Laboratory. Berkeley: University of California Press. ISBN 0-520-06426-7.

- David Kestenbaum, Massive Particle Accelerator Revving Up NPR's Morning Edition article on 9 April 2007

- Ragnar Hellborg (ed.), ed. (2005). *Electrostatic Accelerators: Fundamentals and Applications*. Springer. ISBN 978-3-540-23983-3.

- Fred's World of Science

- Annotated bibliography for particle accelerators from the Alsos Digital Library for Nuclear Issues

- Accelerators-for-Society.org, to know more about applications of accelerators for Research and Development, energy and environment, health and medicine, industry, material characterization.

Chapter 25

Low-energy electron diffraction

Low-energy electron diffraction (**LEED**) is a technique for the determination of the surface structure of single-crystalline materials by bombardment with a collimated beam of low energy electrons (20–200 eV)[1] and observation of diffracted electrons as spots on a fluorescent screen.

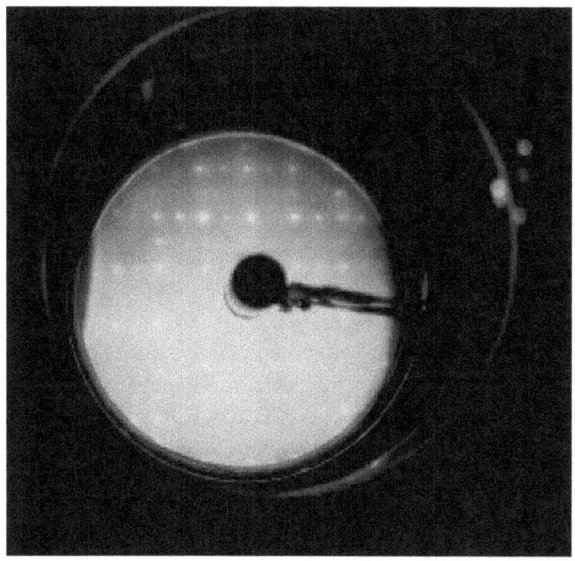

Figure 1: LEED pattern of a Si(100) reconstructed surface. The underlying lattice is a square lattice while the surface reconstruction has a 2x1 periodicity. As discussed in the text, the pattern shows that reconstruction exists in symmetrically equivalent domains which are oriented along different crystallographic axes. The diffraction spots are generated by acceleration of elastically scattered electrons onto a hemispherical fluorescent screen. Also seen is the electron gun which generates the primary electron beam. It covers up parts of the screen.

LEED may be used in one of two ways:

1. Qualitatively, where the diffraction pattern is recorded and analysis of the spot positions gives information on the symmetry of the surface structure. In the presence of an adsorbate the qualitative analysis may reveal information about the size and rotational alignment of the adsorbate unit cell with respect to the substrate unit cell.

2. Quantitatively, where the intensities of diffracted beams are recorded as a function of incident electron beam energy to generate the so-called I-V curves. By comparison with theoretical curves, these may provide accurate information on atomic positions on the surface at hand.

25.1 Historical perspective[2]

25.1.1 Davisson and Germer's discovery of electron diffraction

The theoretical possibility of the occurrence of electron diffraction first emerged in 1924 when Louis de Broglie introduced wave mechanics and proposed the wavelike nature of all particles. In his Nobel laureated work de Broglie postulated that the wavelength of a particle with linear momentum p is given by h/p, where h is Planck's constant. The de Broglie hypothesis was confirmed experimentally at Bell Labs in 1927 when Clinton Davisson and Lester Germer fired low-energy electrons at a crystalline nickel target and observed that the angular dependence of the intensity of backscattered electrons showed diffraction patterns. These observations were consistent with the diffraction theory for X-rays developed by Bragg and Laue earlier. Before the acceptance of the de Broglie hypothesis diffraction was believed to be an exclusive property of waves.

Davisson and Germer published notes of their electron diffraction experiment result in Nature and in Physical Review in 1927. One month after Davisson and Germer's work appeared, Thompson and Reid published their electron diffraction work with higher kinetic energy (thousand times higher than the energy used by Davisson and Germer) in the same journal. Those experiments revealed the wave property of electrons and opened up an era of electron diffraction study.

25.1.2   Development of LEED as a tool in surface science

Though discovered in 1927, Low Energy Electron Diffraction did not become a popular tool for surface analysis until the early 1960s. The main reasons were that monitoring directions and intensities of diffracted beams was a difficult experimental process due to inadequate vacuum techniques and slow detection methods such as a Faraday cup. Also, since LEED is a surface sensitive method, it required well-ordered surface structures. Techniques for the reconstruction of clean metal surfaces first became available much later. In the early 1960s LEED experienced a renaissance as ultra high vacuum became widely available and the post acceleration detection method was introduced. Using this technique diffracted electrons were accelerated to high energies to produce clear and visible diffraction patterns on a fluorescent screen.

It soon became clear that the kinematic (single scattering) theory, which had been successfully used to explain X-ray diffraction experiments, was inadequate for the quantitative interpretation of experimental data obtained from LEED. At this stage a detailed determination of surface structures, including adsorption sites, bond angles and bond lengths was not possible. A dynamical electron diffraction theory which took into account the possibility of multiple scattering was established in the late 1960s. With this theory it later became possible to reproduce experimental data with high precision.

25.2   Experimental Setup

In order to keep the studied sample clean and free from unwanted adsorbates, LEED experiments are performed in an ultra-high-vacuum environment (10^{-9} mbar).

The most important elements in an LEED experiment are [2]

1. A sample holder with the prepared sample

2. An electron gun

3. A display system, usually a hemispherical fluorescent screen on which the diffraction pattern can be observed directly

4. A sputtering gun for cleaning the surface

5. An Auger-Electron Spectroscopy system in order to determine the purity of the surface.

A simplified sketch of an LEED setup is shown in figure 2. [3]

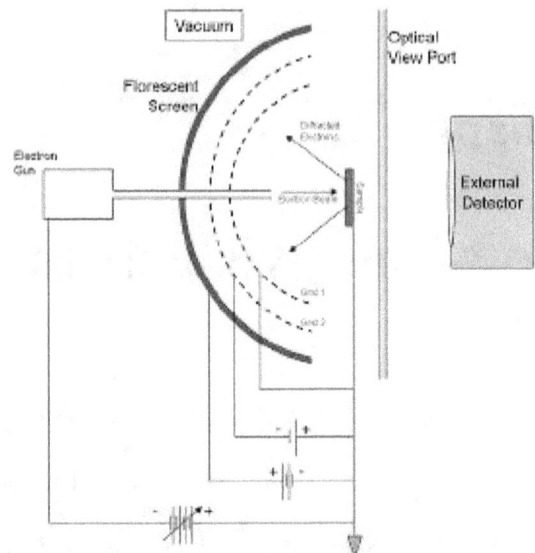

Figure 2 *Diagram of a LEED optics apparatus.*

25.2.1   Sample preparation

The sample is usually prepared outside the vacuum chamber by cutting a slice of around 1 mm in thickness and 1 cm in diameter along the desired crystallographic axis. The correct alignment of the crystal can be achieved with the help of x-ray methods and should be within $1°$ of the desired angle. [4] After being mounted in the UHV chamber the sample is chemically cleaned and flattened. Unwanted surface contaminants are removed by ion sputtering or by chemical processes such as oxidation and reduction cycles. The surface is flattened by annealing at high temperatures. Once a clean and well-defined surface is prepared, monolayers can be adsorbed on the surface by exposing it to a gas consisting of the desired adsorbate atoms or molecules.

Often the annealing process will let bulk impurities diffuse to the surface and therefore give rise to a re-contamination after each cleaning cycle. The problem is that impurities which adsorb without changing the basic symmetry of the surface, cannot easily be identified in the diffraction pattern. Therefore in many LEED experiments Auger Spectroscopy is used to accurately determine the purity of the sample.

25.2.2   Electron gun

In the electron gun, monochromatic electrons are emitted by a cathode filament which is at a negative potential, typically 10-600 V, with respect to the sample. The electrons are accelerated and focused into a beam, typically about 0.1 to 0.5 mm wide, by a series of electrodes serving as electron lenses. Some of the electrons incident on the sample surface are backscattered elastically, and diffraction can be detected if sufficient order exists on the surface. This typically requires a region of single crystal surface as wide as the electron beam, although sometimes polycrystalline surfaces such as highly oriented pyrolytic graphite (HOPG) are sufficient.

25.2.3 Detector system

An LEED detector usually contains three or four hemispherical concentric grids and a phosphor screen or other position-sensitive detector. The grids are used for screening out the inelastically scattered electrons. Most new LEED systems use a reverse view scheme, which has a minimized electron gun, and the pattern is viewed from behind through a transmission screen and a viewport. Recently, a new digitized position sensitive detector called a delay-line detector with better dynamic range and resolution has been developed.

The LEED contains a retarding field analyzer to block inelastically scattered electrons. Because only spherical fields around the sampled point are allowed and the geometry of the sample and the surrounding area is not spherical, no field is allowed. Therefore the first grid screens the space above the sample from the retarding field. The next grid is at a potential to block low energy electrons, it is called the suppressor or the gate. To make the retarding field homogeneous and mechanically more stable this grid often consists of two grids. The fourth grid is only necessary when the LEED is used like a tetrode and the current at the screen is measured, when it serves as screen between the gate and the anode.

25.2.4 Using the detector for Auger electron spectroscopy

To improve the measured signal in Auger electron spectroscopy, the gate voltage is scanned in a linear ramp. An RC circuit serves to derive the second derivative, which is then amplified and digitized. To reduce the noise, multiple passes are summed up. The first derivative is very large due to the residual capacitive coupling between gate and the anode and may degrade the performance of the circuit. By applying a negative ramp to the screen this can be compensated. It is also possible to add a small sine to the gate. A high Q RLC circuit is tuned to the second harmonic to detect the second derivative.

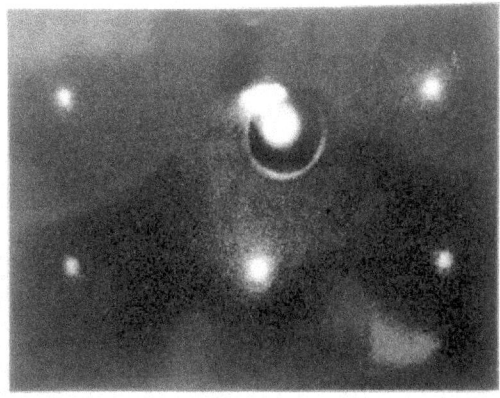

Image 1: LEED pattern of a clean platinum-rhodium (100) (Miller-index) single crystal. Taken in high vacuum using an electron gun with an energy of 85 eV.

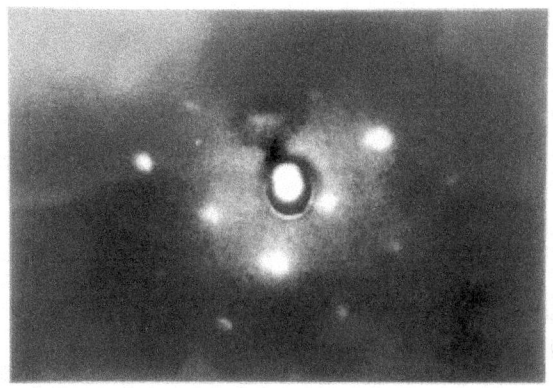

Image 2: LEED pattern of CO on platinum-rhodium (100) (Miller-index) surface of a single crystal. Taken in high vacuum using an electron gun with an energy of 94 eV.

25.2.5 Data acquisition

A modern data acquisition system usually contains a CCD/CMOS camera pointed to the screen for diffraction pattern visualization and a computer for data recording and further analysis.

The shown images are examples of LEED diffraction patterns. The difference between image 1 and 2 is remarkable: where image 1 is of a clean (100) platinum-rhodium single crystal, and image 2 of the same crystal with CO adsorbed on the surface. The original surface order of the clean crystal is clearly visible in image 1, it shows a C(1X1) structure; the extra spots in image 2 are caused by the CO on the surface and are an example of a C(2X2) structure. The diffraction spots are generated by acceleration of elastically scattered electrons onto a hemispherical fluorescent screen, a retarding field analyzer. In the middle one can see the bright spot of the electron gun which generates the primary electron beam.

25.3   Theory

25.3.1   Surface sensitivity

The basic reason for the high surface sensitivity of LEED is that for low-energy electrons the interaction between the solid and electrons is especially strong. Upon penetrating the crystal, primary electrons will lose kinetic energy due to inelastic scattering processes such as plasmon- and phonon excitations as well as electron-electron interactions. In cases where the detailed nature of the inelastic processes is unimportant they are commonly treated by assuming an exponential decay of the primary electron beam intensity, I_0, in the direction of propagation:

$$I(d) = I_0 * e^{-d/\Lambda(E)}$$

Here d is the penetration depth and $\Lambda(E)$ denotes the inelastic mean free path, defined as the distance an electron can travel before its intensity has decreased by the factor $1/e$. While the inelastic scattering processes and consequently the electronic mean free path depend on the energy, it is relatively independent of the material. The mean free path turns out to be minimal (5–10 Å) in the energy range of low-energy electrons (20–200 eV).[1] This effective attenuation means that only a few atomic layers are sampled by the electron beam and as a consequence the contribution of deeper atoms to the diffraction progressively decreases.

25.3.2   Kinematic theory: single scattering

Kinematic diffraction is defined as the situation where electrons impinging on a well-ordered crystal surface are elastically scattered only once by that surface. In the theory the electron beam is represented by a plane wave with a wavelength in accordance to the de Broglie hypothesis:

$$\lambda = \frac{h}{\sqrt{2mE}}, \qquad \lambda[\text{nm}] \approx \sqrt{\frac{1.5}{E[\text{eV}]}}$$

The interaction between the scatterers present in the surface and the incident electrons is most conveniently described in reciprocal space. In three dimensions the primitive reciprocal lattice vectors are related to the real space lattice {$\mathbf{a}$, $\mathbf{b}$, $\mathbf{c}$} in the following way:[5]

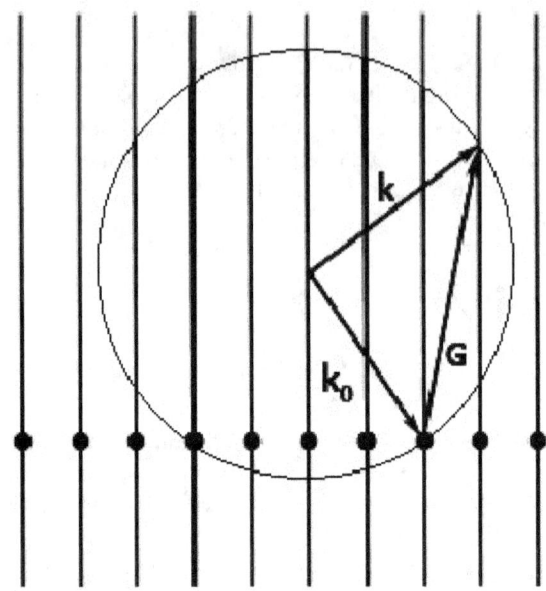

Figure 3: *Ewald's sphere construction for the case of diffraction from a 2D-lattice. The intersections between Ewald's sphere and reciprocal lattice rods define the allowed diffracted beams.*

$$\mathbf{a}^* = \frac{2\pi \mathbf{b} \times \mathbf{c}}{\mathbf{a} \cdot (\mathbf{b} \times \mathbf{c})},$$

$$\mathbf{b}^* = \frac{2\pi \mathbf{c} \times \mathbf{a}}{\mathbf{b} \cdot (\mathbf{c} \times \mathbf{a})},$$

$$\mathbf{c}^* = \frac{2\pi \mathbf{a} \times \mathbf{b}}{\mathbf{c} \cdot (\mathbf{a} \times \mathbf{b})}$$

For an incident electron with wave vector $\mathbf{k}_0 = 2\pi/\lambda_0$ and scattered wave vector $\mathbf{k} = 2\pi/\lambda$, the condition for constructive interference and hence diffraction of scattered electron waves is given by the Laue condition

$$\mathbf{k} - \mathbf{k}_0 = \mathbf{G}_{hkl}, \quad (1)$$

where (h,k,l) is a set of integers and

$$\mathbf{G}_{hkl} = h\mathbf{a}^* + k\mathbf{b}^* + l\mathbf{c}^*$$

is a vector of the reciprocal lattice. The magnitudes of the wave vectors are unchanged, i.e. $|\mathbf{k}_0| = |\mathbf{k}|$, since only elastic scattering is considered. Since the mean free path of low energy electrons in a crystal is only a few angstroms, only the first few atomic layers contribute to the diffraction. This means that there are no diffraction conditions in the direction perpendicular to the sample surface. As a consequence the reciprocal lattice of a surface is a 2D lattice with rods extending perpendicular from each lattice point.

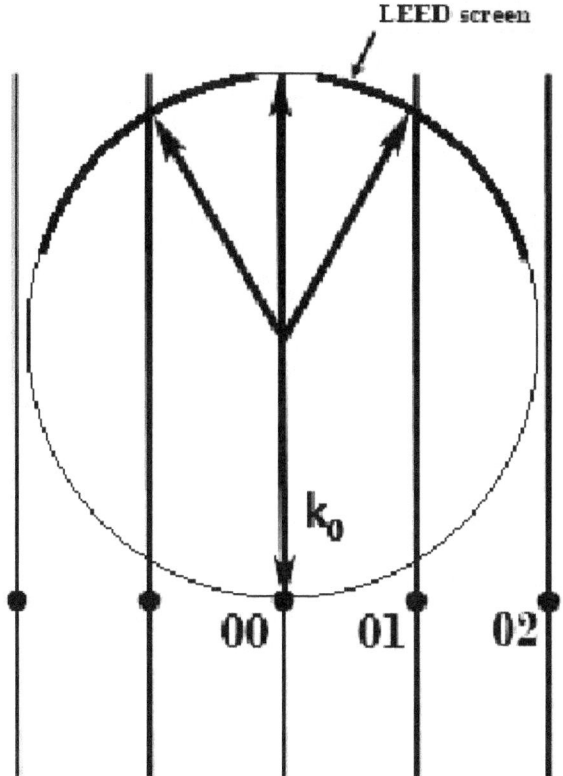

Figure 4: Ewald's sphere construction for the case of normal incidence of the primary electron beam. The diffracted beams are indexed according to the values of h and l.

The rods can be pictured as regions where the reciprocal lattice points are infinitely dense. Therefore in the case of diffraction from a surface equation (1) reduces to the 2D form:[2]

$$\mathbf{k}^{\parallel} - \mathbf{k}_0^{\parallel} = \mathbf{G}_{hk} = h\mathbf{a}^* + k\mathbf{b}^* , (2)$$

where $\mathbf{a}^*$ and $\mathbf{b}^*$ are the primitive translation vectors of the 2D reciprocal lattice of the surface and $\mathbf{k}^{\parallel}$, $\mathbf{k}_0^{\parallel}$ denote the component of respectively the reflected and incident wave vector parallel to the sample surface. $\mathbf{a}^*$ and $\mathbf{b}^*$ are related to the real space surface lattice in the following way:

$$\mathbf{a}^* = \frac{2\pi\mathbf{b} \times \dot{\mathbf{n}}}{|\mathbf{a} \times \mathbf{b}|}$$

$$\mathbf{b}^* = \frac{2\pi\dot{\mathbf{n}} \times \mathbf{a}}{|\mathbf{a} \times \mathbf{b}|}$$

The Laue condition equation (2) can readily be visualized using the Ewald's sphere construction. Figure 4 shows a simple illustration of this principle: The wave vector $\mathbf{k}_0$ of

the incident electron beam is drawn such that it terminates at a reciprocal lattice point. The Ewald's sphere is then the sphere with radius $|\mathbf{k}_0|$ and origin at the center of the incident wave vector.

By construction, every wave vector centered at the origin and terminating at an intersection between a rod and the sphere will then satisfy the Laue condition and thus represent an allowed diffracted beam.

25.3.3 Interpretation of LEED patterns

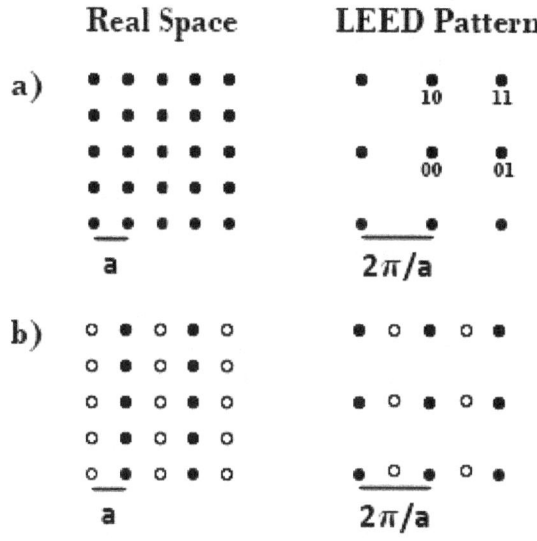

Figure 5: Real space- and reciprocal lattices for the case of a) a (100) face of a simple cubic lattice and b) a (2×1) commensurate superstructure. The white spots in the LEED pattern are the extra spots associated with the adsorbate structure.

Figure 4 shows the Ewald's sphere for the case of normal incidence of the primary electron beam, as would be the case in an actual LEED setup. It is apparent that the pattern observed on the fluorescent screen is a direct picture of the reciprocal lattice of the surface. The size of the Ewald's sphere and hence the number of diffraction spots on the screen is controlled by the incident electron energy. From the knowledge of the reciprocal lattice models for the real space lattice can be constructed and the surface can be characterized at least qualitatively in terms of the surface periodicity and the point group. Figure 5.a shows a model of an unreconstructed (100) face of a simple cubic crystal and the expected LEED pattern. The spots are indexed according to the values of h and k.

Superstructures

We now consider the case of an overlaying superstructure on a substrate surface. If the LEED pattern of the underlying (1×1) surface is known, spots due to the superstructure can be identified as *extra spots* or *super spots*. Figure 5.b shows the simple example of a (2×1) superstructure on a square lattice.

For a commensurate superstructure the symmetry and the rotational alignment with respect to adsorbent surface can be determined from the LEED pattern. This is easiest shown by using a matrix notation.[1] where the primitive translation vectors of the superlattice $\{\mathbf{a}_s,\mathbf{b}_s\}$ are linked to the primitive translation vectors of the underlying (1×1) lattice $\{\mathbf{a},\mathbf{b}\}$ in the following way

$$\mathbf{a}_s = G_{11}\mathbf{a} + G_{12}\mathbf{b},$$
$$\mathbf{b}_s = G_{21}\mathbf{a} + G_{22}\mathbf{b}.$$

The matrix for the superstructure then is

$$G = \begin{pmatrix} G_{11} & G_{12} \\ G_{21} & G_{22} \end{pmatrix}.$$

Similarly, the primitive translation vectors of the lattice describing the *extra spots* $\{\mathbf{a}_s^{*},\mathbf{b}_s^{*}\}$ are linked to the primitive translation vectors of the reciprocal lattice $\{\mathbf{a}^{*},\mathbf{b}^{*}\}$

$$\mathbf{a}_s^{*} = G_{11}^{*}\mathbf{a}^{*} + G_{12}^{*}\mathbf{b}^{*},$$
$$\mathbf{b}_s^{*} = G_{21}^{*}\mathbf{a}^{*} + G_{22}^{*}\mathbf{b}^{*}.$$

G^{*} is related to G in the following way

$$G^{*} = (G^{-1})^{T}$$
$$= \frac{1}{det(G)} \begin{pmatrix} G_{22} & -G_{21} \\ -G_{12} & G_{11} \end{pmatrix}.$$

Domains

An essential problem when considering LEED patterns is the existence of symmetrically equivalent domains. Domains may lead to diffraction patterns which have higher symmetry than the actual surface at hand. The reason is that usually the cross sectional area of the primary electron beam (~1 mm^2) is large compared to the average domain size on the surface and hence the LEED pattern might be a superposition of diffraction beams from domains oriented along different axes of the substrate lattice.

However, since the average domain size generally is larger than the coherence length of the probing electrons, interference between electrons scattered from different domains can be neglected. Therefore the total LEED pattern emerges as the incoherent sum of the diffraction patterns associated with the individual domains.

Figure 6 shows the superposition of the diffraction patterns for the two orthogonal domains (2x1) and (1x2) on a square lattice, i.e. for the case where one structure is just rotated by 90° with respect to the other. The (2x1) structure and the respective LEED pattern are shown in figure 5.b. It is apparent that the local symmetry of the surface structure is twofold while the LEED pattern exhibits a fourfold symmetry.

Figure 1 shows a real diffraction pattern of the same situation for the case of a Si(100) surface. However, here the (2x1) structure is formed due to surface reconstruction.

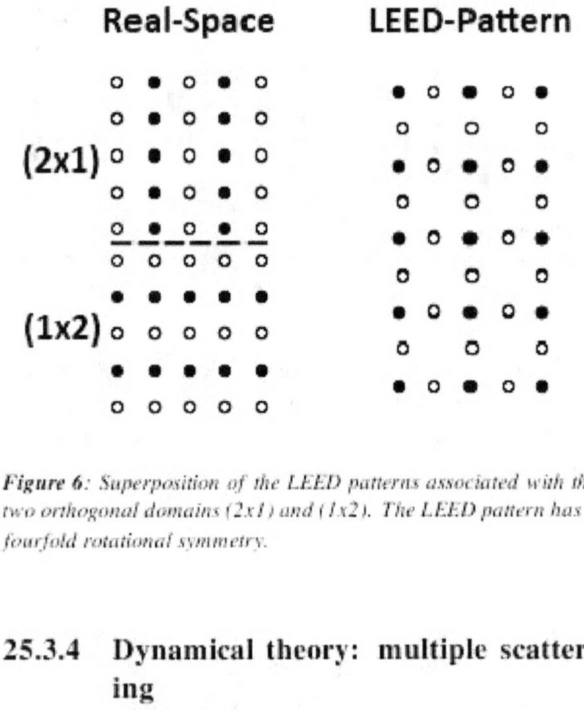

Figure 6: Superposition of the LEED patterns associated with the two orthogonal domains (2x1) and (1x2). The LEED pattern has a fourfold rotational symmetry.

25.3.4 Dynamical theory: multiple scattering

The inspection of the LEED pattern gives a qualitative picture of the surface periodicity i.e. the size of the surface unit cell and to a certain degree of surface symmetries. However it will give no information about the atomic arrangement within a surface unit cell or the sites of adsorbed atoms. For instance if the whole superstructure in figure 5.b is shifted such that the atoms adsorb in bridge sites instead of on-top sites the LEED pattern will be the same.

A more quantitative analysis of LEED experimental data can be achieved by analysis of so-called I-V curves, which are measurements of the intensity versus incident electron energy. The I-V curves can be recorded by using a camera connected to computer controlled data handling or by direct measurement with a movable Faraday cup. The experimental curves are then compared to computer calculations based on the assumption of a particular model system. The model is changed in an iterative process until a satisfactory agreement between experimental and theoretical curves is achieved. A quantitative measure for this agreement is the so-called *reliability-* or R-factor. A commonly used reliability factor is the one proposed by Pendry.[6] It is expressed in terms of the logarithmic derivative of the intensity:

$$L(E) = I'/I.$$

The R-factor is then given by:

$$R = \sum_g \int (Y_{gth} - Y_{gexpt})^2 dE / \sum_g \int (Y_{gth}^2 + Y_{gexpt}^2) dE.$$

where $Y(E) = L^{-1}/(L^{-2} + V_{oi}^2)$ and V_{oi} is the imaginary part of the electron self-energy. In generally $R_p \leq 0.2$ is considered as a good agreement, $R_p \simeq 0.3$ is considered mediocre and $R_p \simeq 0.5$ is considered a bad agreement. Figure 7 shows examples of the comparison between experimental I-V spectra and theoretical calculations.

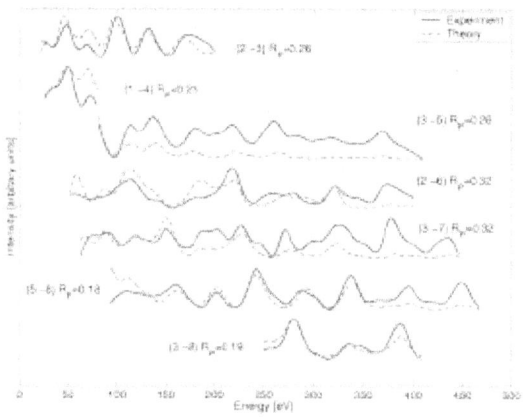

Figure 7: *Examples of the comparison between experimental data and a theoretical calculation (an AlNiCo quasicrystal surface). Thanks to R. Diehl and N. Ferralis for providing the data.*

25.3.5   Dynamical LEED calculations

The term *dynamical* stems from the studies of X-ray diffraction and describes the situation where the response of the crystal to an incident wave is included self-consistently and multiple scattering can occur. The aim of any dynamical

LEED theory is to calculate the intensities of diffraction of an electron beam impinging on a surface as accurately as possible.

A common method to achieve this is the self-consistent multiple scattering approach.[7] One essential point in this approach is the assumption that the scattering properties of the surface, i.e. of the individual atoms, are known in detail. The main task then reduces to the determination of the effective wave field incident on the individual scatters present in the surface, where the effective field is the sum of the primary field and the field emitted from all the other atoms. This must be done in a self-consistent way, since the emitted field of an atom depends on the incident effective field upon it. Once the effective field incident on each atom is determined, the total field emitted from all atoms can be found and its asymptotic value far from the crystal then gives the desired intensities.

A common approach in LEED calculations is to describe the scattering potential of the crystal by a "muffin tin" model, where the crystal potential can be imagined being divided up by non-overlapping spheres centered at each atom such that the potential has a spherically symmetric form inside the spheres and is constant everywhere else. The choice of this potential reduces the problem to scattering from spherical potentials, which can be dealt with effectively. The task is then to solve the Schrödinger equation for an incident electron wave in that "muffin tin" potential.

25.4   Related Techniques

25.4.1   Tensor LEED

In LEED the exact atomic configuration of a surface is determined by a trial and error process where measured I-V curves are compared to computer-calculated spectra under the assumption of a model structure. From an initial reference structure a set of trial structures is created by varying the model parameters. The parameters are changed until an optimal agreement between theory and experiment is achieved. However, for each trial structure a full LEED calculation with multiple scattering corrections must be conducted. For systems with a large parameter space the need for computational time might become significant. This is the case for complex surfaces structures or when considering large molecules as adsorbates.

Tensor LEED[8][9] is an attempt to reduce the computational effort needed by avoiding full LEED calculations for each trial structure. The scheme is as follows: One first defines a reference surface structure for which the I-V spectrum is calculated. Next a trial structure is created by displacing some of the atoms. If the displacements are small the trial structure can be considered as a small perturbation of the reference structure and first-order perturbation theory can be used to determine the I-V curves of a large set of trial structures.

25.4.2 Spot Profile Analysis Low-Energy Electron Diffraction

A real surface is not perfectly periodic but has many imperfections in the form of dislocations, atomic steps, terraces and the presence of unwanted adsorbed atoms. This departure from a perfect surface leads to a broadening of the diffraction spots and adds to the background intensity in the LEED pattern.

SPA-LEED*[10] is a technique where the intensity of diffraction beams is measured in order to determine the diffraction spot profiles. The spots are sensitive to the irregularities in the surface structure and their examination therefore permits more-detailed conclusions about some surface characteristics. Using SPA-LEED may for instance permit a quantitative determination of the surface roughness, terrace sizes or surface steps.*[10]

25.4.3 Other

- Spin-Polarized Low Energy Electron Diffraction
- Inelastic Low Energy Electron Diffraction
- Very Low Energy Electron Diffraction
- Reflection high-energy electron diffraction

25.5 See also

- List of surface analysis methods

25.6 References

[1] K. Oura, V.G. Lifshifts, A.A. Saranin, A. V. Zotov, M. Katayama (2003). *Surface Science*. Springer-Verlag, Berlin Heidelberg New York. pp. 1–45.

[2] M.A. Van Hove, W.H. Weinberg, C. M. Chan (1986). *Low-Energy Electron Diffraction*. Springer-Verlag, Berlin Heidelberg New York. pp. 1–27, 46–89, 92–124, 145–172. doi:10.1002/maco.19870380711. ISBN 3-540-16262-3.

[3] Zangwill, A.. "Physics at Surfaces". Cambridge University Press (1988), p.33

[4] Pendry (1974). *Low-Energy Electron Diffraction*. Academic Press Inc. (London) LTD. pp. 1–75.

[5] C. Kittel (1996). "2". *Introduction to Solid State Physics*. John Wiley, US.

[6] J.B. Pendry (1980). "Reliability Factors for LEED Calculations". *J. Phys. C* **13**: 937. Bibcode:1980JPhC...13..937P. doi:10.1088/0022-3719/13/5/024.

[7] E.G. McRae (1967). "Self-Consistent Multiple-Scattering Approach to the Interpretation of Low-Energy Electron Diffraction". *Surface Science* **8** (1-2): 14–34. Bibcode:1967SurSc...8...14M. doi:10.1016/0039-6028(67)90071-4.

[8] P.J. Rous J.B. Pendry (1989). "Tensor LEED I: A Technuique for high speed surface structure determination by low energy electron diffraction." *Comp. Phys. Comm.* **54** (1): 137–156. Bibcode:1989CoPhC..54..137R. doi:10.1016/0010-4655(89)90039-8.

[9] P.J. Rous J.B. Pendry (1989). "The theory of Tensor LEED." *Surf. Sci.* **219** (3): 355–372. Bibcode:1989SurSc.219..355R. doi:10.1016/0039-6028(89)90513-X.

[10] M. Henzler (1982). "Studies of Surface Imperfections". *Appl. Surf. Sci.* 11/12: 450. Bibcode:1982ApSS...11..450H. doi:10.1016/0378-5963(82)90092-7.

- P. Goodman (General Editor), Fifty Years of Electron Diffraction, D. Reidel Publishing, 1981

- D. Human et al., Low energy electron diffraction using an electronic delay-line detector, Rev. Sci. Inst. 77 023302 (2006)

25.7 External links

- LEED programm packages

Chapter 26

Electron microscope

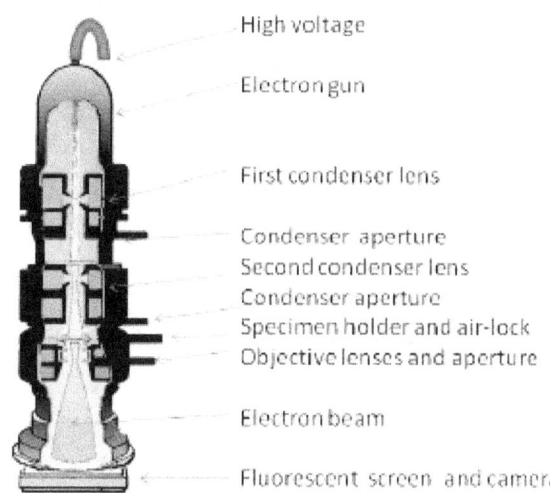

Transmission Electron Microscope

Diagram of a transmission electron microscope

An **electron microscope** is a microscope that uses a beam of accelerated electrons as a source of illumination. Because the wavelength of an electron can be up to 100,000 times shorter than that of visible light photons, the electron microscope has a higher resolving power than a light microscope and can reveal the structure of smaller objects. A transmission electron microscope can achieve better than 50 pm resolution [1] and magnifications of up to about 10,000,000x whereas most light microscopes are limited by diffraction to about 200 nm resolution and useful magnifications below 2000x.

The transmission electron microscope uses electrostatic and electromagnetic lenses to control the electron beam and focus it to form an image. These electron optical lenses are analogous to the glass lenses of an optical light microscope.

Electron microscopes are used to investigate the ultrastructure of a wide range of biological and inorganic specimens including microorganisms, cells, large molecules, biopsy samples, metals, and crystals. Indus-

A 1973 Siemens electron microscope, Musée des Arts et Métiers, Paris

trially, the electron microscope is often used for quality control and failure analysis. Modern electron microscopes produce electron micrographs using specialized digital cameras and frame grabbers to capture the image.

26.1 History

The first electromagnetic lens was developed in 1926 by Hans Busch. [2]

According to Dennis Gabor, the physicist Leó Szilárd tried

Electron microscope constructed by Ernst Ruska in 1933

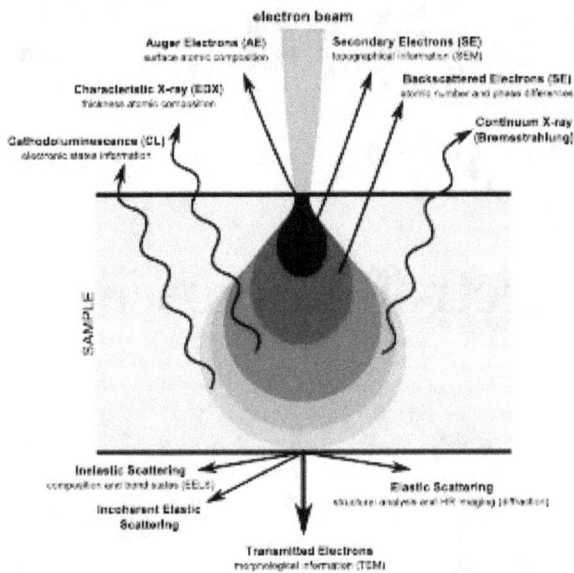

Diagram illustrating the phenomena resulting of the interaction of highly energetic electrons with matter.

plications for the microscope, especially with biological specimens.[4][6] Also in 1937, Manfred von Ardenne pioneered the scanning electron microscope.[7] The first *practical* electron microscope was constructed in 1938, at the University of Toronto, by Eli Franklin Burton and students Cecil Hall, James Hillier, and Albert Prebus; and Siemens produced the first *commercial* transmission electron microscope (TEM) in 1939.[8] Although contemporary electron microscopes are capable of two million-power magnification, as scientific instruments, they remain based upon Ruska's prototype.

26.2 Types

26.2.1 Transmission electron microscope (TEM)

Main article: Transmission electron microscope

The original form of electron microscope, the transmission electron microscope (TEM) uses a high voltage electron beam to create an image. The electron beam is produced by an electron gun, commonly fitted with a tungsten filament cathode as the electron source. The electron beam is accelerated by an anode typically at +100 keV (40 to 400 keV) with respect to the cathode, focused by electrostatic and electromagnetic lenses, and transmitted through the specimen that is in part transparent to electrons and in part scatters them out of the beam. When it emerges from the specimen, the electron beam carries information about the

in 1928 to convince Busch to build an electron microscope, for which he had filed a patent.[3]

German physicist Ernst Ruska and the electrical engineer Max Knoll constructed the prototype electron microscope in 1931, capable of four-hundred-power magnification; the apparatus was the first demonstration of the principles of electron microscopy.[4] Two years later, in 1933, Ruska built an electron microscope that exceeded the resolution attainable with an optical (light) microscope.[4] Moreover, Reinhold Rudenberg, the scientific director of Siemens-Schuckertwerke, obtained the patent for the electron microscope in May 1931.

In 1932, Ernst Lubcke of Siemens & Halske built and obtained images from a prototype electron microscope, applying concepts described in the Rudenberg patent applications.[5] Five years later (1937), the firm financed the work of Ernst Ruska and Bodo von Borries, and employed Helmut Ruska (Ernst's brother) to develop ap-

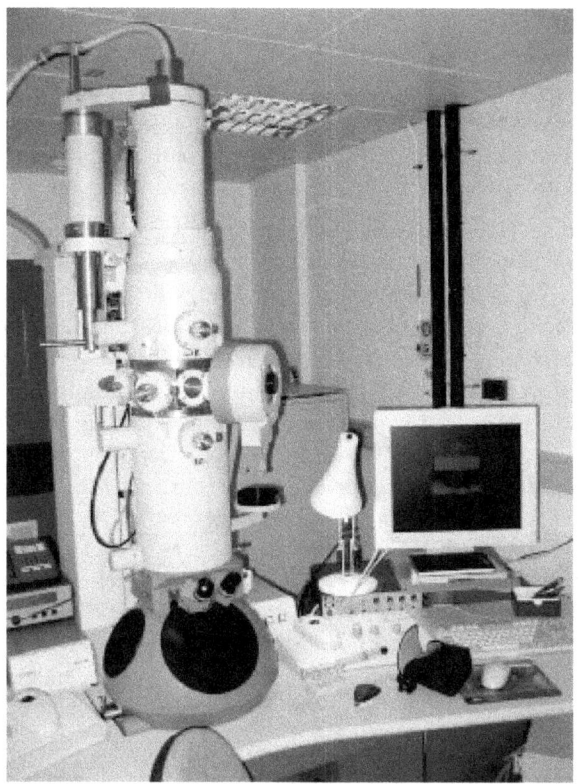

A modern transmission electron microscope

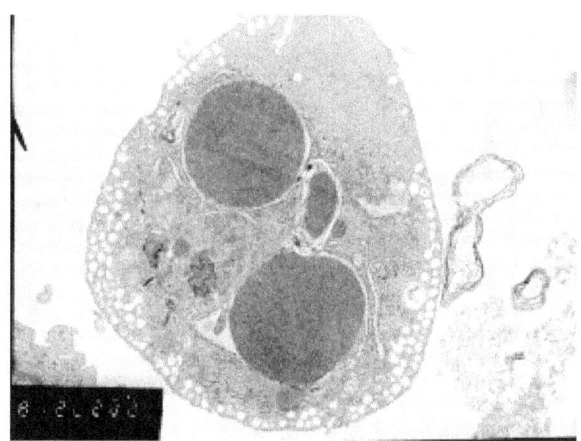

Neonatal cardiomyocytes ultrastructure after anoxia-reoxygenation.

structure of the specimen that is magnified by the objective lens system of the microscope. The spatial variation in this information (the "image") may be viewed by projecting the magnified electron image onto a fluorescent viewing screen coated with a phosphor or scintillator material such as zinc sulfide. Alternatively, the image can be photographically recorded by exposing a photographic film or plate directly to the electron beam, or a high-resolution phosphor may be coupled by means of a lens optical system or a fibre optic light-guide to the sensor of a CCD (charge-coupled device) camera. The image detected by the CCD may be displayed on a monitor or computer.

Resolution of the TEM is limited primarily by spherical aberration, but a new generation of aberration correctors have been able to partially overcome spherical aberration to increase resolution. Hardware correction of spherical aberration for the high-resolution transmission electron microscopy (HRTEM) has allowed the production of images with resolution below 0.5 angstrom (50 picometres)[1] and magnifications above 50 million times.[9] The ability to determine the positions of atoms within materials has made the HRTEM an important tool for nano-technologies research and development.[10]

Transmission electron microscopes are often used in electron diffraction mode. The advantages of electron diffraction over X-ray crystallography are that the specimen need not be a single crystal or even a polycrystalline powder, and also that the Fourier transform reconstruction of the object's magnified structure occurs physically and thus avoids the need for solving the phase problem faced by the X-ray crystallographers after obtaining their X-ray diffraction patterns of a single crystal or polycrystalline powder. The major disadvantage of the transmission electron microscope is the need for extremely thin sections of the specimens, typically about 100 nanometers. Biological specimens are typically required to be chemically fixed, dehydrated and embedded in a polymer resin to stabilize them sufficiently to allow ultrathin sectioning. Sections of biological specimens, organic polymers and similar materials may require special treatment with heavy atom labels in order to achieve the required image contrast.

26.2.2 Scanning electron microscope (SEM)

Main article: Scanning electron microscope

The SEM produces images by probing the specimen with a focused electron beam that is scanned across a rectangular area of the specimen (raster scanning). When the electron beam interacts with the specimen, it loses energy by a variety of mechanisms. The lost energy is converted into alternative forms such as heat, emission of low-energy secondary electrons and high-energy backscattered electrons, light emission (cathodoluminescence) or X-ray emission, all of which provide signals carrying information about the properties of the specimen surface, such as its topography and composition. The image displayed by an SEM maps the varying intensity of any of these signals into the image in a position corresponding to the position of the beam on the specimen when the signal was generated. In the SEM image of an ant shown at right, the image was constructed from signals produced by a secondary electron detector, the normal or conventional imaging mode in most SEMs.

Generally, the image resolution of an SEM is at least an order of magnitude poorer than that of a TEM. However, because the SEM image relies on surface processes rather than transmission, it is able to image bulk samples up to many centimeters in size and (depending on instrument design and settings) has a great depth of field, and so can produce images that are good representations of the three-dimensional shape of the sample. Another advantage of SEM is its variety called environmental scanning electron microscope (ESEM) can produce images of sufficient quality and resolution with the samples being wet or contained in low vacuum or gas. This greatly facilitates imaging biological samples that are unstable in the high vacuum of conventional electron microscopes.

An image of an ant in a scanning electron microscope

26.2.3   Color

In their most common configurations, electron microscopes produce images with a single brightness value per pixel, with the results usually rendered in grayscale.[11] However, often these images are then colorized through the use of feature-detection software, or simply by hand-editing using a graphics editor. This is usually for aesthetic effect or for clarifying structure, and generally does not add information about the specimen.[12]

In some configurations more information about specimen properties is gathered per pixel, usually by the use of multiple detectors.[13] In SEM, the attributes of topography and material contrast can be obtained by a pair of backscattered electron detectors and such attributes can be superimposed in a single color image by assigning a different primary color to each attribute.[14] Similarly, a combination of backscattered and secondary electron signals can be assigned to different colors and superimposed on a single color micrograph displaying simultaneously the properties of the specimen.[15]

Some types of detectors used in SEM have analytical capabilities, and can provide several items of data at each pixel. Examples are the Energy-dispersive X-ray spectroscopy (EDS) detectors used in elemental analysis and Cathodoluminescence microscope (CL) systems that analyse the intensity and spectrum of electron-induced luminescence in (for example) geological specimens. In SEM systems using these detectors it is common to color code the signals and superimpose them in a single color image, so that differences in the distribution of the various components of the specimen can be seen clearly and compared. Optionally, the standard secondary electron image can be merged with the one or more compositional channels, so that the specimen's structure and composition can be compared. Such images can be made while maintaining the full integrity of the original signal, which is not modified in any way.

26.2.4   Reflection   electron   microscope (REM)

In the **reflection electron microscope** (REM) as in the TEM, an electron beam is incident on a surface but instead of using the transmission (TEM) or secondary electrons (SEM), the reflected beam of elastically scattered electrons is detected. This technique is typically coupled with reflection high energy electron diffraction (RHEED) and *reflection high-energy loss spectroscopy (RHELS)*. Another variation is spin-polarized low-energy electron microscopy (SPLEEM), which is used for looking at the microstructure of magnetic domains.[16]

26.2.5   Scanning transmission electron microscope (STEM)

Main article: Scanning transmission electron microscopy

The STEM rasters a focused incident probe across a specimen that (as with the TEM) has been thinned to facilitate detection of electrons scattered *through* the specimen. The high resolution of the TEM is thus possible in STEM. The focusing action (and aberrations) occur before the electrons hit the specimen in the STEM, but afterward in the TEM. The STEMs use of SEM-like beam rastering simplifies annular dark-field imaging, and other analytical techniques, but also means that image data is acquired in serial rather than in parallel fashion. Often TEM can be equipped with the scanning option and then it can function both as TEM and STEM.

26.3 Sample preparation

An insect coated in gold for viewing with a scanning electron microscope

Materials to be viewed under an electron microscope may require processing to produce a suitable sample. The technique required varies depending on the specimen and the analysis required:

- *Chemical fixation* – for biological specimens aims to stabilize the specimen's mobile macromolecular structure by chemical crosslinking of proteins with aldehydes such as formaldehyde and glutaraldehyde, and lipids with osmium tetroxide.

- *Negative stain* – suspensions containing nanoparticles or fine biological material (such as viruses and bacteria) are briefly mixed with a dilute solution of an electron-opaque solution such as ammonium molybdate, uranyl acetate (or formate), or phosphotungstic acid. This mixture is applied to a suitably coated EM grid, blotted, then allowed to dry. Viewing of this preparation in the TEM should be carried out without delay for best results. The method is important in microbiology for fast but crude morphological identification, but can also be used as the basis for high resolution 3D reconstruction using EM tomography methodology when carbon films are used for support. Negative staining is also used for observation of nanoparticles.

- *Cryofixation* – freezing a specimen so rapidly, in liquid ethane, and maintained at liquid nitrogen or even liquid helium temperatures, so that the water forms vitreous (non-crystalline) ice. This preserves the specimen in a snapshot of its solution state. An entire field called cryo-electron microscopy has branched from this technique. With the development of cryo-electron microscopy of vitreous sections (CEMOVIS), it is now possible to observe samples from virtually any biological specimen close to its native state.

- *Dehydration* – or replacement of water with organic solvents such as ethanol or acetone, followed by critical point drying or infiltration with embedding resins. Also freeze drying.

- *Embedding, biological specimens* – after dehydration, tissue for observation in the transmission electron microscope is embedded so it can be sectioned ready for viewing. To do this the tissue is passed through a 'transition solvent' such as propylene oxide (epoxypropane) or acetone and then infiltrated with an epoxy resin such as Araldite, Epon, or Durcupan;[17] tissues may also be embedded directly in water-miscible acrylic resin. After the resin has been polymerized (hardened) the sample is thin sectioned (ultrathin sections) and stained – it is then ready for viewing.

- *Embedding, materials* – after embedding in resin, the specimen is usually ground and polished to a mirror-like finish using ultra-fine abrasives. The polishing process must be performed carefully to minimize scratches and other polishing artifacts that reduce image quality.

- *Metal shadowing* – Metal (e.g. platinum) is evaporated from an overhead electrode and applied to the surface of a biological sample at an angle. The surface topography results in variations in the thickness of the metal that are seen as variations in brightness and contrast in the electron microscope image.

- *Replication* – A surface shadowed with metal (e.g. platinum, or a mixture of carbon and platinum) at an angle is coated with pure carbon evaporated from carbon electrodes at right angles to the surface. This is followed by removal of the specimen material (e.g. in an acid bath, using enzymes or by mechanical separation[18]) to produce a surface replica that records the surface ultrastructure and can be examined using transmission electron microscopy.

- *Sectioning* – produces thin slices of specimen, semi-transparent to electrons. These can be cut on an ultramicrotome with a diamond knife to produce ultra-thin sections about 60–90 nm thick. Disposable glass knives are also used because they can be made in the lab and are much cheaper.

- *Staining* – uses heavy metals such as lead, uranium or tungsten to scatter imaging electrons and thus give contrast between different structures, since many (especially biological) materials are nearly "transparent" to electrons (weak phase objects). In biology, specimens can be stained "en bloc" before embedding and also later after sectioning. Typically thin sections are stained for several minutes with an aqueous or alcoholic solution of uranyl acetate followed by aqueous lead citrate.* [19]

- *Freeze-fracture or freeze-etch* – a preparation method particularly useful for examining lipid membranes and their incorporated proteins in "face on" view. The fresh tissue or cell suspension is frozen rapidly (cryofixation), then fractured by breaking or by using a microtome while maintained at liquid nitrogen temperature. The cold fractured surface (sometimes "etched" by increasing the temperature to about −100 °C for several minutes to let some ice sublime) is then shadowed with evaporated platinum or gold at an average angle of 45° in a high vacuum evaporator. A second coat of carbon, evaporated perpendicular to the average surface plane is often performed to improve stability of the replica coating. The specimen is returned to room temperature and pressure, then the extremely fragile "pre-shadowed" metal replica of the fracture surface is released from the underlying biological material by careful chemical digestion with acids, hypochlorite solution or SDS detergent. The still-floating replica is thoroughly washed free from residual chemicals, carefully fished up on fine grids, dried then viewed in the TEM.

- *Ion beam milling* – thins samples until they are transparent to electrons by firing ions (typically argon) at the surface from an angle and sputtering material from the surface. A subclass of this is focused ion beam milling, where gallium ions are used to produce an electron transparent membrane in a specific region of the sample, for example through a device within a microprocessor. Ion beam milling may also be used for cross-section polishing prior to SEM analysis of materials that are difficult to prepare using mechanical polishing.

- *Conductive coating* – an ultrathin coating of electrically conducting material, deposited either by high vacuum evaporation or by low vacuum sputter coating of the sample. This is done to prevent the accumulation of static electric fields at the specimen due to the electron irradiation required during imaging. The coating materials include gold, gold/palladium, platinum, tungsten, graphite, etc.

- *Earthing* – to avoid electrical charge accumulation on a conductive coated sample, it is usually electrically connected to the metal sample holder. Often an electrically conductive adhesive is used for this purpose.

26.4 Disadvantages

False-color SEM image of the filter setae of an Antarctic krill. (Raw electron microscope images carry no color information.)
Pictured: *First degree filter setae with a V pattern of second degree setae pointing towards the inside of the feeding basket. The purple ball is 1 μm in diameter.*

Electron microscopes are expensive to build and maintain, but the capital and running costs of confocal light microscope systems now overlaps with those of basic electron microscopes. Microscopes designed to achieve high resolutions must be housed in stable buildings (sometimes underground) with special services such as magnetic field cancelling systems.

The samples largely have to be viewed in vacuum, as the molecules that make up air would scatter the electrons. One exception is the environmental scanning electron microscope, which allows hydrated samples to be viewed in a low-pressure (up to 20 Torr or 2.7 kPa) and/or wet environment.

Scanning electron microscopes operating in conventional high-vacuum mode usually image conductive specimens; therefore non-conductive materials require conductive coating (gold/palladium alloy, carbon, osmium, etc.). The low-voltage mode of modern microscopes makes possible the observation of non-conductive specimens without coating. Non-conductive materials can be imaged also by a variable pressure (or environmental) scanning electron microscope.

Small, stable specimens such as carbon nanotubes, diatom frustules and small mineral crystals (asbestos fibres, for example) require no special treatment before being examined in the electron microscope. Samples of hydrated materials, including almost all biological specimens have to be prepared in various ways to stabilize them, reduce their thickness (ultrathin sectioning) and increase their electron optical contrast (staining). These processes may result in *artifacts*, but these can usually be identified by comparing the results obtained by using radically different specimen preparation methods. It is generally believed by scientists working in the field that as results from various preparation techniques have been compared and that there is no reason that they should all produce similar artifacts, it is reasonable to believe that electron microscopy features correspond with those of living cells. Since the 1980s, analysis of cryofixed, vitrified specimens has also become increasingly used by scientists, further confirming the validity of this technique.[20][21][22]

26.5 Applications

26.6 See also

- Category:Electron microscope images
- Acronyms in microscopy
- Electron diffraction
- Electron energy loss spectroscopy (EELS)
- Energy filtered transmission electron microscopy (EFTEM)
- Environmental scanning electron microscope (ESEM)
- Field emission microscope
- HiRISE
- In situ electron microscopy
- Microscope image processing
- Microscopy

- Nanoscience
- Nanotechnology
- Neutron microscope
- Scanning confocal electron microscopy
- Scanning electron microscope (SEM)
- Scanning tunneling microscope
- Surface science
- Transmission Electron Aberration-Corrected Microscope
- X-ray diffraction
- X-ray microscope

26.7 References

[1] Erni, Rolf; Rossell, MD; Kisielowski, C; Dahmen, U (2009). "Atomic-Resolution Imaging with a Sub-50-pm Electron Probe". *Physical Review Letters* **102** (9): 096101. Bibcode:2009PhRvL.102i6101E. doi:10.1103/PhysRevLett.102.096101. PMID 19392535.

[2] Mathys, Daniel, Zentrum für Mikroskopie, University of Basel: *Die Entwicklung der Elektronenmikroskopie vom Bild über die Analyse zum Nanolabor*, p. 8

[3] Dannen, Gene (1998) Leo Szilard the Inventor: A Slideshow (1998, Budapest, conference talk). dannen.com

[4] Ruska, Ernst (1986). "Ernst Ruska Autobiography". Nobel Foundation. Retrieved 2010-01-31.

[5] Rudenberg, H Gunther and Rudenberg, Paul G (2010). "Chapter 6 – Origin and Background of the Invention of the Electron Microscope: Commentary and Expanded Notes on Memoir of Reinhold Rüdenberg". *Advances in Imaging and Electron Physics* **160**. Elsevier. doi:10.1016/S1076-5670(10)60006-7. ISBN 978-0-12-381017-5.

[6] Kruger DH; Schneck P; Gelderblom HR (May 2000). "Helmut Ruska and the visualisation of viruses". *Lancet* **355** (9216): 1713–7. doi:10.1016/S0140-6736(00)02250-9. PMID 10905259.

[7] von Ardenne, M and Beischer, D (1940). "Untersuchung von metalloxyd-rauchen mit dem universal-elektronenmikroskop". *Zeitschrift Electrochemie* (in German) **46**: 270–277.

[8] "James Hillier". *Inventor of the Week: Archive*. 2003-05-01. Retrieved 2010-01-31.

[9] "The Scale of Things". Office of Basic Energy Sciences, U.S. Department of Energy. 2006-05-26. Retrieved 2010-01-31.

[10] O'Keefe MA, Allard LF. "Sub-Ångstrom Electron Microscopy for Sub-Ångstrom Nano-Metrology" (pdf). Information Bridge: DOE Scientific and Technical Information – Sponsored by OSTI. Retrieved 2010-01-31.

[11] Burgess, Jeremy (1987). *Under the Microscope: A Hidden World Revealed*. CUP Archive. p. 11. ISBN 0-521-39940-8.

[12] "Introduction to Electron Microscopy" (PDF). FEI Company. p. 15. Retrieved 12 December 2012.

[13] Antonovsky, A. (1984). "The application of colour to sem imaging for increased definition". *Micron and Microscopica Acta* **15** (2): 77–84. doi:10.1016/0739-6260(84)90005-4.

[14] Danilatos, G.D. (1986). "Colour micrographs for backscattered electron signals in the SEM". *Scanning* **9** (3): 8–18. doi:10.1111/j.1365-2818.1986.tb04287.x.

[15] Danilatos, G.D. (1986). "Environmental scanning electron microscopy in colour". *J. Microscopy* **142**: 317–325. doi:10.1002/sca.4950080104.

[16] "SPLEEM". National Center for Electron Microscopy (NCEM). Retrieved 2010-01-31.

[17] Luft, J.H. (1961). "Improvements in epoxy resin embedding methods". *The Journal of biophysical and biochemical cytology* **9** (2). p. 409. PMC 2224998. PMID 13764136.

[18] Juniper, B.E.; Bradley, D.E. (1958). "The carbon replica technique in the study of the ultrastructure of leaf surfaces". *Journal of ultrastructure research* **2** (1): 16–27.

[19] Reynolds, E. S. (1963). "The use of lead citrate at high pH as an electron-opaque stain in electron microscopy.". *Journal of Cell Biology* **17**: 208-212.

[20] Adrian, Marc; Dubochet, Jacques; Lepault, Jean; McDowall, Alasdair W. (1984). "Cryo-electron microscopy of viruses". *Nature* **308** (5954): 32–36. Bibcode:1984Natur.308...32A. doi:10.1038/308032a0. PMID 6322001.

[21] Sabanay, I.; Arad, T.; Weiner, S.; Geiger, B. (1991). "Study of vitrified, unstained frozen tissue sections by cryoimmunoelectron microscopy". *Journal of Cell Science* **100** (1): 227–236. PMID 1795028.

[22] Kasas, S.; Dumas, G.; Dietler, G.; Catsicas, S.; Adrian, M. (2003). "Vitrification of cryoelectron microscopy specimens revealed by high-speed photographic imaging". *Journal of Microscopy* **211** (1): 48–53. doi:10.1046/j.1365-2818.2003.01193.x.

26.8   External links

- An Introduction to Electron Microscope Resources for Teachers and Students

- Science Aid: Electron Microscopy High School (GCSE, A Level) resource
- Cell Centered Database – Electron microscopy data

26.8.1   General

- Nanohedron.com|Nano image gallery beautiful images generated with electron microscopes.
- electron microscopy Website of the ETH Zurich: Very good graphics and images, which illustrate various procedures.
- FEI Image Contest FEI has a microscope image contest every year since 2008.
- Environmental Scanning Electron Microscope (ESEM)
- X-ray element analysis in electron microscope – Information portal with X-ray microanalysis and EDX contents
- Eva Nogales's seminar: "Introduction to Electron Microscopy"
- good introduction to electron microscopy by David Szondy

26.8.2   History

John H L Watson's recollections at the University of Toronto when he worked with Hillier and Prebus:

- Rubin Borasky Electron Microscopy Collection, 1930–1988 Archives Center, National Museum of American History, Smithsonian Institution.

26.8.3   Other

- The Royal Microscopical Society, Electron Microscopy Section (UK)
- Albert Lleal. Natural history subjects at Scanning Electron Microscope SEM

Chapter 27

Free-electron laser

The free-electron laser FELIX at the FOM Institute for Plasma Physics Rijnhuizen (nl), Nieuwegein, The Netherlands.

A **free-electron laser** (FEL), is a type of laser that uses very-high-speed electrons that move freely through a magnetic structure,[1] hence the term *free electron* as the lasing medium.[2] The free-electron laser has the widest frequency range of any laser type, and can be widely tunable,[3] currently ranging in wavelength from microwaves, through terahertz radiation and infrared, to the visible spectrum, ultraviolet, and X-ray.[4]

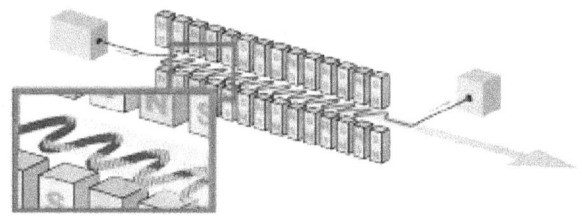

Schematic representation of an undulator, at the core of a free-electron laser.

The term free-electron lasers was coined by John Madey in 1976 at Stanford University.[5] The work emanates from research done by Hans Motz and his coworkers, who built an undulator at Stanford in 1953,[6] [7] using the wiggler magnetic configuration which is the heart of a free electron laser. Madey used a 43-MeV electron beam[8] and 5 m long wiggler to amplify a signal.

27.1 Beam creation

The undulator of FELIX.

To create an FEL, a beam of electrons is accelerated to almost the speed of light. The beam passes through an undulator, a side to side magnetic field produced by a periodic arrangement of magnets with alternating poles across the beam path. The direction of the beam is called the longitudinal direction, while the direction across the beam path is called transverse. This array of magnets is called an undulator or a wiggler, because it forces the electrons in the beam to wiggle transversely along a sinusoidal path about the axis of the undulator.

The transverse acceleration of the electrons across this path results in the release of photons (synchrotron radiation), which are monochromatic but still incoherent, because the electromagnetic waves from randomly distributed electrons interfere constructively and destructively in time, and the resulting radiation power scales linearly with the number of electrons. If an external laser is provided or if the synchrotron radiation becomes sufficiently strong, the transverse electric field of the radiation beam interacts with the transverse electron current created by the sinusoidal wiggling motion, causing some electrons to gain and others to lose energy to the optical field via the ponderomotive force.

This energy modulation evolves into electron density (current) modulations with a period of one optical wavelength. The electrons are thus clumped, called *microbunches*, separated by one optical wavelength along the axis. Whereas conventional undulators would cause the electrons to radiate independently, the radiation emitted by the bunched electrons are in phase, and the fields add together coherently.

The FEL radiation intensity grows, causing additional microbunching of the electrons, which continue to radiate in phase with each other.[9] This process continues until the electrons are completely microbunched and the radiation reaches a saturated power several orders of magnitude higher than that of the undulator radiation.

The wavelength of the radiation emitted can be readily tuned by adjusting the energy of the electron beam or the magnetic-field strength of the undulators.

FELs are relativistic machines. The wavelength of the emitted radiation, λ_r, is given by [10]

$$\lambda_r = \frac{\lambda_u}{2\gamma^2}(1 + K^2)$$

or when the wiggler strength parameter K, discussed below, is small

$$\lambda_r \propto \frac{\lambda_u}{2\gamma^2}$$

where λ_u is the undulator wavelength (the spatial period of the magnetic field), γ is the relativistic Lorentz factor and the proportionality constant depends on the undulator geometry and is of the order of 1.

This formula can be understood as a combination of two relativistic effects. Imagine you are sitting on an electron passing through the undulator. Due to Lorentz contraction the undulator is shortened by a γ factor and the electron experiences much shorter undulator wavelength λ_u/γ. However, the radiation emitted at this wavelength is observed in the laboratory frame of reference and the relativistic Doppler effect brings the second γ factor to the above formula. Rigorous derivation from Maxwell's equations gives the divisor of 2 and the proportionality constant. In an x-ray FEL the typical undulator wavelength of 1 cm is transformed to x-ray wavelengths on the order of 1 nm by $\gamma \approx 2000$, i.e. the electrons have to travel with the speed of 0.9999998c.

27.1.1   Wiggler strength parameter K

K, a dimensionless parameter, tells the wiggler strength as the relationship between the length of a period and the radius of bend.[11]

$$K = \frac{\gamma \lambda_u}{2\pi\rho} = \frac{eB_0\lambda_u}{\sqrt{8}\pi m_e c}$$

where ρ is the bending radius, B_0 is the applied magnetic field and m_e the electron mass.

27.1.2   Quantum effects

In most cases, the theory of classical electromagnetism adequately accounts for the behavior of free electron lasers.[12] For sufficiently short wavelengths, quantum effects of electron recoil and shot noise may have to be considered.[13]

27.2   Large facilities required

Free-electron lasers require the use of an electron accelerator with its associated shielding, as accelerated electrons can be a radiation hazard if not properly contained. These accelerators are typically powered by klystrons, which require a high voltage supply. The electron beam must be maintained in a vacuum which requires the use of numerous vacuum pumps along the beam path. While this equipment is bulky and expensive, free-electron lasers can achieve very high peak powers, and the tunability of FELs makes them highly desirable in many disciplines, including chemistry, structure determination of molecules in biology, medical diagnosis, and nondestructive testing.

27.3   X-ray laser without mirrors

The lack of a material to make mirrors that can reflect extreme ultraviolet and x-rays means that FELs at these frequencies cannot use a resonant cavity like other lasers, which reflects the radiation so it makes multiple passes through the undulator. Consequently, in an X-ray FEL the output beam is produced by a single pass of radiation through the undulator; there must be enough amplification over a single pass to produce an adequately bright beam.

X-ray free electron lasers use long undulators. The underlying principle of the intense pulses from the X-ray laser lies in the principle of self-amplified spontaneous emission (SASE), which leads to the microbunching. Initially all electrons are distributed evenly and they emit incoherent spontaneous radiation only. Through the interaction of this radiation and the electrons' oscillations, they drift into microbunches separated by a distance equal to one radiation wavelength. Through this interaction, all electrons begin emitting coherent radiation in phase. All emitted radiation can reinforce itself perfectly whereby wave crests and wave troughs are always superimposed on one another in the best possible way. This results in an exponential increase of emitted radiation power, leading to high beam intensities and laser-like properties.[14] Examples of facilities operating on the SASE FEL principle include the Free electron LASer in Hamburg (FLASH), the Linac Coherent Light Source (LCLS) at the SLAC National Accelerator Laboratory, the European x-ray free electron laser (XFEL) in Hamburg, the SPring-8 Compact SASE Source (SCSS), the SwissFEL at the Paul Scherrer Institute (Switzerland) and, as of 2011, the SACLA at the RIKEN Harima Institute in Japan.

27.4 Self seeding

One problem with SASE FELs is the lack of temporal coherence due to a noisy startup process. To avoid this, one can "seed" an FEL with a laser tuned to the resonance of the FEL. Such a temporally coherent seed can be produced by more conventional means, such as by high-harmonic generation (HHG) using an optical laser pulse. This results in coherent amplification of the input signal; in effect, the output laser quality is characterized by the seed. While HHG seeds are available at wavelengths down to the extreme ultraviolet, seeding is not feasible at x-ray wavelengths due to the lack of conventional x-ray lasers. In late 2010, in Italy, the seeded-FEL source FERMI@Elettra[15] started commissioning, at the Sincrotrone Trieste Laboratory. FERMI@Elettra is a single-pass FEL user-facility covering the wavelength range from 100 nm (12 eV) to 10 nm (124 eV), located next to the third-generation synchrotron radiation facility ELETTRA in Trieste, Italy. The advent of femtosecond lasers has revolutionized many areas of science from solid state physics to biology.

In 2012, scientists working on the LCLS overcame the seeding limitation for x-ray wavelengths by self-seeding the laser with its own beam after being filtered through a diamond monochromator. The resulting intensity and monochromaticity of the beam were unprecedented and allowed new experiments to be conducted involving manipulating atoms and imaging molecules. Other labs around the world are incorporating the technique into their equipment.[16][17]

27.5 Applications

27.5.1 Medical

Surgery

Research by Glenn Edwards and colleagues at Vanderbilt University's FEL Center in 1994 found that soft tissues including skin, cornea, and brain tissue could be cut, or ablated, using infrared FEL wavelengths around 6.45 micrometres with minimal collateral damage to adjacent tissue.[18][19] This led to surgeries on humans, the first ever using a free-electron laser. Starting in 1999, Copeland and Konrad performed three surgeries in which they resected meningioma brain tumors.[20] Beginning in 2000, Joos and Mawn performed five surgeries that cut a window in the sheath of the optic nerve, to test the efficacy for optic nerve sheath fenestration.[21] These eight surgeries produced results consistent with the standard of care and with the added benefit of minimal collateral damage. A review of FELs for medical uses is given in the 1st edition of Tunable Laser Applications.[22]

Fat removal

Several small, clinical lasers tunable in the 6 to 7 micrometre range with pulse structure and energy to give minimal collateral damage in soft tissue were created. At Vanderbilt, there exists a Raman shifted system pumped by an Alexandrite laser.[23]

Rox Anderson proposed the medical application of the free-electron laser in melting fats without harming the overlying skin.[24] At infrared wavelengths, water in tissue was heated by the laser, but at wavelengths corresponding to 915, 1210 and 1720 nm, subsurface lipids were differentially heated more strongly than water. The possible applications of this selective photothermolysis (heating tissues using light) include the selective destruction of sebum lipids to treat acne, as well as targeting other lipids associated with cellulite and body fat as well as fatty plaques that form in arteries which can help treat atherosclerosis and heart disease.[25]

27.5.2 Biology

Exceptionally bright and fast X-rays can image proteins using a sheet just one molecule thick. This technique allows first-time imaging of proteins that do not stack in a way that allows imaging by conventional techniques, 25% of the total number of proteins. Resolutions of 0.8 nm have been achieved with pulse durations of 30 femtoseconds. To get a clear view resolution of 0.1–0.3 nm is required. The short pulse durations prevented the lasers from destroying the molecules. The bright, fast X-rays were produced at the Linac Coherent Light Source at SLAC. As of 2014 LCLS was the world's most powerful X-ray FEL.[26]

27.5.3 Military

FEL technology is being evaluated by the US Navy as a candidate for an antiaircraft and missile directed-energy weapon. The Thomas Jefferson National Accelerator Facility's FEL has demonstrated over 14 kW power output.[27] Compact multi-megawatt class FEL weapons are undergoing research.[28] On June 9, 2009 the Office of Naval Research announced it had awarded Raytheon a contract to develop a 100 kW experimental FEL.[29] On March 18, 2010 Boeing Directed Energy Systems announced the completion of an initial design for U.S. Naval use.[30] A prototype FEL system was demonstrated, with a full-power prototype scheduled by 2018.[31]

27.6 See also

- Bremsstrahlung

- Cyclotron radiation

- Electron wake

- Gyrotron

- International Linear Collider

- Synchrotron radiation

27.7 References

[1] Huang, Z.; Kim, K. J. (2007). "Review of x-ray free-electron laser theory". *Physical Review Special Topics - Accelerators and Beams* **10** (3). doi:10.1103/PhysRevSTAB.10.034801.

[2] "Duke University Free-Electron Laser Laboratory". Retrieved 2007-12-21.

[3] F. J. Duarte (Ed.), *Tunable Lasers Handbook* (Academic, New York, 1995) Chapter 9.

[4] "New Era of Research Begins as World's First Hard X-ray Laser Achieves "First Light"". SLAC National Accelerator Laboratory. April 21, 2009. Retrieved 2013-11-06.

[5] Hans Motz, W. Thon, R.N. Whitehurst. Experiments on radiation by fast electron beams, *Journal of Applied Physics*, 24(7):826-833, 1953.

[6] Motz, Hans (1951). "Applications of the Radiation from Fast Electron Beams". *Journal of Applied Physics* **22** (5): 527. doi:10.1063/1.1700002.

[7] Motz, H.; Thon, W.; Whitehurst, R. N. (1953). "Experiments on Radiation by Fast Electron Beams". *Journal of Applied Physics* **24** (7): 826. doi:10.1063/1.1721389.

[8] "Phys. Rev. Lett. 38, 892 (1977): First Operation of a Free-Electron Laser". Prl.aps.org. Retrieved 2014-02-17.

[9] Feldhaus, J.; Arthur, J.; Hastings, J. B. (2005). "X-ray free-electron lasers". *Journal of Physics B: Atomic, Molecular and Optical Physics* **38** (9): S799. doi:10.1088/0953-4075/38/9/023.

[10] Neil, G.; Merminga, L. (2002). "Technical approaches for high-average-power free-electron lasers". *Reviews of Modern Physics* **74** (3): 685. doi:10.1103/RevModPhys.74.685.

[11] Robert Soliday (2006-09-05). "WIGGLER". Argon National laboratory.

[12] Fain, B.; Milonni, P. W. (1987). "Classical stimulated emission". *Journal of the Optical Society of America B* **4**: 78. doi:10.1364/JOSAB.4.000078.

[13] Benson, S.; Madey, J. M. J. (1984). "Quantum fluctuations in XUV free electron lasers". *AIP Conference Proceedings* **118**. p. 173. doi:10.1063/1.34633.

[14] "XFEL information webpages". Retrieved 2007-12-21.

[15] "FERMI / HomePage". Elettra.trieste.it. 2013-10-24. Retrieved 2014-02-17.

[16] Amann, J.; Berg, W.; Blank, V.; Decker, F. -J.; Ding, Y.; Emma, P.; Feng, Y.; Frisch, J.; Fritz, D.; Hastings, J.; Huang, Z.; Krzywinski, J.; Lindberg, R.; Loos, H.; Lutman, A.; Nuhn, H. -D.; Ratner, D.; Rzepiela, J.; Shu, D.; Shvyd'ko, Y.; Spampinati, S.; Stoupin, S.; Terentyev, S.; Trakhtenberg, E.; Walz, D.; Welch, J.; Wu, J.; Zholents, A.; Zhu, D. (2012). "Demonstration of self-seeding in a hard-X-ray free-electron laser". *Nature Photonics* **6** (10): 693. doi:10.1038/nphoton.2012.180.

[17] ""Self-seeding" promises to speed discoveries, add new scientific capabilities". SLAC National Accelerator Laboratory. August 13, 2012. Retrieved 2013-11-06.

[18] Edwards, G.; Logan, R.; Copeland, M.; Reinisch, L.; Davidson, J.; Johnson, B.; Maclunas, R.; Mendenhall, M.; Ossoff, R.; Tribble, J.; Werkhaven, J.; O'Day, D. (1994). "Tissue ablation by a free-electron laser tuned to the amide II band". *Nature* **371** (6496): 416. doi:10.1038/371416a0.

[19] "Laser light from Free-Electron Laser used for first time in human surgery". Retrieved 2010-11-06.

[20] Glenn S. Edwards et al., Rev. Sci. Instrum. 74 (2003) 3207

[21] MacKanos, M. A.; Joos, K. M.; Kozub, J. A.; Jansen, E. D. (2005). "Corneal ablation using the pulse stretched free electron laser". *Ophthalmic Technologies XV*. Ophthalmic Technologies XV **5688**. p. 177. doi:10.1117/12.596603.

[22] F. J. Duarte (12 December 2010). "6". *Tunable Laser Applications, Second Edition*. CRC Press. ISBN 978-1-4200-6058-4.

[23] "Efficiency and Plume Dynamics for Mid-IR Laser Ablation of Cornea". 2009-03-18. Retrieved 2010-11-06.

[24] "BBC health". *BBC News*. 2006-04-10. Retrieved 2007-12-21.

[25] "Dr Rox Anderson treatment". Retrieved 2007-12-21.

[26] "Super-bright, fast X-ray free-electron lasers can now image single layer of proteins". KurzweilAI. doi:10.1107/S2052252514001444. Retrieved 2014-02-17.

[27] "Jefferson Lab FEL". Retrieved 2009-06-08.

[28] "Airborne megawatt class free-electron laser for defense and security". Retrieved 2007-12-21.

[29] "Raytheon Awarded Contract for Office of Naval Research's Free Electron Laser Program". Retrieved 2009-06-12.

[30] "Boeing Completes Preliminary Design of Free Electron Laser Weapon System". Retrieved 2010-03-29.

[31] "Breakthrough Laser Could Revolutionize Navy's Weaponry". Fox News. 2011-01-20. Retrieved 2011-01-22.

27.8 Further reading

- Madey, John, "Stimulated emission of bremsstrahlung in a periodic magnetic field". J. Appl. Phys. 42, 1906 (1971)

- Madey, John, Stimulated emission of radiation in periodically deflected electron beam, US Patent 38 22 410, 1974

- Boscolo, et al., *"Free-Electron Lasers and Masers on Curved Paths"*. Appl. Phys., (Germany), vol. 19, No. 1, pp. 46–51, May 1979.

- Deacon et al., *"First Operation of a Free-Electron Laser"*. Phys. Rev. Lett., vol. 38, No. 16, Apr. 1977, pp. 892–894.

- Elias, et al., *"Observation of Stimulated Emission of Radiation by Relativistic Electrons in a Spatially Periodic Transverse Magnetic Field"*, Phys. Rev. Lett., 36 (13), 1976, p. 717.

- Gover, *"Operation Regimes of Cerenkov-Smith-Purcell Free Electron Lasers and T. W. Amplifiers"*. Optics Communications, vol. 26, No. 3, Sep. 1978, pp. 375–379.

- Gover, *"Collective and Single Electron Interactions of Electron Beams with Electromagnetic Waves and Free Electrons Lasers"*. App. Phys. 16 (1978), p. 121.

- *"The FEL Program at Jefferson Lab"*

- Brau, Charles (1990). "Free-Electron Lasers". Boston: Academic Press, Inc.

- Paolo Luchini, Hans Motz, *Undulators and Free-electron Lasers*, Oxford University Press, 1990.

27.9 External links

- Lightsources.org

- FERMI, the new FEL at the ELETTRA synchrotron in Triest

- Free-Electron Laser Open Book (National Academies Press)

- The World Wide Web Virtual Library: Free-Electron Laser research and applications

- European XFEL

- PSI SwissFEL

- SPring-8 Compact SASE Source

- Electron beam transport system and diagnostics of the Dresden FEL

- The Free Electron Laser for Infrared eXperiments FELIX

- W. M. Keck Free Electron Laser Center

- Jefferson Lab's Free-Electron Laser Program

- Free-Electron Lasers: The Next Generation by Davide Castelvecchi New Scientist, January 21, 2006

- Airborne megawatt class free-electron laser for defense and security

- FERMI@Elettra Free-Electron Laser Project

- Center for Free-Electron Laser Science (CFEL)

Chapter 28

Cathode ray tube

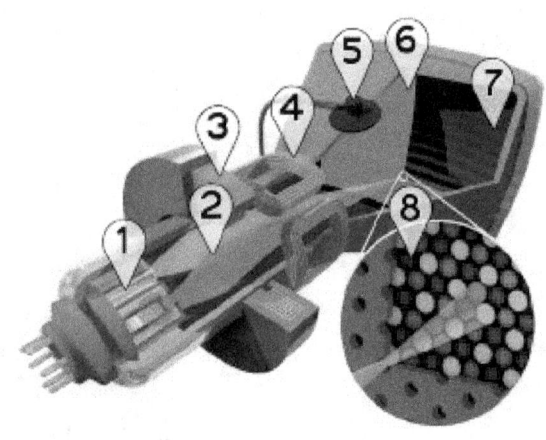

Cutaway rendering of a color CRT:
1. *Three electron guns (for red, green, and blue phosphor dots)*
2. *Electron beams*
3. *Focusing coils*
4. *Deflection coils*
5. *Anode connection*
6. *Mask for separating beams for red, green, and blue part of displayed image*
7. *Phosphor layer with red, green, and blue zones*
8. *Close-up of the phosphor-coated inner side of the screen*

The **cathode ray tube** (CRT) is a vacuum tube containing one or more electron guns, and a phosphorescent screen used to view images.[1] It has a means to accelerate and deflect the electron beam(s) onto the screen to create the images. The images may represent electrical waveforms (oscilloscope), pictures (television, computer monitor), radar targets or others. CRTs have also been used as memory devices, in which case the visible light emitted from the fluorescent material (if any) is not intended to have significant meaning to a visual observer (though the visible pattern on the tube face may cryptically represent the stored data).

The CRT uses an evacuated glass envelope which is large, deep (i.e. long from front screen face to rear end), fairly

heavy, and relatively fragile. As a matter of safety, the face is typically made of thick lead glass so as to be highly shatter-resistant and to block most X-ray emissions, particularly if the CRT is used in a consumer product.

Since the mid 2000s, CRTs have largely been superseded by newer display technologies such as LCD, plasma display, and OLED, which have lower manufacturing costs, power consumption, weight and bulk.

The vacuum level inside the tube is high vacuum on the order of 0.01 Pa[2] to 133 nPa.[3]

In television sets and computer monitors, the entire front area of the tube is scanned repetitively and systematically in a fixed pattern called a raster. An image is produced by controlling the intensity of each of the three electron beams, one for each additive primary color (red, green, and blue) with a video signal as a reference.[4] In all modern CRT monitors and televisions, the beams are bent by *magnetic deflection*, a varying magnetic field generated by coils and driven by electronic circuits around the neck of the tube, although electrostatic deflection is commonly used in oscilloscopes, a type of diagnostic instrument.[4]

28.1 History

Cathode rays were discovered by Johann Hittorf in 1869 in primitive Crookes tubes. He observed that some unknown rays were emitted from the cathode (negative electrode) which could cast shadows on the glowing wall of the tube, indicating the rays were traveling in straight lines. In 1890, Arthur Schuster demonstrated cathode rays could be deflected by electric fields, and William Crookes showed they could be deflected by magnetic fields. In 1897, J. J. Thomson succeeded in measuring the mass of cathode rays, showing that they consisted of negatively charged particles smaller than atoms, the first "subatomic particles", which were later named *electrons*. The earliest version of the CRT was known as the "Braun tube", invented by the German physicist Ferdinand Braun in 1897.[5][6] It was a cold-

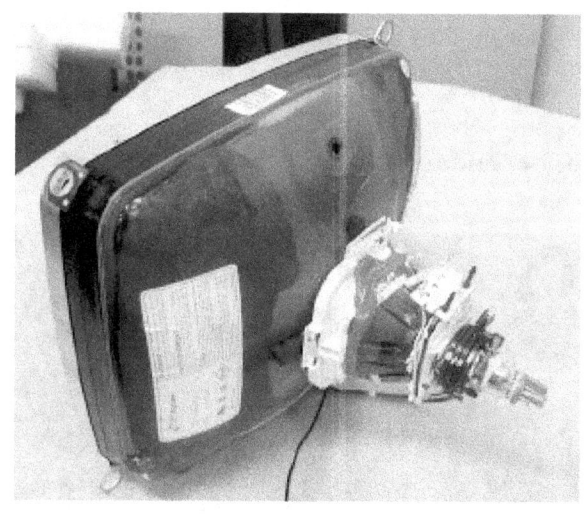

A 14 inch cathode ray tube showing its deflection coils and electron guns

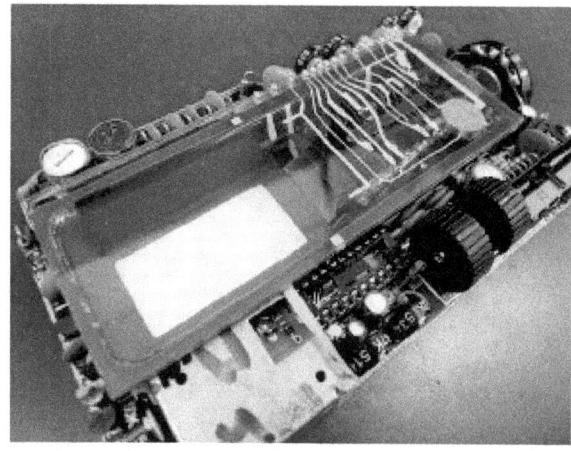

A flat CRT assembly inside a 1984 Sinclair FTV1 pocket TV

Electron gun

Braun's original cold-cathode CRT, 1897

Typical 1950s United States television set

1929.[7] RCA was granted a trademark for the term (for its cathode ray tube) in 1932; it voluntarily released the term to the public domain in 1950.[8]

The first commercially made electronic television sets with cathode ray tubes were manufactured by Telefunken in Germany in 1934.[9][10]

28.2 Oscilloscope CRTs

In oscilloscope CRTs, electrostatic deflection is used, rather than the magnetic deflection commonly used with television and other large CRTs. The beam is deflected horizontally by applying an electric field between a pair of plates to its left and right, and vertically by applying an electric field to plates above and below. Televisions use magnetic rather than electrostatic deflection because the deflection plates obstruct the beam when the deflection angle is as large as is required for tubes that are relatively short for their size.

cathode diode, a modification of the Crookes tube with a phosphor-coated screen.

In 1907, Russian scientist Boris Rosing used a CRT in the receiving end of an experimental video signal to form a picture. He managed to display simple geometric shapes onto the screen, which marked the first time that CRT technology was used for what is now known as television.[1]

The first cathode ray tube to use a hot cathode was developed by John B. Johnson (who gave his name to the term Johnson noise) and Harry Weiner Weinhart of Western Electric, and became a commercial product in 1922.

It was named by inventor Vladimir K. Zworykin in

for the graticule to be illuminated from the side, which improves its visibility.[16]

28.2.4   Image storage tubes

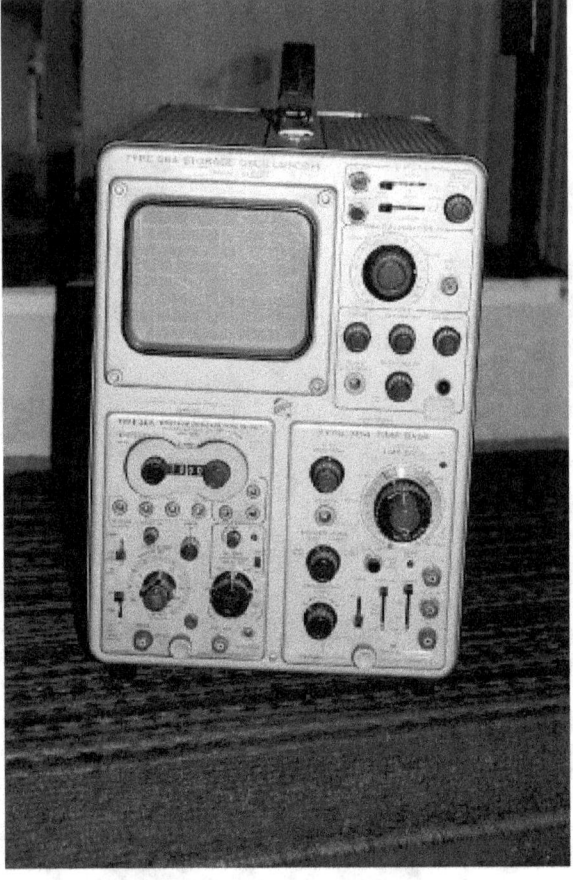

The Model 564 First Mass Produced Analog Phosphor Storage Oscilloscope

28.2.1   Phosphor persistence

Various phosphors are available depending upon the needs of the measurement or display application. The brightness, color, and persistence of the illumination depends upon the type of phosphor used on the CRT screen. Phosphors are available with persistences ranging from less than one microsecond to several seconds.[11] For visual observation of brief transient events, a long persistence phosphor may be desirable. For events which are fast and repetitive, or high frequency, a short-persistence phosphor is generally preferable.[12]

28.2.2   Microchannel plate

When displaying fast one-shot events, the electron beam must deflect very quickly, with few electrons impinging on the screen, leading to a faint or invisible image on the display. Oscilloscope CRTs designed for very fast signals can give a brighter display by passing the electron beam through a micro-channel plate just before it reaches the screen. Through the phenomenon of secondary emission, this plate multiplies the number of electrons reaching the phosphor screen, giving a significant improvement in writing rate (brightness) and improved sensitivity and spot size as well.[13][14]

28.2.3   Graticules

Most oscilloscopes have a graticule as part of the visual display, to facilitate measurements. The graticule may be permanently marked inside the face of the CRT, or it may be a transparent external plate made of glass or acrylic plastic. An internal graticule eliminates parallax error, but cannot be changed to accommodate different types of measurements.[15] Oscilloscopes commonly provide a means

These are found in *analog phosphor storage oscilloscopes*. These are distinct from *digital storage oscilloscopes* which rely on solid state digital memory to store the image.

Where a single brief event is monitored by an oscilloscope, such an event will be displayed by a conventional tube only while it actually occurs. The use of a long persistence phosphor may allow the image to be observed after the event, but only for a few seconds at best. This limitation can be overcome by the use of a direct view storage cathode ray tube (storage tube). A storage tube will continue to display the event after it has occurred until such time as it is erased. A storage tube is similar to a conventional tube except that it is equipped with a metal grid coated with a dielectric layer located immediately behind the phosphor screen. An externally applied voltage to the mesh initially ensures that the whole mesh is at a constant potential. This mesh is constantly exposed to a low velocity electron beam from a 'flood gun' which operates independently of the main gun. This flood gun is not deflected like the main gun but constantly 'illuminates' the whole of the storage mesh. The initial charge on the storage mesh is such as to repel the electrons from the flood gun which are prevented from striking the phosphor screen.

When the main electron gun writes an image to the screen, the energy in the main beam is sufficient to create a 'potential relief' on the storage mesh. The areas where this relief is created no longer repel the electrons from the flood gun which now pass through the mesh and illuminate the phosphor screen. Consequently, the image that was briefly traced out by the main gun continues to be displayed after it has occurred. The image can be 'erased' by resupplying the external voltage to the mesh restoring its constant potential. The time for which the image can be displayed was limited because, in practice, the flood gun slowly neutralises the charge on the storage mesh. One way of allowing the image to be retained for longer is temporarily to turn off the flood gun. It is then possible for the image to be retained for several days. The majority of storage tubes allow for a lower voltage to be applied to the storage mesh which slowly restores the initial charge state. By varying this voltage a variable persistence is obtained. Turning off the flood gun and the voltage supply to the storage mesh allows such a tube to operate as a conventional oscilloscope tube.[17]

28.2.5   Data storage tubes

For more details on this topic, see williams tube.

28.3   Color CRTs

Color tubes use three different phosphors which emit red, green, and blue light respectively. They are packed together in stripes (as in aperture grille designs) or clusters called "triads" (as in shadow mask CRTs).[18] Color CRTs have three electron guns, one for each primary color, arranged either in a straight line or in an equilateral triangular configuration (the guns are usually constructed as a single unit). (The triangular configuration is often called "delta-gun", based on its relation to the shape of the Greek letter delta.) A grille or mask absorbs the electrons that would otherwise hit the wrong phosphor.[19] A shadow mask tube uses a metal plate with tiny holes, placed so that the electron beam only illuminates the correct phosphors on the face of the tube;[18] the holes are tapered so that the electrons that strike the inside of any hole will be reflected back, if they are not absorbed (e.g. due to local charge accumulation),

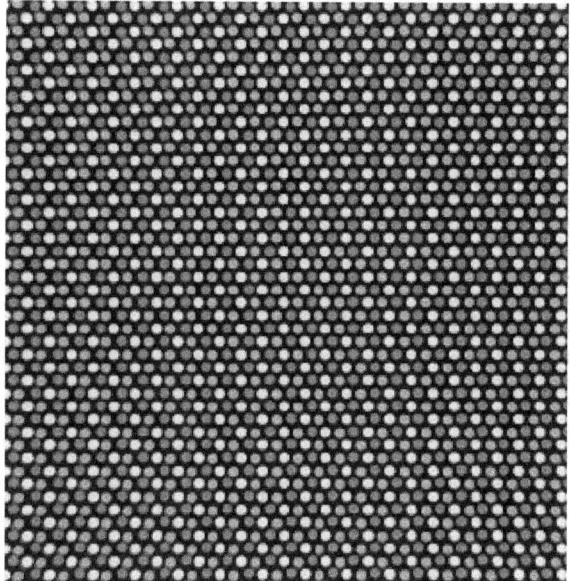

Magnified view of a delta-gun shadow mask color CRT

Magnified view of a Trinitron color CRT

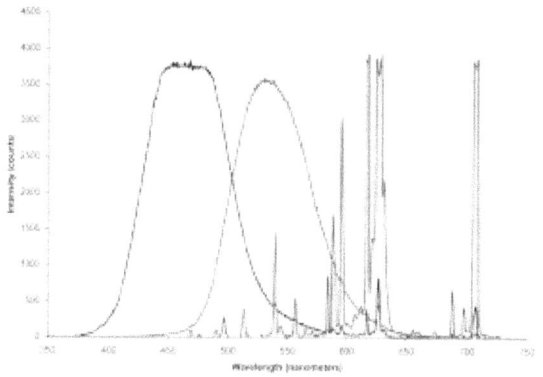

Spectra of constituent blue, green and red phosphors in a common CRT

instead of bouncing through the hole to strike a random (wrong) spot on the screen. Another type of color CRT uses an aperture grille of tensioned vertical wires to achieve the same result.[19]

28.3.1   Convergence and purity in color CRTs

Due to limitations in the dimensional precision with which CRTs can be manufactured economically, it has not been practically possible to build color CRTs in which three electron beams could be aligned to hit phosphors of respective color in acceptable coordination, solely on the basis of the geometric configuration of the electron gun axes and gun aperture positions, shadow mask apertures, etc. The shadow mask ensures that one beam will only hit spots of certain colors of phosphors, but minute variations in physical alignment of the internal parts among individual CRTs will cause variations in the exact alignment of the beams through the shadow mask, allowing some electrons from, for example, the red beam to hit, say, blue phosphors, unless some individual compensation is made for the variance among individual tubes.

Color convergence and color purity are two aspects of this single problem. Firstly, for correct color rendering it is necessary that regardless of where the beams are deflected on the screen, all three hit the same spot (and nominally pass through the same hole or slot) on the shadow mask. This is called convergence.[20] More specifically, the convergence at the center of the screen (with no deflection field applied by the yoke) is called static convergence, and the convergence over the rest of the screen area is called dynamic convergence. The beams may converge at the center of the screen and yet stray from each other as they are deflected toward the edges; such a CRT would be said to have good static convergence but poor dynamic convergence. Secondly, each beam must only strike the phosphors of the color it is intended to strike and no others. This is called purity. Like convergence, there is static purity and dynamic purity, with the same meanings of "static" and "dynamic" as for convergence. Convergence and purity are distinct parameters; a CRT could have good purity but poor convergence, or vice versa. Poor convergence causes color "shadows" or "ghosts" along displayed edges and contours, as if the image on the screen were intaglio printed with poor registration. Poor purity causes objects on the screen to appear off-color while their edges remain sharp. Purity and convergence problems can occur at the same time, in the same or different areas of the screen or both over the whole screen, and either uniformly or to greater or lesser degrees over different parts of the screen.

The solution to the static convergence and purity problems is a set of color alignment magnets installed around the neck of the CRT. These movable weak permanent magnets are usually mounted on the back end of the deflection yoke assembly and are set at the factory to compensate for any static purity and convergence errors that are intrinsic to the unadjusted tube. Typically there are two or three pairs of two magnets in the form of rings made of plastic impregnated with a magnetic material, with their magnetic fields parallel to the planes of the magnets, which are perpendicular to the electron gun axes. Each pair of magnetic rings forms a single effective magnet whose field vector can be fully and freely adjusted (in both direction and magnitude). By rotating a pair of magnets relative to each other, their relative field alignment can be varied, adjusting the effective field strength of the pair. (As they rotate relative to each other, each magnet's field can be considered to have two opposing components at right angles, and these four components [two each for two magnets] form two pairs, one pair reinforcing each other and the other pair opposing and cancelling each other. Rotating away from alignment, the magnets' mutually reinforcing field components decrease as they are traded for increasing opposed, mutually cancelling components.) By rotating a pair of magnets together, preserving the relative angle between them, the direction of their collective magnetic field can be varied. Overall, adjusting all of the convergence/purity magnets allows a finely tuned slight electron beam deflection and/or lateral offset to be applied, which compensates for minor static convergence and purity errors intrinsic to the uncalibrated tube. Once set, these magnets are usually glued in place, but normally they can be freed and readjusted in the field (e.g. by a TV repair shop) if necessary.

On some CRTs, additional fixed adjustable magnets are added for dynamic convergence and/or dynamic purity at specific points on the screen, typically near the corners or edges. Further adjustment of dynamic convergence and purity typically cannot be done passively, but requires active compensation circuits.

Dynamic color convergence and purity are one of the main reasons why until late in their history, CRTs were long-necked (deep) and had biaxially curved faces; these geometric design characteristics are necessary for intrinsic passive dynamic color convergence and purity. Only starting around the 1990s did sophisticated active dynamic convergence compensation circuits become available that made short-necked and flat-faced CRTs workable. These active compensation circuits use the deflection yoke to finely adjust beam deflection according to the beam target location. The same techniques (and major circuit components) also make possible the adjustment of display image rotation, skew, and other complex raster geometry parameters through electronics under user control.

28.3.2 Degaussing

If the shadow mask becomes magnetized, its magnetic field deflects the electron beams passing through it, causing color purity distortion as the beams bend through the mask holes and hit some phosphors of a color other than that which they are intended to strike; e.g. some electrons from the red beam may hit blue phosphors, giving pure red parts of the image a magenta tint. (Magenta is the additive combination of red and blue.) This effect is localized to a specific area of the screen if the magnetization of the shadow mask is localized. Therefore, it is important that the shadow mask is unmagnetized. (A magnetized aperture grille has a similar effect, and everything stated in this subsection about shadow masks applies as well to aperture grilles.)

Most color CRT displays, i.e. television sets and computer monitors, each have a built-in degaussing (demagnetizing) circuit, the primary component of which is a degaussing coil which is mounted around the perimeter of the CRT face inside the bezel. Upon power-up of the CRT display, the degaussing circuit produces a brief, alternating current through the degaussing coil which smoothly decays in strength (fades out) to zero over a period of a few seconds, producing a decaying alternating magnetic field from the coil. This degaussing field is strong enough to remove shadow mask magnetization in most cases.[21] In unusual cases of strong magnetization where the internal degaussing field is not sufficient, the shadow mask may be degaussed externally with a stronger portable degausser or demagnetizer. However, an excessively strong magnetic field, whether alternating or constant, may mechanically deform (bend) the shadow mask, causing a permanent color distortion on the display which looks very similar to a magnetization effect.

The degaussing circuit is often built of a thermo-electric (not electronic) device containing a small ceramic heating element and a positive thermal coefficient (PTC) resistor, connected directly to the switched AC power line with the resistor in series with the degaussing coil. When the power is switched on, the heating element heats the PTC resistor, increasing its resistance to a point where degaussing current is minimal, but not actually zero. In older CRT displays, this low-level current (which produces no significant degaussing field) is sustained along with the action of the heating element as long as the display remains switched on. To repeat a degaussing cycle, the CRT display must be switched off and left off for at least several seconds to reset the degaussing circuit by allowing the PTC resistor to cool to the ambient temperature; switching the display off and immediately back on will result in a weak degaussing cycle or effectively no degaussing cycle.

This simple design is effective and cheap to build, but it wastes some power continuously. Later models, especially Energy Star rated ones, use a relay to switch the entire degaussing circuit on and off, so that the degaussing circuit uses energy only when it is functionally active and needed. The relay design also enables degaussing on user demand through the unit's front panel controls, without switching the unit off and on again. This relay can often be heard clicking off at the end of the degaussing cycle a few seconds after the monitor is turned on, and on and off during a manually initiated degaussing cycle.

28.4 Vector monitors

Main article: Vector monitor

Vector monitors were used in early computer aided design systems and in some late-1970s to mid-1980s arcade games such as *Asteroids*.[22] They draw graphics point-to-point, rather than scanning a raster. Either monochrome or color CRTs can be used in vector displays, and the essential principles of CRT design and operation are the same for either type of display; the main difference is in the beam deflection patterns and circuits.

28.5 CRT resolution

Dot pitch defines the maximum resolution of the display, assuming delta-gun CRTs. In these, as the scanned resolution approaches the dot pitch resolution, moiré appears, as the detail being displayed is finer than what the shadow mask can render.[23] Aperture grille monitors do not suffer from vertical moiré; however, because their phosphor stripes have no vertical detail. In smaller CRTs, these strips maintain position by themselves, but larger aperture grille CRTs require one or two crosswise (horizontal) support strips.[24]

28.6 Gamma

CRTs have a pronounced triode characteristic, which results in significant gamma (a nonlinear relationship in an electron gun between applied video voltage and beam intensity).[25]

28.7 Other types of CRTs

28.7.1 Cat's eye

Main article: Magic eye tube

In better quality tube radio sets a tuning guide consisting of a phosphor tube was used to aid the tuning adjustment. This was also known as a "Magic Eye" or "Tuning Eye". Tuning would be adjusted until the width of a radial shadow was minimized. This was used instead of a more expensive electromechanical meter, which later came to be used on higher-end tuners when transistor sets lacked the high voltage required to drive the device.[26] The same type of device was used with tape recorders as a recording level meter.

28.7.2 Charactrons

Some displays for early computers (those that needed to display more text than was practical using vectors, or that required high speed for photographic output) used Charactron CRTs. These incorporate a perforated metal character mask (stencil), which shapes a wide electron beam to form a character on the screen. The system selects a character on the mask using one set of deflection circuits, but that causes the extruded beam to be aimed off-axis, so a second set of deflection plates has to re-aim the beam so it is headed toward the center of the screen. A third set of plates places the character wherever required. The beam is unblanked (turned on) briefly to draw the character at that position. Graphics could be drawn by selecting the position on the mask corresponding to the code for a space (in practice, they were simply not drawn), which had a small round hole in the center; this effectively disabled the character mask, and the system reverted to regular vector behavior. Charactrons had exceptionally long necks, because of the need for three deflection systems.[27][28]

28.7.3 Nimo

Main article: Nimo tube
 Nimo was the trademark of a family of small specialised CRTs manufactured by Industrial Electronics Engineers. These had 10 electron guns which produced electron beams in the form of digits in a manner similar to that of the charactron. The tubes were either simple single-digit displays or more complex 4- or 6- digit displays produced by means of a suitable magnetic deflection system. Having little of the complexities of a standard CRT, the tube required a relatively simple driving circuit, and as the image was projected on the glass face, it provided a much wider viewing angle than competitive types (e.g., nixie tubes).[29]

Nimo tube BA0000-P31

28.7.4 Williams tube

Main article: Williams tube

The Williams tube or Williams-Kilburn tube was a cathode ray tube used to electronically store binary data. It was used in computers of the 1940s as a random-access digital storage device. In contrast to other CRTs in this article, the Williams tube was not a display device, and in fact could not be viewed since a metal plate covered its screen.

28.7.5 Flood beam CRT

Flood beam CRT's are small tubes that are arranged as pixels for large screens like Jumbotrons. The first screen using this technology was introduced by Mitsubishi Electric for the 1980 Major League Baseball All-Star Game. It differs from a normal CRT in that the electron gun within does not produce a focused controllable beam. Instead, electrons are sprayed in a wide cone across the entire front of the phosphor screen, basically making each unit act as a single light bulb.[30] Each one is coated with a red, green or blue phosphor, to make up the color sub-pixels. This technology has largely been replaced with light emitting diode displays. A similar device has been proposed by one manufacturer as a lamp.

28.7.6 Zeus thin CRT display

In the late 1990s and early 2000s Philips Research Laboratories experimented with a type of thin CRT known as the *Zeus* display which contained CRT-like functionality in a flat panel display.[31][32][33][34][35] The devices were demonstrated but never marketed.

28.8 The future of CRT technology

28.8.1 Demise

Although a mainstay of display technology for decades, CRT-based computer monitors and televisions constitute a dead technology. The demand for CRT screens has dropped precipitously since 2000, and this falloff had accelerated in the latter half of that decade. The rapid advances and falling prices of LCD flat panel technology, first for computer monitors and then for televisions, has been the key factor in the demise of competing display technologies such as CRT, rear-projection, and plasma display.[36]

The end of most high-end CRT production by around 2010[37] (including high-end Sony and Mitsubishi product lines) means an erosion of the CRT's capability.[38][39] In Canada and the United States, the sale and production of high-end CRT TVs (30-inch screens) in these markets had all but ended by 2007; just a couple of years later, inexpensive combo CRT TVs (20-inch screens with an integrated VHS or DVD player) have disappeared from discount stores. It has been common to replace CRT-based televisions and monitors in as little as 5–6 years, although they generally are capable of satisfactory performance for a much longer time.

Companies are responding to this trend. Electronics retailers such as Best Buy have been steadily reducing store spaces for CRTs. In 2005, Sony announced that they would stop the production of CRT computer displays. Samsung did not introduce any CRT models for the 2008 model year at the 2008 Consumer Electronics Show and on 4 February 2008 Samsung removed their 30" wide screen CRTs from

their North American website and has not replaced them with new models.[40]

The demise of CRT, however, has been happening more slowly in the developing world. According to iSupply, production in units of CRTs was not surpassed by LCDs production until 4Q 2007, owing largely to CRT production at factories in China.

In the United Kingdom, DSG (Dixons), the largest retailer of domestic electronic equipment, reported that CRT models made up 80–90% of the volume of televisions sold at Christmas 2004 and 15–20% a year later, and that they were expected to be less than 5% at the end of 2006. Dixons ceased selling CRT televisions in 2007.[41]

28.8.2 Causes

CRTs, despite recent advances, have remained relatively heavy and bulky and take up a lot of space in comparison to other display technologies. CRT screens have much deeper cabinets compared to flat panels and rear-projection displays for a given screen size, and so it becomes impractical to have CRTs larger than 40 inches (102 cm). The CRT disadvantages became especially significant in light of rapid technological advancements in LCD and plasma flat-panels which allow them to easily surpass 40 inches (102 cm) as well as being thin and wall-mountable, two key features that were increasingly being demanded by consumers.

By 2006, although the price points of CRTs were generally much lower than LCD and plasma flat panels, large screen CRTs (30-inches or more) were as expensive as a similar-sized LCD.[42]

28.8.3 Slimmer CRT

Some CRT manufacturers, both LG Display and Samsung Display, have innovated CRT technology by creating a slimmer tube. Slimmer CRT has a trade name Superslim and Ultraslim. A 21-inch flat CRT has 447.2 millimeter depth. The depth of Superslim is 352 millimeters and Ultraslim is 295.7 millimeters.

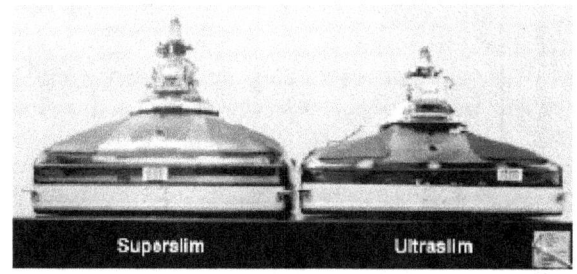

A comparison between 21 inch Superslim and Ultraslim CRT

28.8.4   Resurgence in specialized markets

In the first quarter of 2008, CRTs retook the #2 technology position in North America from plasma, due to the decline and consolidation of plasma display manufacturers. DisplaySearch has reported that although in the 4Q of 2007 LCDs surpassed CRTs in worldwide sales, CRTs then outsold LCDs in the 1Q of 2008.[43][44]

CRTs are useful for displaying photos with high pixels per unit area and correct color balance. LCDs, as currently the most common flatscreen technology, have generally inferior color rendition (despite having greater overall brightness) due to the fluorescent lights commonly used as a backlight.[45]

CRTs are still popular in the printing and broadcasting industries as well as in the professional video, photography, and graphics fields due to their greater color fidelity, contrast, and better viewing from off-axis (wider viewing angle). CRTs also still find adherents in vintage video gaming because of their higher resolution per initial cost, lowest possible input lag, fast response time, and multiple native resolutions such as 576p. Many retro gamers also prefer to use older CRTs with lower resolutions and scan rates due to the distinct scan line effect created when displaying the 240p analog signal generated by most retro consoles.

CRT monitors are still widely used in the study of the brain's visual processing (e.g. in psychophysics). The speed and fidelity of their response, combined with the simplicity of their design, makes them well-suited for experiments where scientists need to have very fine control over stimuli which are presented to an observer.[46]

28.9   Health concerns

28.9.1   Ionizing radiation

CRTs can emit a small amount of X-ray radiation as a result of the electron beam's bombardment of the shadow mask/aperture grille and phosphors. The amount of radiation escaping the front of the monitor is widely considered unharmful. The Food and Drug Administration regulations in 21 C.F.R. 1020.10 are used to strictly limit, for instance, television receivers to 0.5 milliroentgens per hour (mR/h) (0.13 µC/(kg·h) or 36 pA/kg) at a distance of 5 cm (2 in) from any external surface; since 2007, most CRTs have emissions that fall well below this limit.[47]

28.9.2   Toxicity

Older color and monochrome CRTs may contain toxic substances, such as cadmium, in the phosphors.[48][49][50] The rear glass tube of modern CRTs may be made from leaded glass, which represent an environmental hazard if disposed of improperly.[51] By the time personal computers were produced, glass in the front panel (the viewable portion of the CRT) used barium rather than lead, though the rear of the CRT was still produced from leaded glass. Monochrome CRTs typically do not contain enough leaded glass to fail EPA TCLP tests. While the TCLP process grinds the glass into fine particles in order to expose them to weak acids to test for leachate, intact CRT glass does not leache (The lead is vitrified, contained inside the glass itself, similar to leaded glass crystalware).

In October 2001, the United States Environmental Protection Agency created rules stating that CRTs must be brought to special recycling facilities. In November 2002, the EPA began fining companies that disposed of CRTs through landfills or incineration. Regulatory agencies, local and statewide, monitor the disposal of CRTs and other computer equipment.[52]

In Europe, disposal of CRT televisions and monitors is covered by the WEEE Directive.[53]

28.9.3   Flicker

Main article: Flicker (screen)

At low refresh rates (60 Hz and below), the periodic scanning of the display may produce a flicker that some people perceive more easily than others, especially when viewed with peripheral vision. Flicker is commonly associated with CRT as most televisions run at 50 Hz (PAL) or 60 Hz (NTSC), although there are some 100 Hz PAL televisions that are flicker-free. Typically only low-end monitors run at such low frequencies, with most computer monitors supporting at least 75 Hz and high-end monitors capable of 100 Hz or more to eliminate any perception of flicker.[54] Non-computer CRTs or CRT for sonar or radar may have long persistence phosphor and are thus flicker free. If the persistence is too long on a video display, moving images will be blurred.

28.9.4   High-frequency audible noise

50 Hz/60 Hz CRTs used for television operate with horizontal scanning frequencies of 15,734 Hz (for NTSC systems) or 15,625 Hz (for PAL systems).[55] These frequencies are at the upper range of human hearing and are inaudi-

ble to many people; however, some people (especially children) will perceive a high-pitched tone near an operating television CRT.*[56] The sound is due to magnetostriction in the magnetic core and periodic movement of windings of the flyback transformer. This problem does not occur on 100/120 Hz TVs and on non-CGA computer displays, because they are working on much higher frequencies (22 kHz to >100 kHz) compared to the low-frequency noise (50 Hz or 60 Hz) of mains hum.

28.9.5 Implosion

High vacuum inside glass-walled cathode ray tubes permits electron beams to fly freely — without colliding into molecules of air or other gas. If the glass is damaged, atmospheric pressure can collapse the vacuum tube into dangerous fragments which accelerate inward and then spray at high speed in all directions. The implosion energy is proportional to the evacuated volume of the CRT. Although modern cathode ray tubes used in televisions and computer displays have epoxy-bonded face-plates or other measures to prevent shattering of the envelope, CRTs must be handled carefully to avoid personal injury.*[57]

28.10 Security concerns

Under some circumstances, the signal radiated from the electron guns, scanning circuitry, and associated wiring of a CRT can be captured remotely and used to reconstruct what is shown on the CRT using a process called Van Eck phreaking.*[58] Special TEMPEST shielding can mitigate this effect. Such radiation of a potentially exploitable signal, however, occurs also with other display technologies*[59] and with electronics in general.

28.11 Recycling

As electronic waste, CRTs are considered one of the hardest types to recycle.*[60] CRTs have relatively high concentration of lead and phosphors (not phosphorus), both of which are necessary for the display. There are several companies in the United States that charge a small fee to collect CRTs, then subsidize their labour by selling the harvested copper, wire, and printed circuit boards. The United States Environmental Protection Agency (EPA) includes discarded CRT monitors in its category of "hazardous household waste"*[61] but considers CRTs that have been set aside for testing to be commodities if they are not discarded, speculatively accumulated, or left unprotected from weather and other damage.

Leaded CRT glass is sold to be remelted into other CRTs, or even broken down and used in road construction.*[62]

28.12 Advantages and disadvantages

Further information: Comparison CRT, LCD, Plasma

Pros

- High contrast ratio (over 15,000:1),*[63] excellent color, fairly wide color gamut and low black level.

- No native resolution; the only current display technology capable of true multisyncing (displaying many different resolutions and refresh rates without the need for scaling).

- No input lag.

- No ghosting and smearing artifacts during fast motion due to sub-millisecond response time, and impulse-based operation.

- Near zero color, saturation, contrast or brightness distortion.

- Excellent viewing angle.

- Allows the use of light guns/pens.

- Can be used or stored in both extreme hot and cold temperature conditions without harm to the system.*[64]

Cons

- Large size and weight, especially for bigger screens (a 20-inch (51 cm) unit weighs about 50 lb (23 kg)).

- Geometric distortion caused by variable beam travel distances.

- High power consumption. On average, a CRT monitor consumes 2–10× the power that an identically sized LCD monitor would consume, depending on the type of backlight used in the LCD screen, and its brightness setting.*[65]

- A lot of heat can be emitted during operation due to relatively high power consumption, which can mean a short product lifespan.

- Can suffer screen burn-in, though not as quickly as Plasma displays.

- Produces noticeable flicker at refresh rates lower than 85 Hz.

- Hazardous to repair/service without proper training and precautions taken.

- Maximum size for direct-view displays is limited to about 40 inches due to practical and manufacturing restrictions (a CRT display of this size can weigh about 300 pounds), though the sizing can be increased with an array of separate displays, such as the original Jumbotron used at sports arenas.

- The glass envelopes contains toxic lead and barium as X-ray radiation shielding. The phosphors can also contain toxic elements such as cadmium. Many countries treat CRTs as toxic waste and prohibit their disposal in landfills or by incineration.

- Purity and convergence in color tubes, affected by the Earth's magnetic field, usually roughly factory preset (biased) for operation in either the northern hemisphere, the southern hemisphere, or the equatorial area, but may require trimming at final location. Adjustment at final location requires a high degree of technical skill, as well as safety precautions associated with opening the display housing. [66]

- Sensitive to magnetic interference, which can cause the image to shimmer (e.g. if a transformer or other electro-magnetic source is too close to the screen) or the colors to shift (e.g. if an unshielded speaker is too close to the screen).

- Image edges are slightly diffuse (blurred), compared to the razor sharp stationary image an LCD, PDP, or other screen with fixed predefined pixels can produce. For text displays, computer-generated diagrams and line drawing, and other images composed mostly of sharp edges, razor sharp pixel definition is often preferred. However, this may be seen as an advantage for continuous-tone images such as photographic television, as it yields more natural images with no pixellation and it introduces a natural anti-aliasing effect to image information that is pixelated by its origin.

- A "halo" may appear around bright objects on a mostly dark screen.

- They have been phased out in favor of plasmas, LCDs and OLEDs.

28.13   See also

- Computer monitor

- Comparison of CRT, LCD, Plasma, and OLED

- Comparison of display technology

- Crookes tube

- CRT projector

- History of display technology

- Direct-view bistable storage tube

- Flat panel display

- Image dissector

- LCD television / LED-backlit LCD display

- Monitor filter

- Monoscope

- Overscan in Television

- Penetron

- Photosensitive epilepsy

- Surface-conduction electron-emitter display

- TCO Certification

- Trinitron

28.14   References

[1] "History of the Cathode Ray Tube". *About.com*. Retrieved 4 October 2009.

[2] Topic 7 | The Cathode-Ray Tube. aw.com. 2003-08-01

[3] repairfaq.org – Sam's Laser FAQ – Vacuum Technology for Home-Built Gas Lasers. repairfaq.org. 2012-08-02

[4] "How Computer Monitors Work". Retrieved 4 October 2009.

[5] Ferdinand Braun (1897) "Ueber ein Verfahren zur Demonstration und zum Studium des zeitlichen Verlaufs variabler Ströme" (On a process for the display and study of the course in time of variable currents), *Annalen der Physik und Chemie*, 3rd series, **60** : 552-559.

[6] "Cathode Ray Tube". *Medical Discoveries*. Advameg, Inc. 2007. Retrieved 27 April 2008.

[7] Albert Abramson, *Zworykin, Pioneer of Television*, University of Illinois Press, 1995, p. 84. ISBN 0-252-02104-5.

[8] "RCA Surrenders Rights to Four Trade-Marks," Radio Age, October 1950, p. 21.

[9] Telefunken, Early Electronic TV Gallery, Early Television Foundation.

[10] 1934–35 Telefunken, Television History: The First 75 Years.

[11] Doebelin, Ernest (2003). *Measurement Systems*. McGraw Hill Professional. p. 972. ISBN 978-0-07-292201-1.

[12] Shionoya, Shigeo (1999). *Phosphor handbook*. CRC Press. p. 499. ISBN 978-0-8493-7560-6.

[13] Williams, Jim (1991). *Analog circuit design: art, science, and personalities*. Newnes. pp. 115–116. ISBN 978-0-7506-9640-1.

[14] Yen, William M.; Shionoya, Shigeo and Yamamoto, Hajime (2006). *Practical Applications of Phosphors*. CRC Press. p. 211. ISBN 978-1-4200-4369-3.

[15] Bakshi, U.A. and Godse, A.P. (2008). *Electronic Devices And Circuits*. Technical Publications. p. 38. ISBN 978-81-8431-332-1.

[16] Hickman, Ian (2001). *Oscilloscopes: how to use them, how they work*. Newnes. p. 47. ISBN 978-0-7506-4757-1.

[17] The Great Soviet Encyclopedia, 3rd Edition (1970–1979)

[18] "How CRT and LCD monitors work". *bit-tech.net*. Retrieved 4 October 2009.

[19] "The Shadow Mask and Aperture Grill". *PC Guide*.

[20] Norton, Thomas J. (March 2005). "Picture This". *UltimateAVmag.com*. Archived from the original on 2009-11-26.

[21] "Magnetization and Degaussing". Retrieved 4 October 2009.

[22] Van Burnham (2001). *Supercade: A Visual History of the Videogame Age, 1971–1984*. MIT Press. ISBN 0-262-52420-1.

[23] "Moiré Interference Patterns". *DisplayMate Technologies website*. Retrieved 4 October 2009.

[24] "What causes the faint horizontal lines on my monitor?". *HowStuffWorks*. Retrieved 4 October 2009.

[25] Robin, Michael (1 January 2005). "Gamma correction". *BroadcastEngineering*. Archived from the original on 31 May 2009. Retrieved 4 October 2009.

[26] "Tuning-Eye Tubes". vacuumtube.com. Archived from the original on 23 April 2009. Retrieved 1 December 2009.

[27] "CATHODE RAY APPARATUS". Retrieved 4 October 2009.

[28] "INPUT". Retrieved 4 October 2009.

[29] "IEE Nimo CRT 10-gun readout tube datasheet" (PDF). *tube-tester.com*. Retrieved 1 December 2009.

[30] "Futaba TL-3508XA 'Jumbotron' Display". *The Vintage Technology Association: Military Industrial Electronics Research Preservation*. The Vintage Technology Association. 2010-03-11. Retrieved 2014-12-19.

[31] Beeteson, John Stuart (21 November 1998). "US Patent 6246165 – Magnetic channel cathode". Archived from the original on 9 January 2015.

[32] Van Hal, Henricus A. M.; et al. (18 May 1990). "US Patent 5905336 – Method of manufacturing a glass substrate coated with a metal oxide".

[33] Van Gorkom, G.G.P. (1996). "Introduction to Zeus displays". *Philips Journal of Research* **50** (3–4): 269. doi:10.1016/S0165-5817(97)84675-X.

[34] Lambert, N.; Montie, E.A.; Baller, T.S.; Van Gorkom, G.G.P.; Hendriks, B.H.W.; Trompenaars, P.H.F.; De Zwart, S.T. (1996). "Transport and extraction in Zeus displays". *Philips Journal of Research* **50** (3–4): 295. doi:10.1016/S0165-5817(97)84677-3.

[35] Doyle, T.; Van Asma, C.; McCormack, J.; De Greef, D.; Haighton, V.; Heijnen, P.; Looymans, M.; Van Velzen, J. (1996). "The application and system aspects of the Zeus display". *Philips Journal of Research* **50** (3–4): 501. doi:10.1016/S0165-5817(97)84688-8.

[36] Wong, May (22 October 2006). "Flat Panels Drive Old TVs From Market". AP via USA Today. Retrieved 8 October 2008.

[37] "The Standard TV" (PDF). Veritas et Visus. Retrieved 12 June 2008.

[38] "End of an era". The San Diego Union-Tribune. 20 January 2006. Retrieved 12 June 2008.

[39] "Matsushita says good-bye to CRTs". engadgetHD. 1 December 2005. Retrieved 12 June 2008.

[40] "SlimFit HDTV". Samsung. Archived from the original on 10 January 2008. Retrieved 12 June 2008.

[41] "The future is flat as Dixons withdraws sale of 'big box' televisions". London Evening Standard. 26 November 2006. Retrieved 3 December 2006.

[42] Wong, May (22 October 2006). "Flat panels drive old TVs from market". *USA Today*.

[43] "Worldwide LCD TV shipments surpass CRTs for first time ever". engadgetHD. 19 February 2008. Retrieved 12 June 2008.

[44] "LCD outsells plasma 8-to-1 in Q1 2008". engadgetHD. 22 May 2008. Retrieved 12 June 2008.

[45] "X-bit's Guide: Contemporary LCD Monitor Parameters and Characteristics (page 11)". Retrieved 4 October 2009.

[46] LRZ: Der URL ist umgezogen. Lrz.de. Retrieved on 26 August 2013. Archived October 8, 2013 at the Wayback Machine

[47] "SUBCHAPTER J —RADIOLOGICAL HEALTH (21CFR1020.10)". U.S. Food and Drug Administration. 1 April 2006. Retrieved 13 August 2007.

[48] "Toxic TVs". Electronics TakeBack Coalition. Archived from the original on 27 February 2009. Retrieved 13 April 2010.

[49] Peters-Michaud, Neil; Katers, John and Barry, Jim. "Occupational Risks Associated with Electronics Demanufacturing and CRT Glass Processing Operations and the Impact of Mitigation Activities on Employee Safety and Health" (PDF). *Cascade Asset Management, LLC*. Basel Action Network. Archived from the original (PDF) on 26 July 2011. Retrieved 20 January 2011.

[50] "Cadmium". American Elements. Retrieved 13 April 2010.

[51] "CHARACTERIZATION OF LEAD LEACHABILITY FROM CATHODE RAY TUBES USING THE TOXICITY CHARACTERISTIC LEACHING PROCEDURE (TCLP)" (PDF). Archived from the original (PDF) on 22 September 2013. Retrieved 4 October 2009.

[52] "Final Rules on Cathode Ray Tubes and Discarded Mercury-Containing Equipment". Retrieved 4 October 2009.

[53] "WEEE and CRT Processing". Retrieved 4 October 2009.

[54] "CRT Monitor Flickering?". Retrieved 4 October 2009.

[55] Netravali, Arun N.; Haskell, Barry G. (1995). *Digital pictures: representation, compression, and standards*. Plenum Publishing Corporation. p. 100. ISBN 978-0-306-44917-8.

[56] "The monitor is producing a high-pitched whine". Retrieved 4 October 2009.

[57] Bali, S.P. (1994). *Colour Television: Theory and Practice*. Tata McGraw-Hill. p. 129. ISBN 978-0-07-460024-5.

[58] "Electromagnetic Radiation from Video Display Units: An Eavesdropping Risk?" (PDF). Retrieved 4 October 2009.

[59] Kuhn, M.G. (2004). "Electromagnetic Eavesdropping Risks of Flat-Panel Displays" (PDF). *4th Workshop on Privacy Enhancing Technologies*: 23–25.

[60] WEEE: CRT and Monitor Recycling. Executive-blueprints.com (2 August 2009). Retrieved on 26 August 2013.

[61] Morgan, Russell (21 August 2006). "Tips and Tricks for Recycling Old Computers". SmartBiz. Retrieved 17 March 2009.

[62] Weitzman, David. The CRT Dilemma: Cathode Ray Tube Or Cruel Rude Trash. RRT Design & Construction

[63] Soneira, Raymond M. Display "Technology Shoot-Out: Comparing CRT, LCD, Plasma and DLP Displays", DisplayMate Technologies.

[64] No CRT data sheet bothers to mention min or max temperature because any temperature limit usually comes from other parts of the assembly. For example: "Typical high end Oscilloscope tube"

[65] Tom's Hardware: Power Consumption Benchmark Results for CRT versus TFT LCD "Benchmark Results: Different Brightness Testing"

[66] "Monitors: Earth's Magnetic Field Affects Performance". *Apple Support Knowledgebase*. Apple. Retrieved 21 June 2012.

28.15 Selected patents

- U.S. Patent 1,691,324: Zworykin Television System

28.16 External links

- Electromagnetic Deflection in a Cathode Ray Tube, Part 1 Interactive tutorial from the National High Magnetic Field Laboratory

- Electromagnetic Deflection in a Cathode Ray Tube, Part 2 Interactive tutorial from the National High Magnetic Field Laboratory

- The Cathode Ray Tube site

- PCTechGuide: Cathode Ray Tubes

- CRTs at the Virtual Valve Museum

- Samuel M. Goldwasser TV and Monitor CRT (Picture Tube) Information

28.17 Text and image sources, contributors, and licenses

28.17.1 Text

- **Electron** *Source:* https://en.wikipedia.org/wiki/Electron?oldid=689706211 *Contributors:* AxelBoldt, CYD, Mav, Bryan Derksen, AstroNomer~enwiki, Ap, Ed Poor, Andre Engels, Ryrivard, William Avery, SimonP, Peterlin~enwiki, Heron, Camembert, Stevertigo, Bdesham, Patrick, D, JohnOwens, Michael Hardy, Tim Starling, Ixfd64, Fruge~enwiki, Arpingstone, PingPongBoy, Egil, NuclearWinner, Ahoerstemeier, Suisui, Jebba, JWSchmidt, Kingturtle, Aarchiba, Glenn, Scott, Kwekubo, Andres, Jordi Burguet Castell, Mxn, Agtx, Timwi, Wikiborg, Reddi, Rednblu, Markhurd, Maximus Rex, E23~enwiki, Omegatron, Secretlondon, Jusjih, BenRG, Jeffq, Donarreiskoffer, Gentgeen, Robbot, Sanders muc, Vespristiano, Merovingian, Pingveno, Blainster, Hadal, Wikibot, Wereon, Widsith, HaeB, Diberri, Dmn, Dina, Giftlite, Christopher Parham, Ferkelparade, Fastfission, Zigger, Herbee, Dissident, Xerxes314, Curps, Michael Devore, Bensaccount, Ssd, Gilgamesh~enwiki, Vadmium, Gdr, Knutux, Slowking Man, Yath, Gzuckier, Pearbonn, Joizashmo, Karol Langner, Anythingyouwant, RetiredUser2, Bbbl67, Elroch, lcairns, JohnArmagh, JimQ, Mike Rosoft, Mindspillage, Patrick L, Goes, Discospinster, Brianhe, Rich Farmbrough, Guanabot, Hidaspal, Vsmith, Deh, Ardonik, Roybb95~enwiki, Xezbeth, Zazou, Mani1, SpookyMulder, Dmr2, Bender235, ZeroOne, Kjoonlee, Goplat, Calair, Nabla, Brian0918, RJHall, Pt, Jaques O. Carvalho, El C, Huntster, Edward Z. Yang, Susvolans, Art LaPella, RoyBoy, ~K, Bobo192, Army1987, Asierra~enwiki, Flxmghvgvk, AtomicDragon, Evgeny, AllyUnion, Bert Hickman, Deryck Chan, PeterisP, Beetle B., Obradovic Goran, (aeropagitica), Pearle, Mpulier, HasharBot~enwiki, Confusedmiked, Mote, Jumbuck, Gary, ChristopherWillis, Ricky81682, Benjahbmm27, Riana, AzaToth, DonJStevens, BernardH, Malo, David Hochron, Bart133, EagleFalconn, Schapel, Omphaloscope, RainbowOfLight, RichBlinne, H2g2bob, DV8 2XL, Gene Nygaard, Redvers, StuTheSheep, Linas, Mindmatrix, GrouchyDan, StradivariusTV, Uncle G, BillC, Kurzon, Jeff3000, HeorEric X, Eleassar777, Ozielke, Wayward, Palica, Omega21, FreplySpang, Enzo Aquarius, Rjwilmsi, Shaadow, Strait, Mike Peel, Chekaz, Bubba73, Dar-Ape, Yamamoto Ichiro, FlaBot, RobertG, Latka, DannyWilde, Nihiltres, RexNL, Kolbasz, Thecurran, Srleffler, Physchim62, Chobot, DVdm, Unclevortex, Eric B, YurikBot, Wavelength, RobotE, Bambaiah, AcidHelmNun, Jimp, Peter G Werner, Wolfmankurd, Wigie, Ventolin, JabberWok, SpuriousQ, Lucinos~enwiki, Akamad, Ori Livneh, Gaius Cornelius, Shaddack, Eleassar, Rsrikanth05, Salsb, Hawkeye7, Spike Wilbury, Jaxl, Welsh, DarthVader, Dlugosz, BirgitteSB, SCZenz, Retired username, Ravedave, PhilipO, Adam Rock, Mlouns, Chichui, BOT-Superzerocool, Gadget850, Bota47, Kkmurray, James Trotter~enwiki, Dna-webmaster, Ms2ger, Light current, Lycaon, Imaninjapirate, Josh3580, Kriscotta, JoanneB, Peyna, Lpm, JLaTondre, Heavy bolter, RG2, GrinBot~enwiki, Sbyrnes321, ChemGardener, Itub, SmackBot, Zazaban, Incnis Mrsi, KnowledgeOfSelf, Royalguard11, Melchoir, J.Sarfatti, KocjoBot~enwiki, Stepa, Pandion auk, Jrockley, JoeMarfice, ZerodEgo, Edgar181, Yamaguchi 先生, Skizzik, Dauto, JSpudeman, Kurykh, Rajeevmass~enwiki, Persian Poet Gal, Pieter Kuiper, Jprg1966, Acrinym, Miquonranger03, MalafayaBot, Droll, Complexica, DHN-bot~enwiki, Sbharris, RAlafriz, Vladislav, Vanished User 0001, Darthgriz98, Voyajer, Addshore, Percommode, Krich, DavidStern, Theonlyedge, Nakon, Nreprm2026, DMacks, Daniel.Cardenas, Zeamays, Jonnyapple, Sadi Carnot, Bdushaw, Wilt, TriTertButoxy, Chymicus, UberCryxic, Bagel7, Mattfoot, Heimstern, Jaganath, Ocatecir, Mr. Lefty, Ckatz, 16@r, Omnedon, Owlbuster, Waggers, SandyGeorgia, Spiel496, Funnybunny, HappyVR, Iridescent, Newone, NativeForeigner, J Di, Amakuru, Tawkerbot2, Chetvorno, Thermochap, CmdrObot, Ale jrb, Megaboz, RedRollerskate, Ruslik0, MrZap, McVities, WMSwiki, Bakanov, RobertLovesPi, Equendil, Cydebot, Acelor, Reywas92, Cantras, Bvcrist, LouisBB, Travelbird, Llort, David edwards, Tawkerbot4, Christian75, Narayanese, Ssilvers, Thijs!bot, Epbr123, Mbell, Dougsim, Nonagonal Spider, Headbomb, Yzmo, Marek69, West Brom 4ever, Tellyaddict, Cool Blue, Greg L, Sean William, VictorP, KrakatoaKatie, AntiVandalBot, WinBot, Skymt, Voyaging, Opelio, Tyco.skinner, Gef756, Chill doubt, Naturalnumber, Gdo01, Spencer, Leuko, CosineKitty, J-stan, Smith Jones, Acroterion, Magioladitis, WolfmanSF, Bennybp, Bongwarrior, VoABot II, A4, Nyq, JNW, JamesBWatson, الحسني, Drondent, Slartibartfast1992, Jackal irl, Animum, Dirac66, 28421u2232nfenfcene, Hveziris, User A1, Maliz, PoliticalJunkie, DerHexer, GregU, PEBill, MartinBot, BetBot~enwiki, Mermaid from the Baltic Sea, WizendraW, Xantolus, Thereen, CommonsDelinker, AlexiusHoratius, J.delanoy, DrKay, Rgoodermote, Numbo3, Acalamari, TheChrisD, Dispenser, LordAnubisBOT, JayMars, Lathrop, AntiSpamBot, TomasBat, NewEnglandYankee, Nwbeeson, SmoothK, Sunderland06, MetsFan76, Joshmt, Cometstyles, STBotD, RB972, Treisijs, D-Kuru, Dineshextreeme, Martial75, CardinalDan, Idioma-bot, Sheliak, Bondslave777, FeralDruid, X!, VolkovBot, ABF, Thisisborin9, Jacroe, Ryan032, Philip Trueman, DoorsAjar, TXiKiBoT, Oshwah, GimmeBot, Kriak, Hqb, GDonato, Anonymous Dissident, Crohnie, Monkey Bounce, Voorlandt, Mr. Hallman, Michael H 34, TBond, Wikiisawesome, Suriel1981, Rbdebole, Graymornings, Synthebot, Enviroboy, Rurik3, Generalguy11, !dea4u, Insanity Incarnate, Ceranthor, Yoos~enwiki, AlleborgoBot, Kalivd, EmxBot, Neparis, Swimalldor, Ponyo, EJF, SieBot, Graham Beards, Scarian, CircafuciX, BotMultichill, Jauerback, Dawn Bard, Joncam, Caltas, Sergeanthuggy, Bentogoa, RadicalOne, Arbor to SJ, Prestonmag, Thadaddy3233, Oxymoron83, Antonio Lopez, KPH2293, Lightmouse, WingkeeLEE, Ealdgyth, BenoniBot~enwiki, Stustjohn, Dabomb87, PlantTrees, Dolphin51, Nergaal, Tomdobb, Muhends, WikipedianMarlith, ClueBot, Trojancowboy, GorillaWarfare, Artichoker, PipepBot, UniQue tree, The Thing That Should Not Be, Hongthay, Unbuttered Parsnip, GreenSpigot, Liekmudkipz, Mild Bill Hiccup, Correcting nonesense, NovaDog, Blanchardb, Richerman, RandomTREES, Rotational, Piledhigheranddeeper, Inala, DragonBot, Almeaeobtac, Jusdafax, MEJG, Gtstricky, Rhododendrites, Brews ohare, NuclearWarfare, Lunchscale, Jotterbot, PhySusie, Tonyfey, Lkruijsw, Kaiba, SchreiberBike, Stepheng3, Thingg, Jamyricks, Aitias, Melibarr05, Kurtcobain321, Scalhotrod, Versus22, Johnuniq, MasterOfHisOwnDomain, DumZiBoT, TimothyRias, Sjodenenator, XLinkBot, Maky, Rror, Avoided, Mitch Ames, Ilikepie2221, WikHead, Mgaarafan, SkyLined, Addbot, Chizkiyahuavraham, AVand, Some jerk on the Internet, Hurleymann1, Uruk2008, DOI bot, Tenev, Booba5, AkhtaBot, Jessepfrancis, Ronhjones, Jncraton, Moosehadley, CanadianLinuxUser, WFPM, LaaknorBot, Chamal N, CarsracBot, FiriBot, Omnipedian, LinkFA-Bot, Ehrenkater, Pnacitum, Tide rolls, Lightbot, Potekhin, UPS Truck Driver, VP-bot, Luckas-bot, Yobot, Nergalty, Kan8eDie, THEN WHO WAS PHONE?, Eric-Wester, Tonyrex, AnomieBOT, Shootbamboo, DemocraticLuntz, Rubinbot, Götz, Jim1138, IRP, Piano non troppo, lcalanise, Kingpin13, Mydickishuge24, Materialscientist, The High Fin Sperm Whale, Citation bot, Neurolysis, ArthurBot, LovesMacs, Mrhellcool, Rightly, Xqbot, IrishChemistPride, IrishChemistPride2, GeometryGirl, Restu20, Gap9551, Srich32977, S0aasdf2sf, John5955, Alan8, ProtectionTaggingBot, Omnipaedista, RibotBOT, TonyHagale, Phillycheesesteaks, LyleHoward, A. di M., RyanOrdemann, Peter470, Thehelpfulbot, Al Wiseman, FrescoBot, Surv1v4l1st, Eadon-com, Paine Ellsworth, Tobby72, Gauravdce07, Steve Quinn, C.Bluck, Citation bot 1, MarB4, Galmicmi, Gil987, Pinethicket, HRoestBot, Voltron Hax, Raen79, Hooman, Yos233, Allthingstoallpeople, MastiBot, Kuririmo, Noel Streatfeild, Ezhuttukari, Swifterthenyou, Noisalt, Jujutacular, Euchanels, Lissajous, Dude1818, December21st2012Freak, IJBall, Jauhienij, Utility Monster, FoxBot, Sheogorath, Jdlawlis, Diannaa, Odatus, Bestcallumuk, Sampathsris, DARTH SIDIOUS 2, Mean as custard, RjwilmsiBot, TjBot, MinicheddarsandelephantsFTW, Benjadow, Mcmonsterbrothers, Priceracks, Csilcock, Sohaib360, Androstachys, Techhead7890, EmausBot, Optiguy54, GoingBatty, Jjasharpe, Pcorty, TuHan-Bot, Hhhippo, HiW-Bot, John Cline, Harddk, Fæ, Josve05a, StringTheory11, Wackywace, Quondum, GianniG46, Fizicist, Wayne Slam, Raynor42, Amaugir, Jacksccsi, Brandmeister, Donner60, Negovori, RockMagnetist, Mni9791, ClueBot NG, HLachman, Hermajesty21, Jacobkh, Letoya123, Samsau ninjaguy, Ggonzalm, Moritz37, Braincricket, Helpful Pixie Bot, Geo7777, SzMithrandir, Bibcode Bot, Dfbowsmountainer, Ymblanter, Vagobot, Paolo Lipparini, Wzrd1, Lk00la1dl, JacobTrue, Socal212, Begman5, Mark Arsten, Cadiomals, Jikepaddy, Caterpillar111, Macymae, 06seagsa, BEEPTHENOOB, ItzzRevolution, Shawn Worthington Laser Plasma, Duxwing, Klilidiplomus, Uopehem251, Joe0x7F, BattyBot, Justincheng12345-bot, Cyberbot II, ChrisGualtieri, GoShow, Ankap-enwiki, Glenzo999, Barant2, BrightStarSky, Dexbot, Astromango2215, Webclient101, Mogism, 331dot, Spray787, Vanquisher.UA, Lugia2453, Kondormari, Reatlas, JellyBean4.1, Prof.Professer, The User 111, Bluemanyoung, Rohitgunturi, Ugog Nizdast, The Herald, Jwratner1, Zahid2233, MorshusApprentice, 2005-Fan, Phub Dorji, Epic Failure, Gindor, Ian98989898, Monkbot, SkateTier, Waldmannevan, Wiki1098, Wulfiedude14, Mario Castelán Castro, Fleivium, Crystallizedcarbon, Mewikigeek, Jesus is the Light of my life, Acesoli, Soumilm, Lemmegetyou, Flying g shot, SirLagsalott, Tetra quark, Skipfortyfour, Stim 2.0, KasparBot, Fazbear7891, Rewdon1 and Anonymous: 939

- **Subatomic particle** *Source:* https://en.wikipedia.org/wiki/Subatomic_particle?oldid=690494552 *Contributors:* The Anome, Tarquin, Michael Hardy, FrankH, Ixfd64, CesarB, NuclearWinner, Looxix-enwiki, Ahoerstemeier, LittleDan, Glenn, Kwekubo, Schneelocke, Bevo, Chrisjj, Donarreiskoffer, Baldhur, Romanm, Anthony, Alan Liefting, Giftlite, DocWatson42, Awolf002, Mintleaf-enwiki, Dissident, Xerxes314, Everyking, Bensaccount, Vadmium, Antandrus, Rdsmith4, Kenny TM--enwiki, Discospinster, ElTyrant, Vsmith, ESkog, RJHall, El C, Bobo192, Shenme, PiccoloNamek, Stephen G. Brown, Alansohn, Ctande, MarkGallagher, Wtshymanski, Egg, DV8 2XL, Ceyockey, Adrian.benko, Oleg Alexandrov, Mindmatrix, JarlaxleArtemis, Duncan.france, Isnow, SeventyThree, Dysepsion, Rjwilmsi, Strait, Klassykittychick, Scorpiuss, Boccobrock, Erkcan, Naraht, DannyWilde, SouthernNights, King of Hearts, Chobot, Wavelength, Bambaiah, Jimp, Phantomsteve, Loom91, Stephenb, PoorLeno, Bachrach44, Spike Wilbury, Syrthiss, Gat0r, Wknight94, Light current, KGasso, GraemeL, Rlove, Sitenl, Asterion, SmackBot, Incnis Mrsi, JohnRussell, Darkgod, Jordan.ambra, Chris the speller, Bluebot, Persian Poet Gal, DHN-bot-enwiki, Sbharris, Can't sleep, clown will eat me, Drkirkby, Vladislav, Voyajer, Pax85, Radagast83, Edwtie, Drphilharmonic, Thinkingman, Lambiam, Doug Bell, Rigadoun, Ortho, 041744, Ckatz, RandomCritic, Ginkgo100, Esurnir, Tawkerbot2, JForget, Megaboz, Johnlogic, Myasuda, Christian75, Thijs!bot, Epbr123, Guyla, Mbell, Nonagonal Spider, Headbomb, Mjollnir783, Weasel5i2, Escarbot, Ssr, Mentifisto, AntiVandalBot, Luna Santin, Refried, JAnDbot, Instinct, Acroterion, Bongwarrior, VoABot II, Ling.Nut, Glen, Geboy, Mike6271, Davburns, J.delanoy, Yonidebot, Acalamari, Jakey665, McSly, TomasBat, Benito001, Juliancolton, Logic20, Idioma-bot, ArchetypeRyan, VolkovBot, Philip Trueman, Amother, TXiKiBoT, Geht, Jhannah, GcSwRhIc, Seraphim, Martin451, Optigan13, Krazywrath, Lamro, Insanity Incarnate, Monty845, Lkleinow, Alleborgobot, Freependulum, Borne nocker, Hazel77, S8333631, Doclecticwiki, SieBot, Gerakibot, 4RM0-enwiki, Keilana, Flyer22 Reborn, Tiptoety, The Evil Spartan, Sohelpme, AlexWaelde, Wombatcat, Lisatwo, Antman123, Nimbusania, DarkCatalyst, ClueBot, The Thing That Should Not Be, Drugieuk, Arakunem, Drmies, DragonBot, Djr32, Excirial, Monobi, WikiZorro, SpikeToronto, Peter.C, Darren23, Little Mountain 5, Mifter, Garycompugeek, PicoGils, Addbot, Willking1979, Ronhjones, Knowledgesupreme, Download, Chamal N, Bassbonerocks, CosmiCarl, Barak Sh, IOLJeff, Ehrenkater, VASANTH S.N., Tide rolls, Lightbot, Megaman en m, Luckas-bot, Yobot, Amirobot, Eric-Wester, Kulmalukko, AnomieBOT, Ciphers, ^musaz, Jim1138, Icalanise, Ulric1313, Materialscientist, Citation bot, Maxis ftw, ArthurBot, Xqbot, Phazvmk, Gopal81, Addihockey10, Turk oğlan, Dubravko49, GrouchoBot, АлександрВв, TR4YH4N, Mitraunodo, Sushiflinger, Dave3457, Mark Renier, Doremo, Steve Quinn, Gigigogo, Citation bot 1, Pinethicket, Tom.Reding, Maude Frickert, Bgpaulus, Teamspoad, TobeBot, Mptb3, Koyae, MartinHiggs, Diannaa, Harrasser, Tbhotch, RjwilmsiBot, Narayanan20092009, EmausBot, Ajraddatz, Heyimawesome, Wikipelli, K6ka, Hhhippo, JSquish, Harddk, Shuipzv3, Permenent, Anir1uph, Access Denied, Maschen, Epicstonemason, VictorianMutant, CharlieEchoTango, ClueBot NG, Preon, IfYouDoIfYouDon't, Widr, MerllwBot, Helpful Pixie Bot, Novusuna, IrishStephen, BG19bot, Northamerica1000, Neutral current, Kord Kakurios, Zombiecat181, Jethro B, Ducknish, MadGuy7023, Jmeg82, Webclient101, Mogism, Reatlas, Faizan, Bigdaddysound, Frankalbertson, The Herald, Ginsuloft, AddWittyNameHere, Colecharb, 123gogeta, TheRapeTrainBackFromTheDead.Again, CraigyDavi, Robdistasio, Darkenergydesigner, Tetra quark, Jesus.Like.For.Real, Jennifer1122, Harshita1999, ProprioMe OW and Anonymous: 371

- **Electric charge** *Source:* https://en.wikipedia.org/wiki/Electric_charge?oldid=689850873 *Contributors:* AxelBoldt, Mav, Andre Engels, Roadrunner, Peterlin-enwiki, Heron, JohnOwens, Michael Hardy, Ixfd64, Delirium, Looxix-enwiki, Ellywa, Mdebets, Glenn, Rossami, Nikai, Andres, Raven in Orbit, Reddi, Omegatron, Gakrivas, Lumos3, Rogper-enwiki, Gentgeen, Robbot, Fredrik, Dukeofomnium, Wikibot, Fuelbottle, Wjbeaty, Giftlite, DavidCary, Herbee, Snowdog, Dratman, Valen-enwiki, RScheiber, Jason Quinn, Brockert, OldakQuill, Manuel Anastácio, LiDaobing, Karol Langner, Icairns, Iantresman, GNU, Vincom2, Discospinster, Guanabot, Jpk, Dbachmann, ZeroOne, Laurascudder, Bobo192, Rbj, Giraffedata, Kjkolb, Scentoni, Mdd, Alansohn, Atlant, ABCD, Velella, Wtshymanski, HenkvD, Mikeo, DV8 2XL, Gene Nygaard, HenryLi, Oleg Alexandrov, Nuno Tavares, Cimex, Rocastelo, StradivariusTV, Oliphaunt, BillC, Eleassar777, Cyberman, Palica, BD2412, Demonuk, Edison, SMC, Krash, Dougluce, FlaBush, Psyphen, Nivix, Alfred Centauri, Gurch, Kri, Gdrbot, Manscher, YurikBot, Bambaiah, Lucinos-enwiki, Stephenb, Manop, Pseudomonas, JDoorjam, TDogg310, Chichui, Kkmurray, Wknight94, Light current, Enormousdude, Johndburger, Tesetattr, Pinikas, Reyk, Canley, Geoffrey.landis, JDspeeder1, GrinBot-enwiki, Mejor Los Indios, Sbyrnes321, Marquez-enwiki, Moeron, Vald, Thunderboltz, Dmitry sychov, HalfShadow, Gilliam, Oscarthecat, Andy M. Wang, Chris the speller, Lenko, DHN-bot-enwiki, Dual Freq, Hallenrm, Rrburke, The tooth, MichaelBillington, Hgilbert, Drphilharmonic, Daniel.Cardenas, Springnuts, Yevgeny Kats, Andrei Stroe, DJIndica, Naui-enwiki, Nmnogueira, SashatoBot, Lambiam, Richard L. Peterson, Slowmover, Cronholm144, Mgiganteus1, Bjankuloski06en-enwiki, Nonsuch, Ben Moore, RandomCritic, MarkSutton, Stikonas, Dicklyon, Levineps, Igoldste, Tawkerbot2, Chetvorno, JForget, CmdrObot, Kehrli, Jsd, Myasuda, Cydebot, Fl, Bverist, Meno25, Gogo Dodo, WISo, Christian75, Ssilvers, Thijs!bot, Epbr123, Barticus88, N5iln, Mojo Hand, Headbomb, Gerry Ashton, Escarbot, Aadal, AntiVandalBot, Seaphoto, Prolog, DarkAudit, Lyricmac, Tim Shuba, WikifingHelper, Asgrrr, JAnDbot, Acroterion, Bongwarrior, VoABot II, J2thawiki, Sstolper, Jjurik, Bubba hotep, User A1, DerHexer, InvertRect, Robin S, MartinBot, M. Bilal Shafiq, LedgendGamer, Pharaoh of the Wizards, Numbo3, Hans Dunkelberg, NightFalcon90909, Uncle Dick, Ginsengbomb, Katalaveno, DarkFalls, NewEnglandYankee, QuickClown, Juliancolton, ACBest, Treisijs, Lseixas, Jefferson Anderson, Sheliak, Philip Trueman, TXiKiBoT, The Original Wildbear, Ayan2289, Nickipedia 008, LuizBalloti, Monty845, Jpalpant, Biscuittin, Demmy100, SieBot, Gerakibot, Caltas, Gastin, Wing gundam, Msdaghd, JerrySteal, Jojalozzo, Oxymoron83, Faradayplank, Avnjay, Anchor Link Bot, Neo., Loren.wilton, ClueBot, The Thing That Should Not Be, Arakunem, Termine, Mild Bill Hiccup, Stephaninator, LeoFrank, Excirial, Kocher2006, Jusdafax, Brews ohare, Cenarium, Jotterbot, PhySusie, SchreiberBike, Wuzur, JDPhD, Versus22, Thinking Stone, Rror, Cernms, Truthnlove, Addbot, Some jerk on the Internet, CanadianLinuxUser, NjardarBot, LaaknorBot, Scottyferguson, LinkFA-Bot, Naidevinci, Oewaldron, Tide rolls, Lightbot, JDSperling, Legobot, Luckas-bot, Yobot, CinchBug, Duping Man, AnomieBOT, DemocraticLuntz, Sertion, Jim1138, IRP, Pyrrhus16, Kingpin13, Bluerasberry, Materialscientist, Geek1337-enwiki, ImperatorExercitus, Xqbot, TheAMmollusc, Phazvmk, Addihockey10, Capricorn42, Nnivi, ProtectionTaggingBot, RibotBOT, Srr712, A. di M., Constructive editor, Frozenevolution, Ryryrules100, Jc3s5h, Drunauthorized, Mithrandir, Steve Quinn, Davidteng, Fast kartwheels, BenzolBot, DivineAlpha, AstaBOTh15, Pinethicket, I dream of horses, Jivee Blau,

Calmer Waters, Tinton5, MastiBot, Serols, Meaghan, Lalrang2007, Logical Gentleman, FoxBot, TobeBot, SchreyP, Jonkerz, Ndkartik, Vrenator, Taytaylisious09, Ammodramus, Jamietw, DARTH SIDIOUS 2, Eshmate, Irfanyousufdar, EmausBot, John of Reading, GoingBatty, K6ka, Darkfight, Hhhippo, JSquish, Harddk, Stephen C Wells, Liam McM, Sonygal, L Kensington, Donner60, Peter Karlsen, Sven Manguard, Planetscared, ClueBot NG, Jack Greenmaven, Cking1414, Ihwood, Ulflund, CocuBot, MelbourneStar, O.Koslowski, Brickmack, AvocatoBot, Ushakaron, Rm1271, Altafr, F=q(E+v^B), Snow Blizzard, Brad7777, Bhaskarandpm, Eduardofeld, GoShow, Dexbot, JoshyyP, Brandonsnaegregor, Reeceyboii, Frosty, Reatlas, I am One of Many, Eyesnore, Tentinator, Germeten, Nablaedy, Spyglasses, Freddyboi69, 20M030810, SpecialPiggy, Marizperoj, Peterfreed, Rigid hexagon, Jiteshkumar727464, Dyeith, Podayeruma, Oleaster, Layfi, BlueDecker, GeneralizationsAreBad, Pritam kumar Barik, KasparBot, Ramprakashsfc and Anonymous: 429

- **Fermion** *Source:* https://en.wikipedia.org/wiki/Fermion?oldid=689264604 *Contributors:* AxelBoldt, Chenyu, Derek Ross, CYD, Mav, Bryan Derksen, The Anome, Ben-Zin~enwiki, Alan Peakall, Dominus, Deljr, Looxix~enwiki, Glenn, Nikai, Andres, Wikiborg, David Latapie, Phys, Bevo, Stormie, Olathe, Donarreiskoffer, Robbot, Merovingian, Rorro, Wikibot, HaeB, Giftlite, Fropuff, Xerxes314, Vivektewary, JoJan, Karol Langner, Tothebarricades.tk, Icairns, Hidaspal, Vsmith, Laurascudder, Lysdexia, Ashlux, Graham87, Magister Mathematicae, Kbdank71, Syndicate, Strait, Protez, Drrngrvy, FlaBot, Srleffler, Chobot, YurikBot, RobotE, Jimp, Bhny, Captaindan, SpuriousQ, Salsb, Lomn, Enormousdude, CharlesHBennett, Federalist51, Tom Lougheed, Unyoyega, Jrockley, MK8, BabuBhatt, Complexica, Zachorious, Shalom Yechiel, QFT, Garry Denke, Daniel.Cardenas, SashatoBot, Flipperinu, Dan Gluck, LearningKnight, Happy-melon, Paulfriedman7, Cydebot, Meno25, Zalgo, Thijs!bot, Mbell, Headbomb, Nick Number, Orionus, Shlomi Hillel, CosineKitty, NE2, Mwarren us, ZPM, Vanished user ty12k189jq10, Joshua Davis, R'n B, Tensegrity, Rod57, Dgiraffes, Alpvax, VolkovBot, TXiKiBoT, Red Act, Anonymous Dissident, Abdullais4u, בני בני, Tanhueiming, Antixt, Haiviet~enwiki, EmxBot, Kbrose, SieBot, Likebox, Jojalozzo, Dhatfield, Oxymoron83, TubularWorld, ClueBot, Seervoitek, Rodhullandemu, Jorisverbiest, Feebas factor, ChandlerMapBot, Nilradical, Wikeepedian, Stephen Poppitt, Addbot, Vectorboson, Luckas-bot, Yobot, Planlips, Dickdock, AnomieBOT, Icalanise, Materialscientist, Xqbot, Br77rino, Balaonair, 老陳, Paine Ellsworth, Blackoutjack, Kikeku, Rameshngbot, Tom.Reding, RedBot, Alarichus, Michael9422, Silicon-28, TjBot, EmausBot, WikitanvirBot, Quazar121, Solomonfromfinland, JSquish, Fimin, Quondum, AManWithNoPlan, EdoBot, ClueBot NG, PBot1, EthanChant, Bibcode Bot, BG19bot, Petermahlzahn, KingKhan85, ChrisGualtieri, BoethiusUK, DerekWinters, Tentinator, JNrgbKLM, Mohit rajpal, KasparBot, Jiswin1992, Even This Is Taken, Wulframm and Anonymous: 120

- **Wave-particle duality** *Source:* https://en.wikipedia.org/wiki/Wave%E2%80%93particle_duality?oldid=690038204 *Contributors:* AxelBoldt, Tobias Hoevekamp, Derek Ross, MarXidad, The Anome, Manning Bartlett, Wayne Hardman, Andre Engels, Josh Grosse, Miguel~enwiki, ChangChienFu, DrBob, Heron, KF, Stevertigo, Michael Hardy, Gabbe, Dgrant, William M. Connolley, Pizza Puzzle, Hike395, Charles Matthews, Timwi, Geoff, Reddi, El~enwiki, ErikStewart, Time, Maximus Rex, Populus, BenRG, Francs2000, Phil Boswell, SJRubenstein, Robbot, Hankwang, Cdang, Owain, Fredrik, Chris 73, Bkalafut, Sverdrup, Roscoe x, Anthony, Giftlite, BenFrantzDale, Lethe, AJim, Jorge Stolfi, Quamaretto, Eequor, Pearbonn, Antandrus, Unquantum, Lumidek, Chadernook, Adashiel, Trevor MacInnis, Eep², Chris Howard, Rich Farmbrough, Filthybutter, Hidaspal, Vsmith, Mjpieters, Bender235, Cyclopia, JustinWick, Pt, El C, Laurascudder, RoyBoy, Spoon!, Army1987, Enric Naval, Atlant, Ricky81682, AzaToth, Fritzpoll, PAR, BRW, Jheald, Gene Nygaard, Afshar, Oleg Alexandrov, Velho, Woohookitty, Linas, Davidkazuhiro, Zealander, Pol098, Ruud Koot, Cbdorsett, Ch'marr, Mandarax, BD2412, Qwertyus, Thierry Dugnolle~enwiki, Drbogdan, Rjwilmsi, KYPark, Strait, Jmcc150, Cjpuffin, Arnero, Musical Linguist, Nihiltres, Alfred Centauri, Fresheneesz, Srleffler, Kri, Snailwalker, CiaPan, Chobot, Siddhant, Vyroglyph, YurikBot, Wavelength, X42bn6, Wolfmankurd, Hede2000, Gaius Cornelius, Salsb, Schlafly, Expensivehat, SCZenz, Jb849, Danlaycock, Chichui, Zwobot, Ospalh, Wknight94, Ott2, Rwxrwxrwx, Smkolins, Light current, Enormousdude, StuRat, Oysteinp, Keithd, Mavaddat, Sbyrnes321, That Guy, From That Show!, Luk, KnightRider~enwiki, SmackBot, Ashenai, Dmccaig, Joonhon, Delldot, Eskimbot, Gilliam, Hmains, Rrscott, Melburnian, Complexica, DHN-bot~enwiki, Colonies Chris, Fiziker, Duncombe, Lesnail, Kittybrewster, ThreeAnswers, King Vegita, Rich.lewis, Turms, DMacks, Bentreuherz~enwiki, SciBrad, Lambiam, ArglebargleIV, Petr Kopač, Feraudyh, Jpawloski, SMasters, Phancy Physicist, Dicklyon, Spiel496, Hetar, Wizard191, Newone, IRevLinas, Pathosbot, Garretteobb, JR-Spriggs, Chetvorno, Hairyfairycarpetfluff, CRGreathouse, BeenAroundAWhile, Linus M., John courtneidge, L.Person, Myasuda, Act333, Yzphub, Krauss, Nick Y., Bverist, Peterdjones, Michael C Price, Viridae, Abtract, Raoul NK, Letranova, Thijs!bot, Michael D. Wolok, Mbell, Headbomb, JustAGal, GordonRoss, Thadius856, Nipisiquit, JAnDbot, MER-C, .anacondabot, VoABot II, SHCarter, Doeduke, AllenDowney, Tserton, R'n'B, Andrej.westermann, Rieffel, J.delanoy, AstroHurricane001, Maurice Carbonaro, Kevin aylward, TimLong2001, Azus~enwiki, Hakufu Sonsaku, Tarotcards, Aqm2241, Fountains of Bryn Mawr, Laurenpass, SirHolo, Lseixas, TraceyR, Sheliak, Club house, Maniaphobic, DParlevliet, AlnoktaBOT, Maespaunday, Philip Trueman, Revilo314, TXiKiBoT, The Original Wildbear, Red Act, Tom239, Voorlandt, RedAndr, Robert1947, Kpedersen1, Rshob, YohanN7, SieBot, ShiftFn, Hertz1888, Caltas, Yuefairchild, Julianva, Anfieldman, Android Mouse, Flyer22 Reborn, Oxymoron83, Janfri, Szalagloria, Hamiltondaniel, Velvetron, MenoBot, Martarius, ClueBot, Binksternet, Razimantv, Mild Bill Hiccup, Isirr, DanielDeibler, Djr32, TheUNOFFICIALvandalpolice, Tyler, NuclearWarfare, Iohannes Animosus, Zilliput, MelonBot, Mchaddock, James Kanjo, Ost316, Wyatt915, Addbot, DOI bot, Captain-tucker, Moshamoot, WMdeMuynck, CarsracBot, AnnaFrance, Favonian, LinkFA-Bot, ATOE, Luckas-bot, Yobot, TaBOT-zerem, THEN WHO WAS PHONE?, Silea678, AnomieBOT, Piano non troppo, EryZ, Andaza, Danno uk, Citation bot, KT-2500, Peterdx, J JMesserly, Vonharris, RibotBOT, Flaviusvulso, LazyMapleSunday, 老陳, Chjoaygame, Tank.hasmukh, Machine Elf 1735, Ysyoon, Citation bot 1, Vanzac11, Trappist the monk, Wdanbae, Vrenator, Fivedoughnut, RjwilmsiBot, Jonlegere, EmausBot, John of Reading, Dewritech, Tommy2010, Slawekb, AvicBot, 1howardsr1, H3llBot, Fizicist, Chrisman62, Maschen, Bulwersator, Chewings72, Sudozero, Will Beback Auto, ClueBot NG, Jostikas, Andybiddulph, Widr, Helpful Pixie Bot, Verberate, Jubobroff, Bibcode Bot, Bm gub2, BG19bot, Guy vandegrift, Vokesk, PhnomPencil, MusikAnimal, F=q(E+v^B), Jivey81, Elemenat, BattyBot, Samanthaclark11, Mdann52, Thojuf, Dexbot, Hmainsbot1, Purpleare, CuriousMind01, Nishantarya98, Pdecalculus, Iztwoz, Perseus.3.14, Jsresearch, My name is not dave, Mdominguez611, Johnfranciscollins, Monkbot, PlaidPolarity, Master Pok, Rhynhardtk, Pasten2, Vespro Latuna, Knife-in-the-drawer, DireNeed and Anonymous: 359

- **Pauli exclusion principle** *Source:* https://en.wikipedia.org/wiki/Pauli_exclusion_principle?oldid=689632323 *Contributors:* Chenyu, CYD, Andre Engels, Graham Chapman, XJaM, PierreAbbat, Roadrunner, Maury Markowitz, Montrealais, Stevertigo, Michael Hardy, Tim Starling, EddEdmondson, Alan Peakall, Dominus, Shellreef, Graue, Ahoerstemeier, Stevenj, Glenn, Andres, Emperorbma, Charles Matthews, Wikiborg, ElusiveByte, Fibonacci, Robbot, Owain, Modeha, Jheise, Tobias Bergemann, Enochlau, Giftlite, Harp, BenFrantzDale, HorsePunchKid, Gunnar Larsson, Karol Langner, Tsemii, Iwilcox, FT2, Bender235, RJHall, Kaszeta, Liberatus, Army1987, Smalljim, John Vandenberg, Cherlin, Obradovic Goran, Voltagedrop, RJFJR, Pol098, Isnow, Crucis, Graham87, ElCharismo, Enzo Aquarius, Eyu100, Jehochman, FlaBot, RexNL, Fresheneesz, JohnMarkStrain, Srleffler, Chobot, Sharkface217, DVdm, YurikBot, Wavelength, Wolfmankurd, RJC, JabberWok, Okedem, Ravindrala, Rsrikanth05, Wiki alf, Grafen, Ripper234, Enormousdude, RG2, Phr en, Mejor Los Indios, That Guy, From That Show!,

SmackBot, Thorseth, David G Brault, Shai-kun, Georgelulu, Complexica, DHN-bot~enwiki, Canice, Gurevichar, Otis182, BTDenyer, Maximum bobby, Michalchik, Sadi Carnot, Vina-iwbot~enwiki, Andrei Stroe, DJIndica, WhiteHatLurker, Meco, Ambuj.Saxena, D Hill, T.O. Rainy Day, Wfructose, Blehfu, Zipz0p, JRSpriggs, Pseudospin, Xcentaur, Mattbr, Sohum, Vyznev Xnebara, Icek~enwiki, Cydebot, UncleBubba, Michael C Price, Gromonger-17, Headbomb, FourBlades, CharlotteWebb, Greg L, Escarbot, Gioto, Seaphoto, Orionus, Strami, Glennwells, CosineKitty, Wasell, Trugster, Mrfunkyostrich, Aryabhatta, Tanvirzaman, Animum, Dirac66, TristramBrelstaff, Custos0, Melamed katz, Maurice Carbonaro, Mzhsj, Cpiral, Katalaveno, P.wormer, Gombang, Zoedill, Ontarioboy, Keenman76, Sheliak, VolkovBot, John Darrow, Kriak, Nxavar, Anonymous Dissident, LeaveSleaves, UnitedStatesian, Pishogue, Spiral5800, Riick, AlleborgoBot, SieBot, Soler97, Adabow, Likebox, Henry Delforn (old), KoshVorlon, Iain99, Jakeng, Jonlandrum, Anchor Link Bot, ClueBot, Trojancowboy, Bryangv, Dvorsky~enwiki, Dlabtot, Chief buffalo chip, DragonBot, IEROslippersBRYAR, Brews ohare, Thingg, RQG, BodhisattvaBot, SilvonenBot, Quaint and curious, Addbot, Narayansg, CanadianLinuxUser, Chamal N, AgadaUrbanit, Lightbot, Zorrobot, Luckas-bot, Yobot, Heisenbergthechemist, IW.HG, AnomieBOT, Rubinbot, Materialscientist, Citation bot, ArthurBot, Xqbot, Nickkid5, Beeline23, GrouchoBot, Omnipaedista, Nathanielvirgo, 老陳, Craig Pemberton, Cognitivelydissonant, Relke, RedBot, Tjlafave, Hickorybark, Halteres, Korepin, EmausBot, Ethereal-Blade, RA0808, H3llBot, Quondum, ChuispastonBot, Mikhail Ryazanov, ClueBot NG, Gareth Griffith-Jones, Asalrifai, Helpful Pixie Bot, Bibcode Bot, Krishnaprasaths, BG19bot, GKFX, Alexander1102, BattyBot, JYBot, Tony Mach, Frosty, Serten, Septate, Yakamashi, Sp20136761, Kfitzell29 and Anonymous: 198

- **Matter wave** *Source:* https://en.wikipedia.org/wiki/Matter_wave?oldid=687006378 *Contributors:* Heron, Stevertigo, Tim Starling, Deljr, Paul A. Drxenocide, Gentgeen, Cdang, Dratman, Edcolins, DemonThing, Zeimusu, Jossi, Anythingyouwant, DragonflySixtyseven, Brianjd, Pjacobi, Bender235, Nabla, Dataphile, Aranel, El C, Kwamikagami, Laurascudder, Robotje, Reinyday, Jag123, Blinken, Haham hanuka, Free Bear, Keenan Pepper, PAR, Gene Nygaard, Siafu, Jtauber, Linas, Rjwilmsi, Winkels, Ian the younger, FlaBot, Meeve, DVdm, Korg, YurikBot, Ugha, Nmondal, JabberWok, Artur Lion~enwiki, Mingshey~enwiki, Chichui, Enormousdude, Fram, Migdejong, Carlosguitar, Teply, KasugaHuang, SmackBot, Eskimbot, Gaff, Robin Whittle, Amatulic, Pieter Kuiper, Complexica, Tobywheeng, Jbergquist, Daniel.Cardenas, Andrei Stroe, Attys, Calam MacÜisdean, JorisvS, Hemmingsen, Nijdam, Melody Concerto, BillFlis, DI2000, UncleDouggie, Domitori, Rhetth, Zipz0p, Chetvorno, Linus M., Myasuda, A876, RZ heretic, Quibik, DumbBOT, Iliank, Mckinlayr, Adechau, Thijs!bot, Epbr123, Headbomb, Second Quantization, Erik Baas, Qwerty Binary, Steelpillow, JAnDbot, Deflective, Shayno, Igodard, Micahnewman, Dirac66, Warren Dew, STBot, Obscurans, Jqar, Pulsarphysics, TXiKiBoT, Mathwhiz 29, Hhkaviani, Aymatth2, Anandramanathan, SieBot, Graham Beards, Ljagerman, Yintan, Patamia, ClueBot, The Thing That Should Not Be, Niceguyedc, Djr32, Excirial, Alexbot, Thingg, Tjako, DumZiBoT, WeOwnTheNight, Phidus, Addbot, DOI bot, Fgnievinski, EjsBot, MagnusA.Bot, Physicsgrl, Numbo3-bot, Lightbot, Wammes Waggel, Luckas-bot, Yobot, AnomieBOT, 1exec1, Galoubet, Piano non troppo, Csigabi, Life Hurts, Citation bot, Xqbot, Capricorn42, Sfgiants906, Srich32977, Elduquerpi, Samppi111, Kyng, Thehelpfulbot, Chjoaygame, LucienBOT, Craig Pemberton, Ysyoon, Tom.Reding, Gryllida, Citibob, Peabody fly, EmausBot, Az29, Hhhippo, JSquish, ZéroBot, SporkBot, Chrisman62, Maschen, Scientific29, RockMagnetist, ClueBot NG, Muon, Lekrecteurmasque, Funckyfizz, Helpful Pixie Bot, Bibcode Bot, Trunks ishida, BG19bot, B wik, Ragnarstroberg, F=q(E+v^B), Himanshu dtu, Dexbot, Makecat-bot, Tilakdp, Monkbot, Rulonegger, Mitoza123, Dallas a lee, Josefhorwath and Anonymous: 147

- **Electricity** *Source:* https://en.wikipedia.org/wiki/Electricity?oldid=689808820 *Contributors:* AxelBoldt, Marj Tiefert, Derek Ross, Robert Merkel, Zundark, Koyaanis Qatsi, Rjstott, Alex.tan, Ted Longstaffe, Mirwin, Josh Grosse, Youssefsan, Fredbauder, William Avery, Peterlin~enwiki, DavidLevinson, Waveguy, Heron, Ryguasu, Jaknouse, Stevertigo, Patrick, RTC, PhilipMW, Michael Hardy, Tim Starling, Jarekadam, Nixdorf, Ixfd64, Cameron Dewe, Delirium, Tiles, Egil, Looxix~enwiki, Mdebets, Ahoerstemeier, Mac, JayTau, Darkwind, Glenn, Bogdangiusca, Andres, Rob Hooft, Hashar, Mulad, Mbstone, Bemoeial, Reddi, Zoicon5, Timc, Maximus Rex, Grendelkhan, Itai, Paul-L-~enwiki, Omegatron, Tjdw, Raul654, Jerzy, Lumos3, Jni, Robbot, Sander123, Kizor, Ee00224, Goethean, Nurg, Henrygb, Academic Challenger, Rasmus Faber, Hadal, Wikibot, Michael Snow, Fuelbottle, Dina, Wjbeaty, Ancheta Wis, Dominick, Giftlite, DocWatson42, Dkeeper, BenFrantzDale, Zigger, Monedula, Xerxes314, Ds13, Wwoods, Average Earthman, Everyking, Bensaccount, Ssd, Gecko~enwiki, Ryguillian, Yekrats, Xwu, Matthead, Brockert, SWAdair, Bobblewik, Utcursch, Antandrus, Beland, Ot, Rdsmith4, Icairns, CesarFelipe, Sayeth, Kramer, Robin Hood~enwiki, Imjustmatthew, Askewchan, Klemen Kocjancic, Picapica, Demiurge, Corti, Mike Rosoft, Monkeyman, Poccil, Archer3, EugeneZelenko, Discospinster, Rich Farmbrough, Vsmith, Florian Blaschke, Ivan Bajlo, Mjpieters, Dbachmann, Mani1, MarkS, Stbalbach, Edgarde, Bender235, JoeSmack, Nabla, Brian0918, MBisanz, Ruyn, Chairboy, Sietse Snel, RoyBoy, Femto, Jpgordon, Bobo192, Fir0002, Smalljim, Shenme, Man vyi, Bert Hickman, Bawolff, NickSchweitzer, Sam Korn, Haham hanuka, Krellis, Nsaa, Ranveig, Beyondthislife, Storm Rider, Alansohn, Anthony Appleyard, Mark Dingemanse, Rd232, Hipocrite, Andrewpmk, Pauldavidgill, Riana, AzaToth, SlimVirgin, Bokkibear, Bart133, Fasten, Bucephalus, L33th4x0rguy, BBird, BRW, Saga City, Wishymanski, Lumberjack steve, King Bowser 64, RainbowOfLight, BDD, DV8 2XL, HenryLi, Oleg Alexandrov, Tariqabjotu, Smark33021, Stemonitis, Velho, Zudduz, Woohookitty, Henrik, Thewob~enwiki, LOL, PoccilScript, BillC, Drostie, Robert K S, Ruud Koot, MONGO, Damicatz, SCEhardt, Wayward, Prashanthns, Gimboid13, Shanedidona, Dysepsion, Graham87, Deltabeignet, BD2412, MC MasterChef, FreplySpang, RxS, Jshadias, Canderson7, Sjakkalle, Rjwilmsi, Coemgenus, Phileas, Amire80, Hiberniantears, Quiddity, Tawker, Mbutts, Durin, Bensin, Krash, The wub, DoubleBlue, Ttwaring, Dar-Ape, Matt Deres, AySz88, Yamamoto Ichiro, Titoxd, Platyk, CAPS LOCK, Mishuletz, Doc glasgow, Latka, Nihiltres, Crazycomputers, CarolGray, Alfred Centauri, RexNL, Gurch, Wongm, OrbitOne, Lmatt, Alphachimp, Srleffler, Imnotminkus, Azitnay, Butros, King of Hearts, Chobot, SirGrant, DVdm, Gdrbot, Korg, Bgwhite, Mike5904, Elfguy, UkPaolo, Roboto de Ajvol, The Rambling Man, YurikBot, Wavelength, TexasAndroid, John Stumbles, Sceptre, Brandmeister (old), RussBot, WAvegetarian, Bergsten, Netscott, Stephenb, Polluxian, Shell Kinney, Gaius Cornelius, CambridgeBayWeather, Alvinrune, Wimt, Mpa, Ethan, Wiki alf, Justin Eiler, Nader85021, Neoguy9090, Irishguy, Nick, Matticus78, Raven4x4x, Nick C, Syrthiss, Dbfirs, T. Scottfisher, BOT-Superzerocool, Aaron Mendelson, Bota47, Private Butcher, Jhinman, Evrik, Dingy, Zelikazi, Wknight94, Sandstein, Light current, Melca, Lt-wiki-bot, Covington, Theda, Closedmouth, Jwissick, CZero, Steventrouble, Christophercem, JuJube, Petri Krohn, GraemeL, Zuwiki, Alvaro.jaramillo, ArielGold, Kungfuadam, NeilN, Asterion, Mejor Los Indios, Sbyrnes321, DVD R W, Sarah, SmackBot, Aim Here, Mattarata, Unschool, Smitz, Tarret, KnowledgeOfSelf, Primetime, C J Cowie, Pgk, Vald, Blue520, KocjoBot~enwiki, Jagged 85, Davewild, Matthuxtable, Synflame, Jrockley, Delldot, Onebravemonkey, Edgar181, ChristopherEdwards, Xaosflux, Cuddlyopedia, Gilliam, Ohnoitsjamie, Betacommand, Oscarthecat, Skizzik, Carbon-16, Anwar saadat, Teemu Ruskeepää, Improbcat, Stevenwrose, BrownBean, The X, SlimJim, Persian Poet Gal, JackyR, Algumacoisaqq~enwiki, Oli Filth, Miquonranger03, Fluri, Papa November, SchfiftyThree, Deli nk, Whispering, Ctbolt, DHN-bot~enwiki, Sb617, Sbharris, Dual Freq, Hallenrm, Darth Panda, Steve0913, John Reaves, Rheostatik, Can't sleep, clown will eat me, Chlewbot, Onorem, Sub zero133, Thisisbossi, RHJesusFreak40, Rrburke, Addshore, Edivorce, DumLoco, Jmlk17, Fuhghettaboutit, Khukri, Nibuod, Nakon, Steve Pucci, Elmicker, SnappingTurtle, Shadow1, Dreadstar, Danielkwalsh, X-Tron 13, Daniel.Cardenas, Er Komandante, Mion, Bidabadi~enwiki, Sadi Carnot, Dogears, IGod, Nmnogueira, The undertow, SashatoBot, Palmerston–enwiki, Pahles, Vanished user 9i39j3, Kuru, Scientizzle, Zslevi, Heimstern, Gobonobo, Disavian, Timelare, Rundquist, Hadrians, CaptainVindaloo, Sceteaux, Joshua Scott, Shattered, Ckatz, Deadcode, JHunterJ, MarkSutton, Shangrilaista, X2RADialbomber, Joeylawn, Clw, Jon186, Vedexent, Nwwaew, Qyd, Citicat, MTSbot–enwiki, Zapvet, Jose77, Dacium, Sifaka, ShakingSpirit, Levineps, Esoltas, Iridescent, K, BobbyLee, Morrowulf, Tony Fox, CapitalR, Esn, Tawkerbot2, Barometer–enwiki, SkyWalker, CmdrObot, Porterjoh, Sir Vicious, Rawling, Wutime, Drinibot, DeLarge, Zureks, Foraneagle2, Jsd, MarsRover, Shizane, Casper2k3, HonztheBusDriver, Myasuda, Equendil, Mattyh190, Clappingsimon, Atomaton, Ghou–enwiki, Ryan, Gtxfrance, Codice1000.en, VashiDonsk, Gogo Dodo, ST47, Happinessiseasy, Nojika, Tubbyalonso, Oamaro, Tkynerd, Igjav, Karafias, Hispalois, DumbBOT, Chrislk02, Ameliorate!, Quadrius, Ssilvers, Usnerd, JodyB, Zalgo, UberScienceNerd, Eubulide, Thijs!bot, Epbr123, Barticus88, Jedibob5, Kablammo, Gralo, Mojo Hand, Jb.schneider-electric, Headbomb, Lugifan, Marek69, NorwegianBlue, Gerry Ashton, Welzen, Bluerfn, Amitprabhakar, Leon7, J. W. Love, Whoda, SusanLesch, Jeblo, Escarbot, Oreo Priest, KrakatoaKatie, Cyclonenim, AntiVandalBot, Luna Santin, Widefox, Kramden4700, EarthPerson, Tpth, Quintote, Eltanin, Tylerbot, Tlabshier, Blair Bonnett, Dictyosiphonaceae, Elaragirl, Myanw, Choiboi22, JAnDbot, Harryzilber, MER-C, Arch dude, Dagnabit, Sitethief, Hut 8.5, Maias, Kerotan, Suede–enwiki, LittleOldMe, Filtay, Samjohnson, .anacondabot, Acroterion, Freshacconci, Akuyume, Magioladitis, Pedro, Theunicyclegirl, Bongwarrior, VoABot II, Pyr, Dekimasu, DonVander, Praveenp, CTF83!, Brain40, Singularity, WODUP, SparrowsWing, Hifrommike65, Randolph02, Catgut, Nposs, 28421u2232nfenfcenc, Hveziris, Beagel, Rauljdelgado, DerHexer, Khalid Mahmood, InvertRect, Pax:Vobiscum, Oroso, Hdt83, MartinBot, Aidanpugh, NAHID, Rettetast, Mschel, R'n'B, Snozzer, PrestonH, Pomte, RockMFR, Zarathura, Slash, Yjwong, J.delanoy, EscapingLife, Ali, Stephanwehner, Eliz81, Puckett34, Geomanjo, Jerry, GodIsGreater, WarthogDemon, Hippi ippi, Rc3784, M C Y 1008, Jlechem, Katalaveno, PedEye1, DarkFalls, Nemvocalist, McSly, Crezelnuts93, Doomavenger1, Rocket71048576, Pyrospirit, PandoraX, Zrogerz69, 97198, Linuxmatt, NewEnglandYankee, M6060, Fountains of Bryn Mawr, Malerin, KCinDC, JHeinonen, P3rs0n, Juliancolton, Cometstyles, Swiftblade6, Boaster, M bastow, Jamesontai, Tbone762, Treisijs, King Toadsworth, Mike V, Useight, CA387, TehNomad, Squids and Chips, MikeLeeds, Luminate, Coolblaze, Idioma-bot, Xnuala, BierHerr, X!, Spaceman13, VolkovBot, *andreweclark, ABF, Brando130, Alexandria, Ryan032, Philip Trueman, TXiKiBoT, Oshwah, Moofy, KevinTR, Chadmilerna, Technopat, Rogator, Qxz, Someguy1221, DavidSaff, Jimmyx90, Piperh, Shoedizzog, Retiono Virginian, Lradrama, Imasleepviking, Dendodge, Szlam, Martin451, ArmyHero, LeaveSleaves, Akorak, Waycool27, Tulls55, Andy Dingley, Y, Eisd, Enviroboy, Spinningspark, WatermelonPotion, Room214, Why Not A Duck, Iritebs, Lily15, Iritebs1, AlleborgoBot, Symane, Big G Ursa, Munci, ZBrannigan, GraybeardThePirot, NHRHS2010, HybridBoy, Hotdogmckgee, Bong69, SieBot, Koolyman, Hiphopdancequeen, PeterCanthropus, Ak47225, Ellbeeecee, SheepNotGoats, Justinritter, Jauerback, Lemonflash, Drew2131, Hbiz777, Wheeeeee, Mman33, RJaguar3, Jason Patton, Zoragotcha, Aristolaos, Megan.rw1, Keilana, RaAnubisOsiris, Iames, Radon210, Mszegedy, Oda Mari, Bookermorgan, Joseph Banks, Oxymoron83, Antonio Lopez, Beast of traal, Steven Crossin, Mexicanpunjab, Tombomp, SH84, Iain99, Fruitytingles, BenoniBot–enwiki, Karlawilloughby, FuturedOrange, Karl2620, Mygerardromance, Randomblue, Ken123BOT, Superbeecat, Pinkadelica, Tomahiv, G.michalis, Jdolan2726, Dalyman, Tomasz Prochownik, Elassint, ClueBot, Seazoleta, PipepBot, Datta.naikwade, The Thing That Should Not Be, IceUnshattered, ImperfectlyInformed, Matsuiny2004, Rahulvohra29, JonDon69, CliffordWest, Mild Bill Hiccup, Jorge Ianis, J8079s, Silverstein210, CounterVandalismBot, Kitty9992, Masterchamp2, Dsdickie, Harland1, Neverquick, Namazu-tron, Jolibrarian, Saulicious, Nick0416, Machoman739–enwiki, Andrewthejesus, DragonBot, Biria64, Anonymous101, Ryanswimstheworld, Erebus Morgaine, Reperspliter, Cockman45, Cockman46, Dedbaby, Akshay11, Depe0807, Children.of.the.Kron, The Founders Intent, Gottadmit, Jotterbot, Markgriz, Promethean, Primasz, Nukeless, Leffmann, Thingg, Subash.chandran007, Kolakowski, PCHS-NJROTC, Sharkie3000, Saenze, Anon126, Skunkboy74, XLinkBot, Delicious carbuncle, Needtoknow31, Mitch Ames, SilvonenBot, MsVanAuken, Akasuna–enwiki, Noctibus, JinJian, Dwilso, Reconday41, Anticipation of a New Lover's Arrival, The, Imperial Star Destroyer, Addbot, Proofreader77, Betterusername, Surfin simo, Binary TSO, Yobmod, Fieldday-sunday, USchick, Glane23, Bassbonerocks, Jelleve, Debresser, Cjohn67, Xev lexx, Lineface, Tassedethe, Koliri, Nickdog33, Muysal, Alan16, Kelly190, Krano, Jarble, Mishkin11, Legobot, Luckas-bot, ZX81, Yobot, TaBOT-zerem, Les boys, Hulek, Willy56.5, KamikazeBot, Lichen from Hell, Dtrak, IW.HG, ScienceMind, Ying123, Eric-Wester, N1RK4UDSK714, AnomieBOT, IRP, JackieBot, Mintrick, EHRice, AdjustShift, Ulric1313, Flewis, Glaze012, Citation bot, Prabhat278205, Lostkey2, BeyondHisYears, Carlsotr, ArthurBot, Bruce Foods, LilHelpa, User2301, Xqbot, Athabaska-Clearwater, TinucherianBot II, Madsmokey, Capricorn42, Bennyboys, PrometheusDesmotes, SlickWillyLovesSex, DSisyphBot, Uarshad82, Zaingay, Andy12983, Prettyponies12343, Piesaretasty345, AmericasPower, Ubcule, J04n, Youngster14, Nebather, Omnipaedista, Mark Schierbecker, Celebration1981, Soundoftheunderground, Smellmyfeet, Angleofdeath, Mufflinstastenice, 老陳, Remshad, Bekus, GliderMaven, Magnagr, Lagelspeil, Je3s5h, KerryO77, Yickbob, Zanthrax, Sjcandyman, Citation bot 1, Pucketteli, Miguelaaron, Me the Third, Jeff hardy3456, Eengined, Beao, Full-date unlinking bot, FoxBot, Thrissel, NeoAdonis, TobeBot, DixonDBot, Mr Mulliner, Sammetsfan, Ripchip Bot, Jackehammond, Inluminetuovidebimuslumen, EmausBot, WikitanvirBot, Cbommann, Hhhippo, Stubes99, JSquish, Sf5xeplus, HugoLoris, H3llBot, Jarodalien, Fizicist, Morgengave, Sam.P.Hollins, Patrolboat, ResidentAnthropologist, Teapeat, Planetscared, ClueBot NG, Fauzan, Tylko, Snotbot, Alanwilliams101, Castncoot, AussieRulez, Lincoln Josh, NuclearEnergy, Helpful Pixie Bot, Electriccatfish2, Tholme, AvocatoBot, Bryanpiezon, Szczureq, Neshmick, ChrisGualtieri, Embrittled, Nusaybah, Dexbot, Reatlas, Joeinwiki, Skeledzija, Achmad Fahri, Kelvinmike09, SkateTier, Filedelinkerbot, Ejrussellim123, KasparBot and Anonymous: 1103

- **Magnetism** *Source:* https://en.wikipedia.org/wiki/Magnetism?oldid=689354170 *Contributors:* CYD, Bryan Derksen, Zundark, Ed Poor, Stokerm, Peterlin–enwiki, Ktsquare, Waveguy, Heron, Isis–enwiki, Modemac, Stevertigo, Edward, Lir, Michael Hardy, Tim Starling, Llywrch, Ixfd64, Delirium, Egil, Ellywa, Ahoerstemeier, Mac, Theresa knott, Snoyes, Glenn, Mxn, Smack, Quickbeam, Charles Matthews, Reddi, Stone, 4lex, Pedant17, E23–enwiki, Ozuma–enwiki, Donarreiskoffer, Robbot, Fredrik, Jmabel, Texture, Caknuck, Sunray, Fuelbottle, Dina, Tobias Bergemann, Giftlite, Wolfkeeper, Bensaccount, Jfdwolff, Duncharris, Guanaco, Blizzarex, Broekert, DÃ»gosz, Gzuckier, Antandrus, Beland, Ctachme, Piotrus, Karol Langner, Icairns, Karl-Henner, Gseshoyru, DanMatan, Shotwell, D6, Discospinster, ElTyrant, Rich Farmbrough, Pak21, Vsmith, Bender235, ESkog, Nabla, Lankiveil, Joanjoc–enwiki, Edward Z. Yang, Sietse Snel, Femto, CDN99, Bobo192, Meggar, Viriditas, Elipongo, I9Q79oL78KiL0QTFHgyc, Boredzo, MPerel, Haham hanuka, Nsaa, Alansohn, Redfarmer, Spangineer, Hu, Snowolf, Wtmitchell, TheMolecularMan, DV8 2XL, Gene Nygaard, Feline1, Vadim Makarov, Kazvorpal, Schultz.Ryan, Davidkazuhiro, Benbest, Drostie, Duncan.france, Kmg90, M412k, Dyspepsion, Slgrandson, JEB90, Magister Mathematicae, Edison, Koavf, Zbxgscqf, Strait, MZMcBride, EquinoX, Yamamoto Ichiro, Drrngrvy, FlaBot, RobertG, Mathbot, Nihiltres, RexNL, Gurch, SteveBaker, M7bot, Physchim62, King of Hearts, Chobot, Karch, The Rambling Man, Satanael, YurikBot, Oliviosu–enwiki, Hairy Dude, Jeffhoy, Petiatil, Pigman, JabberWok, Chaser, Hellbus, Stephenb, Gaius Cornelius, NawlinWiki, Edinborgarstefan, SEWilcoBot, Wiki alf, Cryptoid, Dlugosz, Hv, Syrthiss, Jeremy Visser, Mistercow, Dan Austin, JustAddPeter, Sinewalker, Sandstein, Light current, Enormousdude, 2over0, Techguru, Closedmouth, Fang Aili, GraemeL, Roberto DR, HereToHelp, Chaiken, Junglecat, RG2, JDspeeder1, GrinBot–enwiki, Zvika, Sbyrnes321, Inthepink, That Guy, From That Show!, SmackBot, Kyle-Cardoza, Dubbin, InverseHypercube, Hydrogen Iodide, Melchoir, David.Mestel, Mjspe1, Unyoyega, Pgk, Proficient, ParkerHiggins, Masonprof, KocjoBot–enwiki, Jagged 85, Renesis, Ozone77, Yamaguchi 先生, Aksi great, Gilliam, Hmains, Skizzik, Kmarinas86, Bluebot, Ferix,

Keegan, Jprg1966, MalafayaBot, Metacomet, J. Spencer, Octahedron80, Colonies Chris, Hongooi, Royboycrashfan, Can't sleep, clown will eat me, Scott3, KaiserbBot, Snowmanradio, Rrburke, Addshore, Celarnor, Wen D House, Fuhghettaboutit, Rolinator, Rajrajmarley, Akriasas, Knutsi, Danikayser84, DMacks, Where, Daniel.Cardenas, Sadi Carnot, Pilotguy, Yevgeny Kats, Charivari, Wikier.ko, GoldenTorc, DJIndica, Nmnogueira, The undertow, SashatoBot, MusicMaker5376, Petr Kopač, Dr. Sunglasses, Dbtfz, Akendall, UberCryxic, Scientizzle, This user has left wikipedia, JorisvS, Jim.belk, IronGargoyle, Postscript07, 16@r, Grapetonix, Special-T, Mr Stephen, Dicklyon, Geologyguy, Mdanziger, Iridescent, Rpb01r, Casull, Filelakeshoe, Chetvorno, Lahiru k, Harold f, CalebNoble, JForget, Thermochap, Ale jrb, Jaeger5432, Woudloper, El aprendelenguas, Whereizben, Nauticashades, Slazenger, Rossf18, Rifleman 82, Gogo Dodo, Xxanthippe, Pinestone, Chrislk02, Ssilvers, Omicronpersei8, A Mom, PoolDoc, Thijs!bot, Epbr123, ProDigit, Nonagonal Spider, Headbomb, Davidhorman, Dfrg.msc, Nick Number, Mr. Trustegious, Natalie Erin, Mentifisto, Hmrox, AntiVandalBot, Gioto, Opelio, Quintote, Gangasudhan, Tim Shuba, Spencer, Sluzzelin, JAnDbot, MER-C, The Transhumanist, Andonic, Mkch, RebelRobot, Kerotan, LittleOldMe, Magioladitis, Bongwarrior, VoABot II, Doodoobutter, C d h, Anthonyramos1, Soulbot, Skew-t, Froid, SwiftBot, Catgut, Allstarecho, LorenzoB, Vssun, Crware, TehBrandon, DerHexer, MartinBot, Halpaugh, Uvainio, Kostisl, CommonsDelinker, Science5, J.delanoy, Pharaoh of the Wizards, JA.Davidson, Lone Skeptic, LordAnubisBOT, Ipigott, GhostPirate, Warut, Jcwf, NewEnglandYankee, Mp50967, Vanilluv30, Mufka, FJPB, Cmichael, Usp, JulianB12, Jamesontai, Vanished user 39948282, Rising*From*Ashes, Mike by, Stuart07, Xenonice, Nmgrad, Deor, Thisisborin9, AlnoktaBOT, Al.locke, WOSlinker, Philip Trueman, Raminmahpour, TXiKiBoT, Paddy-B-Jr, Antoni Barau, Hqb, Z.E.R.O., Anonymous Dissident, Someguy1221, Amog, Tyler Matthews, Pjbcool103, Eubulides, Wenli, Andy Dingley, Spinningspark, Bcfootball, PwncakesN bacon, Ryne1, Insanity Incarnate, Brianga, Fischer.sebastian, TheBendster, ClarkLewis, IndulgentReader, NHRHS2010, EmxBot, Aqwfyj, Biscuittin, Roberdor, SieBot, Tiddly Tom, Caltas, Zoragotcha, Keilana, Happysailor, Flyer22 Reborn, Radon210, Sammyk214, Paolo.dL, Rafonseca, Smilesfozwood, Lightmouse, Techman224, Hobartimus, BenoniBot~enwiki, Werldwayd, Mygerardromance, JCPH, Denisarona, Electrodynamicist, ElectronicsEnthusiast, Martarius, De728631, ClueBot, Trojancowboy, Yoduh2007, The Thing That Should Not Be, Fluidchameleon, IceUnshattered, Jan1nad, Drmies, Blanchardb, Otolemur crassicaudatus, Puchiko, Masonm95, Pointillist, Russell4, Mspraveen, Samsee, Excirial, Jusdafax, Ninjackster, Monobi, Eeekster, Lartoven, Brews ohare, Noor Qasmieh, Davdelehn, Erniesaurus, Ottawa4ever, Thingg, AL2TB, Jocelyne Heys-Gerard, Yuwangswisscom, Versus22, Ryalisaivamsi, SoxBot III, Egmontaz, Flutterman, Party, Editorofthewiki, Spitfire, Gnowor, ChucksGay123456789, Mitch Ames, WikHead, SilvonenBot, Khyranleander, HexaChord, Addbot, Xp54321, Proofreader77, American Eagle, S ortiz, Willking1979, Some jerk on the Internet, DOI bot, Bob sagget jr., Ron B. Thomson, Mr. Wheely Guy, LaaknorBot, Caturdayz, CarsracBot, Bassbonerocks, Chzz, TStein, 5 albert square, Numbo3-bot, VASANTH S.N., Tide rolls, William S. Saturn, Lightbot, Gail, Zorrobot, David0811, Legobot, Luckas-bot, Yobot, Il MusLiM HyBRiD Il, MarcoAurelio, THEN WHO WAS PHONE?, Qui-Gon Jinn, Technobebop, IW.HG, Ykral, Tempodivalse, Synchronism, AnomieBOT, Piano non troppo, Kingpin13, Ulric1313, Flewis, TheFronze05, Materialscientist, Karcih, Granito diaz, The High Fin Sperm Whale, E2eamon, Leif27, Clark89, MauritsBot, Xqbot, Sionus, St.nerol, Capricorn42, Jmundo, Saideepak.budaraju, Abce2, Frosted14, Kyng, Doulos Christos, Shadowjams, Kierkkadon, 老陳, IMNOTARETARDATALL, St. Hubert, Lordyou, George2001hi, FrescoBot, Paine Ellsworth, Tobby72, S colligan, Nicktfx, HJ Mitchell, Silent78, Yoyo2222, Craig Pemberton, Jakhai1000, Citation bot 1, Pinethicket, I dream of horses, Edderso, PrincessofLlyr, 10metreh, Firozmusthafa, Pudgy78685, A8UDI, Flubber88, Robo Cop, Reconsider the static, Jauhienij, FoxBot, Tgv8925, TobeBot, Aliwiki, Dinamik-bot, Vrenator, TBloemink, Dan kelley90, Jeffrd10, Pmbeck, Weedwhacker128, Suffusion of Yellow, Splartmaggot, Jesse V., Mean as custard, Slon02, MELISASIMPSONS, EmausBot, Orphan Wiki, WikitanvirBot, Immunize, Octaazacubane, Karteek987, Racerx11, RenamedUser01302013, Dalegudmunsen, Solarra, Tommy2010, K6ka, Lucas Thoms, Capybara123, Hhhippo, JSquish, John Cline, Fæ, Josve05a, Shuipzv3, Quondum, Crawlbeforeiwalk, EWikist, Wayne Slam, Ownedestroy, Maxrokatanski, Brandmeister, Apocalypse2009, Shrigley, John Aplessed, ChuispastonBot, RockMagnetist, Teapeat, 28bot, Sonicyouth86, Petrb, ClueBot NG, Adawg117, Bchaplucian, Alex-engraver, Cntras, The Troll lolololololololol, Helpful Pixie Bot, Stanlste, Strike Eagle, Bibcode Bot, Cmandouble3, Rijinatwiki, MusikAnimal, Changer-guy, JacobTrue, RaulRavndra, Maxellus, Thekillerpenguin, EntangledSpins, Srikar33, MrBill3, Zedshort, Glacialfox, Eduardofeld, ChrisGualtieri, Embrittled, Yammer68, Fizped~enwiki, MadGuy7023, Bluedot951, SoledadKabocha, Denis Fadeev, Numbermaniac, NinjaNightMare12, Reatlas, Trollarc26, DavidLeighEllis, Tobythedog1, Coz12345, !nnovativ, Manul, SpecialPiggy, JaconaFrere, Skr15081997, SufferinSuckatach, Monkbot, Niuthon, Mrmister6123, MacWarcraft, Scarlettail, Trackteur, Shantygoreman, Suraj 967, Martinpalla123, The Last Arietta, Elite Whitesands Force, RippleStrike, KasparBot, Miskiiy, Mf siddiqui, Creadilo, TheDadeOne and Anonymous: 924

• **Thermal conductivity** *Source:* https://en.wikipedia.org/wiki/Thermal_conductivity?oldid=689974602 *Contributors:* AxelBoldt, Mav, Bryan Derksen, The Anome, AdamW, William Avery, SimonP, Peterlin~enwiki, Jdpipe, Heron, Stevertigo, Patrick, Michael Hardy, SebastianHelm, Alfio, Looxix~enwiki, Mark Foskey, Salsa Shark, Glenn, Rl, Mxn, Charles Matthews, Tantalate, Andy G, Dfeuer, Omegatron, Dmytro, Bearcat, Gentgeen, Robbot, Hankwang, Buster2058, Giftlite, Smjg, DocWatson42, Mat-C, Ido50, Brockert, Bobblewik, Aulis Eskola, Blazotron, Mrdarrett, Dreamtheater, CSTAR, lcairns, Gscshoyru, Trevor MacInnis, Mike Rosoft, Discospinster, Vsmith, Bender235, El C, Mdf, Joanjoc~enwiki, Femto, Smalljim, L33tminion, Kjkolb, Mjager, Matt H, Eric Kvaalen, Riana, Axl, Lightdarkness, Gene Nygaard, Vadim Makarov, Ericl234, Woohookitty, David Haslam, MONGO, Rtdrury, SCEhardt, Bobmilkman, Mandarax, Nirvelli, Yurik, Susten.biz, Rjwilmsi, Wahoofive, Strait, Yamamoto Ichiro, Margosbot~enwiki, RexNL, Tardis, Srleffler, Dougthebug, Roboto de Ajvol, Az7997, Hukseflux, Micah Fitch, RazorICE, Pyrotec, Dhollm, Kittell, Alex43223, Bota47, Mike92591, Jth299, FF2010, Jонанб, Derek1G, Livitup, David Biddulph, Kungfuadam, N3362, Aanidaani, That Guy, From That Show!, TomR, Itub, Doubleplusjeff, SmackBot, Nkrupans, Hydrogen Iodide, Melchoir, CMD Beaker, Chronodm, Brossow, Gilliam, Isaac Dupree, Kmarinas86, Cadmium, Jprg1966, Sadads, Colonies Chris, ToobMug, Frap, Fiziker, Gilloq, Rrburke, Stevenmitchell, -xfi-, Polonium, DavidJ710, SashatoBot, Kuru, Sbmehta, UberCryxic, Loodog, Jaganath, CyrilB, StanBrinkerhoff, Bendzh, Wizard191, IvanLanin, Tawkerbot2, Mikiemike, CmdrObot, Falk Lieder, KyraVixen, Slazenger, Cydebot, Franklinx, Shepplestone, On5deu, Aazn, Tunheim, Mr Ginkgo, Gralo, Headbomb, JdH, Electron9, Uiteoi, Leon7, Orionus, Adams13, Mdkoch84, Somerandom, MER-C, Cerrcerr, Magioladitis, Jswaim, EagleFan, Dan Pangburn, Hbent, Geniescience, MartinBot, Ghostwo, R'n'B, The Anonymous One, Pharaoh of the Wizards, RSRScrooge, Salih, Stan J Klimas, Cperabo, WilfriedC, DorganBot, Craklyn, David H. Flint, BernardZ, Lights, LokiClock, TXiKiBoT, Clarince63, JhsBot, Zolot, Rlsmalling, Microfirmware, Andy Dingley, Temporaluser, Why Not A Duck, EmxBot, Biscuittin, Illinoisavonlady, AquaDTRS, Gknor, Yintan, Scanrod, Radon210, Solar clathrate, Baderimre, Sfan00 IMG, ClueBot, Fyyer, The Thing That Should Not Be, EoGuy, Nickersonl, Cfsenel, Thegeneralguy, C-Therm, SchreiberBike, Munden, Glacier Wolf, DumZiBoT, Darkicebot, XLinkBot, Gonzonoir, Dthomsen8, Sebastian, Ffq, Rxjensen, Addbot, Cxz111, Some jerk on the Internet, Elvire, Montgomery '39, Ronhjones, TutterMouse, Glane23, Barak Sh, Teles, Arbitrarily0, Yobot, Gouryella24, Druiids, Amirobot, Nallimbot, Nummify, AnomieBOT, KDS4444, Metaphase2, Ciphers, Jim1138, Citation bot, Akilaa, Maniadis, LilHelpa, Draxtreme, Nelienke, Wraith69, Logger9, Shadowjams, Bo-Kaj-R, Thehelpfulbot, FrescoBot, Oxonium, Васип Юрий, Pepper, Kdn1982, Proepro, Tavernsenses, HamburgerRadio, OliRG2, Citation bot 1, Þjóðólfr, Nirmos, Pinethicket, NPKResults, Durplub, Nemontemi, Ezzt, Suffusion of Yellow, JesperDoffe, Marie Poise, Pushkar.gaikwad,

DARTH SIDIOUS 2, RjwilmsiBot, MagnInd, DexDor, EmausBot, WikitanvirBot, Dewritech, Enviromet, Carandbike, Hhhippo, Rich 988, John Cline, Pololei, Vramasub, Fizicist, Sahimrobot, Tls60, Finglas, ClueBot NG, Jack Greenmaven, Moses97, School of Stone, Satellizer, Senthilvel32, Ulrich67, Frietjes, Zhoravdb, Rezabot, Zavod219, Jhaupt, Mmarre, Echsecutor, Alex Crikey, Pluma, MerllwBot, Helpful Pixie Bot, Bibcode Bot, BG19bot, Bruekener, Hallows AG, SmellyLilYou, Eio, Zedshort, Hamish59, Kiltanen, NTTFSIT, Minsbot, Evgeny Azarov, Iifar, Stonovic, Jionpedia, AlchemistOfJoy, Adwaele, Mogism, Kap 7, PhantomTech, Everymorning, AresLiam, Johnfranciscollins, BrightonC, TaeYunPark, Dapias, Induleijubilo, Ikkohn, ProprioMe OW, The Quixotic Potato, Poophead2711 and Anonymous: 402

- **Lorentz force** *Source:* https://en.wikipedia.org/wiki/Lorentz_force?oldid=689808959 *Contributors:* Bryan Derksen, The Anome, DrBob, Heron, Tim Starling, SebastianHelm, Dgrant, Looxix~enwiki, Andres, Hollgor, BenRG, Mrdice, Robbot, Smb1001, Modeha, Fuelbottle, Ancheta Wis, Giftlite, Michael Devore, Leonard G., Utcursch, LucasVB, MFNickster, Chris Howard, Rich Farmbrough, Bender235, ESkog, El C, Kwamikagami, Laurascudder, Rbj, Gene Nygaard, Forderud, Rtdrury, Mpatel, BD2412, Ketiltrout, Kinu, RobertG, Goudzovski, Chobot, DVdm, Bgwhite, D.keenan, JabberWok, Rsrikanth05, Salsb, Nick, Mikeblas, Werdna, Tetracube, Petri Krohn, Geoffrey.landis, FyzixFighter, Sbyrnes321, That Guy, From That Show!, Tttrung, SmackBot, InverseHypercube, Qwasty, Jjalexand, Complexica, Metacomet, Colonies Chris, Drkirkby, Jaro.p~enwiki, Fuhghettaboutit, Decltype, Edmundo ba~enwiki, Sadi Carnot, Yevgeny Kats, DJIndica, Nmnogueira, LtPowers, Lwiniarski, Makyen, Dicklyon, JRSpriggs, Mikiemike, Myasuda, Cydebot, Michael C Price, Christian75, Thijs!bot, Headbomb, Electron9, Escarbot, AntiVandalBot, Paclopes, Tim Shuba, Alphachimpbot, Spartaz, JAnDbot, Frobnitzem, JNW, Sfu, Deans-nl, Alexcalamaro, Sunnysite, Cpiral, KylieTastic, Jkeohane, Treisijs, D-Kuru, Sheliak, JohnBlackburne, Thurth, Philip Trueman, Yakeyglee, Sankalpdravid, Ti89TProgrammer, Mihaip, Jackfork, Kiyabg, Lerdthenerd, Falcon8765, Spinningspark, Jradavenport, SieBot, Gerakibot, Jdcanfield, Masgatotkaca, Paolo.dL, Wessmaniac, StaticGull, Uncle Milty, Tharunsr121, Sun Creator, Brews ohare, Alousybum, Boethius65, Rror, WikHead, Khunglongcon, Addbot, George Smyth XI, FDT, Morning277, The mexican boodle, CUSENZA Mario, K Eliza Coyne, TStein, Legobot, Luckas-bot, Yobot, ñ भाषा, Csmallw, AnomieBOT, Archon 2488, Neptune5000, Jcc77, Citation bot, Xqbot, Wavgfkl, Capricorn42, NOrbeck, GrouchoBot, Omnipaedista, GliderMaven, Curtisabbott, BenzolBot, Citation bot 1, Jonesey95, RedBot, Jauhienij, Inbamkumar86, TheBFG, Logichulk, Earthandmoon, Reach Out to the Truth, Olawlor, RjwilmsiBot, TjBot, EmausBot, Lenfreeman, Nitin.i.azam, Solomonfromfinland, Hhhippo, ZéroBot, Ambros-aba, Wikfr, SporkBot, EricWesBrown, BrokenAnchorBot, Maschen, Zueignung, RockMagnetist, ClueBot NG, CocuBot, Movses-bot, Aniruddha22Paranjpye, Helpful Pixie Bot, BG19bot, Amicus of borg, ElphiBot, Acmedogs, John hanley parc, F=q(E+v^B), Khazar2, Monica.alonso.UEM, Quibilia, RaphaeL82, Neilroy1998, NataschaD, Suren.vasilyan, Trainforrest, SkateTier, Lowellbander, Julian ceaser, EoRdE6, Krelcoyne, KasparBot, CanaDeAlba, Robertmorris0 and Anonymous: 181

- **Electric field** *Source:* https://en.wikipedia.org/wiki/Electric_field?oldid=687332898 *Contributors:* Andre Engels, Peterlin~enwiki, Waveguy, Heron, Patrick, Michael Hardy, Tim Starling, GameGod, Modster, Dgrant, Looxix~enwiki, Nikai, Smack, Tantalate, Reddi, Omegatron, BenRG, Lumos3, Donarreiskoffer, Robbot, Nizmogtr, Wjbeaty, Giftlite, Herbee, TomViza, Ssd, Mboverload, Uranographer, Pearbonn, MFNickster, Icairns, AmarChandra, Abdull, Jpk, Mjpieters, Mal~enwiki, Ghitis, Purplefeltangel, El C, Laurascudder, Femto, Smalljim, Shenme, I9Q79oL78KiL0QTFHgyc, Bert Hickman, Larry V, Helix84, Benjah-bmm27, ABCD, Riana, Pion, BRW, Wtshymanski, Gene Nygaard, Netkinetic, Oleg Alexandrov, Tariqabjotu, Nuno Tavares, Woohookitty, Linas, Mindmatrix, Linnea, Quadduc, StradivariusTV, Scjessey, Rtdrury, SCEhardt, Pfalstad, Mandarax, Jorunn, Strait, Jpeham, Margosbot~enwiki, Fresheneesz, Srleffler, DVdm, Bgwhite, Spiderboy, Stephenb, Gaius Cornelius, Salsb, Wimt, NawlinWiki, Grafen, T-rex, DomenicDenicola, Aristotle2600, Light current, Enormousdude, 2over0, Rdrosson, Anclation~enwiki, Mebden, RG2, FyzixFighter, Sbyrnes321, SmackBot, InverseHypercube, Bomac, Pedrose, Gilliam, Bluebot, Complexica, Lenko, PureRED, Hongooi, Can't sleep, clown will eat me, Chlewbot, Wikipedia brown, Fuhghettaboutit, Radagast83, Mwtoews, Xezlee, DJIndica, Nmnogueira, FrozenMan, Ckatz, Frokor, Slakr, Dicklyon, Wjejskenewr, Mfrosz, Courcelles, Ziusudra, JRSpriggs, Tpruane, JForget, Ale jrb, Jackzhp, WeggeBot, Bmk, Funnyfarmofdoom, Scott.medling, Cydebot, Hydraton31, Bverist, Michael C Price, Christian75, Chrislk02, Thijs!bot, Irigi, Headbomb, Second Quantization, Austinenator, Big Bird, Icep, AntiVandalBot, Shirt58, Alphachimpbot, JAnDbot, CosineKitty, Misnardi, Magioladitis, Yurei-eggtart, Bongwarrior, VoABot II, Catslash, Rivertorch, Xanthym, Reinam~enwiki, Edward321, Gludwiczak, DinoBot, Dr. Morbius, TheEgyptian, PrestonH, The Anonymous One, Peter Chastain, Maurice Carbonaro, St.daniel, AntiSpamBot, Qwerty59, Juliancolton, STBotD, Treisijs, Sodaplayer, Lseixas, Robertirwin22, Idioma-bot, Sheliak, Nasanbat, VolkovBot, Doctorkismet, JohnBlackburne, AlnoktaBOT, Philip Trueman, TXiKiBoT, Cosmic Latte, Hqb, OlavN, SheffieldSteel, Andy Dingley, Antixt, Lian1238, HiDrNick, AlleborgoBot, Relilles~enwiki, EmxBot, SieBot, Spartan, Krawi, Gerakibot, JerrySteal, Bentogoa, OKBot, Onlyonefin, ClueBot, André Neves, WriterListener, EoGuy, Jan1nad, Nnemo, Razimantv, Nayafun, Russell4, Excirial, Muro Bot, Corkgkagj, Tvine, MrDeodorant, ZooFari, Wyatt915, Addbot, Tenev, Fgnievinski, Fluxbyte, FDT, LaaknorBot, AndersBot, Numbo3-bot, Qmark42, Tide rolls, Zorrobot, Legobot, Luckas-bot, Yobot, Ptbotgourou, Fraggle81, Sdoregon, KamikazeBot, Qui-Gon Jinn, SwisterTwister, DrTrigon, AnomieBOT, Ipatrol, Geek1337~enwiki, Xqbot, Bdforbes, Renaissancee, DSisyphBot, Raffamaiden, Ruy Pugliesi, RibotBOT, Dreitmen, RGForbes, Atomless, Lookang, Ozhu, Hungryhungarian, Hwong557, I dream of horses, Zoidfather, Mekeretrig, Rausch, Gryllida, TobeBot, عقیل فرانک, SchreyP, TheBFG, Sbgirl54, Ddvche, Vinnyzz, EmausBot, Rokerdud, Immunize, Ajraddatz, Trinibones, Wikipelli, K6ka, Hardestcorest, JSquish, ZéroBot, Quondum, Sonygal, OnePt618, Sahimrobot, Maschen, Benjamink8, Verman1, ClueBot NG, Gareth Griffith-Jones, Jmguerrac, Justlettersandnumbers, Parcly Taxel, Widr, Ryan Vesey, NuclearEnergy, Helpful Pixie Bot, Culo95, Gracielleannep, Chander, Onewhohelps, Bonginkosi zwane, F=q(E+v^B), Ea91b3dd, Pithdillinja, Poilkmn1, Brad7777, Szezureq, IkamusumeFan, SortOfStillCare, BrightStarSky, Dexbot, Webclient101, Cerabot~enwiki, The.ever.kid, Nishant nsp, Frosty, Graphium, Mpov, Jharlanherb, Razibot, Reatlas, DavidLeighEllis, Ciro.Landolfi, Tigraan, Blackbombchu, Shipandreceive, SpecialPiggy, AwesomeEvilGenius, Faekynn, Vieque, Henryalston, Fstevens123, Chevoron, Csemohitagrawal, JC713, Qzekrom, Isambard Kingdom and Anonymous: 373

- **History of electromagnetic theory** *Source:* https://en.wikipedia.org/wiki/History_of_electromagnetic_theory?oldid=690073326 *Contributors:* William Avery, Stevertigo, Arpingstone, Ahoerstemeier, Charles Matthews, Reddi, Topbanana, Lupo, Rik G., Alan Liefting, Sj, Sesel, Beland, Bhugh, ELApro, Rich Farmbrough, LindsayH, Dbachmann, Cladist, I9Q79oL78KiL0QTFHgyc, Wtshymanski, Gene Nygaard, Stemonitis, Simetrical, Woohookitty, Carcharoth, BillC, Ruud Koot, BD2412, Sjö, Rjwilmsi, Hulagutten, Andy85719, Elmer Clark, Gurch, Mallocks, King of Hearts, Gdrbot, Korg, RussBot, Gaius Cornelius, K.C. Tang, Madcoverboy, Aeusoes1, Welsh, Schlafly, Jpbowen, Larsobrien, Evrik, Kkmurray, Dingy, Netrapt, Palthrow, Mais oui!, KNHaw, SmackBot, McGeddon, Jagged 85, Emj, Hmains, Jearroll, Chris the speller, Movementarian, Colonies Chris, Gracenotes, Erzahler, Valenciano, Harryboyles, Sambot, Kuru, Accurizer, IronGargoyle, RandomCritic, Igoldste, RekishiEJ, JRSpriggs, OliverBBurke, Tanthalas39, Ruslik0, Myasuda, Kanags, Gtxfrance, Cyhawk, Doug Weller, Ssilvers, Thijs!bot, Barticus88, Martin Hogbin, Headbomb, Gerry Ashton, James086, D.H, Shirt58, Goldenrowley, TimVickers, Tim Shuba, Pixelface, Emerald Melios, Sluzzelin, Matthew Fennell, Andonic, Mkch, Hut 8.5, Grievous Angel, VoABot II, Mbarbier, Doug Coldwell, Nick Cooper, KConWiki, Dirac66, MartinBot, Keith D, R'n'B, CommonsDelinker, Pharaoh of the Wizards, Adavidb, 5theye, DarkFalls, GhostPirate, NewEnglandYankee, Fountains of Bryn Mawr, ARTE, Tiggerjay, Quacksalber, TXiKiBoT, Plenumchamber–enwiki, Abbyratsolee, Hqb, Rcarey1, PDFbot, Phirosiberia, Billinghurst, Fr33Lanc3r, Insanity Incarnate, Biscuittin, SieBot, Work permit, J.M.Domingo, Keilana, Ambernist, SmallRepair, Arjen Dijksman, Hobartimus, 3rdAlcove, Muhends, ImageRemovalBot, Tomasz Prochownik, ClueBot, SummerWithMorons, Mild Bill Hiccup, J8079s, Misedo, Manishearth, Balrore, Muhandes, Estirabot, Cenarium, SchreiberBike, Audaciter, Ottawa4ever, InternetMeme, Dthomsen8, Mitch Ames, Addbot, Fgnievinski, Letter Ezh, Xantan5, Surferboy244, TStein, Tassedethe, Lightbot, Bartledan, Humphrey Jungle, Middayexpress, PieterJanR, Luckas-bot, Yobot, AnomieBOT, Materialscientist, Citation bot, ArthurBot, Xqbot, EJohn59, DSisyphBot, Omnipaedista, RibotBOT, Ignoranteconomist, Eugene-elgato, Nickbeland, Thehelpfulbot, FrescoBot, Denver26–enwiki, DivineAlpha, JMilty, Citation bot 1, Ivc392, Mikespedia, Full-date unlinking bot, Rzuwig, EmausBot, John of Reading, Never give in, RA0808, Sxoa, Winner 42, Knight1993, Qniemiec, Maschen, RockMagnetist, Sven Manguard, ClueBot NG, Widr, Helpful Pixie Bot, Electriccatfish2, Bibcode Bot, Technical 13, PearlSt82, Davidiad, Tom Pippens, HardBoiledEggs, Rococo1700, BattyBot, Jeremy112233, Khazar2, Sssciencece, Meshed Gears, Dexbot, Mogism, Ashleyleia, So11, Monkbot, 468SM, SQMeaner, LobsterCan and Anonymous: 110

- **Neutron** *Source:* https://en.wikipedia.org/wiki/Neutron?oldid=690594531 *Contributors:* AxelBoldt, Tobias Hoevekamp, Chenyu, Trelvis, Calypso, Mav, Bryan Derksen, The Anome, AstroNomer–enwiki, Malcolm Farmer, Andre Engels, Xaonon, Danny, XJaM, Roadrunner, Jaknouse, Olivier, Patrick, Michael Hardy, Valery Beaud, Ixfd64, TakuyaMurata, NuclearWinner, Looxix–enwiki, ArnoLagrange, Mkweise, Ellywa, Ahoerstemeier, Cyp, Andrewa, Aarchiba, Julesd, Glenn, Nikai, Andres, Stone, Denni, Kbk, Tarosan–enwiki, Maximus Rex, Donarreiskoffer, Gentgeen, Robbot, Fredrik, Romanm, Merovingian, Rursus, Wikibot, Alan Liefting, Dave6, Giftlite, Mikez, Art Carlson, Herbee, Xerxes314, Everyking, Dratman, NeoJustin, Bensaccount, Poupoune5, Jorge Stolfi, Christofurio, Knutux, Karol Langner, Aecarol, Icairns, Zfr, Cglassey, Peter bertok, Frau Holle, M1ss1ontomars2k4, Sparky2002b, Mike Rosoft, Guanabot, Vsmith, Dbachmann, Bender235, Kjoonlee, AlDragon, Geoking66, Neko-chan, RJHall, CanisRufus, El C, Susvolans, Femto, CDN99, Bobo192, O18, Smalljim, SpeedyGonsales, Kjkolb, Obradovic Goran, Sam Korn, Nsaa, Jakew, Eddideigel, Jumbuck, Patsw, Alansohn, Interiot, Riana, Wtmitchell, BRW, NickMartin, Vuo, DV8 2XL, HenryLi, Tchaika, Forteblast, Falcorian, Richard Arthur Norton (1958-), JarlaxleArtemis, WadeSimMiser, Sega381, SDC, Jon Harald Søby, Prashanthns, Abd, LexCorp, Graham87, Magister Mathematicae, Doughboy, Ketiltrout, Rjwilmsi, Nightscream, Zbxgscqf, Strait, AySz88, Oo64eva, Robert Fraser, Rangek, FlaBot, Nihiltres, Goudzovski, Srleffler, Roneborh, King of Hearts, Chobot, DVdm, YurikBot, RobotE, Bambaiah, JWB, TSO1D, Jimp, Phantomsteve, KyleDantarin, Stephenb, Gaius Cornelius, Yyy, Salsb, NawlinWiki, Tupungato, Wiki alf, Complainer, Grafen, Dlugosz, Voidxor, Scottfisher, Kkmurray, Spute, Dna-webmaster, Wknight94, Stefan Udrea, Mike Serfas, Closedmouth, Reyk, Modify, Alchie1, CWenger, RG2, Paul Erik, Triple333, Attilios, SmackBot, Caiyern, Melchoir, Wiki Tiki God, Unyoyega, Jrockley, Dr.Science, Edgar181, Yamaguchi 先生, Kdliss, Wigren, Chris the speller, Rajeevmass–enwiki, Persian Poet Gal, SchfiftyThree, Complexica, DHN-bot–enwiki, Sbharris, Colonies Chris, Brainblaster52, Can't sleep, clown will eat me, DéRahier, Juancnuno, SundarBot, DFriend, Aldaron, KunalKathuria, Nakon, Mwtoews, DMacks, Soarhead77, Bdushaw, Pilotguy, Renafaye77, SashatoBot, Demiex, Tim bates, Mgiganteus1, Slakr, Citicat, Asyndeton, BranStark, Shoeofdeath, Newone, Tawkerbot2, Atonrobot, Mosaffa, CmdrObot, Wafulz, Dycedarg, Rwflammang, Joelholdsworth, Lokal Profil, Karenjc, Myasuda, Safalra, Icek–enwiki, Badseed, Nick Y., Gogo Dodo, Chasingsol, Phydend, Gimmetrow, Thijs!bot, Epbr123, Montazmeahii, Goods21, Tsogo3, N5iln, Oerjan, Headbomb, Marek69, SouthernMan, RoboServien, Escarbot, Aadal, WikiSlasher, AntiVandalBot, Seaphoto, Naturalnumber, Spencer, Astavats, Husond, CosineKitty, Medconn, TheEditrix2, Bongwarrior, VoABot II, Kuyabribri, JamesBWatson, WODUP, Mother.earth, Animum, BatteryIncluded, Dirac66, LorenzoB, DerHexer, JaGa, Hans Moravec, Hyray, Patstuart, MartinBot, Church of emacs, Gnuarm, Mennoblaauw, Andre.holzner, Rettetast, J.delanoy, Dbiel, Extransit, Acalamari, Nemvocalist, TomasBat, MetsFan76, Joshmt, Heavens is the world, Scott Illini, TraceyR, Idioma-bot, Mviduka4197, VolkovBot, Tourbillon, Thedjatclubrock, Jeff G., Mocirne, Seattle Skier, TXiKiBoT, DoctorPiouk, Dev 176, Martin451, ABigGreenHippo, Abdullais4u, FreeFull, Wikiisawesome, Scarymaryfwfc, RadiantRay, Roomyt, W1k13rh3nry, Antixt, Deansinclair, Enviroboy, Burntsauce, Brianga, AlleborgoBot, EmxBot, Neparis, D. Recorder, Ponyo, YohanN7, SieBot, Cwkmail, Yintan, Agesworth, JerrySteal, Keilana, RadicalOne, Toddst1, Tiptoety, JetLover, Arjen Dijksman, Sbowers3, Aruton, Oxymoron83, AnonGuy, Beej175560, Techman224, Anyeverybody, Nergaal, Denisarona, Lord Shivan, Naturespace, ClueBot, RudolfSchmidt, PipepBot, Fasettle, Fyyer, The Thing That Should Not Be, Starkiller88, Industrieman, Mild Bill Hiccup, Polyamorph, Shjacks45, ChandlerMapBot, DragonBot, Gnome de plume, Jusdafax, Ju7kik8ol568r, Cenarium, Jotterbot, Vboobelarus, Subash.chandran007, Plasmic Physics, Versus22, XLinkBot, Dark Mage, PL290, SkyLined, Addbot, Taschna, DOI bot, Ronhjones, Mr. Wheely Guy, LaaknorBot, CarsracBot, JBukon, Favonian, LinkFA-Bot, 5 albert square, AgadaUrbanit, Morgrimm, Numbo3-bot, Ehrenkater, LarryFrank, Tide rolls, Lightbot, Teles, Legobot, Luckas-bot, Yobot, Велесень, 2D, Tohd8BohaithuGh1, Cabb99, AnakngAraw, AnomieBOT, Bsimmons666, Jim1138, AdjustShift, Bluerasberry, Materialscientist, Hdehuer, The High Fin Sperm Whale, Citation bot, Satan's Kitchen, Maxis ftw, Raven1977, ArthurBot, Marshallsumter, Xqbot, Gopal81, Capricorn42, Drilnoth, DSisyphBot, Gilo1969, Paula Pilcher, Faatoafe90, Goostyyy, WaveEtherSniffer, GrouchoBot, Abce2, Amaury, Doulos Christos, Gordonrox24, Shadowjams, A. di M., Samwb123, R8R Gtrs, FrescoBot, LucienBOT, Paine Ellsworth, Cannolis, Citation bot 1, Ecko15, Biker Biker, Pinethicket, HRoestBot, Jonesey95, Nicklcms, Seattle Jörg, Abhinav paulite, Double sharp, Darrell cosare, كاشف عقيل, Mr.98, Diannaa, Ironnickel, Andrea105, Onel5969, TjBot, Magnlnd, Jackehammond, Jimmy be, Robert Johnson 10, EmausBot, Green Day143, WikitanvirBot, Unkenruf, GoingBatty, Illdz, Psturm–enwiki, Pcorty, Wikipelli, Hhhippo, ZéroBot, John Cline, Brazmyth, Quondum, GianniG46, Copper.nanotube, Brandmeister, L. Kensington, Epicstonemason, Sjkimminau, Chris857, VictorianMutant, DASHBotAV, Whoop whoop pull up, ClueBot NG, Nebulosus, CocuBot, Satellizer, Letoya123, TruPepitoM, OverQuantum, Heyheyheyhohoho, Rezabot, Android1188, Widr, Diyar se, Ieditpagesincorrectly, Bibcode Bot, Neutronscattering, Wiki13, Metricopolus, Contact '97, Universaminkeisari, Nathanrohler, Zedshort, Hamish59, Nitrobutane, Hobos-r-us12, Ozniteeki, BattyBot, MeowMeowArf, Dansalmo, ChrisGualtieri, GeorgEhlers, Ducknish, Gladiator222, Dexbot, Mogism, 331dot, TwoTwoHello, Lugia2453, Graphium, FaerieChilde, Fossilsnout, Morg00, Cldorian, Xuanmingzi, DihllonJessie, Jesse.johns, The Herald, Zenibus, Darkch2, Jwratner1, Javierha, My name is not dave, Cytokinetics, EtymAesthete, DudeWithAFeud, Abitslow, AspaasBekkelund, Bballbro62, Mahusha, Light on the wall, Monkbot, Profesionalpretzels, Jayakumar RG, Haftswinch532, Selmatoced50, Istillcant, HMSLavender, Petahr, Orduin, Kethrus, Pulkit 4325, DiscantX, TSchonfeldt, Matan Kovac, KasparBot, Kafishabbir, Lord Wingus The Third and Anonymous: 558

- **Bohr model** *Source:* https://en.wikipedia.org/wiki/Bohr_model?oldid=689019301 *Contributors:* Damian Yerrick, AxelBoldt, Trelvis, Zundark, The Anome, Taw, Andre Engels, Arvindn, Stevertigo, D, JohnOwens, Michael Hardy, Tim Starling, Looxix–enwiki, Ellywa, Ahoerstemeier, Cyp, Glenn, Complex Analysis, Hashar, Charles Matthews, Tantalate, RickK, The Anomebot, Furrykef, Fibonacci, Omegatron, BenRG, Robbot, Vespristiano, Pmineault, Elysdir, Rsduhamel, Enochlau, Decumanus, Giftlite, BenFrantzDale, Lethe, MathKnight, Bensaccount, Dmmaus, Yath, Antandrus, Darksun, Grunt, Discospinster, Vsmith, ArnoldReinhold, Gianluigi, Mani1, Paul August, Kaszeta, El C, MrMarshmallow, CDN99, Army1987, Whosyourjudas, Smalljim, Viriditas, Cmdrjameson, R. S. Shaw, Evgeny, Matt McIrvin, MPerel, Nsaa, Jjron, Mdd, Jumbuck, Storm Rider, Alansohn, TheParanoidOne, Munchkinguy, Fritzpoll, PAR, Redfarmer, Snowolf, Velella, HenkvD, Sciurinæ, Bsadowski1, Linas,

Mindmatrix, Benhocking, Mpatel, Mreult~enwiki, Schzmo, Terence, MFH, Ian**, SeventyThree, Christopher Thomas, Graham87, Galwhaa, DePiep, Mendaliv, Rjwilmsi, Vary, Salix alba, Tawker, Scorpiuss, Yamamoto Ichiro, Sanbeg, Nivix, Krackpipe, RexNL, Gurch, Goudzovski, Srleffler, Chobot, DVdm, Gwernol, YurikBot, Wolfmankurd, Postglock, JabberWok, Rsrikanth05, Spike Wilbury, Aeusoes1, Buster79, CAPS lOCK, Moe Epsilon, Zwobot, Dna-webmaster, Wknight94, Sperril, PTSE, Closedmouth, KGasso, Petri Krohn, Fram, HereToHelp, Spliffy, GrinBot~enwiki, SkerHawx, SmackBot, Thorseth, Blue520, Jacek Kendysz, WookieInHeat, Frymaster, Timotheus Canens, Gilliam, Chris the speller, Pieter Kuiper, OrangeDog, Complexica, DHN-bot~enwiki, Sbharris, Darth Panda, Can't sleep, clown will eat me, Wen D House, Nakon, Localzuk, GoldenBoar, Adam Schloss, DMacks, O RLY?, Ligulembot, Mion, Sadi Carnot, Kukini, Eliyak, John, Gobonobo, Tktktk, Lestatdel.ioncourt, Hemmingsen, 3897515, Goodnightmush, IronGargoyle, PseudoSudo, Smith609, Tase, Beetstra, Martinp23, Domino42, Mets501, Doczilla, Ryulong, Jonhall, Lee Carre, Iridescent, Zootsuits, Joseph Solis in Australia, Morrowulf, Igoldste, Tony Fox, Beve, Blehfu, Courcelles, Profjohn, Tawkerbot2, Chetvorno, JForget, Frovingslosh, Olaf Davis, BeenAroundAWhile, CWY2190, Harej bot, Myasuda, Dept of Alchemy, Cydebot, WillowW, Gogo Dodo, Corpx, Xndr, Shirulashem, DumbBOT, Chrislk02, Sp, Pinky sl, FrancoGG, Epbr123, Wikid77, Goods21, Kablammo, Headbomb, John254, Martin Hedegaard, Pfranson, Dawnseeker2000, Mentifisto, AntiVandalBot, Majorly, Tlabshier, Spartaz, JAnDbot, Harryzilber, MER-C, Andonic, Belg4mit, Hut 8.5, Tstrobaugh, Bkpsusmitaa, R27182818, Ó, Acroterion, Casmith 789, Magioladitis, VoABot II, Wikidudeman, JamesBWatson, Dirac66, 28421u2232nfenfcenc, Schumi555, Cpl Syx, SlamDiego, Mikerobertsn, Vssun, TheRanger, Robin S, Geboy, Kpxxbladexx415, Jemijohn, Mont95, ChemNerd, Slash, J.delanoy, Melamed katz, Uncle Dick, Jonpro, Cpiral, It Is Me Here, Bot-Schafter, AntiSpamBot, Andraaide, NewEnglandYankee, Pez2, Juliancolton, Copsi, Jarry1250, JavierMC, Dextercioby, Cuzkatzimhut, Hugo999, VolkovBot, Philip Trueman, TXiKiBoT, The Original Wildbear, Sarenne, Anonymous Dissident, Piperh, JhsBot, Leafyplant, Jackfork, LeaveSleaves, Itemirus, Thunderbird2, MrChupon, Murkee, DrJunge, EvilBunnyHead, SieBot, Servant Saber~enwiki, Coffee, Ivan Štambuk, GrooveDog, Keilana, Likebox, Flyer22 Reborn, Tiptoety, Grimey109, Oxymoron83, Antonio Lopez, Scorpion451, Jdaloner, Pac72, Lightmouse, Christovac, Nandobike, The Stickler, Mike2vil, Anchor Link Bot, Pinkadelica, Dolphin51, Elassint, ClueBot, Ferred, GorillaWarfare, The Thing That Should Not Be, ArdClose, Swedish fusilier, Lantay77, Neverquick, Auntof6, Djr32, Robert Skyhawk, Excirial, SpikeToronto, Willthedrill, Danmichaelo, Nengscoz416, Iohannes Animosus, Bite Size Monkeys, W.GUGLINSKI, Zerxan, Thehelpfulone, Kakofonous, Thingg, BlueDevil, DumZiBoT, Life of Riley, Jmanigold, Nukeh, Avoided, Skarebo, NellieBly, Mm40, ZooFari, MystBot, RyanCross, Some jerk on the Internet, DOI bot, WMdeMuynck, YUR12008, CanadianLinuxUser, Fluffernutter, Glane23, Bassbonerocks, Glass Sword, LinkFA-Bot, Hainetron, Apteva, Gail, Jarble, Alfie66, Jackelfive, Legobot, Luckas-bot, Yobot, Cflm001, IW.HG, TestEditBot, Tempodivalse, N1RK4UDSK714, AnomieBOT, DemocraticLuntz, Jim1138, Piano non troppo, Sz-iwbot, Materialscientist, The High Fin Sperm Whale, Citation bot, E2eamon, GB fan, ArthurBot, Ammubhave, Jonathan321, Jeffrey Mall, Coretheapple, Omnipaedista, Ilovenickjay, Tgervaisphd, FrescoBot, Qalander, Steve Quinn, Machine Elf 1735, Citation bot 1, Pinethicket, I dream of horses, Teamdojo, BRUTE, RandomStringOfCharacters, Ifritnile, Jordgette, Javierito92, Zink Dawg, Auscompgeek, Redskins247, DARTH SIDIOUS 2, WikitanvirBot, Nuujinn, Super48paul, RA0808, Solarra, Tommy2010, Wikipelli, K6ka, Hhhippo, JSquish, John Cline, Lateg, Quondam, Zloyvolsheb, QEDK, Wagino 20100516, Thine Antique Pen, Dilwala314, WikiPidi, Nick9876, RockMagnetist, Peter Karlsen, CharlotteMab, DASHBotAV, Mtlee7, Billylovespiethethird, Spicemix, Imuwithu2, Rocketrod1960, ClueBot NG, Pruegz778, 12nichja, Satellizer, A520, Widr, Shivsagardharam, Titodutta, Bibcode Bot, BG19bot, Wiki13, Shalom25, MusikAnimal, Eio, Klilidiplomus, Ihateicairns, Achowat, Riley Huntley, Samanthaclark11, Pratyya Ghosh, ChrisGualtieri, Martinkupilas, La marts boys, Bethechangeyouhopetosee, Sdk16420, Makecat-bot, Cerabot~enwiki, Lugia2453, 6033CloudyRainbowTrail, JustAMuggle, Reatlas, Epicgenius, Howicus, Melonkelon, Tentinator, Ihatepauldirac, Ihatedirac, MKCarriegirl, Ugog Nizdast, BruceBlaus, Mourici, Konveyor Belt, Adamharsh, Williamsmith29, Ugotpowned12, LucaMoro, Amortias, Joeyransom, Adam tunard creator of science, DallasSama, GeneralizationsAreBad, YashGarg10, Leawesomepotato, KINGSUFI, MKZG, Pazycraft and Anonymous: 722

- **Effective mass (solid-state physics)** *Source:* https://en.wikipedia.org/wiki/Effective_mass_(solid-state_physics)?oldid=678313077 *Contributors:* Tim Starling, Mac, Stevenj, Tantalate, Bevo, Wereon, Rubber hound, MuDavid, Keenan Pepper, Minority Report, Gene Nygaard, Grnch, Linas, Nanite, Margosbot~enwiki, Kri, Chobot, Jaraalbe, Roboto de Ajvol, Wavelength, Black Falcon, Modify, Chaiken, Sbyrnes321, That Guy, From That Show!, SmackBot, Bluebot, Papa November, Colonies Chris, CosineKitty, CommonsDelinker, Leyo, DavidCBryant, Izno, Felix00, Yomach, RH&ST, Logan, StewartMH, Tai Chi Tech, Tizeff, Etha84, Brews ohare, Kolyma, Dthomsen8, Gekco, Addbot, PV=nRT, Luckas-bot, Yobot, Nallimbot, AnomieBOT, Materialscientist, Citation bot, ArthurBot, Adrignola, Depictionimage, Doraemonpaul, Pinethicket, EmausBot, Beatnik8983, AvicAWB, Frietjes, Dexbot, Martenseemann, Xolroc and Anonymous: 63

- **Double-slit experiment** *Source:* https://en.wikipedia.org/wiki/Double-slit_experiment?oldid=688838250 *Contributors:* AxelBoldt, Sodium, Bryan Derksen, The Anome, Tarquin, Koyaanis Qatsi, Fredbauder, Deb, Nate Silva, Hhanke, Roadrunner, DrBob, Bth, Stevertigo, Mrwojo, Lir, Chas zzz brown, Michael Hardy, Kwertii, Looxix~enwiki, Ahoerstemeier, J-Wiki, Kevin Baas, Glenn, Jouster, Charles Matthews, Reddi, Gutza, Patrick0Moran, Spikey, Finlay McWalter, Carbuncle, Jeffq, Phil Boswell, Robbot, Arkuat, Sverdrup, SchmuckyTheCat, Hadal, Anthony, Ancheta Wis, Giftlite, Ryanrs, 0x0077BE, Laudaka, BenFrantzDale, Fastfission, Dratman, Zhen Lin, Eequor, Andycjp, Alexf, Karol Langner, CSTAR, Lumidek, Thorwald, Rfl, Samboy, Zazou, SpookyMulder, Pt, Dnwq, Laurascudder, SS~enwiki, Lyght, Jpgordon, Bobo192, L33tminion, Julleras~enwiki, Conor~enwiki, Haham hanuka, Hooperbloob, Jakew, Danski14, ABCD, Mac Davis, DV8 2XL, Afshar, Oleg Alexandrov, Darked~enwiki, Joriki, Linas, StradivariusTV, Lofor, DrChinese, Cleonis, Pol098, Someone42, Mr. Qwert, Pfalstad, Rnt20, Graham87, Rjwilmsi, Miserlou, Ohanian, Ttwaring, Wikiliki, Ian Pitchford, Nihiltres, Shooter468, Kolbasz, Srleffler, DVdm, YurikBot, Daverocks, JabberWok, Ansell, Gaius Cornelius, Anomalocaris, Tom Edwards, Haikz, E2mb0t~enwiki, Syrthiss, Kyle Barbour, Olleicua, Dnawebmaster, Jeremyzone, Enormousdude, Closedmouth, Petri Krohn, CWenger, Garion96, John Broughton, FyzixFighter, Ttrung, SmackBot, Dissembly, Unschool, Lemonas', KnowledgeOfSelf, Septegram, Gilliam, Donama, Dauto, Izehar, Jprg1966, Complexica, Redattore, Bbq332, Voyajer, Andy120290, Cybercobra, Lostart, Eratosthenes, FilippoSidoti, Yevgeny Kats, DJIndica, Afiq980, Kimholder, Vincenzo.romano, Camilo Sanchez, Ekrub-ntyh, UKER, Dicklyon, Naumz, Clarityfiend, Masoninman, Tomia, UncleDouggie, CapitalR, Neoking, Chetvorno, George100, Domanix, Van helsing, Megannnn, Mr. Science, Jakelove, Banedon, Cydebot, Peterdjones, Carlamichelle, Gogo Dodo, Deepdreamer, Michael C Price, Quibik, My Flatley, Etycc, Guenael, Mbell, Headbomb, Jojan, Jonny-mt, Stannered, Widefox, Jayron32, Roundhouse0, BaxterG4, AniRaptor2001, Stellmach, Kornbelt888, Magioladitis, SHCarter, Sarahj2107, Dallascloud, Recurring dreams, Knshk.iitr, Ensign beedrill, Originalname37, Ayerz02, User A1, Deltaneos, Timefly, Mårten Berglund, Vinograd19, Custos0, Kpvats, Maurice Carbonaro, TheTerr, Dadn1011, TheSeven, Gblandst, Jotunn, Andejons, Nwbeeson, Pdcook, Idioma-bot, Sheliak, Larryisgood, DParlevliet, Thurth, Kriak, Dreakes, Gbuchana, Bentley4, Spiral5800, Richwil, Falcon8765, Ajrocke, Mars2035, Ttony21, Gprince007, Jim E. Black, Nay01, Lightmouse, OKBot, Svick, Randyburden, Hamiltondaniel, Dolphin51, Emptymountains, Danthewhale, Martarius, ClueBot, Ideal gas equation, Deanlaw, Jjreicher, Radude7, Optics guy07, The Wild West guy, WestwoodMatt, 6E656F, Djr32, Deepsurvival, Mbcudmore, Alexbot, Winston365, Amoceann, SchreiberBike, 1ForTheMoney, Certes, Trefork, Wikiregsters, Jamesscottbrown, XLinkBot, Underdone, Ost316, Nick84, Drlight11,

Addbot, Lovemuffin333, Toyokuni3, Fgnievinski, WMdeMuynck, Mac Dreamstate, TIAA Is An Acronym, Epzcaw, Favonian, LinkFA-Bot, Barak Sh, Dqbeat, Lightbot, Jhayeur, Luckas-bot, Yobot, Windyhead, Amble, AnomieBOT, Rubinbot, Jim1138, RandomAct, Shminux, Materialscientist, Amp3030, Citation bot, Guersk, ArthurBot, GnawnBot, Xqbot, Corrigendas, Tyyeerr, Abce2, Lwickstr, 老陳, FrescoBot, Nh5h, NSH002, Paine Ellsworth, Ribashka, Lookang, D'ohBot, Ugeorge, Citation bot 1, Intelligentsium, Tom.Reding, Ljerabs, Tenuk, Tonymang, Fartherred, Jordgette, Jbenjos, RjwilmsiBot, Siddharth.kulk, Whywhenwhohow, EmausBot, Roier, RA0808, ZéroBot, Mpc755, Gradatmit, Sammyzac, Ὁ οἶστρος, Mophead2002, Jonathan Whitehouse, Plasmageek, C9h13no3hcl, Z1nemo, Kartasto, Whoop whoop pull up, ClueBot NG, Gilderien, TheNewYork123, Milenko.popa, Imyourfoot, Helpful Pixie Bot, Bibcode Bot, BG19bot, Happyboy2011, MusikAnimal, Mark Arsten, Gqu, Weaktofu, Hamedghj, Stigmatella aurantiaca, ChrisGualtieri, Dexbot, Ne1 4 10s, AHusain314, Zziccardi, Acmeg, Julyancartwright, Reatlas, Bluepost22, Ruby Murray, BerFinelli, Justanotherlearner, Lundstorm, OrbitDive, Anrnusna, FrB.TG, Monkbot, Moremore88, Tetra quark, Isambard Kingdom, Sklab, Scipsycho, Niecethe, Cokelid and Anonymous: 341

- **Wave function** *Source:* https://en.wikipedia.org/wiki/Wave_function?oldid=683809777 *Contributors:* Bryan Derksen, MarXidad, XJaM, Youandme, Stevertigo, Michael Hardy, Dhc529, Dominus, GTBacchus, Alfio, Ahoerstemeier, Rob Hooft, Timwi, Wayne-enwiki, Rednblu, BenRG, Robbot, Fredrik, Bkalafut, Fuelbottle, Ancheta Wis, Giftlite, Highlandwolf, BenFrantzDale, Alison, Philgp, Abu badali, HorsePunchKid, Karol Langner, CSTAR, Freakofnurture, Hidaspal, ArnoldReinhold, Night Gyr, MBisanz, El C, Army1987, Pazouzou, Mdd, Carrasco, Pion, Count Iblis, Egg, Oleg Alexandrov, CygnusPius, Woohookitty, Ae-a, Mpatel, Mandarax, Fleisher, Rjwilmsi, John187, Mathbot, Tacid, Fresheneesz, Srleffler, Chobot, GangofOne, Bgwhite, Algebraist, Roboto de Ajvol, YurikBot, Wavelength, JabberWok, Gaius Cornelius, PoorLeno, Zwobot, Sperril, Light current, Modify, Caco de vidro, Banus, RG2, Benandorsqueaks, Sbyrnes321, That Guy, From That Show!, SmackBot, Lestrade, Incnis Mrsi, McGeddon, Frasor, Gilliam, Complexica, DHN-bot~enwiki, Sbharris, Fuhghettaboutit, StoicObloquy, Adfgvx, Joshua Barr, SashatoBot, John, Extremophile, Neoking, Bruinfan12, BeenAroundAWhile, Mct mht, Dragon's Blood, Cydebot, Wrwrwr, Peterdjones, Michael C Price, Christian75, Thijs!bot, Sagaciousuk, Headbomb, Xuanji, Nick Number, Team physicks, Alphachimpbot, B7582, Swpb, Antipodean Contributor, Tedickey, R'n'B, CommonsDelinker, Andrej.westermann, HEL, MITBeaverRocks, Maurice Carbonaro, Clackmannanshireman, (jarbarf), M-le-mot-dit, Julianeolton, DH85868993, Sheliak, Cuzkatzimhut, LokiClock, Philip Trueman, TXiKiBoT, Anonymous Dissident, Spoisp, Xnquist, EmxBot, Psymun747, YohanN7, SieBot, Keilana, Hxhbot, Physics one, Hoof47, Scorpion451, MiNombreDeGuerra, Dravecky, ClueBot, EoGuy, Niceguyedc, Saraiva.if, MorrisRob, Muhandes, NuclearWarfare, PhySusie, Phenylphenol, SchreiberBike, DumZiBoT, Boleyn, Truthnlove, Ttimespan, Realworth, Addbot, Pyfan, Mac Dreamstate, Numbo3-bot, Tide rolls, Yobot, Turiacus, Kan8eDie, Yngvadottir, Gongshow, Tonyrex, AnomieBOT, Rubinbot, Citation bot, Xqbot, Haljolad, GrouchoBot, Omnipaedista, RibotBOT, Adrignola, 老陳, Chjoaygame, FrescoBot, Zero Thrust, Steve Quinn, Machine Elf 1735, Robo37, I dream of horses, Allthingstoallpeople, Thomas1134, Tiberius Curtainsmith, Orenburg1, Augustus the Pony, Puzl bustr, Miracle Pen, Dandrestor, EmausBot, Beatnik8983, Tpudlik, Dewritech, Socob, Stanford96, Brazmyth, Quondum, Milad pourrahmani, Lukedinor, Maschen, CountMacula, RockMagnetist, Llightex, E. Fokker, One really angry guy, Rememberway, Gilderien, Hermajesty21, Snotbot, TeXnocrat, Helpful Pixie Bot, Bibcode Bot, Krishnaprasaths, BG19bot, F=q(E+v^B), Nvallejo, Davidcpearce, BattyBot, Hebert Peró, Eflatmajor7th, Besprnt, ChrisGualtieri, DonnieSwanson, Mogism, LTWoods, Rick from Richmond, AHusain314, YanikB, Samhg, Hossieni2013, Muhsenphysics, Ant.ton.t, JCMPC, Monkbot, Tigercompanion25, Yaymaths, Heinerj, Akaazhar, Isambard Kingdom, Ethansolly, Pengyulong7, Phonon112358 and Anonymous: 160

- **Virtual particle** *Source:* https://en.wikipedia.org/wiki/Virtual_particle?oldid=681850686 *Contributors:* Bryan Derksen, Andre Engels, Roadrunner, Europrobe, Heron, Michael Hardy, Looxix-enwiki, Julesd, Salsa Shark, Gamma-enwiki, Reddi, Mathus-enwiki, Phys, Astronautics-enwiki, Texture, Intangir, Jheise, Sho Uemura, Giftlite, Wolfkeeper, Lethe, Unconcerned, Ferdinangus, Tothebarricades.tk, Lumidek, Yappakoredesho, Jkl, TedPavlic, Pavel Vozenilek, Sunborn, Livajo, El C, Spoon!, Perfecto, Tjic, Matt McIrvin, Keenan Pepper, Linas, Lofor, Shane Drury, Christopher Thomas, Marudubshinki, BD2412, Qwertyus, Rjwilmsi, Koavf, Arnero, RexNL, David H Braun (1964), Krishnavedala, GangofOne, Jinma, YurikBot, Bambaiah, 4C-enwiki, Hede2000, SCZenz, DomenicDenicola, Dna-webmaster, Enormousdude, Ilmari Karonen, Mebden, Sbyrnes321, Finell, Luk, SmackBot, Amcbride, InverseHypercube, Bendykst, Gilliam, Saros136, Jjalexand, Complexica, Sbharris, Jdlambert, KazuiKier-enwiki, Drphilharmonic, Khazar, JorisvS, Mgiganteus1, Lottamiata, JRSpriggs, Chetvorno, Lavateraguy, Arnavion, Cydebot, Krauss, SolarianKnight, Lo2u, Thijs!bot, Mbell, Headbomb, Turkeyphant, Stannered, AntiVandalBot, Luna Santin, Hpesoj00, Tim Shuba, TK-925, Magioladitis, Wormcast, MartinBot, Nono64, HEL, Maurice Carbonaro, K1Aaze, WinterSpw, Jonthaler, Belsazar, Vsst, Coronellian-enwiki, SieBot, BotMultichill, VVVBot, RadicalOne, Bsdipaolo, Wpac5, ClueBot, Mallodi, CounterVandalismBot, DragonBot, Brews ohare, KenDenier, Kakofonous, SoxBot III, DumZiBoT, Terry J. Carter, SkyLined, Addbot, Uruk2008, Deamon138, 5 albert square, Mindlapse, Lightbot, OlEnglish, Yobot, Crispmuncher, Linket, AnomieBOT, Prometheus phys, Materialscientist, Ekolid, Ortho Normal-enwiki, RibotBOT, Waleswatcher, Nerdseeksblonde, Mnmngb, Chjoaygame, FrescoBot, Forward Unto Dawn, SEVEREN, Антон Гліністы, Björn-Bergman, EmausBot, John of Reading, Guevara's Revenge, Jazzalex, RockMagnetist, Curb Chain, Neon, Matt Chase, Stzwz, Ragnarstroberg, Martin.uecker, Khazar2, Metalmikebot, Ardehali, K0RTD, St170e, Yikkayaya, Offy284, Serevix and Anonymous: 109

- **Cathode ray** *Source:* https://en.wikipedia.org/wiki/Cathode_ray?oldid=687586118 *Contributors:* The Anome, SimonP, DavidLevinson, Ellmist, Heron, Bdesham, Reddi, Philopp, Pakaran, Sanders muc, Sunray, Wereon, Leonard G., Foobar, ClockworkLunch, Karol Langner, Rich Farmbrough, Rubicon, Plugwash, Smalljim, Bert Hickman, Alansohn, Lightdarkness, Wtshymanski, LFaraone, Pepepc, Boothy443, BillC, Edison, Rjwilmsi, Lugnad, Lzz, Srleffler, Youssefa, DaGizza, DVdm, Kummi, YurikBot, Gaius Cornelius, Shaddack, Salsb, NawlinWiki, Spike Wilbury, Dogcow, E2mb0t~enwiki, Jeremy Visser, WAS 4.250, Petri Krohn, LeonardoRob0t, Easter Monkey, Katieh5584, SmackBot, Grye, Gilliam, Kmarinas86, Bluebot, FinanceGuru, Chlewbot, Addshore, Zvis-enwiki, Dreadstar, Pwjb, DMacks, Sadi Carnot, Ohconfucius, Wavy G, Lottamiata, Tawkerbot2, Chetvorno, RedRollerskate, Travelbird, Dkronst, Saintrain, Epbr123, Ч, Escarbot, Makipedia, Михајло Анђелковић, Easchiff, Grimlock, I B Wright, Chitownearl, R'n'B, J.delanoy, Nemo bis, ARTE, AlnoktaBOT, Billiska, TXiKiBoT, Jkc0113, Insanity Incarnate, Logan, Neparis, SieBot, Malcolmxl5, Cwkmail, Oxymoron83, Denisarona, Loren.wilton, ClueBot, Ultimatefrisbee92, Polyamorph, Thingg, Darkicebot, HGYAIT, Addbot, Tcncv, FreeLunchBrian, LinkFA-Bot, Lightbot, שי דרך, Zorrobot, Legobot, Luckas-bot, Yobot, 2D, Amirobot, Reindra, AnakngAraw, Messymarcus, AnomieBOT, Materialscientist, Pxlnight13, Xqbot, DSisyphBot, Tad Lincoln, AbigailAbernathy, Pmlineditor, RibotBOT, Techauthor, C.Black, Ysyoon, Citation bot 1, Pinethicket, HRoestBot, Fox Wilson, Wikifan798, ஆஹ், Onel5969, RjwilmsiBot, DexDor, EmausBot, Pete Hobbs, Wikipelli, Auró, Savh, Hhhippo, Wayne Slam, Orange Suede Sofa, 28bot, Nate Guerrette, ClueBot NG, Crtcollector, DieSwartzPunkt, O.Koslowski, Turoktwostep, Helpful Pixie Bot, Argiegeo, Cqdx, Chivesud, ChrisGualtieri, Dexbot, Shubham Majmudar, Thepasta, Leo Spaceman MD, Ruby Murray, Jamesmcmahon0, Vangpace, Electrodryad, Gvprtskvnis, Pranat inder handa, KasparBot, i9dhoughtalen, Thilini C R, Suckitswaggyp and Anonymous: 152

- **Electron-beam lithography** *Source:* https://en.wikipedia.org/wiki/Electron-beam_lithography?oldid=661018184 *Contributors:* Zandark, Timwi, LMB, Altenmann, Niteowlneils, Alexander.stohr, Bobblewik, Fg2, Zro, D6, RJHall, Art LaPella, BRW, Oleg Alexandrov, Rjwilmsi,

DrTorstenHenning, Zotel, Ironside@elec.gla.ac.uk, Gaius Cornelius, Bug42, Anomalocaris, Tony1, Closedmouth, Mike1024, A13ean, Bluebot, DHN-bot-enwiki, Can't sleep, clown will eat me, Guiding light, Fbianco, Vina-iwbot-enwiki, Satish.murthy-enwiki, Chetvorno, CmdrObot, Safalra, Markluffel, Thijs!bot, Electron9, Magioladitis, Mytomi, Electrondevin, Mhesselb, Rettetast, Nigholith, Antony-22, Freekh, Lightmouse, Wikievil666, Raghu07, Alexbot, Rubin joseph 10, SchreiberBike, DumZiBoT, Mrogosky, Addbot, DOI bot, Semachin, Lightbot, Yobot, Lunochod-enwiki, Bunnyhop11, Blm19732008, Amirobot, Gbrake, Materialscientist, Citation bot, Obersachsebot, Xqbot, RGForbes, FrescoBot, OgreBot, Citation bot 1, Matthieu.berthome, Brian24545, Trappist the monk, Vrenator, WikitanvirBot, ChipHogg, ChuispastonBot, Minnsurfur2, ClueBot NG, Drsimonz, Bibcode Bot, BG19bot, Caboz, DarafshBot, Ladygoshasb, Freshgod, Bivas91, Vrmanfri, Anrnusna, Reinchang, Monkbot and Anonymous: 74

- **Electron beam processing** *Source:* https://en.wikipedia.org/wiki/Electron_beam_processing?oldid=678780553 *Contributors:* RJHall, Keltitrout, Rjwilmsi, Bgwhite, Bug42, SmackBot, Kmarinas86, Wizard191, Electron9, Fabrictramp, H1voltage, Squids and Chips, Lamro, Neparis, Hertz1888, Niceguyedc, Akaszynski, DumZiBoT, Doc9871, HGYAIT, Yobot, AnomieBOT, Materialscientist, PigFlu Oink, Ego White Tray, Iotron Industries, Skoot13, 11Andrzej, Vanischenu, Manabeast333, Garuda0001, Telfordbuck, Monkbot, FParkson and Anonymous: 15

- **Linear particle accelerator** *Source:* https://en.wikipedia.org/wiki/Linear_particle_accelerator?oldid=687525587 *Contributors:* Kpjas, SimonP, Heron, Karada, Jiang, Nv8200pa, Bevo, Raul654, Pakaran, Hoss, Pigsonthewing, Michael Devore, Leonard G., Chowbok, Utcursch, Discospinster, Pjacobi, Neko-chan, Laurascudder, Shenme, Brim, Fwb22, Martyman, Jjron, Coma28, Tpikonen, Atomicthumbs, Dirae1933, Lkinkade, Angr, Linas, Graham87, Strait, Arnero, DannyWilde, DaGizza, Gregorik, Korg, YurikBot, David R. Ingham, Welsh, Scottfisher, Blueyoshi321, GrinBot-enwiki, Locke Cole, Renesis, Mad hatter, David.Throop, Sbharris, Pwjb, Huga, Soarhead77, A5b, Zaphraud, Breno, Simkiott, Robert Bond, BethBukata, Drinibot, Myasuda, Thijs!bot, Mojo Hand, Headbomb, CharlotteWebb, AntiVandalBot, K7aay, Albany NY, Magioladitis, MartinBot, Pbroks13, Mattnad, AdrienChen, DAID, DMCer, VolkovBot, TXiKiBoT, Goflow6206, Leafyplant, SieBot, Hiddenfromview, Zuzkafuska, Ivan1984, Destn411, Scaler1112, ClueBot, Mild Bill Hiccup, Gjamesnvda, LSTech, Addbot, Guoguo12, FiriBot, TStein, Eshmo-enwiki, Kurtis, Luckas-bot, Ptbotgourou, Ciphers, Archon 2488, Galoubet, Duvnuj, Materialscientist, LouriePieterse, ArthurBot, Xqbot, RibotBOT, Cooshane, Redrose64, EmausBot, Sgbeer, Steerforth812, Harieoal, BR84, ClueBot NG, Kasirbot, Dimitri Girard, Altair, Sharksrule.alex, YFdyh-bot, TylerDurden8823, Dexbot, Boazhsan, Faizan, Epicgenius, Warrenboling, Tirvane, Mfb, Sunveetsingh, Pyratka and Anonymous: 108

- **Particle accelerator** *Source:* https://en.wikipedia.org/wiki/Particle_accelerator?oldid=690528939 *Contributors:* AxelBoldt, Kpjas, CYD, Bryan Derksen, Rgamble, Rmhermen, Christian List, Michael Hardy, Bewildebeast, Sannse, Alfio, Ahoerstemeier, Emperor, Julesd, Jll, Samw, Tpbradbury, Hellboy1975, Furrykef, Pakaran, David.Monniaux, Finlay McWalter, Hjr, Robbot, Pigsonthewing, Fredrik, Securiger, Lowellian, Merovingian, Sverdrup, Tycho?, Papadope, Xanzzibar, Tobias Bergemann, Giftlite, Harp, Wolfkeeper, Art Carlson, Leonard G., Egomaniac, Bobblewik, Bodhitha, Geni, OverlordQ, Mako098765, Alkivar, FT2, Alistair1978, Bender235, E2m, RJHall, Shad0, El C, Laurascudder, RoyBoy, Femto, Brim, Nk, Physicistjedi, Haham hanuka, Martyman, Gbrandt, HasharBot-enwiki, Jumbuck, Alansohn, Eleland, Mac Davis, Snowolf, Atomicthumbs, Cgmusselman, BRW, TenOfAllTrades, DV8 2XL, Gene Nygaard, Dan100, Vanished user dfvkjmet9jweflkmdkcn234, Ron Ritzman, Linas, Mindmatrix, RHaworth, LOL, Scott.wheeler, Cruccone, WadeSimMiser, Magabund, Tabletop, TreveX, Jugger90, Abd, Christopher Thomas, Magister Mathematicae, Cuchullain, Zzedar, Search4Lancer, Rjwilmsi, Alvinwc, Ae77, Strait, Collard, TheRingess, Crazynas, The wub, Ttwaring, FlaBot, Ground Zero, DannyWilde, Kolbasz, Goudzovski, Chobot, DTOx, DVdm, Digitalme, Gwernol, Krysith, Uk-Paolo, Roboto de Ajvol, EricCHill, YurikBot, Huw Powell, Kafziel, Widdma, Splash, Pigman, Rsrikanth05, Wimt, David R. Ingham, NawlinWiki, Dlugosz, SCZenz, Nick, Voidxor, Scottfisher, Olleicua, DeadEyeArrow, Alarob, Closedmouth, Mais oui!, ViperSnake151, Kgf0, Jesvj, SmackBot, KnowledgeOfSelf, Mrcoolbp, Unyoyega, PeterSymonds, Chris the speller, Jprg1966, Thumperward, O keyes, Dlenmn, Colonies Chris, Firetrap9254, Onorem, Jumping cheese, Khukri, Funky Monkey, Lamikae-enwiki, Kanji-enwiki, Jóna Þórunn, Poobah 4, UberCryxic, Breno, Severoon, Ckatz, Simkiott, Slakr, Mets501, MTSbot-enwiki, Hu12, Mgummess, Joseph Solis in Australia, Newone, Mr Chuckles, Chetvorno, Cryptic C62, Cloudguitar, Drinibot, Runningonbrains, Myasuda, MC10, Kozmik Pariah, HPaul, Difluoroethene, Dancter, HermanFinster, PseudoNym, Chrislk02, Kozuch, Abtract, Mtpaley, Thijs!bot, Epbr123, Wikid77, Headbomb, Davidhorman, AntiVandalBot, Johnny Sumner, David Shankbone, MaXiMiUS, JAnDbot, Husond, ProjectPlatinum, Z22, VoABot II, Pixel :-), PaulAitken, Bubba hotep, Fabricebaro, Cardamon, Web-Crawling Stickler, LookingGlass, Inhumandecency, Kiore, Rettetast, R n'B, Lilac Soul, Terrek, MrBell, 12dstring, Brykupono, Acalamari, Jedd the Jedi, NewEnglandYankee, DAID, BigHairRef, Cometstyles, Kenneth M Burke, Useight, VolkovBot, Danny252, Sohowcome, Themel, Philip Trueman, Irishnightwish, Amaryllis25, Comparat, Andy Dingley, Bill W Ca, Burntsauce, Sesshomaru, Aee is away, SieBot, TJRC, Jlivaudais, PlanetStar, Radon210, Chase92, Oxymoron83, Lightmouse, Tombomp, JsePrometheus, Jbenger, StaticGull, Jacob.jose, Scaler1112, Sfan00 IMG, ClueBot, Fasettle, Murchy, Fyyer, The Thing That Should Not Be, Antivolt, Wwheaton, Richerman, Phenylalanine, Mspraveen, Polaroids4x5, Flaming, DragonBot, Jdrice8, Jusdafax, Bullsrock478, Lartoven, Cenarium, BOTarate, Buckethed, Thingg, Jonverve, DumZiBoT, TimothyRias, LSTech, Pichpich, PervyPirate, Swift as an Eagle, Deltawk, NellieBly, Addbot, Willking1979, Friginator, AkhtaBot, Delos1970-enwiki, Zarcadia, Meander112, LinkFA-Bot, Jonnysonthespot, Tide rolls, Bfigura's puppy, Lightbot, Scientryst, Loupeter, Zorrobot, David0811, Yobot, Fraggle81, JetKing2, Naudefjbot-enwiki, Proxyman1337, Fmrauch, Azcolvin429, DemocraticLuntz, Piano non troppo, Justme89, Apachuri, RandomAct, Duvnuj, Materialscientist, The High Fin Sperm Whale, Citation bot, LilHelpa, Witguiota, Hanberke, Gilo1969, GrouchoBot, ProtectionTaggingBot, Franco3450, Mr.clever2499, Stratocracy, January2009, Wildw79, Krodonnell, FrescoBot, Oneironaut99, Steve Quinn, LuteMidnight, Jmahon01, FoxBot, TobeBot, Trappist the monk, LilyKitty, TheGrimReaper NS, Jeffrd10, Mttcmbs, Reach Out to the Truth, EmausBot, Docjudith, Dewritech, Racerx11, Ferocious osmosis, RenamedUser01302013, Tommy2010, Challisrussia, Wikipelli, 6zeta2tothehalf, Hhhippo, ZéroBot, Josve05a, Wikfr, TGBX, VictorFlaushenstein, John KB, L Kensington, Donner60, Chewings72, QuantumSquirrel, BR84, ClueBot NG, Jack Greenmaven, Frietjes, Moritz37, What is my name silas, Helpful Pixie Bot, Bibcode Bot, Vagobot, Northamerica1000, Neutral current, Lupito123456, LHCpioneer, ChrisGualtieri, SD5bot, 786b6364, Futurist110, Dexbot, Kevinfrank17, Garuda0001, TwoTwoHello, Ghostgammer, Tentinator, Ninar Haneris, Tyronium47, Chemengkiddies, Ct-tiara2013, Monkbot, Mefifm, Leelakrishna86, Curedwales, Trackteur, Jarould, Hhm8, AuxZane, BuddyBJ, EB88625, Tetra quark, JonathanPereiraB, Gorzhn, SuperCarnivore591, Trashboat59 and Anonymous: 362

- **Low-energy electron diffraction** *Source:* https://en.wikipedia.org/wiki/Low-energy_electron_diffraction?oldid=678981987 *Contributors:* Michael Hardy, LMB, Quadell, Chowells, Jak86, Camw, Arnero, Siddhant, Gaius Cornelius, Kkmurray, SmackBot, Kmarinas86, Chris the speller, OrphanBot, Georg-Johann, Owlbuster, Rwmccauley, PSU PHYS514 F06, Hyleelyh, Lantonov, Skier Dude, Jewf, SoCalSuperEagle, Larryisgood, TXiKiBoT, Raymondwinn, Johnny1926, Furious.baz, ClueBot, Piastu, Mild Bill Hiccup, Tevmen-enwiki, Sun Creator, Tanvash, Dthomsen8, Addbot, OlEnglish, Yobot, Citation bot, Killkoll, Mkosterv, Jatosado, Atomless, Citation bot 1, Tom.Reding, Mywtfmp3, RjwilmsiBot, GoingBatty, Hhhippo, ZéroBot, ClueBot NG, Widr, Bibcode Bot, Pakritar, Ilithyia, GabeIglesia, Skilving, Goblinshark17 and Anonymous:

34

- **Electron microscope** *Source:* https://en.wikipedia.org/wiki/Electron_microscope?oldid=689551249 *Contributors:* Magnus Manske, Kpjas, Sodium, The Anome, Stokerm, Andre Engels, Novalis, Enchanter, Heron, Camembert, D, JohnOwens, Michael Hardy, Dgrant, Docu, Snoyes, Kils, Nikai, Tacvek, Dcoetzee, Selket, Itai, JorgeGG, Chuunen Baka, Robbot, Frank A, Astronautics~enwiki, Wikibot, Pengo, Jeremiah, Giftlite, DocWatson42, Mark.murphy, Risk one, Average Earthman, Bensaccount, Sludtke42, Jorge Stolfi, Utcursch, Quadell, Kusunose, Karol Langner, Kesac, Chai~enwiki, Mschlindwein, Deglr6328, ELApro, JTN, Discospinster, Rich Farmbrough, Guanabot, Bender235, Ground, STHayden, CanisRufus, Spearhead, Svdmolen, Perfecto, Bobo192, Stahlkocher1, Rajah, Alansohn, Gary, Lectonar, Hu, Wtmitchell, RainbowOfLight, Sciurinæ, Gene Nygaard, Netkinetic, Ceyockey, Dennis Bratland, Mhazard9, Richard Arthur Norton (1958-), Karnesky, TigerShark, Simon Shek, Kzollman, Hurricane Angel, Frankatca, Steinbach, GregorB, Waldir, MarcoTolo, MoogleFan, Rjwilmsi, George Burgess, DoubleBlue, Ian Pitchford, Nihiltres, Vossman, Terrace4, Chobot, DaGizza, Gregorik, DVdm, Alex Klotz, YurikBot, Wavelength, Borgx, Sceptre, Jimp, Pip2andahalf, RussBot, Fabricationary, Gaius Cornelius, Rsrikanth05, David R. Ingham, NawlinWiki, RUL3R, Nate1481, Bota47, FF2010, Zzuuzz, Imaninjapirate, Oysteinp, Joshua.morgan, Staxringold, Banus, DVD R W, SmackBot, KnowledgeOfSelf, Hydrogen Iodide, Davewild, Grey Shadow, Chych, Jab843, Srnec, Fueled~enwiki, Zephyris, Gilliam, Skizzik, Kissavos, Kmarinas86, Hugo-cs, Chris the speller, Bluebot, SchfiftyThree, DHN-bot~enwiki, Darth Panda, Cgarber, Riflemann, Can't sleep, clown will eat me, Snowmanradio, Dexarouskies, EvelinaB, LouScheffer, SundarBot, Steff, Aaronsharpe, RobHarding, Drphilharmonic, Mwtoews, ILike2BeAnonymous, Acdx, Kukini, Andrei Stroe, Nmnogueira, SashatoBot, Kuru, Bydand, KristianMolhave~enwiki, Dicklyon, Funnybunny, Kvng, WahreJakob, JoeBot, Adijr, Stephen-Buxton, Igoldste, Adambiswanger1, Courcelles, Dlohcierekim, George100, Evan1991, INkubusse, Rustavo, Thermochap, CmdrObot, KyraVixen, Til.Bartel, El aprendelenguas, Joelholdsworth, Myasuda, Cydebot, Gogo Dodo, Travelbird, Corpx, Tahirulislam, Myscrnnm, Tawkerbot4, Carstensen, Shirulashem, Christian75, DumbBOT, JRodeng, Chrislk02, Thijs!bot, Epbr123, Hitmanjon, Daniel, N5iln, JNighthawk, Marek69, Speedyboy, Frank, Tapir Terrific, A3RO, Doyley, Edal, Sikkema, MichaelMaggs, Sean William, AntiVandalBot, Luna Santin, Shirt58, Mary Mark Ockerbloom, Activist, TimVickers, Tmopkisn, Tjmayerinsf, Gh5046, Alphachimpbot, Fireice, JAnDbot, Leuko, MER-C, Plantsurfer, Avaya1, IanOsgood, Albany NY, Acroterion, Bencherlite, S0uj1r0, Bongwarrior, VoABot II, JamesBWatson, Twisted86, 28421u2232nfenfcenc, Chris G, DerHexer, JaGa, DGG, MartinBot, EyeSerene, Simonmn, Anaxial, PrestonH, J.delanoy, Pharaoh of the Wizards, Maurice Carbonaro, Irfanhasmit, 5Q5, Nath87, DarkFalls, McSly, Monkeyknife, Dexter prog, Jasonasosa, Plasticup, NewEnglandYankee, Pliable, STBotD, Doctoroxenbriery, Alan012, Pdcook, VolkovBot, Thedjatclubrock, ABF, DSRH, Jeff G., LokiClock, Cbj77, LeilaniLad, Adam Mihalyi, Philip Trueman, Jaccardi, Ulfbastel, Anna Lincoln, Ferengi, Broadbot, Konsu~enwiki, BotKung, Saturn star, Ephram Shizgal, Andy Dingley, Dr-MikeF, Enviroboy, Rurik3, R.rommel, Brianga, Logan, Closenplay, Golfguy220-, SieBot, Mayapur, Graham Beards, Davidbaca, Gerakibot, Liferulez, Calabraxthis, LeadSongDog, DeltaMicroscopyStudent, Prince Max (scientist), Keilana, Aillema, Flyer22 Reborn, A. Carty, Oxymoron83, Antonio Lopez, Steven Crossin, Maelgwnbot, C0nanPayne, Chem-awb, Esem0, Martarius, ClueBot, The Thing That Should Not Be, Paulhaiti, Dvratnam~enwiki, Mild Bill Hiccup, Spbrandom, DrFO.Jr.Tn~enwiki, Cetcher1159, Blanchardb, MindstormsKid, DragonBot, Jusdafax, Aurelius173, Arjayay, Diaa abdelmoneim, Mikaey, Thingg, Aitias, Horselover Frost, Amaltheus, SoxBot III, Apparition11, Editor2020, DumZiBoT, XLinkBot, Jytdog, Rror, Elearning2000, Ncemer, Subversive.sound, Nickfor1, Csown, Srdomingue, Oceano30~enwiki, Jacopo Werther, Willking1979, DOI bot, Captain-tucker, Blechnic, Ronhjones, Fieldday-sunday, Cst17, Jim10701, MrOllie, Favonian, Santosh sthy, Apteva, Zorrobot, Tijadr, CountryBot, Legobot, Drpickem, Luckas-bot, Yobot, II MusLiM HyBRiD II, Amirobot, ArchonMagnus, Mmxx, ꦒꦗ, SwisterTwister, Mhmolitor, Eric-Wester, Backslash Forwardslash, 007 n1, Jim1138, Hat'nCoat, Piano non troppo, AdjustShift, Kingpin13, Materialscientist, The High Fin Sperm Whale, Citation bot, OllieFury, Bci2, Frankenpuppy, Neurolysis, LilHelpa, Кодекс, Xqbot, Capricorn42, Jjmatt33, Hedgemonkey, Tyrol5, Maro91eg, Fallboy, GrouchoBot, Darkest tree, FrescoBot, Dogposter, Tetraedycal, Citation bot 1, Sub-Angstrom, Isaac yagmoor, Serols, ContinueWithCaution, Horst-schlaemma, TobeBot, Lotje, MrX, SkyLineRxx, Jared lap, Minimac, Hornlitz, Jungy111, Fearstreetsaga, Ripchip Bot, Becritical, Sbertazzo, DASHBot, EmausBot, Grottenolm42, Tylersosmart, Hhhippo, Mz7, ZéroBot, Makingwiththeparticles, A2soup, Doddy Wuid, Felipeavb, Akbar.luck, Gz33, Mayur, Drago27218, BR84, ClueBot NG, MelbourneStar, This lousy T-shirt, Satellizer, Baryonman, Snotbot, Twillisjr, O.Koslowski, Widr, Helpful Pixie Bot, Irisoratoria, Bibcode Bot, Krenair, Wasim14ahmed, MusikAnimal, Orangutans, Claudionico, Snow Blizzard, Vniizht, Worldfactsandpoosandfarts, Chivesud, Gdfusion, Jdoggydog4, Lifetolive.kisor, Frosty, Telfordbuck, Patterdale99, Reatlas, Sevmti faiv, My-2-bits, Sam Sailor, LibraryStudent24, TCMemoire, Floglol, AlmaMer, M nasab, Monkbot, BethNaught, Prakash Narayanan Nair, Ilovetodostuff, Jolly111, MicroPaLeo, Indiebean-7, KasparBot, Thomas M Bernhard, Omfgimcalledjeffrite, Milku3459, Claudionico-commonswiki, Ltumanovskaya and Anonymous: 645

- **Free-electron laser** *Source:* https://en.wikipedia.org/wiki/Free-electron_laser?oldid=679392442 *Contributors:* Bryan Derksen, Julesd, Reddi, Hankwang, Academic Challenger, GreatWhiteNortherner, Alan Liefting, DocWatson42, Jpatej, Micru, Peter bertok, Deglr6328, Deleteme42, Pjacobi, Xezbeth, Kghose, DaveGorman, Kjkolb, Pearle, Hooperbloob, Gene Nygaard, Tylerni7, J M Rice, Graham87, Tommcnabb, Rjwilmsi, China Crisis, KaiMartin, Arnero, Nwatson, Lmatt, David H Braun (1964), Meawoppl, DVdm, Roboto de Ajvol, Shaddack, Eleassar, David R. Ingham, Welsh, RabidDeity, Santaduck, Engineer Bob, Erik J, Saikiri, SmackBot, Mjspe1, Gregjgrose, Kmarinas86, Oatmeal batman, Hgrosser, Hawkwings31, Giancarlo Rossi, Tomatoman, JorisvS, Hu12, Chetvorno, CmdrObot, Thijs!bot, Headbomb, Dtgriscom, Second Quantization, Noclevername, Guy Macon, Lfstevens, Albertvillanovadelmoral, Freddy011, Schmloof, Pagw, Alexander Patrakov, R'n'B, Mjgullans, Stuffysour, Lantonov, DadaNeem, VolkovBot, Landisdesign, TXiKiBoT, Enozkan, Hqb, Revansx~enwiki, Jerryobject, Lightmouse, KJG2007, Polyamorph, Alexbot, Gabella, Davismargaret, XLinkBot, NellieBly, Dyuku, Bonewith, Luckas-bot, AnomieBOT, Obersachsebot, Xqbot, Tomdo08, J04n, Trurle, Xfig, Pozharnikar, Cooshane, Kyteto, Citation bot 1, Pinethicket, Jonesey95, JMMuller, Michael9422, RjwilmsiBot, Paratwa, EmausBot, John of Reading, Karim osama1, WikitanvirBot, ZéroBot, Wikfr, RaptureBot, Wakebrdkid, BR84, ClueBot NG, Powersjcb, ArbHH, Marieto60, MrBill3, Jannick88, Comfr, Dexbot, Ziggy1986 TS, Tony Mach, GravRidr, FizykLJF, Maderthaner, Seneszenz, Ggf4t and Anonymous: 97

- **Cathode ray tube** *Source:* https://en.wikipedia.org/wiki/Cathode_ray_tube?oldid=689660635 *Contributors:* Damian Yerrick, Uriyan, Robert Merkel, Alex.tan, JeLuF, Aldie, Robert Foley, Heron, Edward, RTC, Michael Hardy, Norm, Dominus, Wapcaplet, Penmachine, JerryG, Arwel Parry, Theresa knott, Glenn, Andres, Coren, Reddi, Stone, Maximus Rex, Saltine, Bartosz, Omegatron, Ed g2s, Wernher, Bloodshedder, Francs2000, Rogper~enwiki, Donarreiskoffer, Jeremybornstein, Robbot, Paranoid, Altenmann, SEKIUCHI, Jzhang, Academic Challenger, Pepijn Schmitz, Spike, Auric, Victor, Alan Liefting, Giftlite, Gtrmp, BenFrantzDale, Reub2000, Tom harrison, Koyn~enwiki, Everyking, Curps, Leonard G., Mboverload, Glenn Koenig, Macrakis, Bobblewik, Edcolins, Andycjp, Beland, MisfitToys, Jossi, Sam Hocevar, Nerd65536, JulieADriver, Aknorals, Sonett72, MementoVivere, Grm wnr, Deglr6328, SYSS Mouse, Discospinster, Rich Farmbrough, Rupertslander, ArnoldReinhold, Smyth, Michael Zimmermann, Paul August, Night Gyr, ESkog, Livajo, Willemdd, Femto, Smalljim, John Vandenberg, BrokenSegue, Viriditas, Cmacd123, IDX, Zetawoof, Haham hanuka, Hooperbloob, Nsaa, Jumbuck, Sheehan, Interiot, Keyser Söze, Somebody in the WWW, Atlant, M7, Axl, Zsero, Wtmitchell, Velella, TaintedMustard, XB-70, Rebroad, Wtshymanski, Stephan Leeds, Suruena, H2g2bob, Bsadowski1,

28.17.2 Images

- **File:Egun.jpg** *Source:* https://upload.wikimedia.org/wikipedia/commons/a/ae/Egun.jpg *License:* Public domain *Contributors:* Transferred from en.wikipedia to Commons. *Original artist:* Roychai at English Wikipedia

- **File:Einstein_patentoffice.jpg** *Source:* https://upload.wikimedia.org/wikipedia/commons/a/a0/Einstein_patentoffice.jpg *License:* Public domain *Contributors:* Transferred from en.wikipedia; transferred to Commons by User:Guerillero using CommonsHelper. *Original artist:* Lucien Chavan *[#cite_note-author-1 [1]] (1868 - 1942), a friend of Einstein's when he was living in Berne.

- **File:Electric-eel2.jpg** *Source:* https://upload.wikimedia.org/wikipedia/commons/0/03/Electric-eel2.jpg *License:* CC-BY-SA-3.0 *Contributors:* No machine-readable source provided. Own work assumed (based on copyright claims). *Original artist:* No machine-readable author provided. Stevenj assumed (based on copyright claims).

- **File:Electric_field_point_lines_equipotentials.svg** *Source:* https://upload.wikimedia.org/wikipedia/commons/9/96/Electric_field_point_lines_equipotentials.svg *License:* Public domain *Contributors:* Own work *Original artist:* Sjlegg

- **File:Electric_motor_cycle_3.png** *Source:* https://upload.wikimedia.org/wikipedia/commons/5/59/Electric_motor_cycle_3.png *License:* CC-BY-SA-3.0 *Contributors:* ? *Original artist:* ?

- **File:Electrical_potential_and_field_lines_between_two_wires.png** *Source:* https://upload.wikimedia.org/wikipedia/commons/f/f3/Electrical_potential_and_field_lines_between_two_wires.png *License:* CC BY-SA 4.0 *Contributors:* Own work *Original artist:* Bret Mulvey

- **File:Electromagnet.gif** *Source:* https://upload.wikimedia.org/wikipedia/commons/3/3e/Electromagnet.gif *License:* CC BY-SA 3.0 *Contributors:* This file was created with Blender. *Original artist:* Anynobody

- **File:Electromagnetism.svg** *Source:* https://upload.wikimedia.org/wikipedia/commons/9/91/Electromagnetism.svg *License:* CC-BY-SA-3.0 *Contributors:* Image:Electromagnetism.png *Original artist:* User:Stannered

- **File:Electron_Beam_scattering.svg** *Source:* https://upload.wikimedia.org/wikipedia/commons/c/cd/Electron_Beam_scattering.svg *License:* CC BY-SA 4.0 *Contributors:* <a href='//commons.wikimedia.org/wiki/File:Electron_beam_scattering.jpg' class='image'><img alt='Electron beam scattering.jpg' src='https://upload.wikimedia.org/wikipedia/commons/thumb/9/9f/Electron_beam_scattering.jpg/100px-Electron_beam_scattering.jpg' width='100' height='59' srcset='https://upload.wikimedia.org/wikipedia/commons/thumb/9/9f/Electron_beam_scattering.jpg/150px-Electron_beam_scattering.jpg 1.5x, https://upload.wikimedia.org/wikipedia/commons/thumb/9/9f/Electron_beam_scattering.jpg/200px-Electron_beam_scattering.jpg 2x' data-file-width='651' data-file-height='384' /></a> *Original artist:* Fred the Oyster

- **File:Electron_Interaction_with_Matter.svg** *Source:* https://upload.wikimedia.org/wikipedia/commons/4/49/Electron_Interaction_with_Matter.svg *License:* CC BY-SA 4.0 *Contributors:* Own work *Original artist:* Claudionico~commonswiki

- **File:Electron_Microscope.jpg** *Source:* https://upload.wikimedia.org/wikipedia/commons/c/c5/Electron_Microscope.jpg *License:* CC BY-SA 2.0 *Contributors:* Tecnai 12 Electron Microscope *Original artist:* David J Morgan from Cambridge, UK

- **File:Electron_Microscope.png** *Source:* https://upload.wikimedia.org/wikipedia/commons/2/24/Electron_Microscope.png *License:* Public domain *Contributors:* Wikipedia, from Dr Graham Beards *Original artist:* Dr Graham Beards

- **File:Electroscope.svg** *Source:* https://upload.wikimedia.org/wikipedia/commons/f/fa/Electroscope.svg *License:* CC-BY-SA-3.0 *Contributors:*

- Electroscope.png *Original artist:* Electroscope.png: Stw

- **File:Enrico_Fermi_1943-49.jpg** *Source:* https://upload.wikimedia.org/wikipedia/commons/d/d4/Enrico_Fermi_1943-49.jpg *License:* Public domain *Contributors:* This media is available in the holdings of the National Archives and Records Administration, cataloged under the ARC Identifier (National Archives Identifier) **558578**. *Original artist:* Department of Energy. Office of Public Affairs

- **File:Ernst_Ruska_Electron_Microscope_-_Deutsches_Museum_-_Munich-edit.jpg** *Source:* https://upload.wikimedia.org/wikipedia/commons/1/1f/Ernst_Ruska_Electron_Microscope_-_Deutsches_Museum_-_Munich-edit.jpg *License:* CC BY-SA 3.0 *Contributors:* originally posted to Flickr as Electron Microscope Deutsches Museum *Original artist:* J Brew, uploaded on the English-speaking Wikipedia by en:User:Hat'nCoat.

- **File:Ewaldsphere1.png** *Source:* https://upload.wikimedia.org/wikipedia/commons/1/16/Ewaldsphere1.png *License:* GFDL *Contributors:* Own work *Original artist:* Killkoll

- **File:EwaldsphereLEED.png** *Source:* https://upload.wikimedia.org/wikipedia/en/5/53/EwaldsphereLEED.png *License:* CC-BY-SA-3.0 *Contributors:* ? *Original artist:* ?

- **File:ExperimentCouder-Young.png** *Source:* https://upload.wikimedia.org/wikipedia/commons/9/90/ExperimentCouder-Young.png *License:* CC BY-SA 3.0 *Contributors:* Own work *Original artist:* Krauss

- **File:External_beam_radiotherapy_retinoblastoma_nci-vol-1924-300.jpg** *Source:* https://upload.wikimedia.org/wikipedia/commons/5/59/External_beam_radiotherapy_retinoblastoma_nci-vol-1924-300.jpg *License:* Public domain *Contributors:* National Cancer Institute via Stanford University, image number AV-5700-3472 *Original artist:* Unknown photographer/artist

- **File:FELIX.jpg** *Source:* https://upload.wikimedia.org/wikipedia/commons/1/14/FELIX.jpg *License:* CC BY-SA 3.0 *Contributors:* Own work *Original artist:* China Crisis

- **File:FEL_principle.png** *Source:* https://upload.wikimedia.org/wikipedia/commons/3/3e/FEL_principle.png *License:* CC-BY-SA-3.0 *Contributors:* Transferred from de.wikipedia to Commons. *Original artist:* Selbst erstellt (Horst Frank)

- **File:Faraday-Millikan-Gale-1913.jpg** *Source:* https://upload.wikimedia.org/wikipedia/commons/5/5b/Faraday-Millikan-Gale-1913.jpg *License:* Public domain *Contributors:* Opposite p. 290 of Millikan and Gale's *Practical Physics* (1922) *Original artist:* Probably albumen carte-de-visite by John Watkins

- **File:Feature_stitching_across_fields.png** *Source:* https://upload.wikimedia.org/wikipedia/en/7/7f/Feature_stitching_across_fields.png *License:* CC-BY-SA-3.0 *Contributors:*

Graphic editor

Original artist:

Guiding light

- **File:Plasmonic_Young'{}s_double_slits_interference.png** *Source:* https://upload.wikimedia.org/wikipedia/commons/a/a1/Plasmonic_Young%27s_double_slits_interference.png *License:* CC BY-SA 4.0 *Contributors:* Own work *Original artist:* Sklab

- **File:Portal-puzzle.svg** *Source:* https://upload.wikimedia.org/wikipedia/en/f/fd/Portal-puzzle.svg *License:* Public domain *Contributors:* ? *Original artist:* ?

- **File:Propagation_of_a_de_broglie_wave.svg** *Source:* https://upload.wikimedia.org/wikipedia/commons/2/21/Propagation_of_a_de_broglie_wave.svg *License:* Public domain *Contributors:* Own work *Original artist:* Maschen

- **File:QuantumHarmonicOscillatorAnimation.gif** *Source:* https://upload.wikimedia.org/wikipedia/commons/9/90/QuantumHarmonicOscillatorAnimation.gif *License:* CC0 *Contributors:* Own work *Original artist:* Sbyrnes321

- **File:Quantum_dot.png** *Source:* https://upload.wikimedia.org/wikipedia/commons/6/68/Quantum_dot.png *License:* CC BY-SA 3.0 *Contributors:* http://commons.wikimedia.org/wiki/File:QuantumDot_wf.gif *Original artist:* Saumitra R Mehrotra & Gerhard Klimeck

- **File:Quantum_mechanics_standing_wavefunctions.svg** *Source:* https://upload.wikimedia.org/wikipedia/commons/2/27/Quantum_mechanics_standing_wavefunctions.svg *License:* Public domain *Contributors:* Own work *Original artist:* Maschen

- **File:Quantum_mechanics_travelling_wavefunctions.svg** *Source:* https://upload.wikimedia.org/wikipedia/commons/3/3e/Quantum_mechanics_travelling_wavefunctions.svg *License:* Public domain *Contributors:* Own work *Original artist:* Maschen

- **File:Quantum_mechanics_travelling_wavefunctions_wavelength.svg** *Source:* https://upload.wikimedia.org/wikipedia/commons/f/f1/Quantum_mechanics_travelling_wavefunctions_wavelength.svg *License:* CC0 *Contributors:* Own work *Original artist:* Maschen

- **File:Quark_structure_neutron.svg** *Source:* https://upload.wikimedia.org/wikipedia/commons/8/81/Quark_structure_neutron.svg *License:* CC BY-SA 2.5 *Contributors:* No machine-readable source provided. Own work assumed (based on copyright claims). *Original artist:* No machine-readable author provided. Harp assumed (based on copyright claims).

- **File:Question_book-new.svg** *Source:* https://upload.wikimedia.org/wikipedia/en/9/99/Question_book-new.svg *License:* Cc-by-sa-3.0 *Contributors:*
Created from scratch in Adobe Illustrator. Based on Image:Question book.png created by User:Equazcion *Original artist:*
Tkgd2007

- **File:RCA_Model_EMT3_desktop_electron_microscope_2003_033_006.tf.TIF** *Source:* https://upload.wikimedia.org/wikipedia/commons/7/7e/RCA_Model_EMT3_desktop_electron_microscope_2003_033_006.tf.TIF *License:* CC BY-SA 3.0 *Contributors:* Chemical Heritage Foundation, Photograph by Gregory Tobias *Original artist:* Gregory Tobias

- **File:Regla_mano_derecha_Laplace.svg** *Source:* https://upload.wikimedia.org/wikipedia/commons/6/6a/Regla_mano_derecha_Laplace.svg *License:* CC BY-SA 3.0 *Contributors:* Uses samples from: Image:Electromagnetism.svg *Original artist:* Jfmelero

- **File:Robert_Boyle_0001.jpg** *Source:* https://upload.wikimedia.org/wikipedia/commons/b/b3/Robert_Boyle_0001.jpg *License:* Public domain *Contributors:* http://www.bbk.ac.uk/boyle/Issue4.html *Original artist:* Johann Kerseboom

- **File:Sasahara.svg** *Source:* https://upload.wikimedia.org/wikipedia/commons/c/cc/Sasahara.svg *License:* CC-BY-SA-3.0 *Contributors:* Transferred from en.wikipedia *Original artist:* Original uploader was JWB at en.wikipedia

- **File:Schematic_showing_basic_components_and_operation_of_electron_beam_materials_processing.png** *Source:* https://upload.wikimedia.org/wikipedia/commons/8/8b/Schematic_showing_basic_components_and_operation_of_electron_beam_materials_processing.png *License:* CC BY-SA 3.0 *Contributors:* http://www.worldscientific.com/worldscibooks/10.1142/7745 *Original artist:* Donald E. Powers

- **File:Shen_Kua.JPG** *Source:* https://upload.wikimedia.org/wikipedia/commons/4/40/Shen_Kua.JPG *License:* CC-BY-SA-3.0 *Contributors:* Transferred from en.wikipedia to Commons. The picture is drawn after the likeness of a modern sculpted bust of Shen's head, from the sitewww.phil.pku.edu.cn for the Department of Philosophy at the University of Peking, China. The specific image is found here. *Original artist:* Wikimachine at English Wikipedia

- **File:Si100Reconstructed.png** *Source:* https://upload.wikimedia.org/wikipedia/commons/c/c4/Si100Reconstructed.png *License:* CC BY-SA 3.0 *Contributors:* Transferred from en.wikipedia to Commons. *Original artist:* Killkoll at English Wikipedia

- **File:Siemens-electron-microscope.jpg** *Source:* https://upload.wikimedia.org/wikipedia/commons/b/b0/Siemens-electron-microscope.jpg *License:* CC BY-SA 3.0 *Contributors:* Own work *Original artist:* Edal Anton Lefterov

- **File:Silicon_conduction_band_ellipsoids.JPG** *Source:* https://upload.wikimedia.org/wikipedia/commons/0/0c/Silicon_conduction_band_ellipsoids.JPG *License:* CC BY-SA 3.0 *Contributors:* Own work *Original artist:* Brews ohare

- **File:SinclairFTV1frontPCB6.jpg** *Source:* https://upload.wikimedia.org/wikipedia/commons/a/a2/SinclairFTV1frontPCB6.jpg *License:* CC BY-SA 3.0 *Contributors:* Own work *Original artist:* Binarysequence

- **File:Single_slit_and_double_slit2.jpg** *Source:* https://upload.wikimedia.org/wikipedia/commons/c/c2/Single_slit_and_double_slit2.jpg *License:* CC BY-SA 3.0 *Contributors:* Own work *Original artist:* Jordgette

- **File:Slide4.PNG** *Source:* https://upload.wikimedia.org/wikipedia/en/9/9c/Slide4.PNG *License:* PD *Contributors:*
Group of R Diehl in Penn State
Original artist:
N Ferralis at Penn State.

- **File:Sommerfeld_ellipses.svg** *Source:* https://upload.wikimedia.org/wikipedia/commons/7/75/Sommerfeld_ellipses.svg *License:* Public domain *Contributors:* Own work *Original artist:* Pieter Kuiper

- **File:Standard_Model_of_Elementary_Particles.svg** *Source:* https://upload.wikimedia.org/wikipedia/commons/0/00/Standard_Model_of_Elementary_Particles.svg *License:* CC BY 3.0 *Contributors:* Own work by uploader, PBS NOVA [1], Fermilab, Office of Science, United States Department of Energy, Particle Data Group *Original artist:* MissMJ

- **File:Waves_in_Box.JPG** *Source:* https://upload.wikimedia.org/wikipedia/commons/4/48/Waves_in_Box.JPG *License:* CC BY-SA 3.0 *Contributors:* Own work *Original artist:* Brews ohare

- **File:Weizmann_Institute_particle_accelerator.jpg** *Source:* https://upload.wikimedia.org/wikipedia/commons/0/08/Weizmann_Institute_particle_accelerator.jpg *License:* CC-BY-SA-3.0 *Contributors:* Own work *Original artist:* David Shankbone (attribution required)

- **File:Wiener_process_3d.png** *Source:* https://upload.wikimedia.org/wikipedia/commons/f/f8/Wiener_process_3d.png *License:* CC-BY-SA-3.0 *Contributors:* The description as originally from Wikipedia. *Original artist:* Original uploader was Sullivan.t.j at English Wikipedia.

- **File:Wikibooks-logo-en-noslogan.svg** *Source:* https://upload.wikimedia.org/wikipedia/commons/d/df/Wikibooks-logo-en-noslogan.svg *License:* CC BY-SA 3.0 *Contributors:* Own work *Original artist:* User:Bastique, User:Ramac et al.

- **File:Wikisource-logo.svg** *Source:* https://upload.wikimedia.org/wikipedia/commons/4/4c/Wikisource-logo.svg *License:* CC BY-SA 3.0 *Contributors:* Rei-artur *Original artist:* Nicholas Moreau

- **File:Wiktionary-logo-en.svg** *Source:* https://upload.wikimedia.org/wikipedia/commons/f/f8/Wiktionary-logo-en.svg *License:* Public domain *Contributors:* Vector version of Image:Wiktionary-logo-en.png. *Original artist:* Vectorized by Fvasconcellos (talk · contribs), based on original logo tossed together by Brion Vibber

- **File:Wolfgang_Pauli_young.jpg** *Source:* https://upload.wikimedia.org/wikipedia/commons/4/43/Wolfgang_Pauli_young.jpg *License:* Public domain *Contributors:* ? *Original artist:* ?

- **File:WorldsFairTeslaPresentation.png** *Source:* https://upload.wikimedia.org/wikipedia/commons/8/82/WorldsFairTeslaPresentation.png *License:* Public domain *Contributors:* Transferred from en.wikipedia *Original artist:* Original uploader was Reddi at en.wikipedia

- **File:XRL_Currents.svg** *Source:* https://upload.wikimedia.org/wikipedia/commons/c/c4/XRL_Currents.svg *License:* CC BY-SA 4.0 *Contributors:* <a href='//commons.wikimedia.org/wiki/File:Xrl_currents3.gif' class='image'><img alt='Xrl currents3.gif' src='https://upload.wikimedia.org/wikipedia/commons/thumb/8/86/Xrl_currents3.gif/100px-Xrl_currents3.gif' width='100' height='102' srcset='https://upload.wikimedia.org/wikipedia/commons/thumb/8/86/Xrl_currents3.gif/150px-Xrl_currents3.gif 1.5x, https://upload.wikimedia.org/wikipedia/commons/thumb/8/86/Xrl_currents3.gif/200px-Xrl_currents3.gif 2x' data-file-width='653' data-file-height='664' /></a> *Original artist:* Fred the Oyster

- **File:Young_Diffraction.png** *Source:* https://upload.wikimedia.org/wikipedia/commons/8/8a/Young_Diffraction.png *License:* Public domain *Contributors:* ? *Original artist:* ?

- **File:Ørsted.jpg** *Source:* https://upload.wikimedia.org/wikipedia/commons/7/79/%C3%98rsted.jpg *License:* Public domain *Contributors:* Original artist: Christoffer Wilhelm Eckersberg

28.17.3   Content license

www.ingramcontent.com/pod-product-compliance
Lightning Source LLC
Chambersburg PA
CBHW080759180526
45168CB00006B/2268